Rational Animals?

Edited by

Susan Hurley

*Department of Philosophy, University of Bristol; and
All Souls College, Oxford, UK*

and

Matthew Nudds

Department of Philosophy, University of Edinburgh, UK

OXFORD
UNIVERSITY PRESS

OXFORD

UNIVERSITY PRESS

Great Clarendon Street, Oxford OX2 6DP

Oxford University Press is a department of the University of Oxford.
It furthers the University's objective of excellence in research, scholarship,
and education by publishing worldwide in

Oxford New York

Auckland Cape Town Dar es Salaam Hong Kong Karachi
Kuala Lumpur Madrid Melbourne Mexico City Nairobi
New Delhi Shanghai Taipei Toronto

With offices in

Argentina Austria Brazil Chile Czech Republic France Greece
Guatemala Hungary Italy Japan Poland Portugal Singapore
South Korea Switzerland Thailand Turkey Ukraine Vietnam

Oxford is a registered trade mark of Oxford University Press
in the UK and in certain other countries

Published in the United States
by Oxford University Press Inc., New York

British Library Cataloguing in Publication Data

Data available

Library of Congress Cataloging in Publication Data

Data available

Typeset by Newgen Imaging Systems (P) Ltd., Chennai, India
Printed in Great Britain
on acid-free paper by
Biddles Ltd., King's Lynn

ISBN 978–0–19–852826–5 (Hbk.)
ISBN 978–0–19–852827–2 (Pbk.)

10 9 8 7 6 5 4 3 2

Contents

List of contributors

Elsa Addessi
Istituto di Scienze e Tecnologie
della Cognizione, CNR,
Rome, Italy

Colin Allen
Department of History and Philosophy
of Science and Program in
Cognitive Science,
Indiana University,
Bloomington, IN, USA

José Luis Bermúdez
Philosophy-Neuroscience-Psychology
Program,
Washington University in St Louis,
MO, USA

Sarah T. Boysen
Department of Psychology,
Ohio State University,
Columbus, OH, USA

Josep Call
Max Planck Institute for Evolutionary
Anthropology,
Leipzig, Germany

Nicola Clayton
Department of Experimental Psychology,
University of Cambridge, UK

Richard Connor
Department of Biology,
University of Massachusetts,
Dartmouth, MA, USA

Gregory Currie
Department of Philosophy,
University of Nottingham, UK

Anthony Dickinson
Department of Experimental Psychology,
University of Cambridge, UK

Fred I. Dretske
Department of Philosophy,
Duke University,
Durham, NC, USA

Nathan Emery
Sub-department of Animal Behaviour,
University of Cambridge, UK

William M. Fields
Department of Biology and Language
Research Center,
Georgia State University,
Atlanta, GA, USA

Louis M. Herman
Department of Psychology,
University of Hawaii,
and The Dolphin Institute,
Honolulu, HI, USA

Cecilia Heyes
Department of Psychology,
University College London, UK

Susan Hurley
Department of Philosophy, University of
Bristol, and All Soul's College,
Oxford, UK

Alex Kacelnik
Department of Zoology,
University of Oxford, UK

Janet Mann
Department of Psychology and Biology,
Georgetown University,
Washington, DC, USA

Ruth Garrett Millikan
Department of Philosophy,
University of Connecticut,
Storrs, CT, USA

Matthew Nudds
Department of Philosophy,
University of Edinburgh, UK

David Papineau
Department of Philosophy,
King's College London, UK

Irene M. Pepperberg
Department of Psychology,
Brandeis University,
Waltham, MA, USA

Daniel Povinelli
Cognitive Evolution Group,
University of Louisiana at Lafayette,
LA, USA

Joëlle Proust
Institut Jean-Nicod,
Centre National de la Recherche
Scientifique,
Ecole des Hautes Etudes en Sciences
Sociales et Ecole Normale Supérieure,
Paris, France

Duane M. Rumbaugh
Great Ape Trust of Iowa,
Des Moines,
IA, USA

E. Sue Savage-Rumbaugh
Department of Biology and Language
Research Center,
Georgia State University,
Atlanta, GA, USA

Sara J. Shettleworth
Department of Psychology,
University of Toronto,
Canada

Kim Sterelny
Philosophy Program,
Research School of
Social Sciences,
Australian National University,
Canberra,
Australia and Philosophy Program,
Victoria University,
Wellington, New Zealand

Jennifer E. Sutton
Department of Psychology,
University of Western Ontario,
Canada

Michael Tomasello
Max Planck Institute for Evolutionary
Anthropology,
Leipzig, Germany

Alain J.-P. C. Tschudin
Corpus Christi College,
Cambridge, UK

Elisabetta Visalberghi
Istituto di Scienze e Tecnologie della
Cognizione, CNR,
Rome, Italy

Jennifer Vonk
Cognitive Evolution Group,
University of Louisiana at Lafayette,
LA, USA

Chapter 1

The questions of animal rationality: Theory and evidence[1]

Editors' introduction

Susan Hurley and Matthew Nudds

Are any non-human animals rational? What issues are we raising when we ask this question? Are there different kinds or levels of rationality, some of which fall short of full human rationality? Should any *behaviour* by non-human animals be regarded as rational? What kinds of tasks can animals successfully perform? What kinds of *processes* control their performance at these tasks, and do they count as rational processes? Is it useful or theoretically justified to raise questions about the rationality of animals at all? Should we be interested in whether they are rational? Why does it matter?

The contributors to this volume approach these questions from a variety of theoretical and empirical perspectives. Contributors include distinguished philosophers as well as scientists, who report and reflect on their work with such impressive animals as Kanzi the bonobo, Betty the New Caledonian crow, Sheba the chimpanzee, Sweetie-Pie the scrub jay, Akeakamai the bottlenose dolphin, and Alex the African grey parrot. The volume aims both to bring leading empirical work with different species together for comparison, and to bring philosophical arguments about rationality into contact with empirical evidence of the behavioural and cognitive capacities of animals.

Section 1.1 of this chapter provides a landscape of theoretical issues and distinctions that bear on attributions of rationality to animals; it can be read independently of the synopses of chapters, which follow in the remaining sections. These summarize the arguments of the chapters in each of the volume's six parts, on: types and levels of rationality (Section 1.2); rational versus associative processes (Section 1.3); metacognition

[1] We are grateful to the contributors to this volume, and to Nicholas Shea, for helpful comments on the introduction. We would also like to acknowledge and thank the sponsors of the Oxford 2002 conference of the same name, from which this volume originates, for their generous support of this interdisciplinary project: The McDonnell Centre for Cognitive Neuroscience, Oxford; All Souls College, Oxford; the British Academy; the Mind Association; Mind and Language; and the British Society for the Philosophy of Science. We are also grateful to Jill Taylor for her careful and efficient editorial assistance.

(Section 1.4); social behaviour and cognition (Section 1.5); mind reading and behaviour reading (Section 1.6); and behaviour and cognition in symbolic environments (Section 1.7). The introduction as a whole aims to provide a substantive and self-standing survey of the topic of animal rationality that is accessible to students and researchers in different disciplines, including philosophy and various sciences.

1.1 Background: theoretical questions and distinctions

Before considering more closely what is meant by 'rationality', it is helpful to draw some rough contrasts between rationality and certain other traits and capacities and to consider the relationships between them.

1.1.1 Rationality and intelligence

We may be ready to admit that non-human animals are intelligent, especially in specific domains, but we tend to regard rationality as a special feature of human beings. Aristotle thought that what was distinctive about the human animal was its rationality. To assess the human claim to monopolize rationality, we need a clearer idea of what the difference is between intelligence and rationality (see Chapter 20 by Herman and Chapter 21 by Pepperberg).

Intelligence seems in some ways to be broader: it can be expressed in memory, learning, perceptual, and imaginative abilities. Some might describe dogs as intelligent because they can learn complex obedience or tracking tasks, or say that 2-year-old children display their intelligence in learning language. Others might prefer to say that an individual dog or 2-year old is intelligent on the basis that its capacities in certain domains are exceptional for its species. Such attributions of intelligence need not carry implications of rationality. Nor need rationality imply a high degree of intelligence; some savants might be regarded as rational but not very intelligent.

Should we distinguish intelligence from rationality in terms of certain kinds of *behavioural* tendency or capacity as opposed to certain kinds of *processes* underlying and sustaining behaviour? After all, we might remark, 'That behaviour was intelligent, but was it the result of rational processes?' But while the behaviour/process distinction is very important (and we return to it below), it doesn't coincide with the distinction between intelligence and rationality, since it applies to both. For we could just as well remark, 'That behaviour was rational, but was it the result of intelligent processes?'

1.1.2 Rationality, generalization, and decentring

Both intelligence and rationality require some degree of flexibility or ability to generalize from one context to another. But rationality seems to involve a more explicit use or recognition of more abstract similarities or patterns. There are differences of degree and of grain here: compare the ability to transfer an existing foraging technique to a new type of food with the ability to transfer transitive inference from social to non-social contexts—say from social dominance rankings to rankings of foraging opportunities. The latter transfer seems to require a more abstract ability, one that is less context-bound.

Moreover, we are inclined to recognize rationality as opposed to intelligence when the ability to generalize involves decentring from me-here-now, and entertaining alternative possibilities: taking into account the past, future, or counterfactual possibilities, other places, different possible actions, and other creatures' perspectives. Whether non-human animals have such capacities is discussed in several chapters in this volume.[2]

1.1.3 **Rationality and normativity**

Rationality in a sense that contrasts with intelligence also requires what philosophers of mind call *normativity*: there has to be a difference between doing the right thing and doing the wrong thing.[3] Suppose an animal appears to be trying to achieve some goal and to be making an error in relation to that goal. What basis is there for saying that the animal has made a mistake in the way it attempts to achieve its goal, rather than that it wasn't attempting to achieve *that* goal in the first place? Ironically, rationality requires the possibility that the animal might err. It can't be automatically right, no matter what it does; the possibility of mistakes and errors must make sense. A purely normative theory of rationality has no descriptive ambitions: it simply says what an agent *should* do, or believe, in order to be rational. But even descriptive theories of rationality—the kind relevant to non-human animals, so of concern here—must have a normative aspect: when we say that an agent has acted rationally, we imply that it would have been a mistake in some sense to have acted in certain different ways.[4] It can't be the case that anything the agent might do would count as rational. This is normativity in a quite weak sense.

The requirement of normativity can be interpreted in a stronger, reflective, way, as the demand that a rational animal can in some way *recognize* that it can make mistakes. On this view, it's not enough for there to be a possibility of mistakes only relative to, say, an evolutionary function that we theorists attribute to some behaviour. Rather, there has to be a possibility of mistake *for the animal* (see and cf. Dennett 1996, Ch. 2). Otherwise, although the animal's behaviour can be *described as if* it were rational or irrational from some external perspective, the animal won't really itself have acted for

---

[2] See and cf. the chapters by Millikan, Hurley, Clayton, *et al.*, Call, Shettleworth and Sutton, Currie, Tomasello and Call, Povinelli and Vonk, Tschudin, and Herman; see also Bermudez 2003, Ch. 6, on the distinction between what he calls level 1 and level 2 rationality.

[3] See and cf. the chapters by Hurley, Proust, and Sterelny.

[4] As Bermudez explains (2003, Ch. 6), rationality presupposes a space of alternatives; a rational behaviour contrasts with some other behaviour that could have been performed. The contrast can be between alternative types of behaviour (as in Bermudez' level 0 rationality) or between alternative tokens—particular instances of behaviour (as in his level 1 and level 2 rationality).

reasons. The capacity to recognize that one can make mistakes can be viewed as part of what distinguishes rationality from mere adaptedness, mere 'as if' rationality.[5]

1.1.4 Rationality and consciousness

If rationality makes this stronger reflective requirement, of a capacity to recognize the possibility of mistakes, that may suggest a relationship between rationality and consciousness. Is either necessary for the other? Here it is important to distinguish two aspects of consciousness.[6]

First, there is reflective consciousness, consciousness of or about one's mental states—a kind of higher-order mental state, or access to one's mental states. It is tempting to find links between rationality and reflective consciousness (see Chapter 10 by Call). It can be argued that recognizing the possibility of mistake, or decentring from self to recognize the possibility of other minds, requires higher-order mental states—mental states about other mental states. If such arguments are right, then reflective consciousness may fulfil this requirement of rationality.

Second, there is phenomenal consciousness: 'what it is like' to be in a given conscious state, say, of seeing red, or experiencing hunger. How is phenomenal consciousness related to rationality? Is the presence of information to phenomenal consciousness either necessary or sufficient for it to be available as a reason for acting? Arguably, if information is not available to the animal as a reason for acting then it is not present to phenomenal consciousness either. But could information provide an animal's reason for acting even if it is not phenomenally conscious? It could be argued that a sophisticated robot might, in principle, be a rational agent even if it lacked phenomenal consciousness. However, some theories try to explain phenomenal consciousness in terms of higher-order mental states, in effect linking phenomenal to reflective consciousness. If they are right, a role for reflective consciousness in rationality may also give phenomenal consciousness a role in rationality.

1.1.5 Rationality and conceptual and linguistic abilities

We began with a contrast between rationality and intelligence. Two features—the capacity to generalize in a relatively abstract way including to decentre from me-here-now, and the

---

[5] This reflective sense of normativity is necessary, though not sufficient, for a capacity to justify what one does—a capacity that animals appear to lack (see Bermúdez, 2003 and cf. Proust's chapter). A still stronger sense of 'normativity' would additionally require a capacity for justification (see and cf. Proust's chapter herein; Bermudez 2003).

[6] On consciousness in animals, see Allen and Bekoff 1997, Ch. 8; Dawkins 1998; Gomez 2004, Ch. 10. Further questions arise about the relationships of reflective and phenomenal consciousness to self-consciousness, and of self-consciousness to rationality, which we cannot pursue here. Note that 'self-consciousness' is applied to animals is quite different senses by different researchers; see and cf. Herman's discussion herein; Gallup *et al.* 2002; Gruen 2002.

capacity to make (and perhaps to recognize that one can make) mistakes—appear to be central to what is distinctive about rationality. If this is roughly right, it immediately raises several further issues about the relationship of rationality to conceptual abilities, to language, and to reasoning. Must an animal have concepts and conceptual abilities in order to generalize and to recognize mistakes? Indeed, must it have language?[7]

For example can an animal decentre from itself and take account of the perspective of another animal by simulation—by using its own practical abilities off-line—or does such decentring require theorizing about the other's mind and so require having both concepts of, and a theory about, mental states? Can simulation underwrite a sense in which animals can be rational despite lacking conceptual and theoretical abilities?[8] Must an animal have the concept of a mistake or of a reason in order to recognize that it has made or is likely to make a mistake—or could this again be registered and manifested in practical terms?[9] In the absence of the fine-grained distinctions that language makes available, how can we design experiments to assess such capacities for decentring or recognizing mistakes?[10] What kinds of non-verbal behaviour, under what conditions, might provide evidence for these capacities? How might symbols support or enable rationality?[11]

1.1.6 Rationality and reasoning; behavioural versus process rationality

The above questions about the roles of reflective consciousness and of conceptual and linguistic abilities in rationality in turn point to an underlying issue about the relationship between rationality and reasoning. Reasoning is a special kind of behaviour-generating process, which might be thought to involve reflective consciousness, or conceptual or linguistic abilities. Must rational behaviour be generated by reasoning, or, more generally, by a rational process? Is rationality a feature of behavioural patterns and capacities, or of the processes that sustain behaviour? This latter question is of central importance in considering how and in what ways animals may be rational.

An animal's behaviour can be described as rational from an external perspective without implying that the processes that generate that behaviour are rational. For example behaviour might be described as rational because it maximizes utility, or increases genetic fitness (see Kacelnik's discussions of E-rationality and B-rationality in Chapter 2). Any animal, if it is to satisfy its goals and to survive and reproduce, must act in the

---

[7] See and cf. Chater and Heyes 1994; Bermudez 2003; Allen 1999; Allen and Hauser 1991; Glock 2000; Cook 2002.

[8] See the chapters by Millikan, Hurley, and Proust.

[9] See the chapters by Millikan, Bermudez, and Hurley.

[10] See especially the chapters by Shettleworth and Sutton and by Currie.

[11] See especially the chapters by Herman, Boysen, Pepperberg, and Savage-Rumbaugh *et al.*

world: to do the appropriate thing, given its circumstances, and to adjust its behaviour as its circumstances change. We can ask whether an animal's behaviour is rational in this sense.

But an animal is rational in a different sense if its behaviour is rationally selected by the animal because it is appropriate to its circumstances. There is a sense in which the genetic fitness of certain behaviour may explain it; but it doesn't give a rational explanation of the behaviour: fitness doesn't show that the processes that produced the behaviour were rational. When we ask whether animals are rational, we are interested not just in whether their behaviour is rational, but also in whether rational processes explain their behaviour. That is, we want to know not just whether its behaviour is rational, but if so, why that is so: we want to know *how* the animal selects behaviour appropriate to its circumstances. In particular, we want to know whether rational processes explain how an animal manages to behave rationally.

In application to human beings, different disciplines use 'rationality' in different senses; some focus on rational behaviour and others on rational processes, in line with their different assumptions and purposes (as Kacelnik further elaborates in Chapter 2). As a result, there is a danger of talking at cross-purposes in interdisciplinary discussion. Some further background may help to avoid this danger by showing how the distinction between *behavioural rationality* and *process rationality* arises in various contexts.

1.1.7 Ends versus means

The rationality of ends can be distinguished from the rationality of means, or instrumental rationality. Instrumental rationality is concerned with finding the best means to a given end or goal—something we desire or need or want—rather than with the selection of ends.[12] It concerns whether, given certain ends, an agent is appropriately sensitive to information about means/ends contingencies.[13] 'Appropriate sensitivity' can be understood in different ways, in terms of behaviour or the processes that generate it (see below). An important question here is whether the sensitivity to means/ends information appropriate to instrumental rationality requires, or is enhanced by, representations of causality (see Call's Chapter 10) or of the intentional states of others (see Chapter 9 by Clayton *et al.* and those in Part V).

---

[12] The distinction between ends and means can be understood as relative rather than absolute; means may be more or less proximal, ends more or less ultimate. An intentional bodily movement may be the means to a physical effect, which is in turn the means to signal an intention, which is in turn the means to a social move, which is in turn the means to some further end. Similar differences of degree along a spectrum hold for intentions, which may relate to ends or to means along the spectrum. There are ultimate intentions relating to one's ultimate ends and other less ultimate background or prior intentions. There are also more or less proximate intentions about how to achieve one's ends.

[13] See and cf. the chapter by Clayton *et al.*, herein; Bermudez 2003, Ch. 6, on level 2 rationality.

The rationality of ends themselves is more problematic to assess. Do we have any basis for holding certain human goals to be irrational? (We'll return to this question below, in discussing the relationships between formal and substantive rationality.) In the animal case, if we appeal to biological functions to assess an animal's goals, are we addressing their rationality at all, as opposed to their adaptiveness?

1.1.8 Instrumental rationality as behavioural rationality

Instrumental rationality can be understood as either behavioural or process rationality; classical conceptions of human rationality in different disciplines contrast in this respect. Instrumental rationality is classically conceived by economists in terms of consistent patterns of behaviour (what Kacelnik calls 'E-rationality') and by philosophers and many psychologists in terms of the processes that generate behaviour (what Kacelnik calls 'PP-rationality').

A reasoning process may be reflected in the theory of why the behaviour it would lead to counts as rational. But behavioural rationality does not itself require that reasoning be rehearsed by the agent or that any particular process actually generate her behaviour. Behavioural rationality is thus, in principle, open to the possibility that the processes that actually generate an agent's rational behaviour do not correspond to the theoretical account of why the behaviour counts as rational. In this sense, behavioural rationality is, in principle, compatible with the use of rules of thumb or *heuristics*. Heuristic are decision-making processes that can reliably generate rational behaviour in specific contexts or environments, even though they do not implement ideal reasoning processes and so may not generate rational behaviour in other contexts or environments.

A further question is just *how* reliable the use of heuristics must be, across what range of variation in possible environments, in order to be compatible with rational behaviour. Some perspective on this question can be provided by a potted four-stage history of conceptions of behavioural rationality.

(a) *Classical behavioural rationality*. In economics, idealized normative theories of behavioural rationality are exemplified by orthodox expected utility theory (EUT), on which orthodox game theory builds (see below for more on the difference between individual rationality, as in utility theory, and strategic social rationality, as in game theory). EUT says that the rational choice from a range of alternatives is the one that maximizes the agent's expected utility, that is, the sum of the probability-weighted utilities of each possible outcome of a possible action. This is a conception of instrumentally rational behaviour: possible choices are assessed in terms of their likely consequences, as means to the agent's ends. Alternative means by which an agent's ends might be pursued are assessed in terms of the degree to which their consequences are likely to fulfil the agent's ends. EUT is infinitely flexible, and *domain-general* rather than *domain-specific*. That is, it is not tied to environments with any particular structure, but should in principle work equally well in any environment (*if*—a big 'if'—we abstract from the costs of applying it). Regardless of whether such classical theories of rationality describe the actual processes that generate behaviour, they were regarded as broadly predictive of human behaviour— though of course lapses here and there were admitted.

(b) *Heuristics and biases.* A wave of debunking experimental research was ushered in by Kahneman and Tversky's seminal work, which showed that the norms and/or predictions of classical theories of behavioural rationality were violated systematically by human decision-makers, in a variety of specific, well-charted ways. Decision-makers' patterns of choice were better described in terms of various heuristics and biases,[14] some domain-specific, than by classical conceptions of behavioural rationality such as EUT. These findings were generally taken to be bad news for rationality: to show that human behaviour was less rational than might have been expected or hoped (rather than that rationality had been misunderstood by EUT). From the point of view of behavioural rationality, the bad news was not that much human behaviour is generated by the use of heuristics rather than rational processes; it was the systematic irrationality of the behaviour. Heuristics would have been acceptable had they reliably generated rational behaviour. The problem was that they didn't.[15]

(c) *Ecological rationality.* A more positive reaction to the debunking evidence eventually emerged, represented by Gigerenzer *et al.*'s (1999) slogan that 'simple heuristics make us smart' and accompanying scepticism about the unrealistic 'demonic' character of classical conceptions of rationality, with their disregard for decision-making costs and for the practical limitations on real decision-makers. Further experimental work on ecological rationality demonstrated that in the right environments reliance on simple heuristics, such as the familiarity heuristic, or the one-good-reason heuristic, produces behaviour with better consequences than reliance on more sophisticated and flexible but costly domain-general decision-making processes. Evolution, it was argued, won't build the costly flexibility of demonic processes that produce rational behaviour in all possible environments into decision-makers. Rather, it will sensibly fit them out with a toolkit of heuristics adapted to and triggered by interaction with environments they are likely actually to face.[16] If relevant environments reliably include certain informational

---

[14] See Kahneman, Slovic and Tversky 1982, Kahneman and Tversky 2000a, b; Nisbett and Ross 1980, Bargh 1997, 2005; Bargh and Chartrand 1999; Chartrand and Bargh 1999, and so on. For present purposes, heuristics and biases include confirmation bias, the representativeness heuristic, the base-rate fallacy, uncertainty aversion, the Wason effect, attribution errors, framing effects; loss aversion; chameleon effects, and so on.

[15] Human decision makers were also shown often to depart from the norms of game theory, for example by co-operating more than they rationally 'should' in PDs and by turning down unfair ultimatum offers. The normative import of these social 'failures' of rationality was less clear: perhaps bad news for rationality, but good news for morality?

[16] For the animal version of the adaptive toolkit, see Hauser 2000.

resources, evolution may not duplicate them internally (see and cf. Brooks 1999; Clark 2001; Wilson 2004). On this view, rational behaviour is environmentally situated; the processes that generate it are, in effect, distributed across agents' heuristics and the environments with which agents are interactively coupled. The structure of specific environments and information contained in specific environmental processes does not just contribute to the effects of behaviour, but can contribute to the processes that generate rational behaviour.

(d) *Machiavellian social rationality.* Recently Kim Sterelny[17] has argued that it is no accident that proponents of ecological rationality tend to focus on behaviour in non-social environments. Sterelny explains that natural non-social environments are transparent, or at least translucent, in relation to information about an agent's most effective means to his ends; by contrast, social environments are often informationally opaque and hostile, as a result of strategic thinking, deception, mimicry, and informational manipulation. Nature doesn't try to predict and outwit an individual agent deciding on the best means to achieve his ends;[18] but other agents often do, in order to better achieve their own ends. Evolutionary pressures favouring deceptive mimicry and other ways of manipulating information in social environments contribute significantly to the complexity of social information. As a result, social environments place high demands on cognition. It is implausible that these demands can be met in a way that would generate rational behaviour by simple heuristics that rely on information in environmental structures, given that such information is subject to manipulation and deception. This position draws on the Machiavellian Intelligence Hypothesis:[19] that social life, with its potential for deception and manipulation of information, drives the development of advanced cognitive capacities, which may include capacities for social cognition and metacognition.

If simple heuristics cannot cope with informationally hostile social environments, what is the next bounce for conceptions of instrumental behavioural rationality? Such conceptions might retreat from ecological optimism and revert to a classical view of normative behavioural rationality after all—though one tempered by the realization that human beings fall systematically short of rationality and so are liable to behave irrationally in challenging social environments to which their simple heuristics are inadequate. Or, is there a plausible way for an ecological view of rationality to go social? Are there social heuristics—heuristics well-adapted to social environments—that reliably generate behaviour fit to achieve one's ends in social environments and that are unlikely

---

[17] For example by Sterelny 2003, 25; see also his chapter and Proust's chapter in this volume; cf. Gigerenzer 1997 for scepticism about claims of social complexity.

[18] Natural selection may do plenty of 'outwitting' in phylogenetic time, but not in decision-making time.

[19] Byrne and Whiten 1988; Whiten and Byrne 1997; sometimes also called the 'social intelligence hypothesis'.

to be undermined by deception and the manipulation of information?[20] Can the processes that generate rational behaviour be distributed across an agent's social heuristics and her social environment (including other agents), despite the informational hostility of social environments?

1.1.9 Instrumental rationality as process rationality

By contrast with human rationality as understood in economics, human rationality as conceived in philosophy and parts of psychology is process rationality (as in Kacelnik's PP-rationality). Instrumental process rationality is a feature of the way we select the means to achieve our ends. It is not sufficient for process rationality that someone's behaviour is appropriate to his ends, because this may result from coincidence or accident. A rational process must lead to the right results reliably, not accidentally. Behavioural rationality necessarily concerns behaviour and is in that sense practical; but process rationality can concern the processes that lead to beliefs as well as the processes that lead to behaviour. That is, process rationality includes *theoretical* as well as *practical* rationality (see below for more on this distinction). In the practical instrumental case, process rationality reliably selects the action that is most likely to achieve an agent's ends; in the theoretical case, process rationality leads to beliefs that are true or likely to be true. The essential idea is that of getting things right in the right ways. Behaviour is rational in a sense that derives from the rationality of the process that generates it: rational behaviour is non-accidentally appropriate to my goals, because it results from the right kind of process, a rational one.

When an animal's behaviour on some task is functionally similar to human behaviour on a similar task, it may be tempting to explain the animal's behaviour in the same ways we would explain the human behaviour (see Chapter 11 by Shettleworth and Sutton)—say, in terms of processes of inference and reasoning, or metacognition, or mind reading. But different processes can lead to similar behaviour (even among human beings, for example, in recovery of function after brain damage). Without language, how can we determine whether to attribute a rational process to explain an animal's behaviour? More fundamentally, what kinds of processes are the 'right kind' to count as rational?

(a) *Classical process rationality: reasoning.* A classical conception of process rationality views a rational process as a reasoning process:

> If someone acts with an intention then he must have attitudes and beliefs from which, had he been aware of them and had he the time, he could have reasoned that his act was desirable . . . If we can characterise the reasoning that would serve we will, in effect, have described the logical relations between descriptions of beliefs and desires and the description of the action, when the former gives the reasons with which the latter was performed. We are to imagine, then, that the agent's beliefs and desires provide him with the premises of an argument. (Davidson 1980, pp. 85–86.)

---

[20] See and cf. Gigerenzer 1997; Hurley 2005.

On this view, being rational is a matter of working out the means to one's ends and of reaching decisions and judgements by a normative process of reasoning: by carrying out explicit probabilistic, logical, or decision-theoretic inferences. Both theoretical and practical instrumental rationality are seen to demand reasoning; reasoning can be aimed at solving a problem, making a decision, planning a course of action, arriving at a judgement or prediction. In practical instrumental reasoning we reason about how to achieve our ends; in theoretical reasoning we reason about what is true or probable.

What is characteristic of reasoning, and why is it a good thing to make decisions in this way? Reasoning, arguably, has the core features of flexible generality and normativity—which, as we've seen, are associated with rationality—in rather strong forms.

Reasoning requires a capacity to generalize and to go beyond me-here-now. For example means–ends reasoning requires the agent to think about different non-actual actions and their possible results, to think counterfactually. And reasoning arguably makes this requirement of flexible generality in a strong form. Reasoning is an entirely general capacity; its application is not restricted to particular topics or environments, and extends fully counterfactually. If someone can reason about one thing then they can reason about anything they can think about. The domain-generality and content-neutrality of reasoning reflects its logical structure: you get the right answer irrespective of the content of the premises so long as the reasoning is valid. Given an end and some beliefs about the best means to achieve it, reasoning gives the agent a general-purpose way to work out what to do. If her beliefs are correct, and she makes no mistakes in the working out, the resulting action will achieve the end, or be more likely to do so than any alternative.

Reasoning also is normative in the sense that there is a difference between reasoning correctly and making a mistake in reasoning. It arguably requires thinking processes to fall under certain cognitive norms or rules, in the sense that thinking *should* satisfy these rules, though it may fail to do so. This does not itself require that the agent consciously apply and follow the rules, trying to avoid mistakes; his thinking processes just may follow the rules, fairly reliably. Stronger forms of normativity (as explained earlier) also require a capacity to recognize that one can make mistakes, or reflective consciousness, or even a capacity to justify one's decisions and judgements.

Reasoning is often thought to involve some such stronger form of reflective normativity: to requires not just thinking in accord with rules but also a capacity for reflection on the processes of thinking that lead to a decision, for some form of metarepresentation (see Bermudez 2003, Ch. 8 and 9). Not only should it be no accident that one's behaviour is rational, but it should be no accident that the thinking that leads to one's behaviour is rational. It might be held that normativity in the stronger sense enables reasoners to *know* that the processes that lead to their decisions are likely to lead to the right results. This is because it requires them not just to select the right decisions in the right way, but also to select the right processes for selecting those decisions in the right (reflective) way.

Moreover, the ability to reflect on their thinking processes enables reasoners to correct their mistakes. On the other hand, a strong, reflective requirement of normativity threatens a regress of rules for applying rules for applying rules . . . (as in the tale of Achilles and the Tortoise; Carroll 1895). This may mean that the supposed advantages of some stronger requirements of normativity are illusory.

Classical reasoning processes, then, are usually thought of as requiring a capacity to generalize in the strong form involving domain-neutrality. They can be thought of as taking weaker or stronger forms in respect of normativity: weaker forms require rule-following and the possibility of mistake, while stronger forms require reflective normativity and/or a capacity to provide justifications. It may be attractive to think of rational processes to require reasoning, though not to require the stronger reflective forms of normativity.

(b) *Process rationality without reasoning?* Processes with the unrestricted generality and flexibility of reasoning can be costly, as proponents of bounded and ecological rationality argue. Perhaps the costs outweigh the benefits. Perhaps fully domain-general reasoning is too much to require of rational processes. Can a more inclusive characterization of a rational process be given, which retains the aspect of generality (as well as of normativity) in a weaker form? A number of the contributors to this volume consider how processes other than explicit classical reasoning processes might nevertheless count as rational processes explanatory of animal behaviour.[21]

What kinds of processes other than classical reasoning processes might count as rational, and why? Can the use of heuristics count? Must rational processes at the personal (or animal) level be enabled by subpersonal processes with a corresponding or isomorphic structure, in order genuinely to provide a rational explanation of behaviour? Does explanation of behaviour in terms of associative mechanisms displace explanation in terms of rational processes? Can rational processes or their subpersonal enabling processes be extended—widely distributed across neural, bodily, and situating environmental processes—or must they be contained in the animal?

(c) *Use of heuristics.* While reasoning is a paradigmatic rational process, it may not be as prevalent as we tend to assume, even among human beings. It might be thought that people often arrive at beliefs and perform actions based on perceptual experiences, in ways that do not involve reasoning (even if they reason about the relations among their beliefs). Moreover, as we've seen, there is evidence that people tend systematically to use heuristics and display biases that are not consistent with domain-general reasoning. While the success of heuristics in generating behaviour that achieves our ends may be restricted to specific domains, if those include the domains that matter most to an agent, then perhaps such success is sufficient for rationality. And successful heuristics

---

[21] See the chapters by Dretske, Millikan, Clayton *et al.*, and Proust; cf. the chapters by Call and by Bermudez on animal reasoning, and see Bermudez 2003 on rationality without reasoning.

can be fast and frugal, so less costly than reasoning. Can rationality be understood in terms of an adaptive toolbox of such domain-specific heuristics, layered together by evolution so as collectively to cope flexibly with a wide range of relevant problems?[22] Might arriving at decisions in accord with such heuristics count as a rational process, even if not a process of reasoning?

There are two ways to understand this suggestion, as involving either a less radical or a more radical departure from classical reasoning. On the less radical view, although people can reason about what to do in the classical way, it is often rational to use heuristics instead in certain contexts to decide what to do, because it is known that they are just as likely to be successful in those contexts, and they are easier to use. In this sense, the use of heuristics may be selected reflectively and on domain-general principles, even if heuristics aren't themselves domain-general. We may use reasoning to select which heuristics to use, even if using heuristics is not itself a kind of reasoning; such metacognition may be enough to make the use of heuristics a rational process, or part of one.

On the more radical view, metacognition is not needed for rational processes; using heuristics can be a rational process even if their use is not in turn selected by a process of reasoning. Perhaps we don't decide reflectively to use heuristics, on the general grounds that they are just as likely to lead to success in certain contexts and are less costly. Rather, using them may just be an evolved feature of the way our minds work, perhaps implemented by associative mechanisms. Such processes generate actions that are likely to achieve our goals reliably (in the right environments), not by accident, even though there is no higher-order reasoning process, reflective and domain-general, that explains why we use heuristics. The more radical suggestion is that the use of heuristics could still qualify as a rational process. It may meet weaker requirements of normativity and generality than does reflective, domain-general reasoning. On this view, classical reasoning is too restrictive as a conception of process rationality, even in the human case.

This would make more room for process rationality in animals. Many animals have problem-solving capacities that are specific to a particular task or environment, which they cannot generalize and apply to other tasks. Arguably, domain-general reasoning requires logical concepts and linguistic capacities that animals lack.[23] But if domain-general reasoning is too restrictive as a conception of process rationality in the human case, then it shouldn't be required for process rationality in animals either.

(d) *Personal level versus subpersonal level processes, and associative processes.* Following Dennett (1969, 1978, 1987, 1991; cf. Bermúdez 2000), many philosophers distinguish

---

[22] See the chapter by Hurley herein; see also Shettleworth 1998, 43, 56, and passim; Santos *et al.* 2002; Hare and Wrangham 2002; Bermudez 2003, 185 ff; Hauser 2000.

[23] It may be language that eventually allows domain-general reasoning to emerge; see and cf. the chapter by Hurley; Bermúdez' chapter on non-linguistic analogues of reasoning that might be attributed to animals; the chapter by Savage-Rumbaugh *et al.* on the language and rationality in apes.

descriptions of contentful mental states and processes attributed to persons from sub-personal descriptions of information being processed and passed between subsystems (see Hurley's Chapter 6 for a similar distinction for animals). Cognitive science shows us how the mental lives of persons are enabled by subpersonal information processes (McDowell 1994a). Subpersonal processes can be described functionally or in terms of their neural implementations. Some subpersonal processes are described in cognitive terms, involving representations; some in terms of symbolic representations; others merely in terms of associative mechanisms.[24]

Reasoning is a personal-level process; it seems natural to assume that rational processes more generally are also personal-level processes. But it might be held that a genuinely rational process, one that provides a rational explanation of behaviour, must be enabled by subpersonal processes that correspond to it structurally in certain ways. For example, the subpersonal cognitive processes enabling rational thought processes might be required to have a structure isomorphic to the structure of rational thought.[25] This may be regarded as part of what it is for a rational process to be the right kind of process. It wouldn't be enough, on this view, for subpersonal cognitive processes to enable people reliably to arrive at the right answers, if those subpersonal processes bear no intelligible relationship to personal-level processes.

Related issues also arise for animals. It may be tempting to explain the successful performance of certain complex tasks by animals in terms of cognitive processes such as inference, following abstract rules, or metacognition.[26] But often sophisticated performances can also be explained in terms of associative mechanisms (see Chapter 8 by Papineau and Heyes). How are associative explanations related to cognitive explanations (explanations in terms of representations), and to rational explanations 'at the animal level'? Does some version of an associative/ cognitive distinction hold up? Do associative

---

[24] See Heyes 2000 for a characterization of cognitive processes as theoretical entities that provide a functional characterization of the operations of the central nervous system, which need not be objects of consciousness, and which receive inputs from perception and other cognitive processes and have outputs to other cognitive processes and behaviour. She contrasts how-rules for carving cognitive processes into kinds, based on how information is processed, with what-when rules, based on the content of cognitive processes and the time in development when they typically operate. See also and cf. Shettleworth 1998, 5–6, 567–569.

[25] See Fodor 1975, 1987; Davies 2000. In her chapter herein, Hurley rejects interlevel isomorphism requirements; see also and cf. Hurley 1998a; Bermudez 2003, 22–25, 111, 192; Sterelny 2003. Related familiar debates in philosophy concern mental causation (when is behaviour caused or explained by mental events or properties, as opposed to by the physical events or properties in which they inhere?) and eliminativism (does it matter if non-symbolic connectionist networks or dynamic systems give a better account of cognitive processes than do symbolic representations?).

[26] See the chapters by Allen, Clayton *et al.*, Call, and Shettleworth and Sutton; Bermúdez' chapter argues for the possibility of cognitive processes of inference that are neither associative nor involve following abstract rules.

mechanisms implement explanatory rational processes, or do they undercut them, leaving them with no real explanatory work to do? If associative explanations exclude rational explanations, should we always prefer the former in accord with Morgan's Canon, on the grounds that they are simpler or 'lower' processes ?[27] What kind of behaviour, if any, firmly resists explanation in associative terms, and does it matter?

(e) *Widely distributed processes: extended rationality?* It is a familiar idea that representations and cognitive processes may be distributed. This is often assumed to mean distributed within the head, as in distributed neural networks. However, the distribution may be wider than that: the situated cognition, extended mind, and ecological rationality movements suggest the possibility that the cognitive processes explanatory of behaviour can be distributed across agents and the information-carrying environments with which they interact. This view in turn raises the questions of where rational processes can be located, and how they are bounded. Must rational processes be wholly internal to the rational animal's brain or body, or can they be distributed across the animal's brain, body, and environment? Must the external portions of animal—environment interactions merely stand in causal relationships with rational processes, or can they be part of what constitutes a rational process? On what principles should such a causal/ constitutive boundary be drawn? Can we regard an animal itself as rational if its ongoing behaviour is best explained by such extended processes, not just by internal processes? If so, are the boundaries of extended rational processes provided by biological evolution, or by culture? What is the role of interactions with social and symbolic environments in human rationality? Can reasoning processes extend to include discussion and argument, or the use of pen and paper in proofs? When animals are encultured, raised by and with human beings and trained to use symbols, do they become more rational than they would be 'naturally'?[28]

The explanation of behaviour in terms of widely distributed processes, including processes of language use,[29] begins to blur the distinction between rational processes and rational behaviour. Extended explanations of behaviour are dynamic explanations, of behaviour conceived as extending through time. They explain how ongoing patterns of behaviour are sustained in terms of the dynamic interactions between

---

[27] See Currie's chapter for discussion. Morgan's Canon says that in no case is an animal activity to be interpreted in terms of higher psychological processes, if it can be fairly interpreted in terms of processes that stand lower in the scale of psychological evolution and development. It's plausible that this is just an application to psychology of a general methodological principle in science, viz. prefer simpler explanations. Currie glosses 'higher' processes to enable behavioural capacities that include those enabled by 'lower' processes.

[28] See the chapters by Herman, Pepperberg, Boysen, and Savage-Rumbaugh *et al.* On the extended mind and externalism about the vehicles of mental contents, see for example Clark and Chalmers 1998; Hurley 1998a, b; Rowlands 2003; Clark 2001, 2005a; Wilson 2004 .

[29] See Clark 2005b; Dennett 1996a, Ch. 5.

brain, body, and environment, as one period of movement in a given environment has feedback effects on internal mechanisms that produce the next period of movement, and so on. Bodily movements and their environmental results are an essential part, on the 'embodied, embedded' view, of the processes that explain behaviour. Hence they may be an essential part of the rational subset of these processes—at least the basis for insisting that rational processes must be wholly internally constituted becomes unclear. On this conception, rational processes are on view in the world.

A conception of rational processes that is liberal on all the above issues would admit rational processes other than reasoning processes, which could include the use of domain-specific heuristics, processes implemented by non-isomorphic subpersonal mechanisms including associative mechanisms, and widely distributed processes—if these processes are part of what explains why the person (or other animal) reliably gets the right result. If this is too liberal a view of the processes that could provide a rational explanation, how exactly should it be tightened up, and why? A critical question for animal rationality is how to motivate and characterize a middle ground between a requirement of reflective, domain-general reasoning and what might be viewed as excessive liberality about rational processes.

1.1.10 **Formal versus substantive rationality**

Let's return now to the idea that ends can be assessed for rationality, as well as means. Some light is shed on this idea, as well as on the distinction between behavioural and process rationality, by considering another distinction: that between formal and substantive rationality.

Formal rationality requires that certain formal patterns of preference or expectation, or transitions between them, be respected. For example if A is preferred to B and B to C, then C should not be preferred to A. Substantive rationality by contrast requires that preferences or expectations have certain contents, that they be accurate or enlightened or prudent or virtuous. For example it might be regarded as substantively irrational to desire to take up smoking, even if the pattern of your preferences and expectations was formally consistent. The relationship between formal and substantive rationality has been important for contemporary debates in decision theory. While officially decision theory is concerned with the rules of formal rationality, a problem arises if the contents of preferences and expectations can be endlessly pruned and reinterpreted to make them fit formal requirements.[30] For example perhaps your choices can be regarded as meeting a requirement of formal consistency if your preferences are interpreted to give you the end of avoiding pain on all days of the week other than Thursdays. But the substantive irrationality of such an arbitrary end may argue against this explanation of your behaviour as formally rational, and support instead the view that you are formally

---

[30] See Kahenman and Tversky 2000a, 13; Hurley 1989.

irrational. In the absence of any constraints on the substantive contents of preferences or expectations, formal rationality (whether understood as explanatory or normative) rules nothing out and becomes empty. So formal rationality appears to depend implicitly on assumptions about the substantive contents of preferences and expectations. And such assumptions may in effect include assumptions about the rationality or irrationality of certain ends. How such assumptions could be explicitly defended is a further question. In the case of animals, biological fitness will probably feature in any answer.[31]

The emptiness of a purely formal conception of rationality also puts pressure on a purely behavioural conception of rationality, with no element of process rationality. Our behaviour expresses our intentions and preferences, but it does not wear them on its sleeve. The same observable behaviour could be intentional under quite different descriptions, depending on context, and may express quite different preferences. I may choose the pear rather than the chocolate walnut cake because I am on a diet, because I dislike chocolate, because I am allergic to nuts, because it is more compatible with the wine, etc. The object of my choice is accordingly less independent of the processes that lead to it than a purely behavioural conception of rationality may assume. In order for the theorist of rationality to identify correctly the object of my choice, so as to relate it to the object of my choice in other cases, and hence to determine whether my patterns of behaviour are rational or not, she may need to understand something about my ends, about what intentions and preferences my choices express, and hence about the process that generates my choices. In judging a pattern of behaviour to be rational, we are implicitly judging a pattern of preferences with certain contents, which my choices express, to be rational (see Kahneman and Tversky 2000a, xiv–xv). In the case of human beings, we may overlook this point because we can resolve it using language: in the way we specify the options, or by asking people to say what they prefer (though linguistically specified choices are associated with persistent forms of irrationality, such as framing effects; see Kahneman and Tversky 2000a, b). In the case of animals, language doesn't come to the rescue. But the point remains that behaviour fits different patterns depending on what we take its object to be, and the processes that generate the behaviour will often be relevant to determining that.

1.1.11 Individual versus social rationality

Decision theory concerns individual rationality: what one individual should do who is playing, in the jargon, 'against nature', where nature is not another agent. But the situation is much more complex if we move to game theoretic contexts and to strategic rationality, where the agent is playing against other agents who are in turn playing against him. Here, the agents' decisions are interactive. A basic equilibrium solution concept in game theory is that of 'best reply'. This is a conception of behavioural

---

[31] See Kacelnik's chapter for discussion of related issues about 'biological rationality'.

rationality: my action is rational if it is the best reply for me to what you do, and yours is rational if it is the best reply for you to what I do.

While game theoretic rationality can thus be understood to demand certain behaviours rather than certain decision processes, in the way characteristic of E-rationality, game theory also raises important questions about how we decide what to do in such interactive situations and the function in them of understanding other minds. What is rational for me to do depends on what I predict you will do, and what is rational for you to do depends on what you predict I will do. If we are both rational, we each need to occupy the rational perspective of the other in order to predict the other's behaviour and hence to decide rationally ourselves what to do. Such game theoretic contexts, and social contexts more generally, are a source of pressure to acquire the ability to decentre from self and capacities for social cognition and social rationality: to understand that another agent may have goals and expectations different from one's own, and to use that understanding to predict her behaviour and to perhaps to manipulate her informational environment.[32]

The study of social interactions and cognition in animals is a topic of great current interest. What kinds of understanding of one another do their social interactions involve? Can apparently rational social learning be explained in terms of lower-level processes (see Chapter 15 by Addessi and Visalberghi)? In particular, do non-human animals attribute mental states to other animals: are they mind readers or are they just very smart behaviour readers? What exactly is the difference between genuine mind reading and mere behaviour reading? How can this difference be assessed; what evidence would justify attributing mind reading to non-linguistic animals?[33] Could there be a difference in the functions of mind reading and of behaviour reading, even if the evidence for mind reading is behavioural? Is it helpful to distinguish different kinds or levels of understanding of other minds? There is evidence that children come to understand the desires of others before they understand their beliefs (Rakoczy et al., submitted). Might animals who fail non-verbal versions of tests for the ability to attribute false beliefs to others nevertheless understand the conative states of others and pass other, less demanding, tests of mind reading? Is whatever social rationality of which animals are capable limited to certain domains or activities? Is it best understood in terms of their possession of a rudimentary theory of mind, or rather in terms of a practical ability to simulate the decision-making processes of other agents?

1.1.12 Practical versus theoretical rationality

The distinction between practical and theoretical rationality, the rationality of actions versus the rationality of beliefs, is important in assessing the rationality of non-human

---

[32] See the chapters herein by Proust, Sterelny, Connor and Mann, Tschudin, and Herman; see also Dugatkin and Alfieri 2002; Bermudez 2003, 119–120, 122; Sterelny 2003; Hurley in press.

[33] See the chapters by Tomasello and Call, Povinelli and Vonk, Tschudin, and Savage-Rumbaugh et al.

animals. Theoretical rationality involves evaluating the evidence for and against beliefs, and arguably depends on conceptual or even linguistic abilities. Practical rationality may be less demanding, and be more plausible to attribute to animals.[34] Even if an animal cannot weigh up evidence with due scepticism or assess explanations in a theoretically rational manner, its actions might nevertheless be instrumentally rational, that is, rationally sensitive to information about means–ends contingencies.

The theoretical–practical distinction is relevant to assessing the social rationality of animals. It may be that non-human animals have an ability to understand one another that is more practical than theoretical, in two senses: animals may understand the goals and intentions of others more readily than their beliefs, and they may do so by off-line simulation of practical decision-making capacities rather than by theoretical reasoning. Simulation is another candidate for a rational process that is not a classical reasoning process (though simulation can also be understood as subpersonal enabling process).

The theoretical–practical distinction is also relevant to the idea of rational processes that are not fully domain-general. It is natural, if you think of rationality in theoretical terms, to construe it as a domain-general capacity, not restricted to any specific contexts, such as food gathering or social activities. However, there is evidence that even human rationality may be tuned to specific domains, such as the detection of cheating in social contract situations.[35] If rationality is understood in practical terms, this domain-specificity becomes more understandable, as practical know-how is often implicit and bound to specific procedures and domains. Moreover, what a creature can do on-line may limit what it can do off-line, or simulate doing. So, if the mind-reading abilities of animals are understood in terms of a practical ability to simulate the decision-making processes of others, then we should not be surprised to find that their mind reading skills are limited to specific contexts (see Chapter 6 by Hurley).

1.1.13 The landscape of questions about animal rationality

So far, we've sketched a landscape of distinctions and issues against which animal rationality can be assessed. To summarize:

- Rationality differs from intelligence
  - in requiring a greater capacity for flexible generalization—extending to some capacity for decentring from me here and now, if not to full domain-generality—and
  - in its normativity, which requires the possibility of making mistakes and perhaps some kind of reflective awareness, or metacognition, concerning the possibility of mistake.

---

[34] See the chapters by Millikan and by Hurley.

[35] Cosmides and Tooby 1992, 1997; but cf. Oaksford and Chater 1998; Sperber and Girotto 2003. Domain-specificity should not be confused with innateness. Abilities may be domain-specific but depend on learning and development in, or scaffolding by, specific environments.

◆ Disciplines vary in their focus on rational behaviour as opposed to on rational processes that explain behaviour.

◆ Rational behaviour is usually understood in terms of instrumental practical rationality, concerning the rational means to a given end, rather than the rationality of ends. However, it is arguable that the implicit dependence of purely formal conceptions of rationality on some substantive assumptions puts pressure on behavioural rationality to include some process elements.

◆ Classical conceptions of behavioural rationality, such as EUT, have been challenged by results from the heuristics and biases line of research, and revised and liberalized by the ecological rationality school.

◆ Process rationality can be understood either as practical, concerning the processes that explain actions, or as theoretical, concerning the processes than explain beliefs.

◆ A classical view of process rationality in terms of reflective, domain-general reasoning might be revised and liberalized by allowing other processes to count as rational, such as domain-specific heuristics, processes implemented by associative mechanisms, and widely distributed processes.

◆ Such liberal revisions of classical conceptions of human behavioural and process rationality may make more room for animal rationality.

◆ In social environments, agents play against other agents (rather than nature), who can manipulate information; this creates selection pressure that may drive the development of advanced cognitive capacities, including behaviour reading and mind reading capacities, in social animals.

◆ Practical rationality may be more accommodating than theoretical rationality of the forms of social cognition found in non-human animals, and of the domain-specificity of many of their problem-solving abilities.

With this background, we'll now survey the chapters. They are divided into parts that focus on: types and levels of rationality; the distinction between rational and associative processes; metacognition; social behaviour and cognition; mind reading and behaviour reading; and behaviour and cognition in symbolic environments.

1.2 Types and levels of rationality

The contributors in this section explore the relations between behaviour and the processes that explain behaviour, and the senses in which animal behaviour might be rational in virtue of features other than classical reasoning processes on the human model.

1.2.1 Kacelnik

When different disciplines use a concept in different ways and for different purposes, there is always a danger that they will talk past one another. Zoologist Alex Kacelnik

(Chapter 2) helps to avert this danger by distinguishing three conceptions of rationality, employed in philosophy and psychology, in economics, and in biology. *PP-rationality* is typically adopted by philosophers and cognitive psychologists. On this conception, rationality is exhibited when beliefs or actions are adopted on the basis of appropriate reasons. PP-rationality focuses on the process by which belief or action is arrived at, rather than the outcome of beliefs or actions—in contrast to rationality understood as utility maximization, which is a matter of behaviour appropriate to goals. However, beliefs or actions produced in appropriate ways should, over time, produce appropriate outcomes.

Since PP-rationality is understood in terms of thought processes, Kacelnik suggests that it is not an appropriate conception of rationality with which to address questions about the rationality of animals. From a biologist's perspective, the problem with discovering whether non-linguistic animals are PP-rational is that of finding a way to ascertain what the subject's thoughts are and whether they result from an appropriate process of reasoning. The concepts used in characterizing PP-rationality—belief, thought, reasoning, and so on—are closely tied to language and don't have immediate application to non-linguistic animals. But a more general worry is that the processes that produce behaviour are not observable. And even if some animal behaviour may be best explained by reference to representations that aren't directly observable, the causation of behaviour by representations doesn't entail that the behaviour results from the subject's reasoning. This difficulty is not limited to animals, but applies in the human case also. It is hard to tell whether a human subject's belief has been arrived at on the basis of appropriate reasons: 'the elements that entered into her reasoning process may [although they produce an appropriate belief] have been influenced by the kinds of mechanism that the present definition would explicitly exclude from rationality'.

One response to these problems with the conception of PP-rationality is to switch to a conception of rationality that focuses on outcomes rather than the processes that produce them. There is a conception of rationality prevalent in economics according to which behaviour that maximizes expected utility is rational, no matter how it was produced or selected; Kacelnik calls this *E-rationality*. Rationality in this sense describes a pattern in behaviour.

A problem with this approach is that the utility to be maximized is understood as whatever is in fact maximized by the observed behaviour of the agent; maximizing behaviour is restricted only by formal requirements of consistency. But many peculiar and perverse forms of behaviour can be interpreted as maximizing some common currency, and hence can be regarded as rational. The question of whether an animal's behaviour is E-rational is thus in danger of trivialization, in the absence of some further constraints on what behaviour counts as rational.

Kacelnik suggests that, rather than assessing patterns of behaviour purely in terms of utility-maximization, we can turn to evolutionary biology to provide more substantive constraints on maximization. Evolutionary change results, in part, from natural selection. Natural selection produces phenotypic properties that can be predicted using principles of maximization applied to a defined currency: fitness. Fitness is to biological theory what utility is to economics. But unlike utility, fitness is not revealed by the behaviour of agents; rather, it is defined by the genetic theory of natural selection. The fitness of a biological agent is its degree of success relative to that of other agents in the same population. Genetic material guides the development and functioning of individuals; their behaviour is driven by mechanisms that evolved because they produce biologically rational behaviour. A *B-rational* individual is one whose behaviour maximizes its inclusive fitness across a set of evolutionarily relevant circumstances.[36] In its emphasis on outcomes rather than processes and on consistency, B-rationality is similar to E-rationality (and indeed there is a tradition of fruitful interactions between evolutionary game theory and economics). But B-rationality is far more constrained than E-rationality in its account of what is to be maximized. As Kacelnik explains, a consistent maximizer of fitness is a consistent maximizer of something, and a consistent maximizer of anything counts E-rational. So B-rationality entails E-rationality in relevant circumstances, but goes beyond it to specify that *what* is maximized must be inclusive fitness.

Two theoretical weaknesses of this conception are its under-specification of what counts as an evolutionarily 'relevant' circumstance and its assumption (rather than explanation) of the set of behavioural options from which the fitness-maximizing behaviour is chosen. Both weaknesses limit its explanatory power.[37] Behaviour may not be B-rational in a domain that has not been encountered often enough to influence natural selection—but how often is enough?—or because the animal has not got an adaptive behaviour it its option set—but why hasn't it? Note that lack of a specification of relevant domains can be a problem for B-rationality even if domain-specificity itself is not problematic.

Kacelnik goes on to illustrate how these different conceptions of rationality apply to birds, giving examples of choices of foraging techniques by starlings, choices of flowers by hummingbirds, and the obtaining of inaccessible food by a crow, who makes a tool with which to retrieve it.

B-rationality predicts behaviour in evolutionarily relevant circumstances from models that show which of a limited set of behavioural options would maximize fitness (though models may use various proxies for fitness). For example an explanation of starlings' choices to forage by walking or by flying assumes that starlings are B-rational, but makes no assumptions about whether starlings use reasoning to make their choices, or about what other mechanisms may explain their behaviour.

In another example, Kacelnik shows how apparent violations of E-rationality and hence of B-rationality by hummingbirds in choosing flowers can be understood

---

[36] See also and cf. Bermudez 2003, Ch. 6, on level 0 rationality.

[37] See also the chapter by Sterelny, this volume; Bermudez 2003, Ch. 6, on level 1 rationality.

as rational in these senses after all. Hummingbirds show context-dependent reversals of choice, which are *prima facie* inconsistent. A biologist might respond by claiming that one of the contexts was not evolutionarily 'relevant', or by reinterpreting the context-dependent choices as responses to different problems so avoiding any inconsistency. The availability of such responses to apparent violations of B-rationality may raise concern about how robustly inclusive fitness maximization constrains explanations that assume behaviour is B-rational. Is it still too easy to accommodate apparent violations?

A more general concern is whether, by excluding any element of process rationality, B-rationality tells us any more about behaviour than simply that it is adaptive. It leaves untouched questions about how an animal determines which behaviour is appropriate to its circumstances. How do Kacelnik's foraging birds work out which behaviour satisfies B-rationality? Is this process one of rational decision-making? Is there really nothing useful to be said on these matters?

Kacelnik's final example tempers his pessimism about the applicability of PP-rationality to animals. Betty, a New Caledonian crow, bent a straight piece of wire into a hook, allowing her to retrieve otherwise inaccessible food from the bottom of a vertical plastic well. To do so, she had to leave the site of the problem and find a suitable crack to hold the wire's tip while she bent it. Kacelnik finds it hard to account for Betty's behaviour without attributing to her the capacities to plan ahead, to represent the problem and its solution, and to choose rationally among possible actions. Exercising such capacities seems to amount to something not unlike thinking and perhaps approximates the conditions for PP-rationality. How domain-general these capacities may be in crows is unknown, though Kacelnik doubts that even human reasoning is fully domain-general.

The notion of B-rationality is clearly relevant to many explanatory projects in biology, even though it does not illuminate the processes that enable animals to be B-rational. Many of the contributors to this volume are interested in that further question and in whether those processes can properly be characterized as rational processes. If a positive answer to that question requires animals to engage in reasoning, to have beliefs, desires and other propositional attitudes with conceptual content, or to have linguistic abilities, we may well be sceptical. But there may be other, less demanding ways of understanding the processes that explain animal behaviour as rational.

1.2.2 **Dretske**

Philosopher Fred Dretske (Chapter 3) provides an account of what he calls *minimal rationality,* which is a feature of something that is done for a reason. There is a difference between something's being caused by an event with a certain content and being explained by the fact that the event has that particular content. Minimal rationality requires not only that behaviour be under the causal control of thought or another representational state, but that it be explained by the *content* of that representational state. What is done for a reason in this sense need not be done for *good* reasons, or be the product of *reasoning*, so minimal rationality is not a normative conception in these senses. Minimal rationality cuts across biological rationality; neither entails the other. It is very close to what Clayton, Dickinson and Emery call 'intentional' explanation, in terms of the content of psychological states (see Chapter 9).

This account of minimal rationality provides a sense of process rationality that avoids some of the difficulties with PP-rationality; we don't need to appeal to language or to reasoning to characterize a minimally rational process. But it faces the large question of intentional causation. Representational states are intentional states: they are about something. When is it right to explain a piece of behaviour in terms of what internal representations are about, their content? When does the content of a representation explain its causal function, as opposed to its causal function explaining its content? Given that representations are realized by physical states, when is it possible to explain the causal role of the representational state by appeal to the physical properties of its realization, and when is it necessary to explain its causal role in terms of its content?

Dretske presents his views by means of examples. While thermostats and plants may have inner states that in a sense represent something, and these states cause certain changes or 'behaviour', those changes aren't explained by *what* those states represent. They therefore do not meet the conditions for minimal rationality. For example what explains why a thermostat closes the electrical circuit to the furnace, thus turning on the heat, is the degree of curvature of a metallic strip with certain chemical properties, not what this represents about temperature. And what explains why the thermostat has this representational state in this causal role in the thermostat is not what it represents, but the intentions of the thermostat's designers, who chose to wire it the way they did: they wanted something that functions the way the metallic strip functions. In another of Dretske's examples, a plant, the scarlet gilia, changes from red to white when hummingbirds (who prefer red flowers) depart and hawk moths (who prefer white) arrive. Some chemical state of the plant presumably represents the external conditions in which this colour change becomes adaptive for the plant. But what explains the plant's colour change is its chemical state, not what this state represents about external conditions. And similarly, what explains why the plant has that chemical state in this causal role is not what it represents, but a selection process operating on the states of the plant's remote ancestor's: natural selection 'wanted' something that would function this way.

By contrast, learned behaviour in animals can be minimally rational. A foraging bird, for example, may avoid eating a harmless butterfly because it looks like a butterfly that the bird has learned is bad tasting. Some internal state of the bird represents that the bad-tasting sort of butterfly is present, so it avoids it. The bird's behaviour is explained by the content of what it 'thinks' about the butterfly. The bird's internal state, like the thermostat's and the plant's, represents something as well as having a causal role. But the content of the bird's representational state, unlike those of the thermostat and the plant, explains why that state has the causal role it has in the bird's behaviour. As a result of a learning process, the bird 'thinks' the butterfly tastes bad, and it avoids it because it thinks that. We can't explain why the bird's internal state has the causal role it has—as an element in a process that produces bug-avoidance behaviour—except in terms of what it represents, as a result of learning. Its behaviour thus counts as minimally rational even if it results from associational learning processes.[38] Other examples of what Dretske has in mind as minimal rationality are provided by the birds described by Clayton, Dickinson and Emery (Chapter 9) and by Herman's dolphins (Chapter 20).

Dretske's distinction between being caused by a state with representational content and being explained by the content of that representational state in effect distinguishes merely mechanical processes from intentional or cognitive causal processes (see Chapter 9 by Clayton, Dickinson, and Emery for more on this distinction). This is an important distinction, and provides a sense of 'rational process' that does not require reasoning. But it nevertheless makes *very* minimal demands of rational processes. We may still want to capture a more robust sense in which processes other than reasoning processes can be rational, for which minimal rationality is a necessary but not sufficient condition. There might be a state in the human visual system that, in virtue of its content, plays a role in explaining our ability, say, to recognize different tools; the process of which that state is a part might be minimally rational, in Dretske's sense. Although we might regard such processes as cognitive—as essentially involving representations—we wouldn't think of them as rational processes in a more robust sense.

1.2.3 Millikan

Philosopher Ruth Millikan (Chapter 4) focuses on rational processes that involve what she calls 'trials and errors in the head'. Evolution conducts trial and error explorations over generations; instrumental conditioning improves on this by allowing an animal to learn within its own lifetime from trials and errors in its behaviour. A capacity to conduct trials and errors in the head goes one better: it is quicker and safer than instrumental conditioning or natural selection. Karl Popper noted the benefits of letting one's hypotheses die in one's stead; Popperian animals, as Daniel Dennett (1996) calls them, can do just that, by trying different possible solutions to problems out in their heads. On one reasonable interpretation of rationality, to be rational is to be a Popperian animal, with the capacity for mental trials and errors. Millikan goes on to distinguish two forms of trial and error in the head, a perceptual/practical form that she thinks is plausibly attributed to some non-human animals, and a more theoretical form that she thinks is not.

---

[38] Papineau and Heyes are sympathetic to this claim, in their discussion of associative processes in Chapter 8 herein, but they don't accept Dretske's contention that given cognitive structures count as less rational when they result from evolution than when they result from learning. See also and cf. Dennett 1996; Guzeldere *et al.* 2002; Shettleworth 1998, 13–15; Bermudez 2003, Ch. 6.

Millikan suggests that perceptual representations she calls 'pushmi-pullyu representations' can function in such Popperian processes. These are representations that at once describe the animal's environment and direct its behaviour. They are suitable for immediately guiding action because they represent affordances in the animal's environment: they tell the animal where there are things to be picked up, climbed on, eaten, and so on. Millikan suggests that non-human animals can engage in trials and errors in the head at the level of perceptions of affordances; this would be a kind of practical rationality. She describes a squirrel that she observed solve the problem of how to reach a bird feeder. The squirrel surveyed the situation from various positions and angles before jumping in a way that required a precisely aimed ricochet to reach the feeder. Millikan suggests that the squirrel may have been engaged in a kind of trail and error in perception. It was trying to *see* a way to reach the feeder, to see an affordance.[39] It was not engaging in syllogistic practical inference; but if rationality is the capacity to make trials and errors in one's head, then its behaviour provides evidence of a form of practical rationality, one shared with human beings. While such behaviour is suggestive, the presence of this rational trial and error process in non-human animals would ultimately be settled, Millikan suggests, by neurology and cognitive science—by describing the mechanisms that subserve it. Note that this perceptual/ practical form of trial and error in the head is a form of simulation. The squirrel seems to be able to use his perceptual/ practical skills off-line to solve a problem, in this case to predict the results of alternative possible actions and select the one most likely to succeed.[40]

The theoretical form of trial and error in the head may be specific to humans. Non-human animals represent primarily affordances and features of the environment with direct practical significance for the animals: the presence of a hazelnut, or of danger, the location of cached food, and so on. Their representations of such features are successful when they lead to success in practical activities. But human beings, unlike non-human animals, also assiduously collect representations of what Millikan calls 'dead facts', which do not pertain directly to practical activity and may not have any practical relevance at all (we never know when a dead fact will come alive). Success in representing dead facts does not turn on practical success, but requires an ability to triangulate different methods of recognizing the same objective state of affairs, and to recognize their consistency or inconsistency. That is, success in representing dead facts requires the capacity to recognize contradictions in thought, which depends on representations with subject–predicate structure that can be internally negated. The capacity to recognize consistency and contradiction is not given in perception. Simple representational systems do not contain contrary representations; for example multiple signals used to alert conspecifics to danger do not contradict one another.

---

[39] See also and cf. Bermudez 2003, Ch. 6, 119–123, on level 1 rationality.

[40] See discussions of simulation in the chapters by Hurley, Proust, and Currie.

Perceptual representations of affordances don't contradict one another, even if they require choices. Non-human animals that primarily represent affordances are unlikely to have this capacity to recognize consistency and contradiction. But it is this capacity that makes possible trials and errors in attempts to represent the world theoretically, in ways decoupled from action, and to test and adjust beliefs until they are consistent through reasoning. Thus, non-human animals are unlikely to share with human beings this more theoretical process of trial and error in the head. Their Popperian rationality may be limited to perceptual/practical processes of trial and error in the head.

1.2.4 **Bermúdez**

A psychological explanation of an animal's action would show why the action made sense from the agent's perspective. In philosopher José Luis Bermúdez' view (Chapter 5), this would require showing how the actions could result from a process of reasoning. He is thus concerned with how to understand the possibility that non-linguistic animals might be rational in the sense of Kacelnik's PP-rationality. But rather than extend rational processes beyond reasoning processes, he takes them to be correlative, and instead offers an extended account of reasoning that could apply to animals (see also Bermundez 2003).

One way of understanding the process of reasoning involved in PP-rationality, or the theoretical process of trial and error in the head, is in terms of logical operations defined over linguistic structures. As Bermúdez explains, language makes types of reasoning possible that are not possible in its absence, by making it possible to think about thoughts. Such metarepresentation or 'intentional ascent' enables both reasoning that exploits the internal structure of a thought and inferences that depend on logical concepts, such as negation or material implication ('if…then…'). But this kind of account can make it very difficult to see how reasoning processes could be present in non-linguistic creatures.

Bermúdez explains how analogues of familiar reasoning processes might be possible in the absence of linguistic structure—that is, how it might be possible for non-linguistic animals to reason without exploiting the internal structure of thought or deploying logical concepts. For example consider a form of non-linguistic reasoning analogous to this instance of reasoning by *modus ponens*: *If the gazelle is at the watering hole, then the lion is not at the watering hole; the gazelle is at the watering hole; so the lion is not at the watering hole.* This would require proto-logical analogues of negation ('not…') and material implication ('if…then…'), which do not require logical concepts, language, or metarepresentation.

Bermúdez explains proto-negation in terms of contrary properties, rather than in terms of a formal operation on whole propositions or thoughts. Proto-negation is available in cases where a creature can think about contrary properties. In Bermúdez' example, the thought that the gazelle is absent from the water hole is contrary to the thought that the gazelle is present at the water hole, which allows it to serve a similar role in reasoning to that played by the denial of the thought that the gazelle is present at the water hole, although it does not result from applying the formal, logical operation of negation. Specific contraries, such as presence and absence, exclude one another, even though they do not do so in virtue of their logical structure. While Bermúdez agrees with Millikan that non-linguistic animals cannot understand the general idea of contrariety, he suggests that they do not need to understand contrariety in order to represent specific contraries (such as presence at versus absence from the water hole) and to reason practically and validly in accord with their contrariety. Note that any resulting reasoning capacities might be expected to be correspondingly domain-specific.

Bermúdez next explains a non-linguistic form of proto-conditional reasoning as a primitive kind of causal reasoning. Conditional reasoning takes one from recognition that there is a conditional dependence between two states of affairs, so that if the first holds, then the second holds, plus recognition that the first state of affairs does indeed hold, to the conclusion that the second holds. Proto-conditional dependence, he suggests, could be represented by non-linguistic animals as causal dependence between events (which may or may not include the agent's own actions, and which may be probabilistic). (While conditionals are often appealed to in analyses of causation, the phylogenetic order of acquisition need not match the order of analysis.) The abilities to represent causal regularities and causal dependencies between events or states of affairs can be of great survival value to animals, and need not involve metarepresentation. Such regularities might include, for example, those between the presence of lions and the absence of gazelles at the watering hole, and between the presence of gazelles and the absence of lions; and the absence of gazelles would depend causally on the presence of lions rather than *vice versa*. As a result, an animal may be able to reason from a proto-conditional, and employing contraries, as follows: *If the gazelle is present at the watering hole, then the lion is absent from the watering hole; the gazelle is present, so the lion is absent.*

Such a proto-inference is analogous to *modus ponens*, but is not valid in virtue of its form in the way logical inferences are, since the transition from premises to conclusion only works because of the specific pair of contraries employed. But despite its domain-specificity, the reasoning is valid in the sense that if the premises are true, then so is the conclusion. And it is clearly of practical value to the animal. Bermúdez concludes that the ability to represent specific contraries—proto-negation and causal dependencies—proto-conditionals—would make reasoning available to non-linguistic animals, which could in principle underwrite psychological explanations of their behaviour. Note that while Bermúdez and Millikan differ over whether contrariety can plausibly have a role in animal rationality, Bermúdez is primarily concerned to show how forms of proto-inference involving contrariety could be of direct practical relevance to animals, while Millikan is sceptical about the capacity of animals to represent 'dead facts', with no evident practical relevance, as inconsistent, hence about their capacity for theoretical rationality.

1.2.5 **Hurley**

Philosopher Susan Hurley (Chapter 6) pursues the possibility of domain-specific practical rationality. She argues that practical rationality does not require conceptual capacities—at least in one common understanding, according to which conceptual capacities require the ability to decompose and recombine the elements of a thought in inferential patterns with unrestricted generality, and hence underwrite a domain-general reasoning capacity, or 'inferential promiscuity'. It may be implausible to attribute conceptual capacities in this sense to non-linguistic animals,[41] but that does not mean that they cannot be correctly understood as acting for reasons. It is possible for animals to act for reasons even though their reasons for acting are bound to specific domains, so that only certain aspects or subsets of their behaviour is rational. Animals can occupy 'islands of practical rationality'.

Hurley follows Daniel Dennett and others in distinguishing processes at the subpersonal level—where we describe or explain an action in terms of causal mechanisms, in neural, structural, or functional terms—from processes at the personal level—where we make sense of actions in terms of holistically related and normatively constrained mental states. At the personal level, we understand intentional agency in terms of the rational interactions between someone's various desires and her beliefs about how best to satisfy them. Hurley agrees with John McDowell (1994a,b) that personal-level explanations of behaviour in terms of reasons need not correspond to or compete with subpersonal explanations in terms of causal mechanisms, even though the latter enable people to behave in ways that make personal-level descriptions appropriate. She also holds that our grounds for applying personal-level descriptions are not the structure of internal subpersonal mechanisms, but patterns of observable behaviour in environments; minds are extended, and rational processes are widely distributed. This is realism about minds and reasons, but of a strongly externalist stripe. Hurley proposes an animal analogue of the personal/ subpersonal distinction, and argues that animal action can sometimes be properly understood at the animal rather than the subanimal level. The weight of her claim that animals can act for reasons rests on the view that animal action can sometimes be properly understood as displaying holism and normativity, despite their lack of fully-fledged conceptual abilities.

To defend this view, Hurley first explains how holism and normativity could obtain despite a lack of domain-general conceptual abilities, and then illustrates this possibility with several examples involving chimpanzees.

Hurley emphasizes that a capacity for flexible generalization is not all or nothing. The holism of intentional agency is an instrumental form of the capacity for flexible generalization that does not require conceptually structured inferential promiscuity. An intentional agent's means and ends can decouple and recombine; a given belief or a

---

[41] But see Cook 2002, for a weaker sense in which conceptual abilities can be attributed to animals.

given means can serve various ends, and a given desire can be served by various beliefs and various means, depending on circumstance. An intentional action results from the interaction of the agent's ends and his perception of means/end contingencies, so that an effective means to his ends in given circumstances is selected from those available. But flexibility in adopting effective means to ends can be a matter of degree; some skills that are used in instrumentally rational ways in one domain may not generalize beyond that domain. For example an animal may be able to act in accord with transitive inferences about social dominance in selecting effective means to its ends, but be unable to act in accord with transitive inferences about foraging opportunities. It may not have the conceptual abilities needed to make this transition.

The normativity of acting for reasons requires the possibility of mistake: that there be a difference between making a mistake in acting for one reason, and acting for a different reason. (Note that this is a weaker sense of normativity than, for example, Dretske's: one can make a mistake in acting on a bad reason.) Hurley notes that this in effect requires a solution to Kripke's (1982) version of the rule-following problem, applied to intentional agency. This difficult problem arises equally for subjects with conceptual, theoretical, and linguistic skills, and Hurley doubts it is more tractable for them than for mere intentional agents: it isn't clear why the finer-grain, domain-free flexible generality of inferential promiscuity secures any advantage here. Rather, Hurley suggests that normativity depends on complex control and simulation capacities embedded in teleological contexts, which leave the possibility of mistake no less problematic for mere intentional agents than for creatures with conceptual abilities (see also and cf. Proust, Chapter 12 herein).

Hurley illustrates the possibility of context-specific reasons for action by reference to experimental work with chimpanzees. The first example illustrates how a symbolic context might make instrumental reasons available to an animal, which it cannot generalize to similar but non-symbolic contexts. (The example is drawn from experimental work by Sarah Boysen, described in Chapter 22 in this volume.) Sheba is faced with two dishes of candy, one of which contains more candies than the other. When she points to one dish, she reliably is given the candies in the other dish. Despite this, she continually points to the dish with more candies, so is given the smaller number of candies. In a variant of the task, different numerals are placed in the dishes instead of candies; Sheba has already learned to use numerals in other tasks. When she points to a dish with one numeral in it, she is given the number of candies signified by the numeral in the other dish. She at once begins pointing to the dish with the smaller numeral, and is thus given the larger number of candies. The symbolic context appears to makes instrumental reasons available to the animal. Early studies suggested that this instrumentally rational behaviour would persist when one dish contained a numeral and the other candies (compare Boysen's chapter in this volume for later work). However, when both numerals are replaced with candy again, she reverts to pointing to the dish with more candies. Sheba appears to point to the dish containing the smaller numeral in an instrumentally rational way, order to get as many candies as possible, but cannot generalize this instrumental reason to the all-candy case. Hurley discusses the interpretation of this task in comparison to a related task given to children the windows task.

The second example illustrates how certain social contexts might make reasons for action available to an agent that it cannot generalize to other contexts in which they are also relevant. Various paradigms have been developed to test for mind reading in non-linguistic animals (these methods are discussed further in the chapters in Part V of this volume). In the hider–communicator task, success requires attribution of a false belief to a helpful communicator who indicates the location of food, in circumstances where the communicator has been 'tricked' by a third party. This test produces results in children that correlate well with the results of standard verbal false belief tests of mind reading. Chimps fail this test of mind reading. In a different paradigm, attributions by a subordinate chimp to a dominant chimp of perceptions concerning the presence of food would enable the subordinate to compete more successfully in acquiring food. Chimps pass this test of mind reading; subordinate chimps appear to use information about what dominants did or did not see in instrumentally rational ways. Why should chimps be able to use information about the mental states of others in instrumentally rational ways in one paradigm but not the other? One possibility is that their ability to treat information about the mental states of others as reasons for action is specific to social contexts involving competition for food, which are ecologically significant, but do not generalize to social contexts involving co-operation over food, which are not ecologically significant.[42]

Hurley concludes by explaining how her view of domain-specific practical reasons shares ground with simulationist accounts of cognitive capacities, and by arguing that the functions and interest of making sense of action are not wholly discontinuous at the boundary between human and non-human animals. We should, she claims, avoid over-intellectualizing what it is to have a mind, or viewing rationality as all or nothing. By allowing that rationality can be disaggregated into domain-specific capacities within which basic features of practical rationality are nevertheless present, we can characterize the animal level in terms that are neither too rich nor too impoverished, and chart various specific continuities and discontinuities between our minds and those of other animals.

Hurley is blithely compatibilist about personal level rational processes and subpersonal processes that don't map tidily onto the former.[43] Such compatibilism can be challenged. Explanatory subpersonal mechanisms, such as associative mechanisms, are often regarded as competing with and threatening to displace rational explanations. If so, then even if rational explanations of human behaviour can be defended as indispensable for many purposes, the same may not be true for non-human behaviour. This issue is discussed in more detail in the contributions to Part II.

....

[42] See the chapter by Tomasello and Call; Hare and Wrangham 2002.

[43] Cf. Bermudez 2003, 192; Davies 2000.

1.3 **Rational versus associative processes**

A powerful form of subpersonal explanation of behaviour in both human beings and other animals appeals to associative processes, including classical and instrumental associative conditioning processes and the associative neural networks that may support them. The authors in this section are all concerned with the contrast between explanations of behaviour in terms of associative processes and explanations in terms of 'higher' processes: cognitive, or rational, or intentional explanations. Do such contrasts hold up, and if so, do associative explanations compete with and displace the higher forms of explanation?

1.3.1 **Allen**

Philosopher Colin Allen (Chapter 7) focuses on transitive inferences, of the form: A > B; B > C; therefore, A > C. Allen compares cognitive and associative accounts of behaviour by animals on transitive inference tasks. In doing so he illustrates a more general methodological dispute between ethological and experimental approaches to animal cognition.

Transitive relationships are often important to animals, especially social animals. Transitive inference would permit, for example, an animal to behave in a way appropriate to the dominance relation between other animals A and C, even if their relationship has not been directly ascertained; this could enable the animal to avoid injury and loss of various resources. Given the ecological significance of transitive inference in certain domains, we might expect to find domain-specific rather than domain-general capacities to behave in accord with such inferences. But even if such behavioural capacities were limited to a specific domain, the question would still arise whether they were best explained cognitively or associatively. Cognitive explanations postulate that animals explicitly represent orderings such as A > B > C > D > ···, and use these representations to infer relationships between non-neighbouring pairs. By contrast, associative explanations appeal to past reinforcement histories of the individual elements, without invoking any explicit representation of the entire series.

Allen describes the ping-pong game between cognitive and associative accounts of apparent transitive inferences. If A is rewarded rather than B and B is rewarded rather than C, pigeons and rats given a novel choice between A and C are likely to choose A. This is easy to explain associatively, since A has always been rewarded and C never. An extended version of the test trains animals on five instead of three elements: A is rewarded rather than B, B rather than C, C rather than D, and D rather than E. Many animals given a novel choice between B and D reliably choose B. The original associative explanation no longer applies, since D has been rewarded in training just as often as B. However, more sophisticated associative explanations appeal to value transfer effects: the sure-thing value of A rubs off by association onto B. There is experimental support for this account, though ambiguity between cognitive and associative accounts of this and other experimental results remains.

Animals in experimental paradigms that test for transitive inference learn the relevant associations slowly and laboriously. Ethologists may wonder how relevant the results of such experiments are to explaining the capacities of animals living in complex natural societies, where dominance hierarchies including many dozens of animals are rapidly learned and changes in dominance are flexibly adjusted to. Allen suggests the slow learning rate in the experimental transitivity tasks may be due to the specific nature of the tasks: they use arbitrary stimuli of no ecological significance—such as shapes shaded with patterns—and the ordering of them doesn't reflects a natural or biologically significant ordering, but simply the consequences of a training paradigm. By contrast, the social dominance relations that monkeys negotiate with ease have clear biological significance. Could capacities for transitive inference be domain-specific?

If so, what might that suggest about the general distinction between cognitive and associative explanations? Are these exclusive alternatives? Are associative explanations, when available, always to be preferred? Is scepticism warranted about the defensibility or utility of the general distinction (as expressed by Papineau and Heyes in Chapter 8)? Allen suggests that discussion of this last question may be too abstract to be instructive. Cognitive processes may bear family resemblances to one another rather than satisfy an analytical definition. And even if the general cognitive/ associative distinction cannot be defended, a distinction between different kinds of explanation of specific capacities, such as the contrast between cognitive and associative accounts of behaviour in accord with transitive inference, is still tenable and useful. An explanation that postulates a representation of an entire ordered series and associated cognitive processes that exploit such a representation is a quite different explanation from one in terms of associations between stimuli and responses of varying strengths. We should keep in mind, however, that cognitive accounts of transitive inference may be correct only in application to specific domains where natural orderings and transitive inference are biologically significant, while associative mechanisms play a role in explaining other behaviours that conform to transitive inference. Moreover, cognitive and associative mechanisms for such behaviours may coexist in one animal, each applying to different domains, and yielding different behavioural capacities in experimental tasks that lack ecological significance as opposed natural social environments.

1.3.2 Papineau and Heyes

Strong scepticism about the heuristic and theoretical importance of a distinction between rational and associative explanations is expressed by philosopher David Papineau and psychologist Cecilia Heyes (Chapter 8). Indeed, they suggest that the rational/ associative dichotomy is a modern descendant of Descartes' soul/ matter dichotomy, and equally unhelpful; it obscures the real issues. There is no Rubicon between associative and rational processes to be crossed; rather, evolution adds specific new cognitive capacities by tinkering with previous mechanisms. What we want to know is how animals work: which specific mechanisms account for which specific cognitive capacities in which animals.

They illustrate their view by reference to experiments on imitation in Japanese quail (by Thomas Zentall and others), describing another ping-pong game between rational and associative explanations. The behaviour to be explained is as follows. Observer quails watched a demonstrator quail push a lever; some demonstrators pushed by pecking and others by stepping. Half of the observers saw a demonstrator rewarded with food for this behaviour, while the other half saw a demonstrator perform the same behaviours without reward. Only those observers who had seen the demonstrator rewarded imitated its behaviour when later given access to the lever themselves: they pecked if their rewarded demonstrator had pecked and stepped if it had stepped. The observers of the unrewarded demonstrators did not differentially peck or step. The observers were thus sensitive to demonstrator reward. They appear to learn imitatively, in the sense of learning how to achieve an end by observing another's actions.

It is tempting to explain this sensitivity in rational terms. That is, the imitator birds want food and believe that doing the same thing as the rewarded demonstrators will provide it:

X's pecking produces food.
My pecking will produce food.
I want food.
Therefore,
I will peck.

And it can also be hard to see how this behaviour could be explained in terms of standard associative learning. To see why, note that there are four elements present: demonstrator behaviour, demonstrator reward, observer behaviour, and observer reward. First, how is demonstrator behaviour associated with similar observer behaviour? What the quail sees and feels when it pecks is very different from what it sees and feels when the other pecks. Perhaps this correspondence problem can be avoided by some associative mechanism or innate tendency for mimicry. But second, that would still not explain why the observer mimics only rewarded behaviour. Neither instrumental nor classical conditioning readily explains why an association between demonstrator reward and demonstrator behaviour, plus an association between demonstrator behaviour and observer behaviour (expressed in mimicry) would yield an association between observer behaviour and observer reward. There is nothing obviously rewarding to the observer in seeing some other bird getting food. For related reasons, a capacity for imitative learning, to absorb new means–ends information by observing others, is widely regarded by scientists as a hallmark of advanced cognitive capacities (despite the erroneous popular view of imitation as a low-level capacity).[44]

The apparent difficulties of explaining the quails' behaviour in associative terms may seem to reinforce the rational explanation. This would not entail unlimited cognitive powers; the quails' ability to act rationally on the basis of their beliefs and desires may be specific to certain domains. But even if domain-specific rational explanation is admitted, the issue is not yet resolved. Associative learning theory has further resources for explaining sensitivity to demonstrator reward. Quails tend to feed together, so a quail's feeding may be associated with observing others feed, with the result that seeing another feed acquires rewarding properties. Add to this the tendency to mimicry, so that seeing another peck is associated with pecking oneself. When the observer sees the demonstrator peck and feed, that secondary reward is associated with its own pecking in mimicry, differentially reinforcing mimicry of rewarded demonstrations.

Can an experiment decide between this improved associative account and the rational account? Papineau and Heyes describe a hypothetical variant of the quail experiment, using some foods that are devalued (by prefeeding the birds to satiety) and other foods that are not, which might seem able to do so. But they go on to explain how 'rational' behaviour in their experiment could also be explained in terms of further associative mechanisms, using a cybernetic model of instrumental learning that associates a response with its outcome, including devalued foods, in a negative feedback loop.

Rational explanations of behaviour are typically distinguished from associative explanations by featuring norm-governed reasoning involving belief-like representations. But Papineau and Heyes doubt that the rational/ associative distinction can be clearly drawn by reference to the processing of representations. If rational processes are defined in terms of representations, there is no good reason not to include associative processes as rational. Associative learning involves modifications to internal states that carry information, and the informational processes involved in sophisticated forms of instrumental conditioning are analogous to those in practical inference such as: doing B in C will lead to D; I am in C; I want D, so I do B. Admittedly, associative informational processes may be limited in the kinds of inferences they can enable. But if rational processes are required to underwrite unlimited inferential powers, they represent a cognitive ideal that is useless as a research tool applied to animals—whether human or non-human—, whose cognitive capacities must be realized by specific mechanisms.

---

[44] In fact, a capacity for imitative learning is phylogenetically very rare and has been very difficult to document outside human beings, though in recent years evidence for it in some species, including some great apes, cetaceans, and birds, has been found; see Hurley and Chater, 2005, vol. 1; discussion in Papineau and Heyes, this volume; Heyes and Huber, 2000, part V.

The message from Papineau and Heyes is: we should refocus on specific explanations of how animals do specific things, rather than on the presence or absence of some general or ideal form of rationality that contrasts with associative mechanisms. Perhaps language makes ideal rationality accessible to human beings— though proponents of bounded and ecological rationality would dispute this also. The recommended refocus may not just change our approach to animal minds, but also to assessing the continuities and discontinuities between animal minds and our own.

1.3.3 Clayton, Dickinson, and Emery

Papineau and Heyes are careful to say that there are no obvious associationist explanations of some of the impressive cognitive feats of scrub jays described by psychologists Nicola Clayton, Nathan Emery, and Anthony Dickinson (Chapter 9). Clayton and colleagues contrast mechanistic explanations of behaviour in terms of associative processes with intentional explanations in terms of rationally flexible interactions between beliefs and desires (as in Kacelnik's PP-rationality, or Dretske's minimal rationality). Explanations in terms of beliefs and desires are intentional in the sense that they depend on the representational content of beliefs and desires, their truth or fulfilment conditions. By contrast, associative explanations depend on mechanistic properties. Associative learning consists in the formation of excitatory or inhibitory connections between nodes activated by various events; excitation is transmitted from one node to another until activation of the terminal node is sufficient to generate the observed behaviour. Clayton *et al.* view these different styles of explanation of a given behaviour as mutually exclusive, although different behaviours of the same animal can be explained in different ways. Rationality is primarily a property of processes causing a particular behaviour; a rational animal, whether human or non-human, is an animal *some* of whose behaviour—not necessarily *all*—warrants intentional explanation.

Associative processes may seem capable of explaining memory for food caches in scrub jays. Nodes activated by visual cues around the cache site become associated with those excited by food stored at the site. Re-exposure to the cache site's stimuli activates food nodes that are associated in turn with nodes controlling the appropriate response. But Clayton *et al.* argue that the food-caching and retrieval behaviour of the birds they study is best explained intentionally. Given a certain desire for food and a belief about where the food is, the rational thing to do is to search in certain cache sites. The rationality of such practical inferences, which depend on the contents of desires and beliefs, distinguish intentional explanations from associative explanations, even if the latter are adaptive (and hence biologically rational, in Kacelnik's sense). While intentional accounts are often intuitively attractive, Clayton *et al.* emphasize that behaviour does not wear its intentionality on its sleeve; rival associative accounts are often available. However, rational explanations of the behaviour of non-linguistic animals are more empirically testable, with the right experimental procedures, than sceptics may suppose.[45]

(a) Philosophers have sometimes argued on very general grounds that intentional states cannot have determinate content in the absence of language—or at any rate that we could not know what their contents are.[46] The first line of experiments described by Clayton *et al.* shows how it is possible in specific cases to determine the content of desires on the basis of non-linguistic behaviour. Their empirical procedure addresses whether a desire has a specific or a general content: is it a desire for food, say, or a desire for peanuts? Having cached two different types of food the birds are, after an interval, fed to satiety on one type and then allowed to search their cache sites. They selectively search sites where they cached the other type of food. This behaviour is better explained in terms of a desire for a specific food rather than general desire for food. While an associative account of this behaviour can be offered, in an experiment with a more complex design the birds behave as predicted by an intentional rather than an associative account. There may never be a once-and-for-all, decisive test of the intentionality of animal behaviour, but if we focus on testing particular predictions of associative versus intentional accounts in specific cases the issue is not empirically intractable.

(b) The second line of experiments described, concern whether jays' caching behaviour should be explained in terms of the flexible interactions of beliefs based on various declarative memories in practical inferences. Human cognitive psychology contrasts a declarative memory system, in which general knowledge and recollections of specific episodes interact rationally and flexibility in guiding action, with procedural memories embodied in skills, responses, and habits that are often domain-specific. Declarative memories in jays might include both general memories, say concerning the location of reliable food sources, and specific episodic-like memories of particular life events, such as what food they cached where and when. The scrub jays studied by Clayton and colleagues eat a variety of foods, but have favourites; they prefer crickets to peanuts, for example, but only when the crickets are fresh. And crickets decay relatively rapidly, while peanuts do not. The jays were tested to see if their behaviour in searching for cached food can be explained by flexible, rational interaction of general declarative knowledge of how rapidly crickets decay with specific episodic-like memories of when and where they cached crickets as opposed to peanuts.

The tests proceeded as follows. Birds were allowed to retrieve food after 1 day and then after 4 days (no other intervals were used in the initial learning trials); they rapidly learned to search for cricket caches after 1 day, when the crickets were still fresh,

---

[45] Cf. for example, Kacelnik, and Papineau and Heyes, this volume.

[46] See and cf. Davidson 1984, Ch. 11; 2001, Ch. 7; Allen and Bekoff 1997, Ch. 5; Dennett 1996, 38, 41, 159; Bermudez 2003, Ch. 5.

and for peanut caches after 4 days, by which time the crickets had decayed. Next the experimenters pilfered all the cached food, and then allowed the birds to search after 1, 2, 3, 4, or 5 days. The jays searched for cricket caches at intervals up to 3 days and after longer intervals switched to searching for peanuts. This behaviour suggests that they had generalized from their experience of crickets fresh after 1 day but not 4 days, and formed a general belief that crickets stay fresh for only 3 days, which interacted in instrumentally rational ways with their memories of specific caching episodes to guide their searches.

But how flexible was this interaction? A process of practical inference operating on the contents of declarative memories would produce an instrumentally rational change in behaviour if the content of general declarative memories changed in certain ways after caching had occurred.[47] To address the issue of rational flexibility, Clayton *et al.* allowed the birds to find cached food for the first time after 3 days and to learn, contrary to their prior generalized belief, that crickets had in fact perished by this time. A revised general belief about decay rates should rationally lead to a reversal of behaviour: namely, to searching for peanuts instead of crickets after 3 days. This is indeed what the jays did. This behaviour reversal can be explained as instrumentally rational, in terms of the flexible integration of the contents of their food preferences and their recollections of specific caching episodes with the contents of their changing general beliefs about decay rates of specific foods.

(c) Recall that the demands of social interactions, with the potential to produce manipulation of information, have often been viewed as driving the evolution of advanced cognitive capacities. The third line of experiments investigates caching behaviour in social contexts. Many birds are known to pilfer the caches of other birds they have observed caching, and corvids have been observed to return alone and recache to new sites if they had been observed by other birds during the initial caching. But such behaviour in the wild may be coincidental, and does not resolve whether recaching should be explained intentionally. In an experiment, jays recached significantly more items in new sites when they had been observed during caching than when they had cached in private. Can this behaviour be explained mechanistically? Perhaps recaching is automatically triggered by the memory of having recently been in the presence of another bird. This mechanistic explanation is ruled out by another experiment. Birds who cache in an 'observed' tray immediately before or after caching in a 'private' tray, and are then allowed to recache from either tray, recache selectively from the observed tray rather than the private tray.

An obvious intentional explanation of the selective recaching is that that birds have general beliefs that being observed during caching tends to lead to loss of food and that such loss can be prevented by recaching, which interact with their memories of specific

---

[47] For relevant discussion, see Bermudez 2003, Ch. 6, on level 2 rationality.

caching episodes and their desires to avoid loss of food. But how might such general beliefs arise? A final study compares recaching by birds who are experienced thieves themselves with recaching by naïve birds who have watched other birds cache but have no experience of pilfering other birds' caches. The experienced thieves selectively recached to new sites when their initial caching had been observed; the naïve birds did not. A rational explanation of this behavioural difference involves a kind of mind reading: the experienced thieves appear to attribute their own belief, that observed caches can be stolen, to other birds who observe caching. On this account, their selective recaching of observed caches to new sites is explained by the rational integration of their own experience of pilfering with memories of specific episodes in which they cached food while observed.

The converging evidence for intentional rational processes from these three lines of experiment is impressive indeed. This work contributes strikingly to the growing body of evidence that 'bird brains' may be small but have been seriously underrated: under rigorous experimental probing, birds can compete with the best in the animal rationality stakes.

1.4 **Metacognition**

One approach to distinguishing rational processes from other cognitive processes focuses on metacognition. If rationality requires that the agent can recognize the possibility of mistake, then metacognition could satisfy this demand by allowing the subject to monitor its own cognitive processes. Metacognitive processes would give agents information about the information on which they act, enabling them to monitor certain behaviour-producing processes to check whether, in the circumstances, they are likely to lead to success (as in Millikan's trial and error in thought; see Chapter 4). Domain-specific forms of metacognition may evolve as adaptations to particular environmental problems, and rational processes may eventually emerge from cumulative metacognitive adaptations.[48] The contrast with associative processes arises again here: when is it right to explain animal behaviour in terms of metacognitive as opposed to associative processes? Can metacognition itself be explained as emerging from associative processes, and hence provide an evolutionary bridge between associative and rational processes? What kinds of non-linguistic behaviour would provide evidence of metacognition? An evolutionary perspective is helpful in understanding how associative, intentional, and metacognitive processes may coexist in a given animal, including in human beings: an animal whose behaviour is best explained intentionally or metacognitively in a given domain may fall back on associative processes in other domains (see Chapter 10 by Call)—conceivably these different processes might even compete to govern a given type of behaviour in some circumstances. What implications

---

[48] See especially the chapters by Proust and by Sterelny.

might such an evolutionary perspective on metacognition have for human rationality? Is the final linguistic adaptation by the human animal just another, if rather special, metacognitive layer, or does it make a qualitative difference in enabling classical domain-free reasoning capacities?

1.4.1 **Call**

Psychologist/ primatologist Josep Call (Chapter 10) presents a variety of behavioural evidence that he argues is best explained either by causal reasoning or by metacognition, and goes on to consider the relationships between capacities for reasoning and for metacognition.

(a) *Causal reasoning.* Call first argues for a view exactly contrary to the widespread view that animals are excellent at making associations between arbitrarily related stimuli, but poor at reasoning and causal inference.[49] Perhaps ironically, the greater ease with which human beings learn arbitrary associations may be linked to human facility with symbols, which are arbitrarily associated with what they represent. Call describes performances by various great apes in a series of experiments that contrast arbitrary associations with causal associations, using tasks with similar superficial cues and reward contingencies but different causal structures. The apes do significantly better on the causal tasks than on the arbitrary tasks.

For example, in a causal task the apes must select from two cups the one that is baited in order to be rewarded; both cups are shaken, and only the baited one makes noise—the noise is caused by the food inside the cup. In the paired arbitrary task, both cups are tapped but only the baited one makes a loud tapping noise—a noise causally unrelated to the presence of food in the cup. Apes select the baited cup above chance in the causal task but not the arbitrary task, suggesting that they infer that food in a shaken cup would make noise and do not simply associate food with noise. In a revealing variant on the causal task, the empty cup was shaken and the baited cup was silently lifted; neither produced noise. A negative causal inference is suggested: if food causes noise in a shaken cup, then the silence of a shaken cup implies that food is in the other cup. Apes acted in accord with this inference: they tended to select the lifted cup in this task (when compared with a control task in which both cups were lifted). Similar results supporting causal inferences rather than associative processes were found in visual causal and arbitrary tasks. In the causal task apes were rewarded for selecting an inclined board propped up by food beneath it instead of a flat board, while in the arbitrary task they were rewarded for selecting a previously examined wedge (with the same angle of inclination and presenting the same visual aspect as the inclined board in the causal test) instead of a flat board. Apes selected the inclined board hiding food in 80 per cent of the trials but the wedge in only 50 per cent of the

---

[49] See also and cf. Heyes and Huber, eds, 2000, part III; Wynne 2001, Ch. 3; Gomez 2004, Ch. 4.

trials, suggesting that they understood that the inclination in the causal task but not the arbitrary task was caused by the presence of food. Strikingly, there is no evidence of learning in these tests: the apes either performed well from the beginning (in the causal tasks) or poorly throughout (in the arbitrary tasks).

These results suggest that apes do not simply associate a cue with the presence of food, but understand that the food is the cause of the cue, and can reason accordingly. They could be regarded as providing empirical evidence of a capacity for proto-logical reasoning similar to that described by Bermúdez, relying on pairs of contraries (for example noise versus silence) and conditional dependence based on causal dependence (for example if a cup containing food is shaken, noise results). Call argues that rival explanations of these results, either in terms of associative learning and reinforcement, or in terms of innate dispositions, are unpersuasive. However, these different processes may coexist.

(b) *Metacognition.* Do animals they know whether or not they know how to solve problems? In a second line of argument, Call presents behavioural evidence for metacognitive processes in certain animals—but not others—describing recent results from two experimental paradigms.

The first paradigm allows animals to escape from tasks if they are uncertain of the correct response, where escaping produces less desirable outcomes than responding correctly. For example an auditory task rewarded discrimination of high from low pitches by pressing 'high' and 'low' keys, respectively, but also presented an 'escape' key which moved the subject on to the next trial (with increasing delays to discourage overusing). Dolphins, like human subjects, used the escape key for difficult intermediate pitches. In a visual delayed match to sample task, monkeys did better when allowed to use the escape key than when they were not given this option, suggesting that they knew when they had forgotten the sample stimulus.

The second paradigm allows animals to seek additional information before responding when they are initially given incomplete information. Subjects—including 2-year-old children, orangutans, chimpanzees, gorillas, and bonobos—faced two opaque tubes, and were rewarded for touching the baited tube. In the visible condition the experimenter placed food into one tube in full view of subjects; in the hidden condition subjects knew one tube was baited but did not know which. Subjects were allowed to look inside the tubes before choosing one, and did so more often when they had not seen the baiting. Moreover, they often chose the other tube as soon as they had looked into an empty tube, without looking to see the food in the other tube. These results suggest both metacognition and inference: apes know that they do not know where the food is, so seek additional information, and when they find that the food is not in one tube they then infer that it is in the other. Dogs, by contrast, failed this test. The results from these two paradigms suggest that apes can monitor whether they have the information needed to solve a problem and either escape the situation or seek more information accordingly.

In conclusion, Call suggests that cognitive capacities for causal reasoning and metacognition should be regarded as evolved adaptations. These cognitive adaptations need not wholly displace associative processes; the latter may represent a fallback mechanism for apes when their capacities for reasoning and reflection cannot be engaged. Call is tempted by the view that reasoning and even linguistic capacities may build on metacognitive capacities in a coevolutionary process.

1.4.2 Shettleworth and Sutton

Can success on behavioural tests for metacognition can be explained in other, deflationary ways? What kinds of controls are needed to rule out alternative explanations? These methodological questions are probed by psychologists Sara Shettleworth and Jennifer Sutton in their review of metacognition paradigms and results with animals (Chapter 11). In human beings, success on metacognitive tests correlates with subjects' report of a 'feeling of knowing'. Accurate metacognitive awareness has obvious functional advantages: if one can monitor information gaps, one can take steps either to avoid situations that require the missing information or take steps to acquire it. While we may never be able to assess whether animals have a feeling of knowing, with the right controls we can define functionally similar behaviour and thus assess whether animals' performance depends on metacognitive information.

Non-verbal metacognitive paradigms generally build on well-established non-verbal paradigms for testing perceptual discrimination and memory. The discrimination or memory tests vary in difficulty, and to them is added an 'escape' option, which carries a reward intermediate between correct and incorrect responses on the discrimination or memory task. But, as Shettleworth and Sutton explain, success on some such tests allows of a deflationary associative explanation. For example, suppose an animal consistently escapes from memory trials that are more difficult because they involve a longer interval between stimulus and memory test, but accepts easier trials with a shorter interval. Is this because the animal knows it has forgotten in the harder trials? Or is it because it has simply learned associatively that its rewards are maximized by taking the escape option after long intervals and the test option after short intervals? Controls must be added to such a behavioural test to define functional similarity to metacognition more tightly and rule out alternative explanations.

One important control in metacognitive paradigms is randomly intermixed forced trials, where the escape option is omitted. The metacognitive account predicts that an animal will do worse on the forced trials than on trials it chose to accept when the escape option was available. A second important control is to transfer the animal to tests with new stimuli, so that it cannot rely on learned associations between particular stimuli and rewards, to see if the results across different tests converge on the metacognitive explanation. For example, instead of varying the difficulty of a memory task by varying the retention interval, the interval of exposure to the stimulus could be varied. The metacognitive account would predict an immediate transfer of behaviour consistent with metacognition, without the need for learned associations.

As explained also by Call, discrimination/ escape tasks with dolphins (auditory discrimination) and monkeys (visual discrimination) produced results consistent with metacognition. The results were similar to those obtained from human subjects in almost identical tasks, in which people reported that they escaped trials when they felt uncertain. However, Shettleworth and Sutton point out that these studies with dolphins and monkeys did not include forced trial or transfer test controls. Thus the animals may simply have learned that certain responses to certain stimuli maximized their rewards. When pigeons were given similar tests that included controls, the results were not what metacognition would predict. The pigeons' accuracy was no worse on forced than chosen trials, and they did not differentially escape difficult trials when transferred to new stimuli. Their behaviour was better explained in terms of learning the most profitable response to various stimuli than in terms of choosing to take the test when they knew that they knew the correct response: the pigeons' behaviour was biologically rational, but not rational in a more demanding sense requiring access to the information on which they act. Thus, the evidence is against metacognition for perceptual discrimination in pigeons, and as yet inconclusive in dolphins and monkeys.

Tests of metacognition for memory, or metamemory, use two different paradigms. Serial probe tasks display a series of images, followed by a single probe image; the task is to say whether it was in the series. In more difficult trials, the probe image occurred earlier in the series, or the series is longer. Monkeys took the escape option more often in difficult trials, and were more accurate on chosen than on forced trials. These results appear to be best explained by metamemory: monkeys have information about whether they recognize the presented probe image. Delayed match-to-sample metamemory tasks are even more demanding. They require that the monkey decides whether or not to escape *after* seeing a sample but *before* being presented with the stimuli from which he chooses a match to sample. That is, the monkey has to decide how well it remembers the sample, or to predict whether it will recognize it, before seeing any further stimuli that it might or might not recognize. Metamemory tests with monkeys using this paradigm included both transfer controls and forced trial controls, and the results were as predicted by a metamemory account. The results of these two sets of tests of metamemory provide strong evidence that monkeys can monitor their own memories. By contrast, negative results on related tests with pigeons indicate that pigeons do not have metamemory.

Shettleworth and Sutton emphasize that metacognition for perceptual discrimination is distinct from metacognition for memory. A given species may have one but not the other; different species may have metacognitive capacities in different domains. Moreover, the extensive training these tests of metacognition require contrasts with the spontaneous uses of metacognition by people in everyday life. Tests for metacognition in animals might do well to focus on biologically relevant contexts in which such abilities might have a natural function. In this respect, they cite approvingly Call's paradigm in which apes can choose whether to obtain more information by looking into opaque tubes for bait before choosing one. While there is strong evidence for metamemory processes as opposed to mere biologically rational behaviour in primates, a challenge for the future is to develop more naturalistic contexts in which animals might display possibly domain-specific capacities to access the information on which they act.

1.4.3 **Proust**

The perspectives on metacognition offered by Call and by Shettleworth and Sutton lead to the thought that evolution may have selected for domain-specific metacognitive capacities, in accordance with their adaptiveness to different animals. A related evolutionary approach to metacognition is given a rich theoretical framework by philosopher Joëlle Proust (Chapter 12). Proust contrasts top–down approaches to animal rationality, which begin with human reasoning and look for something relevantly similar in non-human animals, with bottom–up approaches such as her own, which view human rationality as emerging from evolved mechanisms for control and representation in non-human animals. Her synthesizing discussion charts the adaptive steps that lead from merely biologically rational behaviour to human reasoning processes, and clarifies the critical functions of metacognitive processes in this genesis.

(a) *Rationality as cognitively-operated adaptive control*. Proust finds a powerful clue to the biological sources of rationality in the very idea of bounded rationality: the limited informational resources of simple cognitive systems in variable environments bring with them new selection pressures for flexible behaviour. These favour capacities to extract information from changing environments so as to respond appropriately, and to control and manipulate informational resources in ways that advantage oneself and disadvantage one's competitors. She endorses the view of Peter Godfrey-Smith (1998, 2002) and Kim Sterelny (2003, and Sterelny's Chapter 14, herein) that variable, complex environments exert strong selective pressure for flexible behaviour.

Adaptive control structures evolve in response to this pressure. Information flows in a dynamic loop through control structures: given a target, the system generates output, which in turn generates feedback, which is compared with the target; the comparison of target and feedback governs output in a way that permits disturbances to be neutralized. In 'slave' control systems such as thermostats,[50] this comparator process is implemented inflexibly by the physics of the system; they cannot learn or change their goals. By contrast, adaptive control systems achieve flexibility by the use of internal

---

[50] See and cf. the discussions of thermostats and varieties of control in the chapters by Dretske and by Hurley. The transitions from simple to more complex and flexible control systems has been an important theme in naturalistic approaches to the mind.

models: learning and memory allow them to predict the results of alternative commands and flexibly to select the right commands to achieve their target in varying environments. Simple internal models can be local, specialized, and not refer to anything outside the system. But control becomes more powerful and flexible when internal models employ cognitive representations, which are selected to refer to specific events or properties that can be re-identified over time and which can combine in indefinitely new ways to represent alternative possible courses of action.

In its earliest form, Proust proposes, rationality can be understood as a disposition that tends to be realized by adaptive control systems that are cognitively operated. Her notion of cognitive operated adaptive control is closely related to Dretske's conception of minimal rationality, in which the explanation of behaviour depends on the content of representations, as well as to Clayton *et al.*'s conception of intentional explanation. However, such flexible control has costs as well as benefits, which vary with both the ease with which information can be extracted and its reliability; in some environments, rigid behaviour can bring greater rewards than flexible behaviour, which may involve, for example, higher chances of error. Moreover, in the variable, informationally noisy, environments that select for flexible cognitive control, information itself becomes a good. Organisms face various trade-offs and constraints in discriminating signals from noise and in categorizing signals. They are advantaged if they can manipulate the informational properties of their environments effectively, to increase information quality for themselves and decrease it for their competitors.

(b) *Rationality as metacognitive control.* What Proust calls 'epistemic actions' are actions with the goal of such informational manipulation. This goal can be pursued by physical means, such as squirting ink in the eyes of a predator, or by representational means, such as issuing a deceptive signal. Proust refers to epistemic actions pursued by representational means as 'doxastic actions', and identifies metacognition as a sub-category of doxastic action of particular relevance to the development of rationality: metacognition pursues informational goals in the system itself by informational means. It monitors and controls various of the system's own cognitive processes (as opposed to controlling behaviour directly), functioning to improve cognitive flexibility and informational quality in complex, multivalued control systems. However, metacognitive control processes also bring a new set of costs, such as the need for complementary mechanisms of binding, calibration, inhibition, and so on. Whether metacognition is worth the cost in the sense of biological rationality varies with the needs and environment—including informational environment—of different species.

Metacognition is covert in the sense that it enables a creature to make trials and errors in its head, as Millikan puts it—to entertain possible actions without incurring the potential costs of actual actions. Metacognitive capacities emerged, Proust suggests, in the form of covert simulative processes, which use the action control system itself off-line to represent possibilities (the need for inhibitive mechanisms in off-line simulation is one of its costs). Like Hurley, Proust views simulation in control systems as an efficient and adaptive way of implementing a variety of cognitive functions; there is increasing evidence that nature has used it widely. Note that while simulative metacognition can improve the results of my actions here and now, it also enables various forms of decentring from me-here-and-now, which we (along with various contributors) have suggested is an important aspect of rationality.

(c) *Rationality as explicit metarepresentation of reasons.* Proust next highlights an important distinction between metacognition as a practical capacity for information control exemplified by simulation, and metarepresentation, a theoretical capacity for explicit, conceptual representation of mental states such as beliefs and desires—one's own and others (see also and cf. Bermudez 2003, Ch. 9, 166 ff.). A closely related distinction has played an important role in debates between theory-theory and simulation-theory concerning the basis of mind reading capacities. She surveys empirical work with animals (including work discussed by other contributors) which provides evidence that metarepresentation is not required for metacognitive control, of either information available to others or to oneself. On the other hand, there is no species with capacities for metarepresentation but not for metacognitive control. She explains this asymmetry in terms of a distinction between the implicit, procedural reflexivity of metacognitive control and the explicit, conceptual reflexivity of meta-representation. The continuous dynamic cycles of control architectures, including architectures for metacognitive control, mean that command and feedback refer to one another procedurally or implicitly: metacognitive information is embodied in the dynamic properties of the control loop used in simulative mode. This can, but need not, be articulated as explicit mental content.

How and why is implicit metacognitive information available to some animals transformed into explicit metarepresentation, which is characteristic of the final stage of fully-fledged human rationality? She suggests that in human beings the output of metacognitive control loops, which carry information about information, may be inputs to higher-level structures that control linguistic communications. The latter recode metacognitive information in an explicit, metarepresentational format. This further capacity serves social control functions in a Machiavellian world in which information is manipulated by linguistic (among other) means, by permitting one to report one's reasons and offer justifications for acting. She cites neurophysiological evidence for such a superposition of control structures in human beings; both forms of control are aspects of human rationality, but only the lower level of rationality is shared with other animals.

As we've seen, as well as metacognition, normativity is often regarded as a requirement of rationality. Proust ends her discussion of metacognition and rationality by arguing that normativity depends on explicit metarepresentation. Metacognition without metarepresentation is not prescriptively normative: it does not determine an optimal solution that the system *should* select. Thus non-linguistic animals cannot meet this requirement. Here she takes issue with Millikan's account of normativity in terms of natural evolutionary functions,[51] including cognitive functions, according to which the functions of representations underwrite the normativity of their content. Proust denies that biological functions are natural, causally efficacious properties. Rather they super-vene on the history of causally efficacious physical properties, and are relative to the explanatory concerns of biologists. Metacognition without metarepresentation can indeed support the capacity of control systems to recognize and correct mistakes relative to their purposes; but there is nothing 'necessarily and inherently bad' about such mistakes. This is insufficient for prescriptive normativity in Proust's strong sense. She concludes that prescriptive normativity is only possible when metarepresentation, with its social control functions, enables explicit, public justification and assessment of mistakes (see and cf. Dennett 1996, Ch. 2).

We might raise the question, however, of why the social control functions of reporting reasons and offering justifications are any more naturally prescriptive, or any less relative to the human concerns or the purposes of control systems, than are biological control functions. Does the possibility of explicitly and publicly assessing and recognizing a mistake make the mistake necessarily and inherently bad? It is useful to distinguish social normativity from biological normativity, but it is not clear why social functions and associated norms are more objectively, inherently or naturally prescriptive—hence more normative—than biological functions and associated norms. The difference may be one of kinds of normativity rather than degree of normativity. This suggestion goes naturally with scepticism about the very idea of intrinsic prescriptivity, a scepticism that might undermine the capacity of this idea to support a conception of normativity distinctive to human rationality. But such scepticism is quite compatible with a characterization of the distinctive normativity of human rationality in terms of the social functions of explicit metarepresentation.

1.4.4 **Currie**

We suggested above that we are inclined to recognize rationality as opposed to 'mere' intelligence when the flexibility of behaviour indicates a capacity to decentre from me-here-now. Philosopher Gregory Currie (Chapter 13) suggests that since pretence depends on a capacity to decentre, it is an indication of rationality. In pretence, a creature responds to the world as transformed by imagination.

While decentring may be a form of metacognition in Proust's sense, Currie argues that it does not depend on metarepresentation. He distinguishes the vertical ascent involved in metarepresentation (as in the movement from thinking 'p' to thinking 'She thinks that p') from the horizontal shifts involved in decentring (as in shifts of perspective from me-here-now to me-there-yesterday or me-there-tomorrow or

---

[51] Endorsed by Hurley in her chapter herein.

you-here-now or some fictional character-place-time). The vertical and horizontal dimensions of variation are conceptually independent. Representing the world from another perspective by decentring needn't necessarily depend on representing another perspective—metarepresentation. Conversely, metarepresentation, say of another's thoughts, needn't necessarily involve decentring. Decentring and metarepresentation should be distinguished so that questions about whether they tend to go together and how they interact can be resolved by further empirical work and arguments. Currie doesn't try to sort out these questions in his chapter, but he does urge that issues about decentring as distinct from metarepresentation are also important for characterizing animal cognition.

What forms of behaviour provide evidence of pretence? Play need not involve the imaginative transformation and decentring of pretence. Assessing pretence often requires tracking behaviour over time and looking for elaborations of pretence that distinguish it from mere play. One such form of elaboration is *recognition*: for example, evidence for pretence that globs of mud are pies might be recognition that there are two more globs of mud over there, hence two more pies to be shared out. Another form of elaboration is creative *enrichment*, in which the game is taken to a new stage not demanded by anything so far specified or enacted. Currie takes such recognition and enrichment to be good evidence, though not strictly required, for pretence. They are found in 2-year-old children, and in solitary as well as social play. But are they found in other primates?

To address this question, Currie distinguishes pretence from the overlapping categories of deception and imitation, with which pretence may be confused. Deception, in his view, requires intentional behaviour by the deceiver, directed toward an end which it is likely to achieve only by inducing a false belief, not shared by the deceiver, in the deceived. It thus doesn't require the intentional manipulation of belief, but rather behaviour that depends for its success on affecting another's belief. Deception requires first-order intentionality in deceiver (intentions) and deceived (beliefs), but not metarepresentation (intending to bring about a false belief) or mind reading. While imitation of goal-directed acts requires decentring to the other's perspective, Currie is concerned here with imitation of bodily movements, which doesn't require even first-order intentionality. Understood in these ways, imitation, deception and pretence can occur independently or together.

Various examples of animal behaviour that may seem to involve pretence are better seen as imitation or deception. A gorilla wiping a surface with no evidence of recognition or enrichment is better seen as imitating human cleaning behaviour than pretending to clean. False alarm calls by monkeys that induce desired behaviour in conspecifics are better seen as deception than as pretending there is a threat nearby. What might warrant viewing a false alarm episode as pretence would be recognition or enrichment behaviours—say if the alarm-giver made eye movements apparently tracking the movements of a threatening predator. Eye-covering play in various primates is puzzling, but Currie sees no reason to view it as involving the imagination transformation of the world required for pretence. The few examples of recognition and enrichment behaviour in animals that do seem to warrant attributions of pretence are found in human-reared primates who have been given language training and encouraged to imitate. For example Kanzi the bonobo appeared to eat an invisible fruit and spit out its pips, signing that it was 'bad', and also to grab an imaginary piece of food he had put on the floor when someone else reached out to that spot.

In Currie's view, current evidence suggests that pretence, like language but unlike deception, is beyond the capacities of non-human primates unless they are raised in such humanly enriched environments. He endorses a version of Morgan's Canon, according to which we should attribute lower mental processes that enable only observed behaviours rather than higher processes that would also enable further behaviours—unless we have some evidence of capacity for those further behaviours. Any creature that can respond to the world as imagined can also respond to the world as it is, but the converse does not hold. Since a higher mental process such as pretence enables a creature to do more, to attribute pretence we need evidence of those additional behavioural capacities, such as capacities for recognition and enrichment.

Along with Millikan, Currie allows for the possibility of rationality in perception, and shows how the kind of decentring found in pretence illustrates this possibility. Pretence, he explains, is closely linked with the perceptual phenomenon of 'seeing-in'. When you see something in a picture, you don't confuse the picture and what you see in it. Some apes, in particular human-reared apes, show evidence of seeing-in without any such confusion and can, for example, sort line drawings according to what they depict. Similarly, we can see in simple mimetic acts the depiction of driving a car, or nursing a baby, with no confusion of the mimetic acts and the acts they depict: we don't take them for the real thing. Currie suggests that such seeing-in is an important aspect of pretence, perhaps part of its primitive basis. If so, pretence is a kind of decentring that does not require full-blown conceptual capacities and is perceptually based: a kind of rationality in perception that is available to young children and perhaps to some human-reared non-human primates. There is some evidence that what young children think is depicted or pretended depends on what they themselves see in depictions or acts of pretence.

Currie concludes with a speculation about how the capacity for seeing-in and hence for pretence might have evolved. Behaviour-manipulating signals evolve into belief-manipulating deceptive signals that potentially influence a whole range of behaviours. These are in turn countered by the development of the capacity to recognize deception: for example, to see a warning gesture *in* deceptive behaviour, without seeing it *as* a genuine warning. Seeing-in may thus have resulted from an arms race between deception and deception-detection.

1.5 **Social behaviour and cognition**

1.5.1 **Sterelny**

Philosopher Kim Sterelny's complex argument (Chapter 14) pursues the relationship between metarepresentation and social cognition within an evolutionary perspective on rationality. His argument falls into three broad stages. First, he describes the selective pressures that drive the genetic evolution of capacities for sophisticated information use by animals. Second, he describes the further distinctive pressures on hominid cognitive capacities associated with the social as well as genetic transmission of information. Finally, against the background of this dual view of the genetic and sociocultural roles of cognitive capacities, Sterelny contrasts two conceptions of rationality that might be used to describe and contrast animals and human beings. The first is related to Kacelnik's biological rationality: a measure of overall cognitive efficiency, of the general evolved capacity of an agent to respond adaptively to the informational challenges it faces. Sterelny provides reasons for scepticism about this approach. He recommends instead a second, more specific conception of rationality in terms of metarepresentational folk logic. He views folk logic as an evolved capacity to assess socially (and especially linguistically) transmitted information, with its inherent openness to manipulation and high costs of error.

Sterelny explains, by reference to Peter Godfrey-Smith's views, how animals' capacities to use information are driven by the need to respond flexibly to variable environments. For example in foraging, starlings must trade off the costs of declining efficiency in collecting food as their beaks fill up against the costs of flying back and forth to deliver their loads to waiting offspring. The efficient trade-off depends on variable information about the agent's environment. But adaptive behaviour also depends on a trade-off against the costs of gathering such information and of error: if the costs of information are too great, flexible responses reflecting information about varying environments may not be adaptive on balance. Some decision problems require environmental information that is relatively costless to acquire, but others require costly information. Environmental information can be degraded and become costly to collect in various ways that derive from competitive biological interactions. For example other organisms pollute the informational environment by pre-empting the use of simple single cues for appropriate behaviour: prey hides, disguises itself, mimics poisonous organisms, and so on. And targeted agents often respond subversively to the actions of other agents, to block or thwart them: fast and frugal heuristics may work for catching a baseball, but are unlikely to intercept the non-ballistic, subversive path of fleeing prey. Such informationally hostile environments select for cognitive advances, such as the use of multiple cues rather than single cues to track features of the environment, and the decoupling of stimuli from automatic responses.

Hominid cognition reflects the selective pressures imposed by such informationally hostile biological interactions, as well as those deriving from environmental variation—from unstable climate and migration. The expansion of hominid populations into new environments created selective pressure for adaptive plasticity, as well as for specific adaptations by local groups to local environments. Like many other species, human beings construct important aspects of their own environmental niches. But human niche construction is distinctively cumulative, as one generation inherits the improvements of previous generations, such as cooking techniques and domesticated plants and animals, which in turn induce profound changes in social environments and transform the risks and cognitive demands individuals face. Such changes occur faster than genetic adaptation can solve the specific problems they present, creating selective pressure to improve the cultural transmission of solutions between generations. In response, hominids have evolved distinctive genetic adaptations that facilitate social learning: capacities for imitative learning and teaching, an extended childhood for transmission of information between generations, language, and a division of cognitive labour. Once social learning is genetically enabled, however, it is self-amplifying through a process of cultural evolution. Runaway selection for the social transmission of information has, of course, costs as well as benefits: it is a powerful means of adapting swiftly to a changing environment, but it also risks extreme error costs, which in turn generates pressures for control of these risks.

Against this background of ideas about the evolution of cognition in general and of human cognition in particular, Sterelny sets out a contrast between two approaches to animal rationality. He argues that the evolution of rationality should not be viewed as the evolution of optimal cognitive design, and that it is more fruitfully viewed as the evolution of folk logic.

The first approach conceives of rationality by reference to a general, optimally adaptive design for obtaining and using information. But Sterelny argues it is unlikely to work. Why not? Because of a broader problem: there is no accepted, objective to measure overall fit between organism and environment or to assess the optimality of adaptations in general terms (as Lewontin argues). Fitness and adaptation are local, highly specific relations, between specific organisms and specific environments; definitions of general relations of fitness or adaptiveness have proved elusive. Moreover, explanatory work is done by specific first-order features of phenotype and environment, not by general fitness level. Sterelny argues that this problem with the idea of optimal adaptation carries across when it is applied to cognitive adaptations (see also and cf. Shettleworth 1998, 10, 570). Just as what counts as fitness enhancing is specific to organism and environment, so what counts as efficient use of information is specific to organism and environment. There is no global metric of cognitive efficiency. It is multidimensional and cannot be maximized in every direction simultaneously; the right trade-offs are sensitive to specific contexts. Good cognitive design in some contexts may be unsophisticated, as the costs of information may exceed its benefits. If we were to think of rationality in these terms, it would have no intrinsic link to degree of cognitive sophistication.

Sterelny favours the second approach, which conceives of rationality in terms of the more specific metarepresentational capacity of hominids to *assess* (as well as represent) the thought and talk of others. Such assessments express a *folk logic*: a set of more or less explicit norms about reasoning and the representations it produces. Folk logic presupposes metarepresentation, but involves more than that. Metarepresentation, of the signals of other agents as signals, might be useful in predicting their behaviour, even if it did not assess their accuracy. But folk logic assesses, as well as representing, representations and inferences. What is the function of such normative assessment? Folk logic is not required for inference; a capacity to make valid inferences does not require a capacity explicitly to assess inferences against norms. Do the norms of folk logic promote inferential efficiency? While they have done so in the history of science, folk logic presumably predates science. Like Proust, Sterelny gives normativity a function related to the potential for social manipulation of information. Sterelny thinks that folk logic can be understood as an evolved response to the costs and benefits of using socially transmitted information. Social learning is risky as well as powerful. Socially transmitted ideas can exploit and damage their biological hosts, and put recipients at risk of manipulation. Some forms of social transmission are harder to manipulate than others: the social transmission of information through imitative learning of expert skills is harder to manipulate than the social transmission of information through linguistic signals, which are arbitrary and carry no intrinsic marks of reliability. As Sperber (2000) has suggested, folk logic filters out some of the bullshit from socially transmitted information, especially linguistically transmitted information. Sterelny argues that folk logic also responds to selection for cognitive plasticity in learning to use novel inferential techniques and representational media: to learn about reasoning flexibly, you need to be able to represent and evaluate your own reasoning. Thus, folk logic both amplifies the effects of the social transmission of information, and responds to pressure for control of the risks of costly errors incurred by social transmission.

If the optimal cognitive design approach could be made to work, it might provide a graded conception of evolved rationality applicable in degrees across human and non-human animals. But as we've seen, Sterelny thinks it will not work. The folk logic approach he favours does not provide a conception of rationality applicable to non-human animals, since only human beings appear to have evolved folk logic. Nevertheless, it does provide a naturalistic view of how rationality develops from social intelligence. It also raises questions about genetic and cultural coevolution: is folk logic, hence rationality, to be viewed as a genetically evolved form of control and guidance of cultural evolution, or as itself a product of cultural evolution—or a bit of both?

The chapters by Proust and by Sterelny place rationality in the context of the social intelligence hypothesis, that the cognitive demands of social life drive the evolution of intelligence.[52] Social life and the social transmission of information create

---

[52] See Byrne and Whiten 1988; Whiten and Byrne 1997; Shettleworth 1998, 511–522.

opportunities for animals to gain advantages over others by manipulating information, but also create opportunities to benefit by warranting the information they provide to others and by filtering the information provided by others. Rationality responds to the pressures to warrant and vet information that are generated by social learning. This general approach provides a plausible view of the evolution of rationality in the strong sense of human reasoning processes, but what implications does it have about prior stages of rationality in animals? Proust provides for intermediate stages of rationality in terms of metacognition without metarepresentation, while Sterelny regards rationality in the sense of metarepresentational folk logic as a hominid preserve.

1.5.2 Addessi and Visalberghi

Studies of Capuchin monkeys reported by primatologists Elsa Addessi and Elisabetta Visalberghi (Chapter 15) provide a cautionary perspective on the link between social learning and rationality. Field observations of these monkeys, who forage in social in groups, make it tempting to suppose that they learn from observing others which foods are safe to eat by means of rational processes that would be natural to attribute to human beings in similar contexts. However, Addessi and Visalberghi's experiments undermine the attribution of (what Kacelnik would call) PP-rational social learning to the monkeys, and instead suggest an explanation of their foraging behaviour in terms of lower-level processes operating in ways that are biased by social contexts, which nevertheless produce adaptive, or B-rational, behaviour in the relevant environments. This chapter thus draws together various themes already encountered, as well as that of the relations between social life and rationality, including: the methodological difficulties in attributing PP-rationality to animals, the possibility of explaining apparently rational behaviour in terms of lower-level rather than rational processes, and the possibility of 'outsourcing' to relevant environments some elements of the processes that support B-rational behaviour.

Social learning about food plays a large role in human eating habits, and many researchers have found it plausible to suppose that the feeding habits of other social primates are shaped by similar learning processes. The assumption has often been made that naïve animals observe experienced group members eating and learn from them which foods are good to eat and which are unsafe. Capuchin monkeys eat novel foods less than familiar foods, but eat more novel foods when in a group than when alone. Observations of social foraging in the wild that seem to support an attribution of rationally sophisticated social leaning are better explained, in light of experimental results, in terms of individual preferences and learned aversions, associative and trial and error learning, and social biases on individual learning.

Addessi and Visalberghi allow a monkey to see demonstrator monkeys eat a brightly coloured food, and the observer is offered a novel food of the same appearance and colour to eat, or a novel food of a very different colour. Whether the demonstrators are eating food of the same colour or of a different colour, the observer consumes more of its novel food when they can see demonstrators eating. That is, the observer eats no more of its novel food when its colour matches that of the demonstrators' food than when its colour differs. Interestingly, the observer consumes more of the novel food when there are more demonstrators eating. Other experiments showed that giving either the observers or the demonstrators a choice between two differently coloured foods does not produce any tendency for the observers to choose the food that matches the food the demonstrators eat. Since eating novel foods is socially facilitated regardless of what is eaten and is not directed to specific novel foods, the monkeys do not appear to be learning what is safe to eat by observing what others eat. Nor does the experimental evidence support the hypothesis that monkeys learn which foods to avoid by observing which foods others avoid eating (absence is harder for animals to detect than presence).

What does appear to influence capuchins' choice among foods? The monkeys tend to prefer sweet foods and avoid bitter foods; sweetness reflects energy content and toxic foods are often bitter. They are cautious about novel foods; learning about foods begins with small samples, which are less likely to be lethal even if the food is toxic. If sickness results, the food is associated with it and avoided; if there are no bad consequences it is progressively included. These individual processes go a long way to enabling monkeys to select the right foods. However, social biases still influence an individual's food choices, though in a less 'targeted' and rationalistic way, in which social influences are mediated by an environment with certain properties. Social facilitation works non-specifically, by supporting the synchronization of feeding activities rather than the learning of what to eat. However, since individuals moving together are likely to encounter the same food sources in most environments, feeding synchronization increases the probability that naïve individuals, along with knowledgeable individuals, eat the same foods. This social effect combines with individuals' preferences for certain tastes, their tendency to avoid novel foods, and food aversion learning, to increase the chances that monkeys feed in similar ways, which are biologically rational given their environments.

Thus, a widely assumed interpretation of wild foraging behaviour in terms of rational social learning does not stand up to experimental probing. Rather, the role of social facilitation in adaptive food choice can be understood in terms of lower-level processes and interactions with environmental features (compare the discussion by Papineau and Heyes of rival rational and associative accounts of imitative social learning in quails, Chapter 8 this volume). The authors' methodological moral is that field observations may tempt us to make unwarranted attributions of human-like reasoning and, by themselves, can be an unreliable basis for attributions of process rationality, in the sense of Kacelnik's PP-rationality, to animals.

1.5.3 Connor and Mann

Cetologists Richard Connor and Janet Mann (Chapter 16) are very aware of the difficulty of inferring from complex social behaviours in the wild to the cognitive processes that produce them. The alliance formation behaviours of bottlenose dolphins that their fieldwork meticulously documents are not only dauntingly complex, but appear to be intractable to experimental methods of probing process rationality. Nevertheless, their fascinating observations over many years of the social lives of over 600 bottlenose dolphins in Shark Bay provide a rich basis for theorizing about the biological rationality of dolphin behaviour. Connor and Mann compare the selective pressures imposed by social complexity in marine as opposed to terrestrial environments, and draw attention to the very different adaptations found in different cetacean species, which display variation in brain-to-body size ratios comparable to that between human beings and the great apes. Their work aims to contribute to our understanding of how the cognitive abilities demonstrated in captive bottlenose dolphins (see Herman's Chapter 20 in this volume) function in the wild, and of how the Machiavellian intelligence hypothesis—that the demands of social life select for advanced cognitive capacities—may apply to cetaceans.

Connor and Mann provide a review of dolphin social life, focussing on male alliances (both first-order and higher-order alliances), on female relationships, and on the roles of affiliative interactions, and making revealing comparisons with primates along the way. Like primates, dolphins mature slowly, and their offspring nurse for about four years; unlike primates, dolphins of both sexes maintain their natal territorial range as adults, so that from early life they can begin building social knowledge and negotiating relationships that may be important during their later reproductive years. And like chimps, dolphins live in a fission—fusion society, in which individuals associate in small groups of changing composition. However, while chimp social groups have strong boundaries, dolphin societies are open, with no boundaries between groups or territories but rather a pattern of overlapping home ranges. Primates are likely to know all the other individuals in their social group, but in Shark Bay dolphin A may know B and B know C, while A does not know C.

(a) *Male multilevel alliances and social complexity.* The strongest affiliations among the dolphins are same-sex affiliations, though male–male and female–female affiliations differ in character. A distinctive and unusual feature of dolphin society is the existence of male multilevel alliances within a social group. Females with calves approaching weaning attract aggressive control or 'herding' by first-order alliances of two or three males, which capture females from other alliances for herding and defend them against attack. Alliance partnerships may anticipate female fertility as existing calves mature, but may also function to test and consolidate male–male bonds. Stable first-order alliances in turn work together in second-order alliances that might have a dozen or more members; the more labile the first-order alliances, the larger the second-order alliances.

In general, alliances within social groups make for complexity, as when dolphin B may recruit A in alliance against C by means of affiliative behaviours such as grooming, while C is also trying to recruit A to ally against B. Such triadic interactions mediated by affiliative behaviour are prevalent between individuals within groups but not between groups, in non-human primates. However, social complexity increases dramatically if such triadic interactions are recapitulated between groups of allies as well, so that alliance ABC may compete with alliance GHI to recruit alliance DEF in a second-order alliance: consequences of individual interactions may be relevant at both levels of alliance. Such nested or hierarchical alliances generate a landscape of strategic options and high risks that exacerbate the demands on cognition and place a premium on social intelligence; among primates, they are characteristic only of human beings. They are also characteristic of dolphins.

Connor and Mann illustrate the challenges of nested alliance structures among male dolphins with several case studies: a 17-year history of the relations between dolphins Rea and Luc; a study of the 14-member second-order 'super-alliance' WC; a study of the shifting multilevel alliances of three provisioned males given dead fish by human beings on daily beach visits. A striking observation of interactions among three second-order alliances PD, KS, and WC suggests a third level of alliance formation. While Connor and Mann's interpretation is cautious, it is tempting to describe the interaction as one in which an initial third-order alliance between PD and KS, which were not competing for their respective controlled females, was attacked by WC, which recruited PD to a new third order alliance that expelled KS, in which PD retained possession of its females but WC appropriated KS's female. However, it was difficult to determine whether PD actively 'took sides' or merely remained neutral.

(b) *Female associations and social learning.* Females spend considerably less time than males do with their closest same-sex associations, and their group sizes vary widely; unlike male–male relationships, female–female relationships are very rarely antagonistic. Females may associate for reproductive functions, or to mob sharks. Members of one female group were observed to engage in petting the males in a trio that had been herding another female, while other members of the female group flanked the herded female, concealing her both visually and acoustically from the distracted males, and escorted her away to freedom—an unusual behaviour that may suggest intentional deception (at any rate, the males were fooled). The demands of foraging for single, mobile prey may account for aspects of female behaviour, such as neither stealing nor sharing prey (even with offspring). Moreover, females often specialize in specific, distinctive, foraging techniques, such as sponge-carrying (a form of tool-use) or beaching, which are passed on to their offspring who can be seen 'practicing'. Sponge-carrying, for example, is found in less than 10 per cent of the female population, and only in calves born to sponge-carrying mothers. Connor and Mann's work demonstrates that there is social transmission of several foraging 'traditions' across generations, in addition to the amply-documented social learning of vocal activity.

(c) *Affiliative interactions.* Alliances among both primates and dolphins are mediated by affiliative interactions, such as grooming in primates or petting and synchronized swimming in dolphins, which may function to test the bond between individuals.

In dolphins they are found not just between males in the same first-order alliance but also between males in different first-order alliances but the same second-order alliances. Males prefer to associate with other males from an early age, and homosexual social behaviours involving role exchanges and synchrony are likely to play a role in establishing male bonds. On one occasion males were observed to herd members of a rival second-order alliance as if they were females, directing at the rival males behaviours associated with consortship of females. To observers, emotions appear to play a role in aggressive and affiliative interactions among dolphins; they might function to guide complex social decision-making by providing a timely integration and comparison of the values of past interactions with others.

Connor and Mann end their discussion by reflecting on the selection pressures that could have led to high ratios of brain to body size in dolphins, given the large variation in brain-to-body size ratios among cetaceans. The energy-rich food of dolphins makes a large brain more affordable; but what benefits does it provide? Connor and Mann find unpersuasive the argument that large brains are needed to implement sophisticated echolocation functions for foraging, and support Herman's suggestion that social complexity drives the evolution of large brains in dolphins. Connor and Mann suggest that multilevel alliances among dolphins provide a link between theories of alliances and conflict *within* groups of non-human primates and theories of alliances and conflict—that is warfare—*between* groups of human beings. Warfare may increase the selective pressure for sophisticated social cognition because it requires individuals to co-operate against an enemy with other individuals in their group, with whom they are simultaneously in reproductive competition. Similarly, multilevel alliances demand that individuals base decisions at one level of interaction partly on the impact of those decisions at other levels. But what aspects of life in the open ocean provide the relevant trigger for such interdependence and social complexity? Perhaps cetaceans facing predation from sharks have nothing to hide themselves or their offspring behind except each other. Comparative study of the relationships among brain size, cognitive capacities, and behaviour among diverse cetacean species promises to increase our understanding of these and other issues about the role of social life in driving cognitive adaptations.

1.6 **Mind reading and behaviour reading**

The ability to understand the mental states of others, or mind reading,[53] is widely regarded as the most demanding form of social cognition. Mind reading was originally

---

[53] We here follow current usage in adopting the generic term 'mind reading' for the understanding of the mental states of others. We agree with Nichols and Stich (2003, 2) that other terms (especially 'theory of mind') often are understood to carry substantive theoretical commitments (especially given the rivalry between theory-theory and simulation-theory accounts of mind reading).

conceived as the exercise of a theoretical capacity for reasoning about the mental states of others, but simulation theories of mind reading are now prominent rivals to such 'theory-theories'. In recent years, two further significant developments have occurred in work on mind reading. First, a widespread view that mind reading is an exclusively human capacity has been challenged, and controversy re-emerged about animal mind reading. Second (though relatedly), mind reading has been disaggregated into different aspects, some of which animals may have and others they may lack. For a time, tests for attributions of false beliefs to others were treated as the gold standard in assessments of mind reading. Children under about 4 years of age, and autistic persons, cannot answer correctly questions that require them to attribute to someone else a belief that they themselves believe to be false in order to predict the other's behaviour; instead, they attribute their own belief to the other. But recently, Tomasello (1999) and others have urged that a more graded view of mind reading capacities is needed, rather than an all-or-nothing assessment based on false belief tests (see also Povinelli and Vonk, Chapter 18 this volume). There is evidence that the ability to attribute desires and intentions to others precedes the ability to attribute beliefs to others, both ontogenetically in children (Rakoczy et al., submitted), and phylogenetically (Tomasello 1999; Tomasello and Call 1997; Guzeldere et al. 2002). At the same time, non-verbal experimental paradigms have been developed for assessing not just the capacity for false belief (see Tschudin's Chapter 19), but also various other aspects of mind reading capacity. In particular, paradigms have been developed that focus on whether chimps understanding perceptual states of others, such as seeing, as opposed to epistemic states, such as beliefs (discussed in Chapter 17 by Tomasello and Call and Chapter 18 by Povinelli and Vonk, herein).

An important aspect of the current controversy over animal mind reading concerns how results obtained using some of these non-verbal paradigms should be interpreted, and indeed how informative they are in principle. Here we represent the debate between two distinguished centres for the study of ape cognition, Michael Tomasello's group in Leipzig and Daniel Povinelli's group in Louisiana (see also Gomez 2004, Ch. 8). While apes have been unable to pass standard non-verbal false belief tests, they have passed other non-verbal tests of mind reading for seeing, including those described by Tomasello and Call in Chapter 17. However, Povinelli and Vonk argue that success on such tests can just as well be explained in terms of behaviour reading as mind reading. We also represent inconclusive but suggestive data from non-verbal false belief paradigms applied to dolphins by Alain Tschudin, which casts further light on the difficult methodological issues in play.

In considering what kind of behavioural evidence favours animal mind reading as opposed to behaviour reading, relevant factors include not just methodological factors about the transfer of success to new tasks, the evidence base required for successful performance, and various appropriate controls. Another relevant factor is the conception of mental states that is in play. Given evidence may warrant attributions of mind reading on some conceptions of mental states but not on others. Crudely, some conceptions of mental states and their contents emphasize their internal, inferential connections among one another, while others emphasize their external connections, informational or teleological, to the world. Suppose an animal whose mind-reading capacities are in question behaves in a way that suggests he attributes a given mental state to another animal but not a further, inferentially connected mental state. This may count against the mind reading interpretation less on an informational or teleological conception of the mental than on an inferential role conception of the mental.[54] Differences over evidence for animal mind reading may reflect such implicit differences in the underlying conception of the mental. Moreover, the extent to which animal mind reading counts as an expression of social *rationality* in particular, as opposed to social perception or cognition more generally, may depend on the degree to which attributions of mental states are viewed as carrying with them inferential commitments.

1.6.1 **Tomasello and Call**

Comparative psychologists Michael Tomasello and Josep Call (Chapter 17) distinguish two approaches to interpreting the behaviour in non-human animals. Boosters interpret behaviour in psychologically rich ways; scoffers prefer psychologically lean interpretations. The ultimate boosters think there are no significant differences between human and non-human cognition, while the ultimate scoffers are radical behaviourists, who do not find it useful to talk of cognitive processes at all. Boosters about primate social cognition attribute understanding of others' psychological states to apes; scoffers regard apes as clever behaviourists themselves, who merely read and respond to others' behaviour but lack understanding of others' psychological states. In their chapter, Tomasello and Call compare richer and leaner interpretations of recent data from four experimental paradigms concerning whether chimpanzees know what others can or cannot see, or only what others are looking at. They argue that the booster hypothesis, that apes know what others see, is better supported than the various scoffer hypotheses that would be needed to explain these results.

(a) *Gaze following*. Experiments show that chimpanzees reliably follow the gaze of other chimpanzees and human beings, even when doing so requires looking past and ignoring other novel objects along the way, or moving in order to follow a gaze to a target behind a barrier. If they find nothing of interest, they check back and track the other's gaze again, and eventually stop following the gaze of individuals who repeatedly look to locations where no salient target is found. The booster interpretation of this behaviour is that chimpanzees follow gaze because they want to see what the other sees. The scoffer interpretation avoids attributing to the chimpanzee knowledge that

---

[54] Thanks here to Nick Shea. See also and cf. Sterelny 2000; Allen and Bekoff 1997, 97.

the other is seeing something and instead attempts to explain the behaviour by appeal to biological predispositions and learning: perhaps they are disposed to orient the way others orient, or have learned to look in the direction others are oriented.

(b) *Competing for food.* A series of experiments place a subordinate and a dominant chimpanzee in rooms on opposite sides of a baiting area in which food is placed in various positions in relations to barriers before the chimps are allowed to go for it. The dominant chimp will take any food she can see, but the subordinate can monitor the dominant's visual access to the food. It has been found that subordinates chimpanzees go for food that a dominant chimpanzee cannot see because it is hidden from view behind a barrier much more often than they go for food that the dominant can see.

Tomasello and Call consider and dismiss five 'scoffer' interpretations of this behaviour, some of which are ruled out by further experiments. The subordinates are not merely responding to the initial behaviour of the dominants, since when they are given a head start they still go for the food the dominant cannot see. The subordinates do not merely view the barrier as something that might slow down the dominant's physical approach to the food, since they do not favour food placed behind transparent barriers. The subordinates do not simply prefer to forage near barriers rather than in the open, since when all the food is placed near barriers they still prefer food dominants have not seen. The subordinates do not simply respond to the presence or absence of a dominant during baiting, since they go freely for food seen by one dominant if the first dominant is replaced by a second who has seen nothing; they seem to know what particular individuals have seen or not seen. Finally, the subordinates do not simply avoid food that a dominant has looked at (the 'evil eye' hypothesis); if the subordinate monitors a dominant watching the food being placed initially, but the food is then moved to a new location in view of the subordinate but not the dominant, the subordinates still go for it even though the dominant has looked at it. It is not plausible to suppose that the chimpanzees have had past learning experiences involving transparent barriers, relocated food, and so on, that would explain these results. A better interpretation is that chimpanzees do not just follow the gaze of others, but also know something about what others see, on the basis of which they infer what others are likely to do and hence decide what they themselves should do.

(c) *Begging and gesturing.* In these experiments, chimpanzees can beg for food from one of two trainers in different attentional states. The chimps chose to beg from a trainer facing them rather than one turned away from them. But they did not beg more from a trainer wearing a blindfold over his mouth as opposed to his eyes, or from one with a bucket on his shoulder as opposed to over his head, or from one with hands covering his ears as opposed to his eyes, or from one with eyes open as opposed to closed, or from one turned away but looking over his back at the subject as opposed to turned fully away. The scoffer interpretation of these results is that chimps attend to the body orientation of others, but know little if anything about what others see. However, in a modified design in which chimps faced only one human trainer rather than having to chose between them, they gestured differently depending on where the trainer was looking and the orientation of her face (see the comments by Povinelli and Vonk on these results, in section 3 of the appendix to Chapter 18, this volume). A booster interpretation of these results is that body orientation indicates the trainer's disposition to give the chimp food, while face orientation carries information about whether the trainer can see the chimp's begging gesture. Another line of experiments, in which chimps compete with human trainers for food, supports the view that face orientation gives chimps information about what others can see. When a trainer's body is facing one piece of food but his head is turned away toward another piece of food, chimps attempt to steal the food in front of his body rather than the food the trainer can see. A scoffer interpretation of these and other competition-with-a-human-being results is that chimps do not understand what someone can see, but simply avoid approaching food if they can see a competitor's face. This hypothesis was ruled out by an experiment in which chimps avoided going for food when the chimp could not see the human competitor's face but it was clear that the human competitor could nevertheless see the chimp approaching the food.

(d) *Self-knowledge.* The last of the four paradigms is one also discussed by Call's chapter on metacognition, in which one of two tubes is baited; to get the reward, the chimps have to touch the baited tube. Recall that chimps looked into the tubes before choosing which to touch more often when the tubes had been baited out of their view than when they had watched the baiting. A scoffer explanation of these and related results could appeal to hypothetical past learning of a difficult conditional discrimination. The alternative booster interpretation is simply that chimps know what they themselves have and have not seen, and hence know to look in the tubes when they have not seen them being baited.

Taken together, Tomasello and Call argue, these results make a strong case for the booster hypothesis that chimps understand what others (and they themselves) can and cannot see. The alternative scoffer hypothesis requires a dozen different hypotheses to explain the same phenomena, including various *ad hoc* and unlikely prior learning scenarios for which there is no independent evidence. They aren't impressed by either parsimony or Morgan's Canon as a reason to favour the scoffer view. It isn't clear whether parsimony here favours invoking fewer different hypotheses, as the boosters do, or invoking only behaviouristic learning rather than mind reading, as the scoffers do. Morgan's Canon counsels us to avoid postulating higher-level mechanisms to do explanatory work that can be done by lower level mechanisms, but the higher–lower distinction is problematic. Finally, learning is not incompatible with understanding seeing: they hypothesize that chimps may learn about what others can see more readily than they learn arbitrary associations, just as they can learn causal relations more readily than arbitrary associations (see Call's Chapter 10 for discussion). While Tomasello and Call believe that the evidence that chimps understand seeing is now overwhelming, they caution that this does not mean that the chimps also understand other psychological states. Mind reading is not all-or-nothing; chimps may be able to understand some psychological states but not others, and empirical evidence is needed for each case.

1.6.2 Povinelli and Vonk

Comparative psychologists Daniel Povinelli and Jennifer Vonk (Chapter 18) mount a methodological argument that there is no compelling evidence for the claim that chimps understanding seeing in others because the kinds of experimental paradigms that have been used in support of this claim, such as those discussed by Tomasello and Call in their chapter, are in principle incapable of providing evidence for it. A logical problem, they argue, precludes obtaining empirical evidence for animal mind reading by employing such paradigms; until the logical problem is resolved, the empirical issue of ape mind reading is not joined.

(a) *Louisiana vs. Leipzig: The logical problem with existing experimental paradigms for animal mind reading.* They explain the logical problem as follows. The standard experimental aim is to design an experiment in which a capacity to represent specific behaviours and behavioural invariances in relation to the environment predicts one response, while a capacity to represent mental states in addition predicts a different response. This assumes that it is clear, on folk psychological general principles, that a certain response would *only* be generated given the additional capacity to represent mental states. But this assumption is unjustified. The responses typically taken as evidence for representations of mental states are just as well explained in terms of representations of behaviour and environment. The animal's representations of behaviour and environment would in these cases have to provide the evidence for its inferences to 'intervening variables', that is, to others' mental states. An animal cannot avoid the work of detecting invariances and regularities and nuances in others' behaviour by attributing mental states to them, since the former provide evidence of the latter. So what explanatory power is gained by attributing to the animal representations of mental states in addition to representations of behaviour? Mind reading presupposes smart behaviour reading in these cases, we might say, and there is no evident explanatory work being done by mind reading that smart behaviour reading isn't already doing. Attributions of mind reading in addition, therefore, appear to be explanatorily redundant. For example, in the begging experiments, a chimp's attributions of seeing to the experimenter must be based on the chimp's representations of the experimenter's body and behaviour, for example on whether her eyes are blindfolded or not. That is, the behavioural representations on which mind reading depends must be present in any case. But then, why can't the chimp predict how the experimenter will respond to a begging gesture on the basis of such behaviour reading *by itself* just as well as on the basis of such behaviour reading plus attributions of seeing to the experimenter? And if the chimp can predict the experimenter's behaviour equally well by means of behaviour reading as by means of behaviour reading plus mind reading, then the chimp's own behaviour in such experiments is equally well predicted by attributing behaviour reading to it as by attributing behaviour reading plus mind reading. The chimp's behaviour thus provides no evidence for mind reading; mind reading has no added explanatory value. More generally, the experimental paradigms currently in use cannot, in principle, distinguish between behaviour reading by itself and behaviour reading plus mind reading.

Povinelli and Vonk consider and reject the idea that explanations of behaviour that appeal to mind reading are more parsimonious than explanations that appeal to representations of many different, specific behavioural regularities. Their reason is that mind reading by chimps in these experimental paradigms presupposes that chimps represent the relevant behavioural regularities in any case; there is no increase in parsimony by adding representations of mental states as well. Adding control conditions to rule out reliance on representations of a specific behavioural regularity, as the Leipzig group has done, does not address the point that mind reading in these paradigms must rely on representations of some behavioural regularity or other, which are really doing the work. This is a logical rather than an empirical problem.

(b) *How to avoid the logical problem.* To avoid the logical problem requires finding a category of social behaviour that cannot be explained by attributing representations of behaviour but that can be explained by attributing representations of mental states. There are two broad approaches here (cf. Sterelny 2003, 75).

First, one can focus on the output side: on the functions of representations of mental states. Perhaps it can be shown that there are limits to the functions that representations of behaviour can have in generating social complexity, but that representations of mental states can have further functions. If so, then evidence from animals that such further functions are being performed would be evidence for representations of mental states in addition to representations of behaviour. Although Povinelli and Vonk do not pursue this approach, arguments exist by means of which it might be developed. For example, it can be argued that 'mirror metaheuristics' provide co-operative solutions to certain games, such as one-off Prisoner's Dilemmas, that are not made available by accurate predictions of behaviour alone but are only available to those who can mind read as well as behaviour read.[55] On this view, representations of others' heuristics have further functions than do representations of their behaviour, even if the latter provide evidence for the former. If there were evidence that animals can co-operate effectively in one-off Prisoner's Dilemmas (in which they do not repeatedly play against the same partners—and of course in which solutions are not genetically determined), that might provide evidence that they were doing more than predicting one another's behaviour and were indeed representing the heuristics of others and their similarity or dissimilarity to their own heuristics (see and cf. the

---

[55] Hurley 2005; Danielson 1991, 1992; Howard 1988.

discussion of ecological validity of co-operative versus competitive paradigms, in the appendix to Chapter 18, section 6; Hare and Wrangham 2002; Dugatkin *et al.* 2002).

Povinelli and Vonk develop a second way of avoiding the logical problem, which focuses in effect on the input side: on the evidence for representations of mental states. The aim here is to find tasks that avoid the crucial feature of the logical problem—that is, tasks for which there is no plausible representation of the others' behaviour that is presupposed by the postulated representations of their mental states. The desired tasks should be such that success can be explained by attributing representations of others' mental states, but cannot be explained by attributing representations of others' behaviour—because no relevant representations of others' behaviour are available. Following a lead provided by Cecilia Heyes (1998), Povinelli and Vonk propose pursuing a class of such tasks for which successful performance can only be generated by mapping one's own experience onto that of others: self–other inference tasks. Such inferences are familiar to philosophers who study the problem of other minds under the heading of 'arguments from analogy'.

Povinelli and Vonk put forward an example of an experimental task of a conceptually different nature from those they criticize, which avoids the logical problem and hence succeeds in joining an empirical issue about ape mind reading. Chimps could be given first-person experience of wearing two buckets containing visors. The buckets are different colours; while the visors look the same from the outside, one can be seen through while the other is opaque. When two experimenters don the buckets, will chimps who have had first-person experience of wearing the buckets prefer from the first trial to beg from the experimenter wearing the bucket containing the see-through visor? If chimps represent only others' behaviour, the predicted answer is 'no': since they have never seen anyone wearing either bucket before, no information is available to them about what others wearing buckets do, on which to base such a preference. If chimps can represent others' minds as well, the predicted answer is 'yes' since if they can represent their own experience and then attribute an analogue of it to the other, information is available to them on which to base such a preference.[56]

---

[56] In work in progress, one of the editors, SH, argues as follows. Such first-to-third-person inference tasks may not be conceptually different from the tasks that face the logical problem, as Povinelli and Vonk claim. To resolve this issue it is important to distinguish two dimensions of possible difference. One is the difference between representing self and representing others. The other is the difference between representing behaviour and representing a mental state, such as an experience. These distinctions cut across one another. From the fact that the chimps in the proposed experiment have no representations of *others'* behaviour of wearing buckets it does not follow that they have no representations of behaviour of wearing buckets: they may well have representations of their *own* behaviour while wearing buckets, since they have worn them. Perhaps they infer from nuanced representations of their own behaviour while wearing the buckets to others' analogous behaviour while wearing buckets. The inference from self to other can attribute analogous behaviour to the other instead of analogous experience, as in: when I wear the red bucket but not the blue bucket,

Povinelli and Vonk emphasize that the disagreement between Louisiana and Leipzig does not turn on whether mind reading is an all or nothing capacity; both sides agree that it should *not* be assumed to be all or nothing. Indeed, this view is reflected in the Louisiana group's focus on perception rather than belief in its seeing/not-seeing studies with chimps. They conclude by reviewing and replying to a series of criticisms of this research programme.

This chapter directs its attention to recent non-verbal paradigms for assessing the attribution of seeing to others, given that claims for ape mind reading have been made based on that paradigm. However, the logical challenge laid down is quite general, and in principle also applies to non-verbal false belief paradigms. While the consensus is that apes fail non-verbal false belief tests, the evidence is as yet inconclusive for dolphins.

1.6.3 Tschudin

Comparative psychologist Alain Tschudin (Chapter 19) reviews his work using non-verbal false-belief paradigms with dolphins, which has yielded promising but equivocal results. He illuminates various methodological challenges such work faces and how they can be met, and makes a strong case for the importance of further research with these highly social, cognitively sophisticated animals to resolve the false belief issue.

Tschudin places the question whether dolphins can attribute beliefs to others in the context of both: (a) the hypothesis that social complexity drives the evolution of advanced cognitive capacities; and (b) the correlations between social complexity and social group size with brain-to-body ratios in both primates and dolphins.

....

I engage in instrumentally controlled manipulation of objects in front of me; so when she wears the red bucket but not the blue bucket, she is likely to do so as well. Philosophers have often regarded self-to-other inferences in arguments from analogy as suspect; but it is not obvious why an equivalence between one's own and another's behaviour is any more suspect than an equivalence between one's own and another's experience, which Povinelli and Vonk are willing to attribute. If attributing self-to-other inference is acceptable in explaining chimp behaviour, why isn't it available equally for as an inference from own to others' behaviour as from own to others' experience? But if it is, then it is not clear that the proposed experiment does avoid the logical problem. Inferences from own to others' behaviour appear capable of doing the same explanatory work as inferences from own to others' experiences; what added explanatory value is there in attributing the latter?

This points suggests that the logical problem is hard to contain once it gets a foot in the door, at least by means of the second strategy of finding tasks that require representations of mental states for which no relevant representations of behaviour provide a basis. For representations of one's own behaviour may be available to self-other-inferences, even if representations of others' behaviour are not directly available. And such representations of behaviour may be doing the work. For this reason, we may do better to return to the first strategy for avoiding the logical problem, by focussing on what functions representations of mental states can perform that representations of behaviour cannot. Even if representations of behaviour provide the evidence for attributions of mental states for others, representations of others' mental states may have functions that representations of others' behaviour do not have.

The background features of social complexity and large brains in relation to body size that have prompted investigation of the mind-reading capacities of apes also apply to dolphins. Moreover, dolphins exhibit some of the precursors to mind reading, such as a capacity for joint attention, spontaneous comprehension of referential pointing, and gaze following.

(a) *Pilot study*. Given these supporting considerations, Tschudin and colleagues conducted a pilot study with dolphins using a hider–communicator false-belief paradigm. As he explains, it is easier to distinguish behaviour expressing an agent's own belief from behaviour that expresses the agent's attribution of a belief to another if the other-attributed belief is inconsistent with the agent's own beliefs or 'false'. In the pilot study, the dolphin can see that a hider baits one of two opaque boxes with a fish, but cannot see which one. However, the dolphin can also see that a different trainer, a communicator, can see which box has been baited. After baiting, this trainer tapped on one of the boxes, and the dolphin indicated its choice of one of the boxes and was given its contents: a fish, if it indicated the box containing the fish. During the training trials, the boxes were not switched after baiting, so that the dolphin learned to obtain fish by choosing the box on which the communicator tapped. During the test trials, the communicator departed after baiting, and the dolphin observed the other experimenter switch the boxes. The communicator then returned, and tapped the 'wrong' box. All the dolphins passed this false belief task.

However, there were some methodological problems with the design of the pilot study, which tempered excitement about the result. First, a confound may have been generated by pretest control trials in which the boxes were switched after the tap by the communicator while the dolphin watched, and in which the fish was moved from one box to the other after the tap while the dolphin watched. The dolphins passed these tests, which were intended to ensure that the dolphins understood spatial displacement and could ignore the communicator when appropriate. However, they may have learned from these trials simply to choose the opposite box to the tapped box whenever there is a switch. Work with 4-year-old children underscores this concern. Second, the same experimenter baited the boxes and presented the dolphin with a choice between them, and may inadvertently have cued the correct response to the dolphin. Subsequent experiments were designed to address these problems.

(b) *Experiment 1*. In experiment 1, the pretest controls were omitted and true-belief tasks (the boxes are switched in the presence of the communicator) were interspersed with false-belief tasks to prevent reversal learning. Moreover, different experimenters were used for baiting and presenting. One of four dolphins failed to pass the training phase, perseverating to the left; but the other three succeeded on the first false-belief trial and two of these succeeded on the first true-belief trial. Overall, there was a significant relationship between whether the communicator's indicated 'beliefs' were true or false and whether the dolphins responded in accord with them or not. But when the results were pooled on a given trial for all animals, or for a given animal on all trials, they are of equivocal significance; this may be an effect of the small number of trials. As well as the need for an increased sample size, further methodological concerns arose from this experiment, including a worry that the communicator herself may have inadvertently provided cues other than the tap signal, and that the dolphins, who had also been used in the pilot study, might remember a reversal strategy learned during the pilot study.

(c) *Experiment 2*. Experiment 2 was designed to address these concerns, using a naïve animal and interspersing trials in which the communicator did not know whether the boxes had in fact been switched during her absence or not. Cueing and learning were now thoroughly controlled for. Unfortunately, the naïve dolphin did not pass the training phase but perseverated in various ways, so the experiment was discontinued. This dolphin did not fail the false belief test; rather, he never met the training conditions for taking the test.

The combined results from the pilot study and experiment 1 may make it tempting to ascribe mind reading to dolphins. But this would be premature; alternative explanations need to be ruled out. A reversal learning explanation is possible, as indicated; a larger sample size and naïve dolphins are needed to assess whether dolphins can succeed on first trials with all appropriate controls in place, which reversal learning could not explain but mind reading could. Another alternative explanation would be one of the kind Povinelli and Vonk press, namely, that the dolphins attribute a behavioural state concerning visual access rather than a mental state to the communicator, and base their responses on this. Tschudin argues that a behavioural rule could not account for success on first false belief trials, because it would have to be learned; first trial performance should be at chance in the absence of mind reading. (However, Povinelli and Vonk might reply that mind reading must be based on behaviour reading even in first trials, and that the added value of mind reading in explaining first trial success is therefore unclear.) Another explanation is in terms of inadvertent cueing of correct responses by experimenters; unfortunately, in experiment 2, with the best design to exclude various forms of cuing, was discontinued.

A mind reading interpretation gains support from the background considerations already reviewed, though it would have to explain the failure of two dolphins to pass the training phase. These failures may simply reflect differences between individual animals. Moreover, the naïve dolphin used in experiment 2 was socially isolated in relation to other dolphins, which may have interfered with the development of his normal social capacities despite his interaction with human trainers. To provide points of comparison, experiment 2 was also run with seals and with children. The seals and 4-year-old children failed, and only three out of fourteen 5-year-old children passed false-belief trials on first trial. Moreover, children have been attributed the capacity to mind read based on experimental designs similar to that of the pilot study, with fewer controls than used in experiment 1 with dolphins. Arguably, evidence for mind reading by dolphins is being held to higher standards than evidence for mind reading by children.

Given the methodological lessons of Tschudin's work on dolphin mind reading, the supportive direction of his results and lack of clear failure by dolphins on false belief tests, further research is warranted on the question of mind reading by dolphins and, more generally, on the social cognitive capacities that support their complex social behaviour. If dolphins were to be successful on false belief tasks with all appropriate controls in place, this would raise further questions: in particular, whether success would be equally open to explanations in terms of behaviour reading as in terms of mind reading (see Chapter 18 by Povinelli and Vonk); and, if the latter, what further experiments might reveal whether such success were best understood as an expression of theoretical rationality (as in theory-theory accounts of mind reading) or of practical rationality (as in simulation theory accounts).

1.7 Behaviour and cognition in symbolic environments

1.7.1 Herman

Cetologist and psychologist Louis Herman (Chapter 20) is a founding pioneer of the experimental study of dolphin cognition. Without his work we would have far less understanding than we do of the remarkable intelligence of these animals. His chapter reviews the cognitive accomplishments of his four dolphins, Akeakamai, Phoenix, Elele, and Hiapo, in declarative, procedural, social, and self-related domains, presenting evidence for their rational responses in these domains. Throughout his discussion the dolphins' trained facility in understanding symbolic gestures is apparent, raising questions about how this capacity to understand symbolic gestures is related to the rational responses dolphins display.

A number of factors make dolphins compelling subjects for experimental studies of cognition. Their ratio of brain to body mass is second only to that of human beings; they live to 40 or 50 years and have a protracted period of development; they are highly sociable and naturally inhabit highly complex social environments. While their echolocation abilities are extremely well developed, these do not explain their large brains, which are more plausibly viewed as subserving their intelligence in a variety of domains. In Herman's view, intelligence is manifested in behavioural flexibility, the ability to adapt behaviour to a changed environment to enable effective functioning, in ways that go beyond biologically programmed and learned behaviours. The capacity for intelligently flexible behaviour can be present to various degrees in various dimensions, and provides the foundation for rational behaviour. A rational animal can perceive the structure and function of the world it occupies, and can make inferences that enable it to function effectively, based on its model of that world; moreover, it can incorporate evidence to model a changing world, or even radically different worlds, and adapt its behaviour appropriately, flexibly generalizing, inferring, and innovating. Wild dolphins are able to respond flexibly and to function effectively in laboratory settings that are radically different from their natural environments, in ways that provide evidence that dolphins have created accurate models of their new world.

(a) *The declarative and procedural domains.* The declarative domain concerns abilities to perceive things and their properties, and to understand references to these things; the procedural domain concerns understanding of how to do things and of how things work. Herman discusses these domains together, since some of his experimental paradigms require both declarative and procedural knowledge. Various paradigms provide evidence of rational behaviour in these domains.

Akeakamai responds correctly to communications that use a symbolic language of hand gestures with an inverse grammar: the order of the symbols is not the order in which responses to them are required, and the significance of symbol order depends on the whole sequence. For example, the symbol order *indirect object/direct object/ relational action* requires the dolphin to perform a particular action on a particular object in relation to another object. *Swimmer/surfboard/fetch* requires the dolphin to bring the surfboard to the swimmer, while *surfboard/swimmer/fetch* requires the dolphin to bring the swimmer to the surfboard. Akeakamai responds correctly to novel sequences using this relational syntactic frame, treating objects grammatically correctly as direct or indirect objects according to their place in the order of symbols. Novel sequences that are semantically anomalous, requiring impossible performances, produce no response other than continued attention to the experimenter. In other cases, Ake takes initiatives to enable instructions to be complied with. Innovating in response to the instructions to jump over a surfboard that was resting on the side of the pool, Ake first moved the surfboard into the pool and then jumped over it; asked to put a ball in a basket that was already in a basket, she first removed it and then replaced it. Without specific training, she typically decomposes syntactic anomalies and acts on a syntactically correct part of the string. When modifiers *left* and *right* trained using non-relational commands, such as *swim through left hoop,* are added to the inverse relational syntax distinguishing direct and indirect objects to create a five-place syntax, Ake immediately, with no further training, generalizes and responds correctly. For example she responds correctly to *put the right ball in the left basket* and to *put the left ball in the right basket.* She responds correctly (on over 80 per cent of trials) to queries referring to absent objects, using 'yes' and 'no' paddles to questions about whether certain objects are in the pool after observing objects being thrown over her head into the pool behind her. If asked to transport an absent object, she spontaneously, without training, pressed the 'no' paddle. If asked to transport a present object to an absent object, from first trial and with no training, she pushed the present object to the 'no' paddle, and has continued to respond in this way for any combination of present direct objects and absent indirect objects. While enculturated chimps have shown little initial interest in televised images, Ake responded immediately and with no TV-specific training to televised images of trainers giving gestural commands, even when these images were degraded to include only the hands; and she outperformed some human staff members in responding correctly to still further degraded images. With no TV-specific training, another dolphin Elele generalized to perform match-to-sample tasks from televised images of samples, as accurately as from real samples. All four dolphins generate novel, self-selected behaviours during training sessions, if and only if, they are given a *create* command.

(b) *The social domain*. Evidence of flexible generalization, inference, and innovation by dolphins extends to the social realm. In human beings, pointing is a means of sharing attention to the same object with another. Unlike chimps, Ake spontaneously understands the object of referential pointing by human beings, including cross-body pointing, responding correctly to instructions accompanied by points. Remarkably, without training she integrated points into the relational syntax she had learned, generalizing the inverse grammatical rule to the order of points. Herman suggests that dolphins may understand pointing by generalization from an analogue of pointing provided by their highly focussed echolocation beams. These may provide a natural mechanism for sharing attention; it has been shown that dolphins can tell what object another dolphin is echolocating.

As Connor and Mann explain in their chapter, synchronized swimming is a natural affiliative behaviour. All Herman's dolphins have been trained to respond in synchrony to commands preceded by a *tandem* gesture, and they generalize this command from the few behaviours they are trained on to others. Most remarkably, the first time they are exposed to the combination of *tandem* and *create*, pairs of dolphins respond correctly. Typically, they swim side by side for longer than they would if asked to synchronize on a specific behaviour, and then execute a selected behaviour synchronously, such as a spinning leap while spitting water from their mouths (which requires extra water to be taken into their mouths before they leap). This creative co-ordination is not taught, and Herman has not been able to determine how it is accomplished; close video analysis does not reveal a leader and a mimic. However, dolphins are talented vocal and motor mimics, and also show creative generalization here. Ake was trained to mimic vocally a variety of electronically produced sounds of different waveforms, and when the models were out of her preferred vocal range she spontaneously reproduced the sound contour at an octave above or below the model. She also recognized tunes transposed by an octave; octave generalization is a rare phenomenon among animals. The dolphins respond to a gestural command *mimic* in combination with motor behaviours modelled by human beings as well as by other dolphins. When mimicking human beings, the dolphins spontaneously analogize between human and dolphin body parts, raising a tail, say, when a human being raises a leg; and they generalized their mimicry to televised models.

(c) *The domain of self-knowledge*. As well as displaying mirror self-recognition, dolphins display awareness of their own behaviours and body parts. Phoenix repeats her past behaviours or chooses a different behaviour on request, requiring her to represent her past behaviour and generate a match or a mismatch. Elele associates her various body parts with gestural symbols, and can respond correctly to instructions requiring her to use different body parts in the same way or the same body part in different ways; she can respond to such requests involving novel combinations not in her natural or learned repertoire on the first occasion of presentation.

These and other paradigms that Herman reviews provide multiple instances of rational responses not specifically trained and requiring flexible generalization, inference, and innovation. Correct responses are inferred without specific training when new semantic elements are inserted into familiar syntactic slots, when the syntactic positions of objects are reversed, when novel syntactic structures combine previously familiar syntactic structures, including references to absent objects, when symbolic gestures are represented in novel televised formats, when pointing gestures are incorporated into familiar syntactic slots, when responses require creative coordination with another dolphin or untaught analogies between human and dolphin body parts or generalization across octaves. Initiatives are taken to rearrange objects so that instructions can be complied with. This level and variety of flexible generalizations and innovations suggests not merely 'islands of practical rationality', in Hurley's terms, but significant conceptual and inferential abilities across several domains.

What processes select for evolution of this level of intelligence? Herman makes the suggestion, which Connor and Mann, and Tschudin also find plausible, that the pressures of social complexity play a key role in driving the evolution of dolphin intelligence. Herman suggests that these operate along with various ecological pressures. However, Herman's work raises further compelling questions about the role of symbols in his dolphins' remarkable abilities: does training with symbols support or extend dolphin cognitive abilities and thus enable dolphin rationality, does it express an independently grounded social intelligence, or does it reflect natural skills at symbolic communication that we have not yet been able to evidence in the wild?

1.7.2 Pepperberg

Related questions about the role of symbol training in facilitating remarkable cognitive performances by animals are raised by comparative psychologist Irene Pepperberg's work with African grey parrots (Chapter 21). Pepperberg draws on her work to probe the relationship between animal intelligence and rationality, and between standards of rationality appropriate to human and to animal behaviour. Behaviour that is intelligent, in the sense that it displays sophisticated cognitive abilities that are biologically adaptive, may not count as rational by human standards.

In some cases biological intelligence converges with human standards of rationality, as when the adaptive mating behaviour of female great tits conforms to transitive inferences about the rank of various males. The intelligent behaviour of Pepperberg's parrots also often converges with human standards of rationality. Alex's accomplishments include comprehending and producing English labels for objects, for specific colours, shapes, and numbers, for the categories of material, colour, and shape, and various English commands, requests, and responses. Using these skills, he responds to many tasks in ways that would be regarded as rational in a human child. He produces correct responses in English to recursive, conjunctive questions in English that combine labels in new ways; he correctly answers questions about size, quantity, colour, similarity or difference of attributes in application to new objects; he requests absent objects. For example he can label the quantity of a specific subset of objects in a complex heterogeneous array: when shown a tray containing one blue box, three green boxes, four blue cups, and six green cups, and asked how many green cups there are, he will respond by saying 'six'. This capacity generalizes to other questions about the same objects and to new objects, such as a question about what shape the green paper is, when shown an array of seven objects of different colours, shapes, and materials (Pepperberg 1999, Ch. 7, 8).

However, Pepperberg here focuses not on these well-known accomplishments, but on further labelling behaviours outside the usual training process and on certain 'mistakes' the birds make, for what they reveal about the relationships between biological intelligence and various standards of rationality.

(a) *Transitions to referential labelling and model rival training.* Pepperberg has developed a distinctive training process, the model–rival system. The parrot eavesdrops on two trainers talking about interesting objects (compare Kanzi's eavesdropping on attempts to language-train his mother Matata; see Savage-Rumbaugh, Rumbaugh, and Fields, Chapter 23, this volume). One trainer asks the other questions and rewards correct responses by giving the other the labelled object(s), as well as by praise. Errors are also demonstrated and punished with scolding and withdrawal of the object(s). The second trainer is the parrot's model as well as its rival for attention. The trainers reverse roles, and also begin to include the parrot in their interactions. Pepperberg has demonstrated with juvenile parrots that training is unsuccessful if it omits any element of this three-way social modelling process. Moreover, when the birds are given model–rival training for some labels and other training techniques are used with other labels, the birds 'practice' the former labels but not the latter to themselves after working hours (Pepperberg 2005).

Pepperberg compares early label learning by her parrots with that by children. Label learning is initially slow and difficult for both, but picks up speed, and considerable self-initiated learning appears to occur outside of training for the parrots, as when children play naming games. For parrots as for children, she argues, first labels are qualitatively different from later labels; they are not fully representational, referring to immediately present items and often overextended or underextended, perhaps reflecting lack of memory capacity or capacity for categorical classification. In parrots as in children, she suggests, the transition to representational use of labels parallels the transition from a self-centred orientation to recognition of others as information sources and to the rudiments of social rationality, a transition mediated by the cognitive functions of emotions. Underlying neural developments, especially in mirror systems in human beings and in possible analogous structures in avian brains, may contribute to these transitions by enabling the combinatorial recreation of complex observed structures of speech and action.

The parrots' transition from self-centred early label learning is seen, for example, in sound play outside training sessions, in which they spontaneously recombine or vary labels to produce new sounds: after learning 'grey', Alex produced 'grape', 'grate', 'grain', 'chain', and 'cane'. If such new sounds are rewarded with corresponding objects, they are readily incorporated into the bird's repertoire of labels. A label learned in a specific context, such as 'wool' for a woollen pompon, might be used while pulling at a trainer's sweater. It is possible that by means of these behaviours the birds, in the early stages of forming categories and representations, are rationally probing their trainers for information about the referents of sounds and testing hypotheses. The model–rival technique may facilitate the transition to representational use of labels by using two trainers, to separate the utterance from its referent or the required response, and by providing a model to imitate that generalizes across role reversals. Model–rival training has produced significant improvement in the communicative and social abilities of autistic children; Pepperberg suggests that it engenders exceptional learning, which would not occur given normal input.

(b) *Responses to being tricked and playing tricks.* The parrots demonstrate an appreciation of object permanence by retrieving an object that has been covered by a container and then passed behind various screens. When the birds have witnessed a particularly desirable object being covered and the object they find is not the expected one (the experimenter has 'tricked' them by an unseen swap), the birds do not continue looking for the desirable object, but rather display apparent surprise and anger. While their emotional response to not finding the expected object expresses an intelligent awareness, it is not clear whether this response is more or less rational than continuing to look for the desired object would be.

In another case, a parrot appears to play a trick on the experimenter. When shown a tray of seven items of different colours, shapes, and materials, the parrot may be asked, for example, the colour of the square wooden object. He has already demonstrated mastery of this type of task, but on occasion, instead of providing the correct answer, he provides each of six possible *wrong* answers, and then repeats them. He thereby sacrifices a reward, but also produces a frustrated emotional response from his trainer. No doubt considerable intelligence is needed to perform in this systematically perverse way. But is his behaviour irrational? Or is it a rational means of producing an emotional response from his trainer that he enjoys more than he would enjoy the reward?

Pepperberg concludes by emphasizing the need to find experimental methods of assessing rationality across various taxa that are sensitive to the possibilities of continuities and convergence in rational functioning, but do not rely inappropriately on human-centred values and goal assumptions.

1.7.3 **Boysen**

Primatologist Sarah Boysen (Chapter 22) directly addresses the role of symbol training in facilitating the cognitive performance of enculturated chimps. She argues that immersion in symbol-laden human culture and long-term social relationships with human beings dramatically increases their attentional resources. By permitting an animal to process selected information about objects, she suggests, symbols can minimize interference from low-level dispositions and enable them to respond rationally. Here she surveys work from three paradigms: on the way numerals modulate chimps' evaluative dispositions, on the way numerals and symbols for 'same' and 'different' modulate reaction time, and on chimps' use of scale models to solve problems.

(a) Candies and numerals: interference effects and symbolic facilitation. Boysen explains that, in certain circumstances, multiple evaluative dispositions (which perhaps function at different levels in the cognitive structure of the animal) may be expressed, and that these may conflict. Such conflict is illustrated in her work on chimpanzees' capacity to make quantity judgements. This work showed that chimps were unable to perform in an instrumentally rational way, that is, to point to a smaller quantity of candy in order to obtain a larger quantity; moreover, their performance was poorest when the disparity in quantity of candies was largest, so that they stood to gain the most. However, these chimps had previously been trained to use numerals, and they succeeded immediately on the task when candies were replaced with numerals: they pointed to the smaller of two numerals and were accordingly rewarded with a number of candies corresponding to the larger numeral. Related results have been obtained with children, who respond more rationally to photos of food than to actual food. Boysen suggests that the inherent incentive features of the candy introduce a bias toward the larger quantity that conflicts with an instrumental associative disposition based on the reinforcement contingency (you get the larger quantity if you point to the smaller quantity), which they appeared to have learned. (When different quantities of rocks are substituted for candies, the chimps also respond incorrectly, suggesting that an interference effect can also be based on perceived array mass.) Numerals represent quantity while abstracting from the interfering incentive properties, thus modulating the interaction of the chimps' conflicting dispositions.

To assess whether the presence of numerals fostered an abstract mode of processing that would overcome the incentive effect of the candies, Boysen next conducted a feasibility study in which chimps were presented with mixed arrays of candies and numerals on the same reversed reinforcement contingency. In the feasibility study, the chimps did not simply point to the candies in the mixed array sessions, and were only slightly worse at pointing to the smaller quantity in mixed arrays than in numeral–numeral arrays. The mixed array feasibility study thus supported the hypothesis that numerals facilitate the instrumentally rational response.

However, in a full-scale study that presented chimps with candy–candy, numeral–numeral, and candy–numeral arrays within a single session, the chimps appeared to be unable to decipher the operative rules, and instead to adopt low-level response strategies such as a right-hand bias. Surprisingly, they did not show significant incentive interference effects in candy–candy trials or symbolic facilitation effects in numeral-numeral trials, but appeared to respond randomly. In mixed candy–numeral arrays, they tended to point to the numeral regardless of quantity. In effect, mixing array types within a single session seemed to confuse the chimps, suggesting limits to the extent to which numerals facilitate flexibly rational responses.

A final study was conducted in which different trial types were not mixed within a given test session, to see if the original interference and facilitation effects would again be found. They were: candy–candy sessions produced interference; numeral–numeral sessions produced facilitation. But mixed trial sessions revealed no consistent group pattern. More work is needed to clarify the conditions under which numerals facilitate rational responses by chimps.

(b) *Reaction time effects of combining numerals with symbols for 'same' and 'different'*. In another experiment, Boysen combined numerals with symbols for 'same' and 'different', with which the chimps were also already competent from other work. They were asked to use the symbols for 'same' and 'different' to report whether quantities matched or not, and their reaction times were measured. The two quantities were presented as two arrays of dots, as two numerals, or as a combination of dot array and numeral. The aim was to probe and compare possible counting and subitizing processes. Subitizing is the rapid, accurate, perceptual apprehension of small quantities up a limit of between three and five, as opposed to the slower, serial enumeration involved in counting larger quantities. The chimps could all use numerals '0' to '6', and some could extend to '8', so whether they used the faster or slower process might be expected to vary with quantities compared or with mode of presentation (dots vs. numerals) and to be revealed by reaction times. The animals were first trained on the task dot–dot, numeral–numeral, and dot–numeral pairs but only using the quantities 2 and 5. They were then trained similarly on 1 and 6, and then on 1, 2, 5, and 6. They were then tested on novel combinations using quantities 0, 3, and 4 as well. Their 'same' judgements were faster for smaller arrays of dots than for larger arrays, but their 'different' judgements were significantly faster than their 'same' judgements for both dots and numerals, even for smaller quantity 'same' comparisons. Reaction times tended to be longest for numeral–numeral comparisons and shortest for dot–dot comparisons, but these differences across stimulus types did not reach significance. Thus, the quickest reaction times are for judgements of different quantities, whether presented by means of dots or numerals, rather than for comparisons of smaller quantities or of dots as opposed to numbers. No symbolic facilitation effect is shown, nor is it clear how the subitizing/ counting distinction might explain this intriguing result.

(c) *Solving problems with scale models*. The third set of experiments reported by Boysen investigates chimps' abilities to find hidden objects by using scale models and other representational media, such as photographs and videos. In similar studies with children by De Loache, the child watches as a miniature toy is placed in a scale model of the playroom and the child is then asked to find the real toy in the real playroom, which is hidden in the corresponding place. Three-year-olds were successful, though children only 6 months younger had difficulty; however, both age groups were successful when photographs were used instead. It was suggested that the scale model was harder for the younger children to interpret as a representation of the real world than the photos were, because the scale model had the additional aspect of a separate toy-like object, which the photographs lacked.

Boysen investigated chimps' ability to solve the hiding game using a scale model as well as other representations of the hiding place, including photographs and video. One of her chimps, Sheba, readily solved all variants of the problem, while another, Bobby, failed all variants and perseverated in various ineffective search strategies, such as circling the room clock-wise. When the hiding game was extended to include other chimps, a sex difference emerged: three females were successful even when the hiding sites in the model and its referent were rearranged between trials, while the males performed poorly and perseverated in the way Bobby had. Boysen suggests that these male individuals could not meet the attentional demands of representational processing, so could not control their impulsive, lower-level dispositions by using the representational strategy successfully employed by the females. She speculates that a metarepresentation bridging the scale model and its referent might provide a facilitating symbolic link that would enable the animals who failed the task to inhibit interfering responses, as in the candy–numeral studies, and plans further experiments to address this possibility.

The three lines of experiment together indicate that symbols facilitate rational responses and/or the inhibition of interfering dispositions in some situations, and for some individuals, but not others. While more work is needed to understand how and when symbolic facilitation operates, these results indicate the powerful effects symbols can have in modulating dispositions, and, as Boysen suggests, their adaptive role for our early hominid ancestors.

1.7.4 Savage-Rumbaugh, Rumbaugh, and Fields

Primatologists Sue Savage-Rumbaugh, Duane Rumbaugh, and William Fields (Chapter 23) provide an overview of language research conducted with apes over four decades, aiming to delineate the circumstances that promote linguistic capacities.[57] As animals acquire language, they can display evidence of rationality that is problematic to obtain otherwise. But learning language requires learning symbols that are freed from their context of acquisition to be used in novel ways, not merely learning associations.

---

[57] See also and cf. Gomez 2004, Ch. 10; Shettleworth 1998, 549, 563.

The authors begin by providing a useful comparison of the methodologies of nine different animal language projects. The projects may use sign language, lexigrams on a touch panel, and/or spoken English; they may emphasize symbol production or comprehension; they may use traditional associative training techniques, alternative training techniques (such as the model–rival system), or no explicit training at all but immersion in social activities including language use. The authors argue that the differences among these projects in various methodological dimensions hinder direct comparisons of the skills that result, which also differ in various dimensions.

The authors' research on ape language at the Language Research Center divides into four phases, employing different methodologies that are informed by experience in previous phases.

(a) *Lana*. In the first phase, the chimpanzee Lana was trained using geometric lexigrams embossed on a keyboard, which were displayed in the sequence touched above the keyboard. She was taught 'stock sentences', strings of lexigrams following specific grammatical rules. Real progress began when her caregiver, Tim, entered her chamber and began to interact with her using the keyboard, and turned on processes of social learning rather than the rewarding of motor responses. For example Lana was able to form novel sentences, recombining elements from five different stock sentences, to get Tim to share his coke with her when her vending machine was empty. No previously learned associations or behavioural invariances could explain this behaviour. In contrast with Herman's dolphins, who understand novel symbol combinations in grammatically appropriate ways, Lana was herself producing such novel sentences. Moreover, when first shown an object she didn't know the name of, and no one bothered to tell her, Lana spontaneously asked 'what name of this?' These innovations indicate that Lana understood the limitations of her stock sentences. However, some of her sentences were not interpretable, and her comprehension skills were limited.

(b) *Sherman and Austin*. Phase two focussed on achieving communication between chimps rather than human/chimp interactions, and used single words rather than stock sentences, in order to try to understand how language might have evolved without continual assistance by competent language-users. Using the lexigrams keyboard, and emphasizing both production and comprehension, Sherman and Austin were taught to name items, to request items and to give specific food items to the experimenter in response to requests—which they found difficult initially, as they were strongly biased to attend to the foods they wanted to eat rather than to those symbolically requested. When they'd learned to do so, they were given the opportunity to request foods from one another. They initially failed to communicate, lacking the needed non-verbal co-ordination skills. But they began to succeed in verbal communication as they learned non-verbal gestural, timing, turn taking, and attentional skills that co-ordinate conversation. For example if Austin were looking away when Sherman selected a symbol, Sherman would wait until Austin looked back, point to the symbol, point to the object it referred to, and so on. They appeared to recognize that speakers must monitor listeners' attention and comprehension, and repair conversations as needed.

In another test, one chimp's keyboard was inactivated and he was provided instead with brand labels from food packaging by means of which he could communicate the contents of a container to the second chimp, who could then request it using his keyboard. Both chimps did this from the first trial. They were not trained or shown how to use the brand labels as symbols for food or to associate them with lexigrams, but were merely provided with the materials. No previous experiences explain this innovative use of labels as symbols; rather, the chimps used them to communicate information they recognized the other lacked, indicating understanding of the mental state of the other.

Moreover, Sherman and Austin appeared to recognize that differential mental states existed more generally; they spontaneously began to announce what they were about to do before doing it and to report unusual events selectively to others who had not witnessed them. They also engaged in pretence games, showing what Currie would call enrichment, for example by pretending dolls had bitten their fingers and then nursing their fingers. As their television-watching skills developed, they spontaneously began to differentiate between live and taped images of themselves, watching the live images of themselves monkeying about and using the live images to investigate their own bodies. The authors suggest that these behaviours display an increased capacity for self-reflection as well as for understanding the minds of others.

(c) *Kanzi*. Phase three focussed on the linguistic capacities of bonobos rather than chimpanzees, beginning with an adult female Matata. Her training was the same as that Sherman and Austin had received, but her progress was extremely slow. However, her son Kanzi was allowed to remain with her during training sessions, from age 6 months to 2 years, at which point he was separated from his mother temporarily. Suddenly, his latent linguistic learning was activated; he used the keyboard over 300 times in the first day of separation. In the first month of separation, Kanzi stated his intentions, shared food, combined symbols spontaneously, and replied to requests symbolically. By age 3 he had acquired all of Sherman and Austin's trained skills, though he had never received training himself but only observed attempts to train his mother. Rather than receiving training, Kanzi was immersed in language in ordinary life; he acquired new lexigrams rapidly along with recognition of corresponding spoken words.

By age 8, Kanzi could interpret novel English utterances that were syntactically complex and semantically unusual, such as 'Get the toy gorilla and slap him with the can opener' or 'Feed the doggie some pine needles'. Formal tests of this ability indicated that Kanzi had broken the syntax barrier and was simultaneously processing both syntax and semantics. For example when confronted with a collection of objects including a knife, a mask, an apple, a toy dog, a balloon, a cup of water, a TV, juice in a bowl, and an egg in a bowl, Kanzi responded correctly both when asked to pour the egg into the juice and when asked to pour the juice into the egg. He responded correctly to novel recursive requests such as 'get the tomato that is in the microwave', ignoring both the tomato in front of him and the other items in the microwave in doing so, and to questions such as 'can you feed your ball some tomato?', finding a previously ignored face on his ball and placing the tomato on its mouth. The authors argue that such responses cannot be explained as responses to single words in the order presented or in certain contexts, without an understanding of syntax.

(d) *Panbanisha and Panzee.* Was the ease with which Kanzi acquired language compared with Lana, Sherman, and Austin due to immersion in language use, or to his species? In phase four, this question was addressed by co-rearing a chimpanzee, Panzee, and a bonobo, Panbanisha, from infancy to 4 years in a language immersion environment. Both began to 'babble' at their keyboards at about 6 months of age, but Panbanisha began using lexigrams to communicate at an earlier age than did Panzee. By age 4 years, the bonobo understood 179 words and lexigrams, including many proper names for chimps and people, while the chimp understood only 79, including no proper names. Moreover, the chimp often over-generalized her symbols, failing to mark distinctions that the bonobo did mark. The chimp's comprehension was more closely linked to the here and now and presently displayed objects, while the bonobo responded appropriately to remarks about the future or the past, and was relatively more concerned to communicate about social relationships than about tools.

The authors summarize the most important findings of their research as follows. First, language can be readily acquired by apes under 3 years of age, without any training, by immersion in a social environment where English is spoken normally to provide a narrative of everyday life and of social interactions, and where lexigrams accompany spoken words. Second, lexical production emerges spontaneously from untrained comprehension, but comprehension does not emerge spontaneously from the trained production of words in the right context. The latter tends instead to produce learned associations bound to a specific context. Such training may actually inhibit acquisition of a fully representational language that can be used freely across contexts and to communicate about what is absent or past. Third, when an ape achieves a level of symbolic understanding that enables it to employ symbols freely to state its intentions and to report events, many other rational capacities begin to be displayed as well, including capacities for directing attention by pointing, for pretence, for self-reflection, and for mind reading.

In the authors' view, acquiring language brings with it the understanding that it functions to communicate information, and hence the realization that the contents of others' minds can differ from those of one's own mind. The authors share Povinelli's view that current non-verbal assessments of animal mind reading are inconclusive. But they argue that linguistically competent animals can instead be asked what they are thinking about, under conditions in which the questions and answers are novel and could not be learned responses. The bonobos reared in language immersion environments answer this question appropriately in various contexts and validate their reported thoughts with appropriate behaviour. The authors report, moreover, that Kanzi and Panbanisha pass standard verbal false belief tests normally used with children on first trial. Language acquisition thus makes it possible to assess aspects of animal rationality, evidence for which would otherwise be problematic.

Finally, the authors endorse the theory of Greenspan and Shanker that symbolic communication emerges from emotional signals in interactions between infant and caregiver. They suggest that the paradigms likely to produce the highest levels of linguistic competence in animals are those that foster the closest relationships with their caregivers, a prediction confirmed by ape language results.

1.8 Why does it matter?

Why, then, does it matter whether animals are rational? It matters both for our understanding of other animals and of ourselves. We live with animals, we interact with them and use them in our daily lives, and share a planet with them. Yet we see ourselves as discontinuous from them in important ways. Rationality is one of the main hooks on which human distinctiveness and specialness has been hung. We treat rationality as having intrinsic worth, in addition to sentience. If a creature can feel pain, we feel we ought to avoid making it suffer unnecessarily, but we may not on that account grant it the additional intrinsic value and dignity associated with rationality. Understanding whether and in what ways non-human animals can be rational may prompt us to rethink human rationality, our relations to other animals, and our own irrationalities. Perhaps our rationality is more piecemeal, less theoretical, more embedded in and conditioned by our environments than we realize. The possibility of explaining relatively complex animal behaviour by appealing to a variety of relatively simple, domain-specific processes may lead us to re-evaluate our presuppositions about human rationality and to ask whether we are operating a double standard in assessing human vs. animal rationality (see and cf. Shettleworth 1998, 563). On a disaggregated view of rationality, there is no single boundary distinguishing the human mind from other animal minds in respect of rationality; rather, there are various specific dimensions of comparison. Making comparisons across species and with human beings in this way elucidates the extent to which the abilities of different animals and of humans can be explained by appealing to similar kinds of processes, and in what sense these processes are rational. Should this rethinking debunk human rationality? Not necessarily. Rather, it may help us to understand it better, and to understand the continuities as well as the discontinuities between human and other animals.

References

Allen, C. (1999). Animal concepts revisited: The use of self-monitoring as an empirical approach. *Erkenntnis*, **51**: 33–40.

Allen, C. and Bekoff, M. (1997). *Species of mind*. Cambridge: MIT Press.

Allen, C. and Hauser, M. D. (1991). Concept attribution in non-human animals. *Philosophy of Science*, **58**: 221–240.

Bargh, J. A. (1997). The automaticity of everyday life. In: R. S. Wyer, ed. *The Automaticity of Everyday Life*. Mahwah, New Jersey: Lawrence Erlbaum Associates.

Bargh, J. A. and Chartrand, T. L. (1999). The unbearable automaticity of being. *American Psychologist*, **54**: 462–479.

Bargh, J. (2005). Bypassing the will: towards demystifying the non-conscious control of social behaviour. In: R. Hassin, J. Uleman, and J. Bargh, eds. *The New Unconscious*. New York: Oxford University Press.

Bekoff, M., Allen, C., and Burghardt, G. M., eds (2002). *The Cognitive Animal*. Cambridge: MIT Press.

Bermúdez, J. L. (2000). Personal and sub-personal: A difference without a distinction. *Philosophical Explorations*, **3**: 63–82.

Bermúdez, J. L. (2003). *Thinking Without Words*. Oxford: Oxford University Press.

Brooks, R. (1999). *Cambrian Intelligence*. Cambridge: MIT Press.

Byrne, R.W. and Whiten, A., eds (1988). *Machiavellian Intelligence: Social Expertise and the Evolution of Intellect in Monkeys, Apes, and Humans*. New York: Oxford University Press.

Carroll, L. (1895). What the tortoise said to Achilles. *Mind*, **4**: 278–280.

Chartrand, T. L. and Bargh, J. A. (1996). Automatic activation of social information processing goals: Non-conscious priming reproduces effects of explicit conscious instructions. *Journal of Personality and Social Psychology*, **71**: 464–478.

Chartrand, T. L. and Bargh, J. A. (1999). The Chameleon Effect, *Journal of Personality and Social Psychology* **76(6)**, 893–910.

Chater, N. and Heyes, C. (1994). Animal concepts: Content and discontent. *Mind and Language*, **9**: 209–246.

Clark, A. (2001).Reason, robots and the extended mind. *Mind and Language* **16**: 121–145.

Clark, A. (2005a). Intrinsic content, active memory, and the extended mind. *Analysis*, **65**: 1–11.

Clark, A. (2005b). Word, niche, and super-niche: How language makes minds matter more. *Theoria*, **20**: 255–268.

Clark, A. and Chalmers, D. (1998). The extended mind. *Analysis*, **58**: 7–19.

Cook, R. G. (2002). Same-different concept formation in pigeons. In: M. Bekoff, C. Allen, and G. M. Burghardt, eds. *The Cognitive Animal*, pp. 229–237. Cambridge: MIT Press.

Cosmides, L. and Tooby, J. (1992). Cognitive adaptations for social exchange. In: J. Barkow, L. Cosmides, and J. Tooby, eds. *The adapted mind: Evolutionary psychology and the Generation of Culture*. New York: Oxford University Press.

Cosmides, L. and Tooby, J. (1997). The multimodular nature of human intelligence. In: A. Schiebel and J. W. Schopf , eds. *Origin and Evolution of Intelligence*. Center for the Study of the Evolution and Origin of Life, UCLA.

Danielson, P. (1992). *Artificial Morality: Virtuous Robots for Virtual Games*. London: Routledge.

Danielson, P. (1991). Closing the compliance dilemma: How it's rational to be moral in a Lamarckian world. In: P. Vallentyne, ed. *Contractarianism and Rational Choice*, pp. 291–322. New York: Cambridge University Press.

Davidson, D. (1980). *Essays on Actions and Events*. Oxford: Oxford University Press.

Davidson, D. (1984). *Inquiries into Truth and Interpretation*. Oxford: Oxford University Press.

Davidson, D. (2001). *Subjective, Intersubjective, Objective*. Oxford: Clarendon Press.

Davies, M. (2000). Persons and their underpinnings. *Philosophical Explorations*, **3**: 43–61.

Dawkins, M. S. (1998). *Through Our Eyes Only?* Oxford: Oxford University Press.

Dennett, D. C. (1969). *Content and Consciousness*. London: Routledge and Kegan Paul.

Dennett, D. C. (1978). *Brainstorms: Philosophical Essays on Mind and Psychology*. Montgomery, VT: Bradford.

Dennett, D. C. (1987). *The Intentional Stance*. Cambridge, MA: MIT Press.

Dennett, D. C. (1991). *Consciousness Explained*. Boston: Little Brown.

Dennett, D. C. (1996). *Kinds of Minds*. New York: Basic Books.

Dugatkin, L. A. and Alfieri, M. S. (2002). A cognitive approach to the study of animal cooperation. In: M. Bekoff, C. Allen, and G. M. Burghardt, eds. *The Cognitive Animal*, pp. 413–420. Cambridge: MIT Press.

Fodor, J. A. (1975). *The Language of Thought*. Cambridge: Harvard University Press.

Fodor, J. A. (1987). *Psychosemantics*, Appendix: Why there still has to be a language of thought. Cambridge, MA: MIT Press.

Gallup, G. G., Jr., Anderson, J. R., and Shillito, D. J. (2002). The mirror test. In: M. Bekoff, C. Allen, and G. M. Burghardt, eds. *The Cognitive Animal*, pp. 325–334. Cambridge: MIT Press.

Gigerenzer, G. (1997). The modularity of social intelligence. In: A. Whiten and R. W. Byrne, eds. *Machiavellian intelligence II*. Cambridge: Cambridge University Press.

Gigerenzer, G., Todd, P. M., and the ABC Group, eds (1999). *Simple Heuristics That Make us Smart*. New York: Oxford University Press.

Glock, H. J. (2000). Animals, thoughts, and concepts. *Synthese* , **123**: 35–64.

Godfrey-Smith, P. (1998). *Complexity and the Function of Mind in Nature*. Cambridge, UK: Cambridge University Press.

Godfrey-Smith, P. (2002). Environmental complexity, signal detection, and the evolution of cognition. In: M. Bekoff, C. Allen, and G. M. Burghardt, eds. *The Cognitive Animal*, pp. 135–142. Cambridge: MIT Press.

Gomez, J. C. (2004). *Apes, Monkeys, Children, and the Growth of Mind*. Cambridge: Harvard University Press.

Gruen, L. (2002). The morals of animal minds. In: M. Bekoff, C. Allen, and G. M. Burghardt, eds. *The Cognitive Animal*, pp. 437–442. Cambridge: MIT Press.

Güzeldere, G., Nahmias, E., and Deaner, R. (2002). Darwin's continuum and the building blocks of deception. In: M. Bekoff, C. Allen, and G. M. Burghardt, eds. *The Cognitive Animal*, pp. 353–362. Cambridge: MIT Press.

Hare, B. and Wrangham, R. (2002). Integrating two evolutionary models for the study of social cognition. In: M. Bekoff, C. Allen, and G. M. Burghardt, eds. *The Cognitive Animal*, pp. 363–369. Cambridge: MIT Press.

Hauser, M. (2000). *Wild Minds: What Animals Really Think*. London: Allen Lane.

Heyes, C. (2000). Evolutionary psychology in the round. In: Heyes and Huber, eds. pp. 3–22.

Heyes, C. M. (1998). Theory of mind in non-human primates. *Behavioral and Brain Sciences*, **21**: 101–148.

Heyes, C. and Huber, L., eds (2000). *The Evolution of Cognition*. Cambridge: MIT Press.

Howard, J. (1988). Cooperation in the prisoner's dilemma. *Theory and Decision*, **24**: 203–213.

Hurley, S. L. (1989). *Natural Reasons*. New York: Oxford University Press.

Hurley, S. (1998a). *Consciousness in Action*. Cambridge: Harvard University Press.

Hurley, S. (1998b). Vehicles, contents, conceptual structure, and externalism. *Analysis*, **58**: 1–6.

Hurley, S. (2005). Social heuristics that make us smarter. *Philosophical Psychology*, **18**: 585–61

Hurley, S. and Chater, N. (2005). *Perspectives on Imitation: from Neuroscience to Social Science*, vol. 1: Mechanisms of imitation and imitation in animals. Cambridge, MA: MIT Press.

Kahneman, D. and Amos, T., eds (2000a). Choices, Values, and Frames. Cambridge, UK: Cambridge University Press.

Kahneman, D. and Amos T., eds (2000b). Choices, values, and frames. In: D. Kahneman and A. Tversky, eds. *Choices, Values, and Frames,* pp. 1–16. Cambridge, UK: Cambridge University Press.

Kahneman, D., Slovic, P., and Tversky, A., eds (1982). *Judgment Under Uncertainty: Heuristics and Biases.* Cambridge: Cambridge University Press.

Kripke, S. (1982). *Wittgenstein on Rules and Private Language.* Oxford: Basil Blackwell.

McDowell, J. (1994a). The content of perceptual experience. *Philosophical Quarterly*, **44**: 190–205.

McDowell, J. (1994b). *Mind and World.* Cambridge, MA: Harvard University Press.

Nichols, S. and Stich, S. (2003). *Mindreading: An Integrated Account of Pretense, Self-awareness and Understanding Other Minds.* Oxford: Oxford University Press.

Nisbett, R. E. and Ross, L. (1980). *Human Inference: Strategies and Shortcomings of Social Judgement.* Prentice-Hall.

Oaksford, M. and Chater, N. (1998). *Rationality in an Uncertain World.* Hove, East Sussex: Psychology Press.

Pepperberg, I. M. (1999). *The Alex Studies.* Cambridge, MA: Harvard University Press.

Pepperberg, I. M. (2005). An avian perspective on language evolution: implications of simultaneous development of vocal and physical object combinations by a grey parrot (Psittacus erithacus). In: M. Tallerman, ed. *Language Origins: Perspectives on Evolution.* Oxford: Oxford University Press.

Rakoczy, H., Warneken, F., and Tomasello, M. (submitted). 'This way!', 'No! That way!'—3-year-olds know that two people can have mutually incompatible desires.

Rowlands, M. (2003). *Externalism.* Chesham: Acumen.

Santos, L. R., Hauser, M. D., and Spelke, E. S. (2002). Domain-specific knowledge in human children and non-human primates: Artifacts and foods. In: M. Bekoff, C. Allen, and G. M. Burghardt, eds. *The Cognitive Animal.* Cambridge: MIT Press.

Shettleworth, S. J. (1998). *Cognition, Evolution, and Behavior.* New York: Oxford University Press.

Sperber, D. (2000). Metarepresentations in an evolutionary perspective. In D. Sperber, ed., *Metarepresentations.* Oxford: Oxford University Press. Pp. 117–137.

Sperber, D. and Girotto, V. (2003). Does the selection task detect cheater-detection? In: J. Fitness and K. Sterelny, eds. *From Mating to Mentality: Evaluating Evolutionary Psychology*, pp. 197–226. Macquarie University Series in Cognitive Psychology.

Sterelny, K. (2000). Primate worlds. In: C. Heyes and L. Huber, eds. *The Evolution of Cognition*, pp. 143–162. Cambridge: MIT Press.

Sterelny, K. (2003). *Thought in a Hostile World.* Oxford: Blackwell.

Tomasello, M. (1999). *The Cultural Origins of Human Cognition.* Cambridge, MA: Harvard University Press.

Tomasello, M. and Call, J. (1997). *Primate Cognition.* New York: Oxford University Press.

Whiten, A. and Byrne, R. W., eds (1997). *Machiavellian Intelligence II: Evaluations and Extensions.* Cambridge University Press.

Wilson, R. (2004). *Boundaries of the Mind: The Individual in the Fragile Sciences.* Cambridge: Cambridge University Press.

Wynne, C. (2001). *Animal Cognition: The Mental Lives of Animals.* Basingstoke, Hampshire: Palgrave.

Part I

Types and levels of rationality

Chapter 2

Meanings of rationality

Alex Kacelnik

Abstract

The concept of rationality differs between psychology, philosophy, economics, and biology. For psychologists and philosophers, the emphasis is on the process by which decisions are made: rational beliefs are arrived at by reasoning and contrasted with beliefs arrived at by emotion, faith, authority, or arbitrary choice. Economists emphasize consistency of choice, regardless of the process and the goal. Biologists use a concept that links both previous ideas. Following Darwin's theory of natural selection, they expect animals to behave as if they had been designed to surpass the fitness of their conspecifics and use optimality to predict behaviour that might achieve this. I introduce the terms *PP-rationality*, *E-rationality*, and *B-rationality* to refer to these three different conceptions, and explore the advantages and weaknesses of each of them. The concepts are first discussed and then illustrated with specific examples from research in bird behaviour, including New Caledonian crows' tool design, hummingbirds' preferences between flowers, and starlings' choices between walking and flying. I conclude that no single definition of rationality can serve the purposes of the research community but that agreement on meanings and justifications for each stand is both necessary and possible.

'When I use a word,' Humpty Dumpty said, in rather scornful tone, 'it means just what I choose it to mean—neither more nor less'

'The question is,' said Alice, 'whether you can make words mean so many different things.'

'The question is,' said Humpty Dumpty, 'which is to be master—that's all.'

2.1 Introduction

The main questions that concern the contributors to this volume are:

◆ Are any non-human animals rational?

◆ What are the character and limits of rationality in animals?

◆ Are unobservable processes such as reasoning valid causal accounts of behaviour?

◆ What leads to differences in the kind of rationality exhibited by different species?

These are tough issues in the best of cases, but the real problem, as I see it, is that without a semantic effort we cannot even begin to discuss them: the questions contain words whose meanings cannot be assumed to be shared among those interested in the matter. Even accepting that too much defining inhibits thinking about the real issues, and that (as Humpty Dumpty tells us) definitions are arbitrary, clearly we cannot avoid reflecting on what our central theme, 'rationality', means for different authors. Responding to this need, my modest goal here is to discuss some ways in which this polysemous word is, and perhaps should be, used.

Guided by their differing goals and acceptability criteria, scholars in various disciplines have reached within-field consensus on workable definitions of rationality, and they produce data, reflections, models, theorems, and so on that provide evidence for the presence or absence of rationality and its boundaries as they understand them. These definitions, however, are at best consensual within particular fields. In my experience, a great deal of time is wasted arguing at cross-purposes while holding different understandings of rationality in mind. To mitigate this difficulty, I start by presenting an admittedly idiosyncratic discussion of various conceptions of rationality. In the case of my own field, biology, I will be forced to make up a definition, as none really exists at the moment.

I do not think that it is advisable (or feasible) to use a one-size-fits-all definition. Notions from different fields highlight such different aspects that to propose one overarching definition would be futile because few would follow it. I shall instead subsume all meanings of rationality into three categories, derived from my perception of the main uses in Philosophy and Psychology (PP-rationality), in Economics (E-rationality), and in Evolutionary Biology (B-rationality). I find all these uses necessary and appropriate for specific aims, but as I describe each of them I shall highlight what appears to me to be their virtues and their vices.

2.2 **PP-rationality**

The *Oxford Companion to Philosophy's* entry for 'Rationality' is a good starting point:

> This is a feature of cognitive agents that they exhibit when they adopt beliefs on the basis of appropriate reasons [...] Aristotle maintained that rationality is the key that distinguishes human beings from other animals. [...] A stone or a tree is non-rational because it is not capable of carrying out rational assessment. A being who is capable of being rational but who regularly violates the principles of rational assessment is irrational. [...] Rational beliefs have also been contrasted with beliefs arrived at through emotion, faith, authority or by an arbitrary choice. (Brown 1995, p. 744)

I suspect that this definition would sound acceptable to most non-philosophers, and also, to some extent, to contemporary cognitive psychologists (behaviourists may feel more comfortable with what I call 'E-rationality', discussed in the next section). Hence I will use this entry as a working definition of PP-rationality. Two features are particularly noteworthy.

First, the emphasis is on *process*, not on outcome. We can separate rational from non-rational beliefs depending on how they were arrived at, rather than according to their contents or the pattern of behaviour that results from them. There is clearly a difficulty in distinguishing 'appropriate' from inappropriate reasons, and the criteria for this distinction are likely to depend on cultural context. For example to believe that giraffes result from a cross between panthers and camels did count as PP-rational once upon a time because it was based on what were then appropriate reasons. Indeed this belief was held by champions of rationality such as Aristotle and other Greek scholars. This empirically mistaken belief would not qualify as rational today, but no doubt it is rational today to believe in theories that will prove factually wrong as time goes by and science progresses.

Second, PP-rationality is understood not in terms of observable behaviours but of entities such as *thoughts* and *beliefs*. To judge whether behaviour is PP-rational one needs to establish if it is caused by beliefs that have emerged from a reasoning process. To assess the PP-rationality of non-humans, we would have to devise means to expose not just our subjects' beliefs and the processes by which they were arrived at, but also to find a basis for judging whether these processes include 'appropriate' reasons in the sense discussed in the previous paragraph. This makes it very hard to assess whether, for example, a lion is rational, irrational, or non-rational.

The adherence of cognitive psychologists to this second point is typified by Oaksford and Chater (1998), who point out that, although stomachs may be well adapted to perform their function (digestion), 'they have no beliefs, desires or knowledge, and hence the question of their rationality does not arise' (p. 5). This exclusion of stomachs places them in the same rationality bracket as stones or trees, and seems reasonable within this framework, but it raises the question of which definition of rationality is at issue when questions about rationality are raised about the non-human world.

These two features would appear to place PP-rationality in a wholly unsuitable position to address our brief. Our focus is on non-human animals, whose thoughts, desires, and beliefs are inaccessible in practice and possibly also in principle, and certainly are not the stuff of normal animal research. In dealing with non-verbal subjects, biologists find the notion of using such entities as causes of behaviour problematic, even though the use of some of them are now (after the cognitive revolution) widely accepted. Some kinds of behaviour are best explained by reference to 'concepts' and 'representations' that are observable only indirectly. For instance, if an animal is exposed repeatedly to an interval between two events, such as a flash and a food reward, it will later show the same interval between the flash and performing a food-related action. Since the animal produces the interval, it is fair to say that the interval is represented in the animal and causes its behaviour. However, the fact that a representation causes behaviour does not imply that the subject has used reasoning.

The difficulties with PP-rationality are not limited to research with non-human animals. Many processes that give rise to the beliefs held by human subjects are in fact inaccessible to the holders of these beliefs, making it very hard to determine whether a belief has been arrived at on the basis of appropriate reasons. The hundred or so possibilities that chess masters are aware of examining before each actual move are a small subset of the available legal moves (de Groot 1965; Simon and Schaeffer 1992). It is likely that this subset is determined by unconscious processes that delve into the 50 000 or so positions chess masters remember, and that choices are often made under the irrational influence of emotional or aesthetic factors without the player being aware of their influence or of their access to her full knowledge base of chess positions. Thus, even if the whole process ends in the belief that a given move is best, and if the player feels that she has arrived at this conclusion by reasoning, the elements that entered into her reasoning process may have been influenced by the kinds of mechanism that the present definition would explicitly exclude from rationality. If, say, the player has acquired a Pavlovian aversion to a given position because she saw it while she had a toothache, then she will play so as to avoid it, and in doing so, she will be influenced irrationally by her knowledge base, though this influence and the active parts of her knowledge base may be unconscious.

I am aware that my concerns apply not just to assessments of rationality but to many other aspects of animal experience including welfare, pain, goal-directed behaviour, theory of mind, and so on, as well as to some aspects of human experience. Nevertheless, I think that within our present focus (rationality in non-humans), PP-rationality is particularly hard to assess. The combined weight of these problems would lead me to exclude PP-rationality from my own research, were it not for my (perhaps PP-irrational) desire to 'understand' my avian subjects and my belief that some progress can be made through painstaking experimentation. I shall return to this issue in the last of my empirical examples.

The focus of PP-rationality as I have described it is the rationality of beliefs or of the agents that hold them, or—in the language of cognitive psychologists—the rationality of information processing, rather than the rationality of actions. Yet psychologists, along with economists, are often concerned with the latter. To the extent that action is understood as essentially caused by certain mental processes and beliefs, my preceding comments about PP-rationality apply in similar ways. However, some notions of rationality concern themselves not with the mental processes that lead to beliefs or to behaviour, but with the resulting patterns of behaviour itself, and to this I now turn.

2.3 E-rationality

Economics is not what it once was. Until some time ago economic theory was basically a consistent set of mathematical models developed from rationality assumptions, but nowadays experimental economics (which studies what economic agents actually do without ignoring deviations from what is expected from them) is booming. This is illustrated by the choice of an experimental economist (Vernon Smith) and a cognitive psychologist (Daniel Kahneman) as Economics Nobel laureates for 2002. But many who concede that the assumption of full rationality is unhelpful in describing the actual behaviour of economic agents nevertheless broadly agree with orthodox economists in what they mean by the word. As part of an insightful discussion of the concept of rationality and the reasons why economists had to introduce it, another Economics Nobel laureate wrote:

> It is noteworthy that the everyday usage of the term 'rationality' does not correspond to the economist's definition as transitivity and completeness, that is, maximisation of something. (Arrow 1986, p. S390)

The 'something' to which this definition refers is 'Expected Utility'. Expected utility maximization is itself characterized mathematically in terms of the axioms of completeness and transitivity, and related properties such as independence and regularity (see Mas-Colell *et al.* (1995) for definitions of these terms). These are all properties related to the internal coherence of the agent's choices. Behaviour that is compatible with expected utility maximization counts as rational; in contrast with PP-rationality, the processes that generate behaviour are not the focus. For present purposes, the essential idea is that expected utility maximization constrains patterns of behaviour to be internally coherent in a certain sense, but places no substantive demands on behaviour, such as that choices lead to wealth accumulation, happiness, biological success, or the honouring of commitments.

While this is the prevalent view in economics, there are some important dissenting voices. Critics of the 'internal consistency' approach include, besides those mentioned so far, Amartya Sen, the Economics Nobel laureate for 1998. In Sen's view, a person who behaves according to these principles may well be rational, but if his behaviour is unrelated to happiness or other substantive concerns then he must also be a bit of a fool (Sen 1977). He proposes a view of rationality that does take into account the substantive interests of mankind. Although I have sympathy for Sen's perspective on what constitutes a suitable target for policy, I shall use the 'internal consistency' definition of rationality under the label of E-rationality for this discussion, both because it is the most widely used definition among economic theorists and because it is readily usable in the context of non-human behavioural research.

In contrast with PP-rationality, E-rationality concerns patterns of action rather than beliefs and the cognitive processes that may cause these actions. E-rationality is about observable behaviour rather than about unobservable or private mental states, and about outcomes rather than processes. What matters are the choices made and how they relate to each other rather than whether they originated from emotional impulses or cold reasoning. Indeed, economically rational agents can be institutions (which do not have private mental experiences) rather than persons, opening the possibility that even plants—and perhaps stomachs—may behave rationally.

Despite the apparent epistemological advantages of focussing on observable outcomes rather than unobservable mental processes, E-rationality faces difficulties of its own.

The central concept of expected utility maximization is not by itself accessible to observation, but is defined *post facto* as whatever is (consistently) maximized by the observed behaviour of the agent. While defining utility by reference to preferences as revealed in behaviour has logical advantages, it also limits considerably the class of observations that could be accepted as a violation of rationality. As long as observed choices can be interpreted as maximizing some common currency, however exotic or perverse, the choices are rational, under this definition. As Sen (1977) puts it, 'you can hardly escape maximizing your own utility'. Empirically, very little is ruled out by the mere internal consistency of E-rationality, unless we implicitly revert, as non-economists tend to do, to using an intuitive notion of utility as related to some property of the agent (viz. desire for wealth) that causes him to express those preferences in his behaviour.

The depth of this empirical difficulty can be easily illustrated by reference to state-dependency. For instance, non-transitive preference cycles (preferring a to b, b to c, but c to a) are definite irrationalities, assuming that the subject has remained in a constant state. However, if I choose lamb over ice-cream at 8 pm, ice-cream over coffee at 9 pm, and finally coffee over lamb half an hour later, this is not a serious breach of any principle of consistency because I have changed my state in the intervening period, that is, my preferences are state-dependent and this eliminates the intransitive cycle.[1] The point made by this example is theoretically trivial, but in practice, especially when applied to non-humans, it can lead to misleading conclusions, as I shall explain later.

Another example of the limitations of the revealed preferences approach is provided by lactation. In passing resources to her offspring, a lactating mother incurs a material loss. An observer may think, *prima facie,* that for a rational agent defined as utility maximizer any loss entails disutility and may then wonder if lactation violates utility maximization, but this would be mistaken. According to the revealed preference approach, the mother's choice to pass resources to her offspring simply means that her utility encompasses the well-being of her offspring. The degree to which the child's and her own consumption combine to define the mother's utility can be directly construed from her allocation of available resources. This has been encapsulated in a 'dynastic utility function' that depends on the utilities and number of children of all descendants of the same family line (Becker and Barro 1988). This approach can bring lactation back into the realm of rational behaviour, but it raises the question of how one determines the utilities of all those involved other than tautologically by their observed choices.

---

1 Editors' note: compare this issue about the empirical tractability of the transitivity of revealed preference in E-rationality, given the possibilities of state and context dependencies, with Allen's concerns about transitivity, this volume. Allen raises issues about the empirical tractability of attributions of transitive inference to animals, but these concern the processes involved and in particular the supposed distinction between genuinely rational or merely associative processes, and hence with what Kacelnik would label PP-rationality.

Economists sometimes revert to the intuitive notion of using a substantial criterion and identify utility with an objective observable such as consumption, but this is not helpful either, since consumption maximization for either one agent or a dynasty of them has no external justification as a currency for maximization.

The contrast between PP-rationality and E-rationality is evident: the former deals with causal processes and with unobservable events, such as beliefs, while the latter deals with outcomes and with observable actions. An agent can be 'rational' in either of these senses while being 'irrational' in the other. Biology provides yet another approach, and I discuss this next.

2.4 **B-rationality**

Rationality has never been a primary concern for evolutionary biologists, but ideas that relate to it underlie the logic of optimality modelling of decision-making in animals and plants. Furthermore, there is a rich and thriving tradition of contact between biology and economics: evolutionary game theory has made important contributions to economic theory, experimental economists challenge concepts within evolutionary theory, and experimental animal behaviourists test the E-rationality of their subjects (Shafir 1994; Hurly and Oseen 1999; Bateson *et al.* 2002; Schuck-Paim and Kacelnik 2002; Shafir *et al.* 2002; Fehr 2003). Given this busy exchange, it seems useful to make an attempt at tidying up a definition of rationality from a biologist's viewpoint.

B-rationality is necessarily linked to fitness maximization because it is based on the historical process of evolution of behavioural mechanisms. Evolutionary change is caused by both directional (natural selection) and non-directional processes (genetic drift). Both have some predictability. For instance, random genetic drift results in a predictable rate of accumulation of mutations and this serves as a clock to measure evolutionary distances between species. However, only natural selection generates phenotypic properties that can be anticipated using principles of maximization of a defined currency. For this reason biological rationality is best examined with natural selection (and hence fitness) at the centre.

Fitness is as central to biological theory as utility is to economic theory, but the two concepts are epistemologically very different. Current notions of fitness are grounded not on the revealed preferences of agents but on the genetic theory of natural selection. Broadly speaking, the fitness of a biological agent is its degree of success (growth as a proportion of the population) relative to that of other agents in the same population. Because fitness is relative and not absolute success, the fitness of an agent is always dependent on the population context and is not an intrinsic property of each agent.

Strictly speaking, agents should be alleles, which are different versions of each gene, but with suitable transformations it is possible to discuss the fitness-maximizing behaviour of individuals that carry these alleles, using the 'individual-as-maximizing-agent' metaphor (Grafen 1999; Grafen 2000). The passage from gene to individual level of analysis is not trivial. As an example, I return to lactation, in which the behaviour of the individual-level agent (the mother) compromises her nutritional state for the sake of another individual (her child). From the gene's point of view, one could argue that each allele in the mother's body that has an influence on her behaviour may also be (with a probability of one half) in the body of the child, and hence the allele as an agent is influencing the passage or resources between an older and a younger version of itself. There will be an optimum level of transfer that has to do with the replicating chances of each of the two versions, and this can be examined and predicted. As mentioned earlier, within the utility framework this can be handled by transforming the mother's utility function into a dynastic version that includes consumption of the resource by self and by her descendants, thus building a function that fits the observed behaviour (Becker and Barro 1988).

The biological alternative is to work out the fundamentals from the point of view of the alleles and the mechanics of population genetics, and then identify the appropriate transformation of the predicted allocation of resources to define a function that the individual ought to maximize. (For a rigorous treatment of the individual-as-maximizing-agent see Grafen (1999, 2000).) This concept was developed in detail by William Hamilton (1964) under the name of 'inclusive fitness' and it takes into account that alleles sit in more than one body and the dynamics of evolving populations. Because of the complexities of sexual reproduction and incomplete dominance between alleles, the individual as maximizing agent analogy is not yet fully tied to the genetic theory of natural selection, but theoreticians are busy working on this. Meanwhile, it is through this analogy that behavioural researchers from an evolutionary persuasion pursue their programme.

In summary: Under B-rationality the behaviour of individuals is a function of the genetic material that guides their development and functioning. Natural selection is a consistent (E-rational?) process that determines the distribution of alleles in populations and hence imposes properties on the genes that predominate in the individuals of each species at any given time. As a consequence of natural selection, alleles shape the behaviour of their carriers in ways that promote their own (the alleles') success. Theoretical population genetics is used to define a function (inclusive fitness) that describes how to see the individual as the maximizing agent, and this underlies the logic of optimality in biology. What the individual agent maximizes with its behaviour can never be understood without reference to the fact that the same alleles sit in more than one body.

This deceptively straightforward theoretical picture opens two difficult new problems: how the concept of individual-level rationality derived from inclusive fitness relates to those used by philosophers, psychologists, and economists and how these ideas can be tackled empirically when studying behaviour of real animals.

As a first approximation to addressing the first issue, I suggest that a B-rational individual can be defined as one whose actions maximize its inclusive fitness. This definition is closer to E-rationality than to PP-rationality in that it emphasizes outcome rather than process, it operates with observable behaviour, and it strongly emphasizes consistency. It differs from the former, however, in that what is maximized by B-rational agents (inclusive fitness) is far more constrained than what is maximized by E-rational agents (utility). As we have seen, inclusive fitness is definable and in principle measurable at genetic level, independently of the subject's choices, while utility is constructed from these choices. The cognitive or emotional processes that may accompany or even cause behaviour are not important for B-rationality: the approach is applied equally to bacteria, oak trees, blue whales, and humans in spite of their cognitive differences (but does not apply to stones, because, in contrast with living things, they have not been 'designed' by natural selection.

In terms of the second problem (testing), the main difficulty with B-rationality as defined above is that it is not explicit about the conditions across which the subject can be an inclusive fitness maximizer. It would be unjustified to expect any living creature to be B-rational under all conceivable circumstances. Having no foresight, natural selection only shapes neural mechanisms on the basis of encountered situations, so that as circumstances change evolved mechanisms fall short of fitness maximization. Since no creature can be expected to be universally B-rational, the concept is only useful when relativized to limited sets of circumstances and limited classes of decisions. Individual behaviour is driven by mechanisms evolved because they induce B-rational behaviour and not by the intentional pursuing of fitness maximization. These mechanisms may well include submission to emotions, authority, faith, and false beliefs.

It is worth mentioning, however, that an alternative view is also tenable. Imagine an organism that computes the consequences of each possible action and then uses reason to act in a way that maximizes inclusive fitness. That organism would be globally B-rational. It would also be PP-rational because it arrives at its decisions by reasoning and it would be E-rational because inclusive fitness would be identical to its utility as constructed from revealed preferences. I am not aware of anybody who explicitly defends the existence of such a creature, but in expecting humans to behave under a great variety of present cultural environments according to the maximization of inclusive fitness, some evolutionary psychologists and behavioural ecologists fall only marginally short of assuming that humans operate in this way.

A number of recent biological publications have used the term 'rationality' in an unqualified way that primarily reflects its economic definition, namely with emphasis on self-consistency. Thus, hummingbirds, starlings, jays, and honeybees have been charged with irrationality (Shafir 1994; Hurly and Oseen 1999; Waite 2001a and b; Bateson *et al.* 2002; Shafir *et al.* 2002) because their behaviour violates either transitivity or regularity (a principle of choice that states that the addition of further alternatives to a set of options should never increase the level of preference for a member of the original set (Luce 1977; Simonson and Tversky 1992; Tversky and Simonson 1993)). Counterclaims state that these observations may be compatible with B-rationality when state dependency is considered (Schuck *et al.* in press).

Before I proceed to consider how these three definitions of rationality may apply to birds, it may be helpful to summarize the main points. PP-rationality requires that beliefs or actions be based on reasoning; it focuses on how beliefs or actions are arrived at rather than what they consist of. E-rationality focuses on whether behaviour is consistent in the sense of maximizing some function that is called 'utility'. Utility maximization is not tied to substantive criteria such as fitness or well-being. It focuses on how an individual behaves rather than on how it has arrived at its preferences. B-rationality is the consistent maximization of inclusive fitness across a set of relevant circumstances; the under-specification of this set is perhaps the main weakness of B-rationality. It should be clear by now that individuals can certainly be rational in some of these senses while violating the other two. In the next section I go through examples of empirical research in bird behaviour that relate to the topics discussed so far.

2.5 Rational birds? Some experimental tests

2.5.1 B-rationality: optimal foraging theory

Tests of B-rationality typically start by considering situations assumed to be evolutionarily relevant and then proceed to create models that predict behaviour assuming that the subject chooses the strategy that maximizes inclusive fitness among a limited set of options (Kacelnik and Cuthill 1987; Krebs and Kacelnik 1991). Such models may fail to predict actual behaviour in experimental conditions, but this is not surprising because several elements of the model-making process can lead to failure.

In the first place, the situation considered may in fact not be ecologically relevant either at present or in the past, meaning that the choice may not have been encountered often enough during the evolutionary history of the species to have shaped its behavioural patterns.

Next, there is a serious problem with the description of the strategy set that the models draw on. It is unsatisfactory and somewhat circular to claim that an animal performs the best action among those in its strategy set and that if it fails to act adaptively it is because it has not got the best behaviour in its repertoire. Often, however, there is no escape from such claims. This is obvious when we deal with a physiological or anatomical limitation. For instance, behavioural ecologists can use models to predict the optimal choice between walking, jumping, and flying for locusts but they can only include the first two options for frogs. Clearly, a frog will not fly because flying is not within its strategy set.

However, and more problematically, this limitation also applies when the constraint is psychological. For instance, starlings will forego foraging gains so as to be close to conspecifics (Vásquez and Kacelnik 2000). Overall, this gregarious drive is adaptive because in nature flocking enhances feeding rate through several mechanisms, including sharing of information about both food location and predation danger (Giraldeau and Caraco 2000; Fernández-Juricic et al. 2004). When we expose a starling to a situation where the best feeding patch is not in the greatest vicinity to other starlings, we expose a psychological mechanism that stops the starling from foraging in a way that maximizes foraging yield. This mechanism may have evolved because on average it yields greater benefits, but it does sometimes stop starlings from satisfying optimality predictions for specific situations.

It is impossible to determine precisely which observed features of a creature should be included in the assumptions of an optimality model and which should be left to be predicted by the model itself. This means that testing B-rationality in practice is more dependent on revealed preferences than biologists like myself would like to admit and that some of the circularities of E-rationality afflict this approach as well.

These aspects of the study of B-rationality are evident in any detailed application of optimality in foraging behaviour, and as an example I present one case in some detail.

2.5.2 **B-rationality: To fly or to walk?**

Starlings forage sometimes by walking on the ground (poking the soil with their beaks to dig for hidden grubs) and sometimes by taking short flights (hawking small airborne insects). In one study (Bautista *et al.* 2001) these two foraging modes were taken as given and optimality modelling was used to examine how the birds chose among them in a laboratory situation. Starlings could work for food by walking or by flying. In each foraging mode a number of trips (walks or flights) between a resting place and a food source were necessary to obtain one reward. In 11 different treatments, the flying requirement was fixed as a number of flights between one and 11, and the number of walks that made the two foraging modes equally attractive to each bird was found using a titration technique: the number of walks increased or decreased by one depending on whether the animal's previous choice had favoured walking or flying respectively. This number eventually oscillated around a certain value, and this was taken to be the number of walks that were as attractive as the number of flights in that treatment.

The model was based on the fact that hawking yields more frequent captures but is more expensive, because flying uses more energy per unit of time than walking. I describe below the details of the model. Skipping these details should not obscure my main point.

Let S_w and τ_w be respectively the size (in energy units) and involvement time (in time units) per prey from walking, while S_f and τ_f are the size and involvement time from flying. Involvement time includes the times taken to travel to and to consume a prey item. Getting food using each mode involved time resting, travelling, and handling the food, but the model uses the average metabolic rates throughout all these components. This average differs among modes. Energy gain is known to play a major role in this sort of problem, but several metrics are reasonable possibilities. Foraging theorists use either energy gain over time ('net rate' and its simplification 'gross rate') or energy gains per unit of expenditure ('energetic efficiency').

Net Rate for an option i is defined as $\gamma_i = (\text{Gain}_i - \text{Cost}_i)/\text{time}_i$, where gain is the energy content of a capture, cost is the energy spent in procuring it, and time is the involvement time. Gain is the usable caloric content of a prey and cost is the product of the metabolic rate (in energy per unit of time) times the involvement time. Walking and flying yield equal net rates when $\gamma_w = \gamma_f$ as expressed by:

$$\frac{S_w - m_w \tau_w}{\tau_w} = \frac{S_f - m_f \tau_f}{\tau_f} \tag{1}$$

where m_w and m_f are the average metabolic rates in the walking and flying modes respectively. When reward sizes are equal between the two modes (as in the experiment discussed here) the cost in walking time that yields the same rate as a particular flying time cost is given by solving Equation 1 for τ_w:

$$\tau_w = \frac{1}{\dfrac{1}{\tau_f} + \dfrac{m_w - m_f}{S}} \tag{2}$$

where S is the common size of both rewards. Gross rate is a simplification in which the negative terms in the numerators of Eq. 1 (the costs) are ignored, that is treated as if their values were negligible. If this is done and reward sizes are equal, then Eq. 2 simplifies so that equal gross rates are achieved when $\tau_w = \tau_f$.

Efficiency may be the optimal choice when the main constraint is availability of energy for immediate use, for instance if energy gains do not become immediately usable. When this occurs, it makes sense for an optimal decision maker to maximize the ratio of gains to expenditure regardless of the time involved. With the same notation as before, we can see that efficiency is equalized between modes when

$$\frac{S_w - m_w \tau_w}{m_w \tau_w} = \frac{S_f - m_f \tau_f}{m_f \tau_f}. \tag{3}$$

Then, with equal reward sizes, we get

$$\tau_w = \frac{m_f}{m_w} \tau_f. \tag{4}$$

The predictions of the models and the titration results from the starlings are shown in Fig. 2.1. As the figure shows, net rate maximization (Eq. 2) is extremely successful in predicting the starlings' choices, while the alternatives (gross rate and efficiency) either underestimate or overestimate preference for walking.

This example shows that it is possible to predict animal preferences assuming a consistent (hence E-rational) criterion on the part of the subjects, and that this criterion may be deduced *a priori* from functional considerations, so that B-rationality is tested as well. On the other hand, the example also shows that the *a priori* biological criterion was not inclusive fitness but a proxy, and that this proxy was not unique. At least three alternatives (gross rate, net rate and efficiency) were judged possible, and which one drives

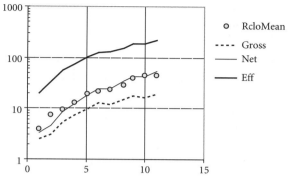

Fig. 2.1 To walk or to fly? Starlings were offered a choice between obtaining a food reward by a number of walks (Rclo Mean) or by a number of flights. The number of fights (abscisa) was experimentally fixed, and the number of walks at indifference (ordinate) was obtained by titration. The symbols show the average indifference point among four starlings. The lines, from top to bottom, show the predictions of choice according to maximum gain per unit of expenditure, maximum net gain per unit of time and maximum gross gain per unit of time. The agreement with the middle line indicates that starlings make these choices in a way that maximizes net rate of energy gain. Modified from Bautista *et al.* 2001.

choice was established by reference to the birds' preferences, thus submitting to some extent to the tautology of utility-based approaches. Nothing in this example links to PP-rationality. There is no reference to the psychological mechanism by which the subjects make choices. There is no suggestion for or against the possibility that the starlings reason their way to choice, and hence the study does not connect with PP-rationality.

2.5.3 **E-rationality**

The economic concept of rationality, with its emphasis on consistency of choice, offers an interesting and radically different source of inspiration for experiments on animal decision-making. At some level E-rationality may be seen as a corollary of B-rationality, but one that leads to different questions and different experiments. If a subject is a consistent maximizer of inclusive fitness, then it is a consistent maximizer of something, and a consistent maximizer of anything is by definition E-rational. It follows that empirical observations of violations of E-rationality pose problems that biologists need to address. Recent years have seen a proliferation of studies with an emphasis on E-rationality (Shafir 1994; Hurly and Oseen 1999; Waite 2001a and b; Bateson 2002; Shafir *et al.* 2002) and here I discuss one such example.

One conceptual violation of E-rationality is found in some forms of preference reversals due to context dependence. Imagine that a subject, facing a choice between A and B, prefers A in one context and B in another. This would be a sign of E-irrationality that should warrant intensive attention and research by biologists. Precisely such an observation was reported by Hurly and Oseen (1999) in a field study of risk preference by wild rufous hummingbirds *Selasphorus rufus*. These authors offered wild, free-living hummingbirds choices between artificial flowers that differed in the level of variance in amount of sugar solution. There were three kinds of flowers, identified by their colour. Type N (for No Variance) always offered the same amount of nectar, type M (for Medium variance) offered a coefficient of variation in volume of 33.3 per cent, and type H (for High Variance) offered a coefficient of variation of 66.6 per cent. If an animal systematically avoids variance in amount (this is a frequent trend; see Kacelnik and Bateson 1996), it should always prefer the option with lowest variance in any set. The hummingbirds complied with this expectation by significantly preferring N over M, M over H, and N over H when facing pairs of alternatives. However, when the birds were offered a choice between the three types of flower presented simultaneously, their maximum preference was for the intermediate level of variance, thus reversing the ranking between medium and no variance. Similar observations have been made in other species and paradigms including honeybees, starlings, and jays.

Now, how might a biologist respond to observations of this kind? One option is to accept the violation of E-rationality at face value, including the implied breach of B-rationality, and to conclude that the hummingbirds are indeed poor choosers in this situation. This is perfectly tenable, as it is possible that although the same choice (say lower variance) is always better, in their evolutionary history they may have only very rarely faced simultaneous encounters of more than two kinds of flowers, and hence even if this situation leads to a suboptimal preference, the cost has not been sufficient to select for mechanisms that avoid this malfunction. Another possibility is to reflect on whether an adaptive explanation may exist to account for the observed reversal. In fact, several suggestions exist for why relative preferences between two options may change adaptively when other options are added.

One idea (Houston 1997) is that the decision maker infers (meaning here: behaves as if it infers) what options will be available in the future from the set of options present at the time of choice. If this were so in the hummingbird example, an animal facing two sources, one with no variance and the other with medium variance, would infer that this is what the future has in store, and then it would deploy the best choice for that scenario. On the other hand, an animal facing the same two options within an extended set (after the inclusion of a third alternative with even greater variance) would infer that the future will make this third option available as well, and this may change the nature of the problem sufficiently to alter the relative values between the original alternatives. There is no evidence that this explains the hummingbirds' reversal, but Houston's idea illustrates how apparent violations of B- and E-rationality may be expressions of a complex context-dependent set of optimal strategies.

A related possibility (Schuck-Paim 2003) looks back into the agent's history rather than forward into how the agent infers future options. This idea postulates that the different contexts during training may result in different energetic states at the time preferences are measured. If this occurs, then the problem faced by the agent has changed, and there is no reason to expect consistency in preference ranking. With a subject in a different state, the benefit accrued from each option may be different. The difference in state of the subject in experiments such as those with hummingbirds may occur because preferences are not exclusive. Since the subject allocates some choices to all available options, then its energetic state reflects the conditions that precede the tests, and animals trained in pairwise choices may be in different energetic states (or have a history of different variance in state) from those trained with simultaneous presentations of three options. If the subject's state is different, then relative preferences may reverse without this being an indication of E-rationality. This remark is similar to the point of the 'lamb vs. ice cream vs. coffee' example, in which, because state varied as the meal progressed, reversals did not violate rationality.

In summary, apparent violations of E-rationality are interesting and pose challenges to B-rationality, but it may often be the case that when a B-rational interpretation has been found, the notion of a violation of E-rationality may become unnecessary. In particular, it is crucial to show that when a subject shows inconsistent preferences, these are measured with the subject in the same state. In Houston's interpretation the state of the subject differs because it possesses different information about the future and in Schuck-Paim's idea the state differs because the recent history of the animal has placed it in a different energetic state. If any of these ideas is supported, then the breach of either kind of rationality is explained away.

Once again, in this section I did not refer to the notion of PP-rationality, as whether hummingbirds' actions resulted from a reasoning process or otherwise could not and was not addressed. To tackle the possibility of saying something about this issue, I turn to my final avian example.

2.5.4 **PP-rationality: New Caledonian crows**

In the introductory description of PP-rationality I exposed my pessimism regarding the possibility of testing of PP-rationality in non-humans. This stems from the fact that testing this kind of rationality requires access to the process by which subjects reach beliefs, rather than dealing with some mapping between the material situation of the decision maker and its behaviour. Very worthy attempts have been made to cross this bridge by researchers working on topics such as mind reading, gaze following, and other forms of social inference. In most cases the results raise very difficult problems of interpretation (Povinelli 2000; Visalberghi 2000; Tomasello et al. 2003; Tomasello and Call, Chapter 17, this volume; Povinelli and Vonk, Chapter 18, this volume). Without any expectation of solving the difficulties, I describe below one example from our work with New Caledonian crows (*Corvus moneduloides*) that goes some way to temper my pessimism.

The case in question (Weir et al. 2002) concerns what a crow named Betty did when faced with an out-of-reach morsel in a small basket at the bottom of a vertical plastic well. On previous occasions Betty had been provided with two wires, one straight and the other hooked, and had been able to lift the basket with the hooked wire. On a crucial trial, however, she only had available a straight wire. After failing to lift the basket with the straight wire, she took it to a fracture in a nearby plastic tray, wedged the tip there and pulled perpendicularly from the proximal side, bending the wire until it formed a hook. She then returned to the well, retrieved the basket and ate the food. Further observations showed that she could bend wires using several different techniques to achieve functional tools.

Many corvids are capable of being trained to use tools (Powell and Kelly 1977), and New Caledonian crows are consummate tool-makers and users (Hunt 1996; Hunt 2000), but before this observation there had been no report of a bird solving a novel problem with such degree of creativeness. New Caledonian crows make and use hook-like and other tools from plant material in the wild and in the laboratory (Hunt and Gray 2004; Chappell and Kacelnik 2004). However, to our knowledge they normally do not have access to pliable material that can be bent into a shape and preserve it. Betty used her motor skills and knowledge of the principle of hooks to devise a new solution to her problem, and this solution required leaving the site of the problem, finding a suitable crack to hold the wire's tip, and modifying the wire appropriately before returning to finish the task. It is hard to account for all of this without reference to some form of planning ahead, some representation of the problem and its solution, and some choice among possible actions leading to the solution. This is what many would call thinking, and with only a little suspension of disbelief might even be accepted as reasoning and as approximating the conditions for PP-rationality.

The attribution of any form of rationality cannot be based on one set of observations, however compelling this set may be. We do not know how domain-general the New Caledonian crows' ability to plan and execute solutions to new problems is. We need to investigate this while bearing in mind that there is evidence that even for humans there is no such a thing as totally domain-independent reasoning abilities (Cosmides and Tooby 1997). In a sufficiently attractive scenario we may hope to show that animals reveal underlying processes that include wide range anticipation and planning. This may help to progress even in the hard field of PP-rationality.

2.6 **Conclusions: Are animals rational after all?**

My purpose in this chapter is to provide a small degree of clarification in the use of the term (and hence the concept) of rationality, because the term is manifestly interesting and worth researching but by virtue of its being used in various fields has diverse and potentially incompatible interpretations. I constrained my narrative to three notions of rationality, and illustrated them with examples from bird research. I am aware that I did not exhaust the discussion of previous uses of the term nor did I necessarily use the most striking available examples. (I realize with some dismay that I did not discuss the achievements of Alex the African grey parrot; see Pepperberg 1999 and this volume.)

As a minimum palliative to this parochialism, I add a few words about a use of the term 'rationality' that I have not discussed so far.

The notion of 'Ecological Rationality' has been fostered through many studies by Gerd Gigerenzer, Peter Todd, and their colleagues (Gigerenzer *et al.* 1999). This concept refers to human subjects' hypothetical use of thinking rules ('fast and frugal heuristics') that achieve satisfactory solutions under the limited sets of circumstances subjects face in their everyday life. Ecological rationality is thus clearly different from the three definitions of rationality used through this paper.

It differs from PP-rationality in that subjects' are not assumed to arrive at beliefs by logical reasoning, from B-rationality in that it does not focus on the evolutionary or developmental origins of each rule, and from E-rationality in that there is no paramount role for internal consistency among choices. The idea that humans follow simple cognitive heuristics to cope efficiently with the problems they face in real life is close to that of 'rules of thumb', which were used in early foraging theory to account for how behaviour that may require complicated calculation to be identified as optimal is performed by insects or birds that could not be expected to perform the calculations. Krebs and McCleery expressed this concept thus:

> It is generally assumed that foraging animals use simple 'rules of thumb' to solve their foraging problems, and that these rules may approximate to the solutions predicted by optimization models. (Krebs and McCleery 1984, p. 118)

This notion was applied to many search problems, including for instance the issue of how birds might find a balance between the advantages and costs of acquiring knowledge (Houston *et al.* 1982).

The use of this approach for animals has had mixed success. In fact, there is a dichotomy between work on behaviour that is directly related to fitness and follows genetically preprogrammed rules (as when parasitoid wasps track chemical gradients to localize hosts or female birds use the size of males' song repertoires to choose among suitors) and work on problems where subjects learn a great deal, including the nature of the problem and the parameters of each alternative. For the first kind of problem the rules-of-thumb approach works well, but for problems that are evolutionarily unpredictable and require individual learning, unravelling the mechanisms of preference acquisition seems a better strategy. Learning mechanisms (pre-eminently associative learning) are species-wide and generate a variety of behavioural rules, each tailored to a given agent's individual history. Criteria for rationality are more likely to apply to the learning mechanisms by which animals acquire preferences than to the rules that result from them.

It seems appropriate to end this chapter by leaving the last word to Lewis Carroll. As he probably would have pointed out, there is no reason to impose a universal set of definitions of rationality, but there are plenty of reasons why, if we are about to claim that animals are or are not rational, we should make a serious effort to define what we mean.

I maintain that any writer of a book is fully authorized in attaching any meaning he likes to any word or phrase he intends to use. If I find an author saying, at the beginning of his book 'Let it be understood that by the word 'black' I shall always mean 'white' and by the word 'white' I shall always mean 'black',' I meekly accept his ruling, however injudicious I may think it. (Carroll 1964/2000, p. 226).

Acknowledgments

I am deeply grateful to the Editors for the very substantial comments and suggestions that improved previous versions of this paper. Thanks also to Martha Klein and Marian Dawkins for thoughtful and useful advice. This paper was supported with financial support from the BBSRC (Grant 43/S13483) and benefited from a partial research leave spent at the Institute for Advanced Studies in Berlin.

References

Arrow, K. J. (1986). Rationality of self and others in an economic system. *Journal of Business*, **59**: S385–399.

Bateson, M. (2002). Context-dependent foraging choices in risk-sensitive starlings. *Animal Behaviour*, **64**: 251–260.

Bateson, M., Healy, S. D., and Hurly, T. A. (2002). Irrational choices in hummingbird foraging behaviour. *Animal Behaviour*, **63**: 587–596.

Bautista, L. M., Tinbergen, J., and Kacelnik, A. (2001). To walk or to fly: how birds choose among foraging modes. *Proceedings of the National Academy of Sciences*, **98**: 1089–1094.

Becker, G. S., and Barro, R. J. (1988). A reformulation of the economic theory of fertility. *The Quarterly Journal of Economics*, **103**: 1–25.

Brown, H. I. (1995). Rationality. In: T. Honderich, ed. *The Oxford Companion to Philosophy*, pp. 744–745. Oxford: Oxford University Press.

Carroll, L. (1964/2000). Symbolic logic and the game of logic. In M. Gardner, ed. *The Annotated Alice*, pp. 225–226. London: Penguin.

Chappell, J., and Kacelnik, A. (2004). Selection of tool diameter by New Caledonian crows *Corvus moneduloides*. *Animal Cognition*, **7**: 121–127.

Cosmides, L., and Tooby, J. (1997). Dissecting the computational architecture of social inference mechanisms. In: G. Bock and G. Cardew, eds. *Characterizing Human Psychological Adaptations*, pp. 208: 132–156. Chichester: Wiley.

de Groot, A. D. (1965). *Thought and Choice in Chess*. The Hague: Mouton.

Fehr, E. (2003). The nature of human altruism. *Nature*, **425**: 785–791.

Fernández-Juricic, E., Siller, S., and Kacelnik, A. (2004). Flock density, social foraging, and scanning: an experiment with starlings. *Behavioral Ecology*, **15**: 371–379.

Gigerenzer, G., Todd, P. M., and The ABC Research Group (1999). *Simple Heuristics that Make us Smart*. New York: Oxford University Press.

Giraldeau, L.-A., and Caraco, T. (2000). *Social Foraging Theory*. Princeton: Princeton University Press.

Grafen, A. (1999). Formal Darwinism, the individual-as-maximising-agent analogy and bet-hedging. *Proceedings of the Royal Society of London B*, **266**: 799–803.

Grafen, A. (2000). A biological approach to economics through fertility. *Economics Letters*, **66**: 241–248.

Hamilton, W. D. (1964). The genetical evolution of social behaviour. *Journal of Theoretical Biology*, **7**: 1–52.

Houston, A. I. (1997). Natural selection and context-dependent values. *Proceedings of the Royal Society of London B*, **264**: 1539–1541.

Houston, A. I., Kacelnik, A., and McNamara, J. M. (1982). Some learning rules for acquiring information. In: D. J. McFarland, ed. *Functional Ontogeny*, pp. 140–191. Boston: Pitman.

Hunt, G. R. (1996). Manufacture and use of hook-tools by New Caledonian crows. *Nature*, **379**: 249–251.

Hunt, G. R. (2000). Human-like, population-level specialization in the manufacture of pandanus tools by New Caledonian crows *Corvus moneduloides*. *Proceedings of the Royal Society of London B*, **267**: 403–413.

Hunt, G. R., and Gray, R. D. (2004). The crafting of hook tools by wild New Caledonian crows. *Proceedings of the Royal Society of London B (Supplement)*, **271**(S3): S88–S90.

Hurly, T. A., and Oseen, M. D. (1999). Context-dependent, risk-sensitive foraging preferences in wild rufous hummingbirds. *Animal Behaviour*, **58**: 59–66.

Kacelnik, A., and Bateson, M. (1996). Risky theories—the effects of variance on foraging decisions. *American Zoologist*, **36**: 402–434.

Kacelnik, A., and Cuthill, I. C. (1987). Starlings and optimal foraging theory: modelling in a fractal world. In: A. C. Kamil, J. R. Krebs, and H. R. Pulliam, eds. *Foraging theory*, pp. 303–333. New York: Plenum Press.

Krebs, J. R., and Kacelnik, A. (1991). Decision making. In: J. R. Krebs and N. B. Davies, eds. *Behavioural Ecology: an Evolutionary Approach*, pp. 105–136. Oxford: Blackwell Scientific Publications.

Krebs, J. R., and McCleery, R. H. (1984). Optimization in behavioural ecology. In: J. R. Krebs and N. B. Davies, eds. *Behavioural Ecology: an Evolutionary Approach*, pp. 91–121. Massachusetts: Sinauer.

Luce, R. D. (1977). The choice axiom after twenty years. *Journal of Mathematical Psychology*, **15**: 215–233.

Mas-Colell, A., Whinston, M. D., and Green, J. R. (1995). *Microeconomic theory*. Oxford: Oxford University Press.

Oaksford, M., and Chater, N. (1998). An introduction to rational models of cognition. In: M. Oaksford and N. Chater, eds. *Rational models of cognition*, pp. 1–17. Oxford: Oxford University Press.

Pepperberg, I. M. (1999). *The Alex Studies: Cognitive and Communicative Abilities of Grey Parrots*. Cambridge, MA: Harvard University Press.

Povinelli, D. J. (2000). *Folk Physics for Apes*. Oxford: Oxford University Press.

Powell, R. W., and Kelly, W. (1977). Tool use in captive crows. *Bulletin of the Psychonomic Society*, **10**: 481–483.

Schuck-Paim, C. (2003). *The Starling as a Rational Decision-Maker*. Unpublished DPhil Thesis, Zoology Department, University of Oxford.

Schuck-Paim, C., and Kacelnik, A. (2002). Rationality in risk-sensitive foraging choices by starlings. *Animal Behaviour*, **64**: 869–879.

Schuck-Paim, C., Pompilio, L., and Kacelnik, A. (In press). State-dependent decisions cause apparent violations of rationality in animal choice. *Public Library of Science Biology*.

Sen, A. (1977). Rational fools: a critique of the behavioral foundations of economic theory. *Philosophy and Public Affairs*, **6**: 317–344.

Shafir, S. (1994). Intransitivity of preferences in honey-bees—support for comparative-evaluation of foraging options. *Animal Behaviour*, **48**: 55–67.

Shafir, S., Waite, T. A., and Smith B. H. (2002). Context-dependent violations of rational choice in honeybees (*Apis melifera*) and gray jays (*Perisoreus canadensis*). *Behavioural Ecology and Sociobiology*, **51**: 180–187.

Simon, H. A., and Schaeffer, J. (1992). The game of chess. In: R. J. Aumann and S. Hart, eds. *Handbook of Game Theory*, pp. 1: 1–17. Amsterdam: Elsevier.

Simonson, I., and Tversky, A. (1992). Choice in context: trade-off contrast and extremeness aversion. *Journal of Marketing Research*, **29**: 281–295.

Tomasello, M., Call, J., and Hare, B. (2003). Chimpanzees understand psychological states—the question is which ones and to what extent. *Trends in Cognitive Sciences*, **7**: 153–156.

Tversky, A., and Simonson, I. (1993). Context-dependent preferences. *Management Science*, **39**: 1179–1189.

Vásquez, R. A., and Kacelnik, A. (2000). Foraging rate versus sociality in the starling *Sturnus vulgaris*. *Proceedings of the Royal Society of London B*, **267**: 157–164.

Visalberghi, E. (2000). Tool use behavior and the understanding of causality in primates. In: E. Thommen and H. Kilcher, eds. *Comparer ou Prédire: Examples de Recherches Comparatives en Psyhologie Aujourd'hui*, pp. 17–35. Fribourg, Switzerland: Les Editions Universitaires.

Waite, T. A. (2001a). Background context and decision making in hoarding gray jays. *Behavioral Ecology*, **12**: 318–324.

Waite, T. A. (2001b). Intransitive preferences in hoarding gray jays (*Perisoreus canadensis*). *Behavioural Ecology and Sociobiology*, **50**: 116–121.

Weir, A. A. S., Chappell, J., and Kacelnik, A. (2002). Shaping of hooks in New Caledonian crows. *Science*, **297**: 981.

Chapter 3

Minimal rationality

Fred I. Dretske

Abstract

In thinking about animal rationality, it is useful to distinguish (what I call) minimal rationality, doing something for reasons, and doing something for good reasons, reasons that (if true) exhibit the behavior as contributing to goal attainment and desire satisfaction. Minimal rationality, though it is less demanding than rationality (it doesn't require good reasons), is, in another way, more demanding than other forms of rationality (e.g. biological rationality). It requires the behavior to not only be under the causal control of thought, but to be explained by the content of these thoughts. It is for this reason that a lot of behavior (especially by plants and machines) that would appear to be rational is not even minimally rational.

When you make a sudden movement towards my eyes, I blink. I cannot help myself. Of course, I do not want your finger in my eye. I also believe one way to keep your finger out of my eye is to close my eye when you poke at it. But though I think these things, and though I close my eye when you poke at it, I do not close my eye because I think these things. I would close my eye whether or not I had these beliefs and desires. The mechanisms for these reflexes are hard-wired. They swing into action well before thought has time to act. I have reasons to close my eyes, but my reasons for closing them are not the reason I close them.

So despite the fact that I do exactly what I think will get me what I want, my behavior is not a purposeful act. It does not exhibit what I will call *minimal rationality*. Though the behavior is in conformity with thought, it is not explained, not governed, by thought. Minimal rationality requires that thought be involved in the process by means of which the behavior is produced. It is, for this reason, more demanding than what Alex Kacelnik (Chapter 2, this volume) describes as *biological* rationality—behavior that can be seen or interpreted as designed by natural selection to achieve a consistent goal of higher fitness. The blink reflex is biologically, but not minimally, rational. It contributes to fitness, but it doesn't do so in a way—by a process involving thought—that earns for the agent the accolade of rationality.

In another sense, though, minimal rationality is *less* demanding than biological rationality. Minimal rationality requires that what is done be done for reasons, but it doesn't require that it be done for *good* reasons. Nor does it require reason*ing*. Although the behavior must be explained by a thought in order to qualify as minimally rational, it needn't be rationalized or rationally justified by the thought that explains it, and the agent needn't have computed (reasoned) his way to that result.[1] The behavior needn't contribute, even in ideal conditions, to an organism's overall fitness. The behavior can, in fact, diminish fitness. It can be the *result* of reasoning but be *contrary* to reason.

An old joke illustrates minimal rationality at its minimum. Clyde is looking for his keys under a street lamp on a dark night.

Jane: What are you doing?
Clyde: Looking for my keys.
Jane: Did you lose them here?
Clyde: No, I lost them over there (pointing to a dark alley).
Jane: Why, then, are you looking here?
Clyde: The light is so much better here.

Assuming Clyde is doing what he's doing for the reason he gives, his behavior is minimally rational. Thought (that the light is better here) is controlling his behavior. Clyde is, nonetheless, irrational. Given the truth of his beliefs, there is no chance of finding his keys here, no way his behavior can achieve his goals. Though the thought that the light is better here is controlling his behavior, it does not rationalize this behavior. It does not increase, let alone maximize, Clyde's utilities (in Kacelnik's economic sense of rationality); just the opposite. The behavior is counter-productive. Nonetheless, it is still minimally rational. It is under the control of thought. What Clyde is doing is explained by what he thinks

Or consider an example mentioned by Sally Boysen.[2] Someone straps on explosives, climbs aboard a bus, and blows himself and 60 other people up. His reason? He was told, and he believed, that the act would help the cause (imagine whatever 'cause' you care to here) and that 40 virgins would await him as his heavenly reward. Is this a reasonable thing to do? Is it rational? It certainly doesn't increase fitness. I assume that most of us think this act cannot be rationally justified; some may even regard it as insane. But is it minimally rational? Yes. The behavior is explained by what the terrorist believes and wants.

---

[1] I take the imitative behavior—at least some of it—of the type described by Richard Byrne (in his talk Who Needs Rationality at the 2002 Oxford conference on rationality in animals; see also Byrne 2005) and Herman's bottlenosed dolphins (this volume) as minimally rational. When the behavior being imitated is observed, a representation of it stored in memory, and then, later, the imitative behavior is elicited by some symbolic gesture or signal, the behavior is being explained by a thought in my sense of 'thought'.

[2] During discussion at the 2002 Oxford conference on rationality in animals.

Whether or not we think the behavior is totally irrational, it is under the control of thought. The behavior is goal directed, however much we may object to the goal.

Minimal rationality does not require rationality in a normative sense of this term. You or I don't have to think behavior B is rational, it doesn't have to accord with our picture of what it makes sense to do, or what it would be best to do, in order for it to be minimally rational. This, indeed, is why I think minimal rationality is a useful concept in thinking about animal rationality. It keeps normative issues at bay; it brackets them, so that we can investigate whether animals are doing things for reasons quite aside from whether we, or anyone, think these are *good* reasons. Why should a chimpanzee or a crow have to be doing what you and I think is rational in order to be doing exactly the same sort of thing (doing something for reasons) you and I are doing when we do what we think is rational? Why can't an animal's behavior, just like ours, be under the control of thought even though we are unable to recognize the point or purpose (the rational justification) of such behavior? If we are going to investigate animal rationality, why not *first* identify minimal rationality in animals, and then, after we are sure what is being done is being done *for* reasons, ask whether the reasons for which it is being done make it a reasonable thing to do. Do the reasons that explain the behavior also make it a reasonable thing to do? One thing at a time.

The same procedure is appropriate in thinking about machine mentality. Instead of immediately asking whether computers are smarter than we are—whether, for instance, Deep Blue (IBM's computer that 'beat' the world champion) is a better chess player than Kasparov—let's first ask whether what machines have is intelligence at all? Do they even play chess? Do they want to win? Do they have purposes? Intentions? Is anything they do explained by what they want? Do they have *minimal mentality*? Do pocket computers add and subtract? Or do *we* add and subtract *with* pocket computers the way we pound nails with hammers? If computers don't even play chess, if they don't have the minimum credentials for this sort of activity, how can they be better or worse at it than we are?

Minimal rationality is a prerequisite for playing the rationality game. No matter how clever you appear to be, if nothing you do is explained by what you think, you are not a rational being, you are an impostor. Somebody *else*—whoever or whatever designed you—may be very clever (Mother Nature?), but not *you*. One could as well say that the sprinkler system is smart because it extinguishes the fire that would otherwise destroy it. If I did that, if I put out a fire because it threatened me, I would be rational. My behavior would be explained by what I think. But that isn't why the sprinkler system does it. It doesn't have minimal rationality. Neither do I display minimal rationality when I perspire—thereby cooling myself—when my body temperature starts to rise. The behavior isn't under the control of thought. It may be a smart thing to do, but I'm not smart for doing it.

But what does it mean to be *governed* by thought? How can thought *explain* behavior? Is it enough if inner representations (thoughts?) cause the behavior? No. There is a difference between being caused by an event that means (represents) M and being explained by the fact that it means (represents) M.[3] This difference is important for understanding what it takes to qualify as minimally rational. It is this difference that explains why machines and plants—and perhaps even some animals—who do exactly what we do don't qualify as rational agents.

Consider an ordinary home thermostat. It turns the heat on and off and in so doing keeps the room at a comfortable temperature. This is typical thermostat behavior. If we did this, our behavior would count as purposeful, as rational. Why is it merely behavior, not a rational action, not even minimally rational, when the thermostat does it? Because the thermostat's behavior is not—in the *relevant* sense—under the control of what it 'thinks' (represents) about temperature.

What governs the behavior of the thermostat when it turns the heat on? Well, in the case of most thermostats, there is a bimetallic strip that functions as both a thermometer—something that represents or stands for room temperature—and an electrical switch. Its degree of curvature represents room temperature. If it gets cold enough in the room, this metallic strip bends enough to make electrical contact with an adjustable point. This closes a circuit to the furnace—thereby turning the heat on. The thermostat's behavior—its turning the heat on and off—is thus controlled by an internal element, the metallic strip that represents temperature. The thermostat 'senses' a drop in temperature and responds by turning the furnace on. If we ignore the fact that it is *our* desires (not the thermostat's) that are being satisfied by this arrangement, the behavior looks remarkably like a minimally rational act: an internal representation of temperature (something we might loosely interpret as *perception* of temperature) causes the device to behave in some appropriate way. The instrument's behavior is under the control of what it thinks (represents) about the temperature.

If we ignore, for the moment, the fact that the thermostat's representation of temperature isn't quite what we had in mind by speaking of *thought* (about temperature) explaining behavior, there is still an important difference between the thermostat's behavior and, say, our behavior when we turn the heat up when it gets chilly. Our behavior is explained by what we think about the temperature—that it's getting chilly. Even if we credit the thermostat with 'thoughts' about temperature, its behavior isn't explained by *what* it thinks about temperature. The thermostat's behavior is caused by the thermostat's 'thoughts' about temperature, yes, but it isn't explained (as it is with us) by what the thermostat 'thinks' about temperature. To see why, think about why we don't regard a microphone as obedient just because it does what we tell it to do, and does it, moreover, *because* we tell it to do it.

---

3 See Clayton *et al.*, this volume, for a similar claim that the rationality of behavior requires more than that the behavior be caused by a belief and desire; the behavior must, they point out, be caused in the *right* way, an intentional way, by the belief and desire.

I say something into a microphone: I say 'Vibrate rapidly'. In response to my command the microphone's diaphragm vibrates rapidly. This instrument's behavior is determined by the frequency and amplitude of the sounds I produce. These sounds have a meaning. They mean *vibrate rapidly*. The microphone, therefore, does exactly what I tell it to do. Despite this fact, and despite the fact that my telling it to vibrate rapidly causes it to vibrate rapidly, the device's behavior is not governed, it is not explained, by what I say. It is not *what I say* that is relevant to—and, thus, controls—the microphone's behavior. It is the sounds I produce in saying it.

So there is an important difference between being caused by an event that means M and being explained by the fact that it means M. This difference is important for understanding what it takes to be minimally rational. For even if an animal (a computer? a thermostat?) has thoughts and even if these thoughts cause the animal (computer, thermostat) to do something, the animal (computer, thermostat) may not do what it does because it thinks what it does. What it thinks, the content or meaning of its thoughts, may be quite irrelevant. This is especially evident in the case of the thermostat. It is the degree of curvature of the metallic strip, not what this curvature means about temperature, that explains why it closes the electrical circuit to the furnace—thus turning on the heat.

Once we distinguish, as we must, between a meaningful event causing behavior and its meaning explaining the behavior, and once we identify thought's governance of behavior with the latter, machines are quickly disqualified as minimally rational agents. They can do a lot of things, things that would be rational if you or I did them, but the machine does not thereby qualify as rational. Their behavior may be caused by internal representations, but it is not (as it is with us) explained by the way these representations represent things. The same is true of plants.

A plant, the scarlet gilia, changes from red to white in mid July each summer. This is something the plant does, a piece of plant behavior. A plant doesn't have thoughts and desires, intentions and plans, but it does do things, sometimes very interesting things, and botanists are interested in explaining why plants behave in the way they do. Why does the scarlet gilia do this? Why does it change from red to white in the middle of July each year? One explanation (the one given by the botanists from whom I take the example) is that the plant does this in order to attract pollinators (Paige and Whitham 1985). Early in the flowering season, hummingbirds are the chief pollinators, and hummingbirds are more attracted to red blossoms. Later in the season the hummingbirds migrate and hawkmoths, preferring whiter blossoms, become the principal pollinators. According to Whitman and Paige, the plant changes colour in order to exploit this seasonal alteration in its circumstances. It sets more fruit by changing colour, and this is why it does it. If this is, indeed, the correct explanation of why it changes colour, the plant is completely rational in Kacelnik's biological sense of rationality. But is it minimally rational? Is anything it does explained by its inner representations?

In order to reap the benefits, this behavior must occur at a particular time. For maximum advantage it has to occur at the time the hummingbirds leave and the hawkmoths arrive. There must, therefore, be something inside the plant, some botanical clock or calendar, that both indicates when the time is right and acts as a chemical 'switch' to initiate processes that result in a change of color. Causally speaking, this botanical clock plays exactly the same role in the behavior of this plant that the bimetallic strip plays in the behavior of the thermostat. It both represents (accurately or inaccurately as the case may be) the external conditions in which the behavior is supposed to occur (for maximum advantage) and it triggers the events that constitute this behavior.

This plant can be 'fooled'. Unseasonably hot weather and a drought make the internal clock run fast. Chemical changes occur in early June that normally do not take place until July—the time when the hummingbirds leave. Since hummingbirds do not like white blossoms, they ignore the plant. The hawkmoths will not arrive for another 6 weeks, so pollination does not occur. The plant suffers for its 'mistakes'.

When behavior is brought about by internal representations of this sort, we are tempted to use intentional language. I spoke above about the plant being fooled and making mistakes. We do it with instruments and we do it with plants. Botanists succumb to this temptation by explaining the plant's behavior in purposive terms: the plant changes color, they say, 'in order' to attract pollinators. Clearly, though, we do not yet have anything like action, nothing like purpose, nothing that would justify 'in order to', nothing that would qualify as minimal rationality. The internal clock in these plants might *mean* it is July, but the fact that it means this is not relevant to why the plant changes color (I'll say exactly why in a moment).

For the purpose of understanding whether—and if so, how and why—some animal behavior differs from machine and plant behavior, and thus qualifies as minimally rational (and, therefore, possibly rational in the fully normative sense of this term), it will prove useful to look at a very simple case of animal learning. Here, I believe, we first see meaning start to play a genuine explanatory role in animal behavior. It is here that minimal rationality first emerges on the evolutionary scene. It is here that intentional (vs. mechanistic) explanations of behavior (see Clayton, *et al.* this volume for this way of putting it) enter the picture.

Consider, then, a simple case of learned behavior. A foraging bird tries to eat a monarch butterfly. This butterfly has been reared on a toxic form of milkweed. Such butterflies are poisonous and cause birds to vomit. After one nasty encounter, the bird avoids butterflies that look like the one that made it sick. A day later our bird sees a tasty viceroy, a butterfly with an appearance remarkably like that of the noxious monarch. The viceroy, though, is not poisonous. It has developed this coloration as a defense from predatory birds. It mimics the appearance of the monarch so that birds will 'think' that it, too, tastes nasty and avoid it. Our bird sees the viceroy and flies away.

A perfectly tasty meal is ignored by a hungry bird. Why? Why didn't the bird eat the viceroy? Was the avoidance of a tasty meal a rational thing for the bird to do? Was it minimally rational?

We all know—or think we know—why the bird didn't eat the viceroy, but we have to choose our words carefully. If the bug it saw happened to be a poisonous monarch, we could have said that it *recognized* the butterfly as one of those nasty tasting bugs and avoided it because it didn't want to get sick again. That would have been a biologically useful thing to do. But what it saw was not a nasty tasting bug. No recognition took place. There was no knowledge. We need a different word. What is it we call some perceptual state that would be recognition or knowledge if only it were true? Belief! Judgment! Thought! The bird *thinks* (falsely as it turns out) that the bug is an M bug, something that tastes bad. That is why it doesn't eat it. So thought is governing its behavior.

The causal structure of the bird's behavior is remarkably similar to the thermostat and the gilia. The bird sees the butterfly, it discriminates it from other objects, and, after learning, this perceptual state (the one induced in the bird by the viceroy) controls behavior. An element inside the bird—a conscious experience of some sort—tells the bird that an M-type bug is present in the same way something inside the thermostat—the curvature of the bimetallic strip—tells the instrument the temperature is too low and something inside the plant—presumably some chemical process—tells it July has arrived. So there is something inside the bird that indicates or *means* a bug-of-type-M is present, and this element functions as a behavioral 'switch' (triggering avoidance behavior) in the same way corresponding elements inside the thermostat and plant act as switches for appropriate behavior.

Unlike the thermostat and plant, though, the meaning of the bird's internal representation of the butterfly is explanatorily relevant to its behavior. Like the thermostat and plant, this internal representation in the bird (call it R) has both a meaning and a causal role, but, unlike the instrument and the plant, its meaning explains its causal role. R is causing avoidance behavior, it was given that job to do as a result of the bird's unpleasant encounter with M, because it means that a butterfly of type M is present, the sort of object the bird, after its unpleasant experience, wants to avoid. So the causal story looks like this: an R in this particular bird means that an M is present, and this R causes the bird to avoid V (the viceroy) because it means that an M is present. A meaningful state, R, not only causes behavior (this was also true in the thermostat and the plant), but its meaning explains why it is causing the behavior it is. A meaningful state also causes behavior in the thermostat and plant, of course, but in the case of the thermostat it is not *its* (the internal state's) meaning that explains its causal role. It is the intentions and purposes of the instrument's designers and builders that explains why the instrument was wired the way it is—why, therefore, it is now turning the furnace on. In the case of the plant, it is, once again, not the meaning of this plant's internal state (the plant whose behavior we are trying to understand) that explains its causal role in the plant but the meaning of similar meaningful states in this plant's ancestors, the ones from whom it inherited its genetic constitution. The fact that R in this plant means that July has arrived does not explain why this plant changes color. The meaning of this plant's internal states does not (as it does in the case of the bird) explain why the plant is (now) wired the way is and, therefore, why the plant (now) behaves the way it does. It is, rather, the meaning of the internal states of the plant's remote ancestors that explains, via some selectional process, why this plant is now wired in the way it is and, therefore, why this plant now behaves the way it does.

This is the key difference between the plant (machine) and the animal. Animals, at least some of them, sometimes have internal circuits that were reconfigured by a learning process in which the meaning of internal elements played an important causal role. The behaviors that result from such processes are behaviors that are, in part at least, explained by the meanings that helped shape these changes. It is for this reason that the meaning of the bird's internal states—that yonder bug is an M-bug—helps explain the bird's behavior while the meaning of the thermostat's and scarlet gilia's internal states, although present, remains explanatorily inert. The bird is minimally rational while the machine and plant are not.

It should be clear from my argument and examples that minimally rational behavior, behavior that is explained by the meaning or content of the internal states that produce it, cuts across some distinctions that others might prefer. Allen (this volume), for instance, contrasts cognitive with associative explanations of behavior; Papineau and Heyes (this volume) discuss a similar distinction between rational and associative psychological processes. Though Papineau and Heyes don't think this distinction is particularly useful, they along with Allen would, I assume, interpret my bird's behavior as a result of straightforward associational processes and, therefore, as something less than fully rational.

If this is, indeed, the way they would classify things, I see no reason to follow them in carving things up this way. It seems to me that although the bird does not (or may not) engage in any reasoning or inference in avoiding the viceroy, it does do something perfectly rational: it avoids eating something not because it tastes bad (the viceroy doesn't taste bad), but because it 'thinks' it tastes bad. Given its reasonable 'beliefs' (the viceroy looks just like a monarch, a bug that tastes bad), that is a sensible, rational thing to do. It is something I would do, and, if I did it for these reasons, it is something I would expect to be deemed rational for doing. This explanation for the bird's behavior is what Clayton *et al.* (this volume) describe as an intentional (as opposed to a mechanistic) explanation, an explanation that adverts to the content or meaning of the bird's psychological states— what the bird believes and desires. I agree with them in their judgment about what is essential to (minimally) rational action. My purpose in this paper has been two-fold: (1) to say why this condition is necessary for rational action; and (2) to show how this condition is actually satisfied in simple learning situations.

Is the bird's behavior really purposeful? Does the bird really *think* (mistakenly) that the bug it sees tastes bad? Is this really why it avoids the bug? All I have argued, I know, is that in this kind of learning process an internal state that indicates or means something about the animal's external environment comes to play a role in the animal's subsequent behavior, and it comes to play that role because of what it means. (The caching behavior of scrub jays being influenced by their past experience is an even more dramatic example of this process; see Clayton *et al.*, this volume.) The informational content or meaning of this internal, causal, element is, thus, genuinely explanatory. This, I concede, is not sufficient to show that *thought* is governing the acquired behavior in the relevant (explanatory) sense since I have not shown that internal states with meaning of this kind are thoughts. Still, we have here, if not thought itself, a plausible antecedent of thought— an internal representation whose meaning or content explains why the system in which it occurs behaves the way it does (for details on this approach to rational action see Dretske 1988). To my ear, that sounds enough like thought not to haggle about what is still missing.

References

Byrne, R. (2005). Detecting, understanding, and explaining animal imitation. In: S. L. Hurley and N. Chater, eds. *Perspectives on Imitation: from Neuroscience to Social Science*, vol. 1, pp. 225–242. Cambridge: MIT Press.

Dretske, F. (1988). *Explaining Behavior: Reasons in a World of Causes*. Cambridge, MA; MIT Press, A Bradford Book.

Paige, K. N., and Whitham, T. G. (1985). Report of research published in *Science. Scientific American*, **252**: 74.

Chapter 4

Styles of rationality

Ruth Garrett Millikan

Abstract

One way to describe rationality is as an ability to make trials and errors in one's head rather than in overt behaviour. I speculate about two different kinds of cognitive capacities of this sort that we humans seem to have, one of which we may share with many of the animals, the other, perhaps, with none. First, there is a certain kind of rationality that may occur on the level of perception, prior to cognition proper. Second, there is the capacity to form subject–predicate judgements sensitive to a negation transformation, hence subject to the law of non-contradiction. This latter capacity may be what allows humans to learn to represent world affairs that are not of immediate practical interest to them, a capacity that we probably don't share with the animals.

By whatever general principles and mechanisms animal behaviour is governed, human behaviour control rides piggyback on top of the same or very similar mechanisms. We have reflexes. We can be conditioned. The movements that make up our smaller actions are mostly caught up in perception–action cycles following perceived Gibsonian affordances. Still, without doubt there are levels of behaviour control that are peculiar to humans. Following Aristotle, tradition has it that what is added in humans is rationality ('rational soul'). Rationality, however, can be, and has been, characterized in many different ways. I am going to speculate about two different kinds of cognitive capacities that we humans seem to have, each of which is at least akin to rationality as Aristotle described it. The first I believe we share with many other animals, the second perhaps with none.[1]

Traditionally, the paradigm of human 'rational behaviour' is taken to be engaging in Aristotelian practical inference. Practical inference is usually described as reasoning in something like the form of a proof: I desire A, doing B will probably lead to A, therefore

---

[1] Since this chapter derives from a session of the conference on rational animals (Oxford 2002) designated as a 'brainstorming' session, I will take philosopher's license, presenting no more than the softest sort of intuitive evidence for these ideas. The suggestions made in this chapter are spelled out in much more detail in Millikan (2004).

I'll do B. In reality, however, that is not the way practical reasoning goes at all. Rather, it is the way conclusions are justified to other people. The core of actual practical reasoning processes is not like a proof but like a *search* for a proof. Just as in mathematical reasoning you are likely to start with something you would like to prove, in practical reasoning you begin with something you would like to do or to have done and then attempt to construct something like a proof. And how do you construct a proof? This is largely a matter of trial and error. You start with what you would like to prove and work backwards, trying to find plausible steps that might lead to that conclusion, and you start also with things you already know to be true and work forwards to see where these things might lead. You try to fill in the gap between what you find going forward and what you find going backward.

Looked at this way, the emergence of reasoning appears as just one among other examples of the evolution of evolvability, in this case the emergence of a new level of trial and error learning beyond operant conditioning. New levels of trial and error yield quicker ways to adapt organisms to their environments than genic selection. Reasoning is just trial and error in thought. Dennett (1996) calls animals capable of trial and error in thought 'Popperian'. The reference is to Popper's remark that it is better to let one's hypotheses die in one's stead. The Popperian animal is capable of thinking hypothetically, of considering possibilities without yet fully believing or intending them. The Popperian animal discovers means by which to fulfil its purposes by trial and error with inner representations. It tries things out in its head, which is, of course, quicker and safer than trying them out in the world. It is quicker and safer than either operant conditioning or natural selection.

One of many reasonable interpretations of what it is to be rational is that being rational is being a Popperian animal. The question whether any non-human animals are rational would then be the question whether any of them are Popperian. I suppose the cash value of the answer to that question is in the neurology. We would need to inspect the mechanics of thought of the various animals, what kinds of inner representations they employ for what purposes. At the moment we have mostly behavioural evidence on these matters, and it is often controversial how much light the behavioural evidence can be made to cast on underlying structures. Still, based on very informal behavioural evidence, I will suggest a certain way in which it seems plausible that both humans and many of the higher animals are indeed Popperian.

So far as I know, rationality has always been assumed to occur only on the level of cognition, as distinguished from the level of perception. However, I suggest that a certain kind of rationality may occur on the level of perception prior to cognition. A difficulty here, of course, is that exactly what the distinction between perception and cognition consists in, indeed, whether there exists such a definite distinction at all, is controversial. For example perception is generally assumed to include object recognition, which may look rather like the applying of concepts, yet concept application is usually thought of as involving cognition. Both the ventral or 'what' visual and auditory systems and the dorsal or 'how' systems are generally supposed to be involved in perception (Jeannerod 1997,

Norman 2002). Yet recognizing *what* one sees or hears is also thought of as involving cognition. For simplicity, I will take as paradigmatic of perception the production of representations suitable for guidance of immediate action, suitable because these represent for the perceiving animal its own present relation to various world affairs as needed for action. Perception tells where in relation to the animal certain affordances are exhibited, such as relations to things to be picked up, things to hide from, places to hide, things to climb up on, things to eat, and so forth. That is, perception of objects, as perception, is for immediate use in practical activity. If perception involves concepts, they are in the first instance practical concepts, repositories for procedural rather than propositional knowledge, for storing know-how rather than factual information.

The perceptual level, so defined, is a level that involves only 'pushmi-pullyu representations' (Millikan 1996). These are representations that are undifferentiated between being indicative and being imperative, between describing and directing. The simplest examples of pushmi-pullyu representations are signals and signs used to co-ordinate behaviour among conspecifics. Danger signals, for example, tell in one undifferentiated breath when there is danger nearby and when to run or take cover. Bee dances tell in one undifferentiated breath where there is nectar and where other worker bees are to go. Similarly, perception of a predator, for most animals, is perception of where the danger is and perception of which way to run or to take cover or, if distinctions among various kinds of predators are recognized, of other ways to handle the situation depending on the predator perceived. Similarly, perception of a precipice, of a prey, of water when thirsty, of edible plants, and so forth, are perceptions of affordances, potential guides to things that must or might be done by the animal. Each has a directive aspect as well as a descriptive aspect.

I want to suggest that there is such a thing as mental trial and error, hence rationality, at the perceptual level, so defined. I haven't got any fancy animals such as chimps or dolphins or African grey parrots in my laboratory, but we do have grey squirrels in our yard. And we have a bird feeder that hangs on a chain from well under the eves above the deck of our house, hung there to keep it out of reach of the grey squirrels. The grey squirrels are not satisfied with this arrangement. Not long ago I watched one eyeing the feeder from the deck. It studied the situation long and hard from one side of the deck, then from the other. It climbed up on the railing to study the situation from there, first from one side, then from the other side, and then from underneath. It eyed the screen on the door that goes out to the deck. Finally it made a try. Starting from a run along the railing, it leapt and ricocheted off the screen toward the feeder but missed. Once again it surveyed the situation from various angles, and finally succeeded by hitting the screen a little higher up, then hanging on tight to the whirligig feeder while it wound up, unwound, wound up again and unwound. I hadn't the heart to shoo it away!

Now what was going on in that squirrel's head as it sat up on its haunches, studying and studying from this angle and that? What was going on, I suspect, was a sort of trial and error in perception. It was trying to *see* a way up, trying to *see* an affordance.

Similarly, when hiking a steep path, you may take a brief moment sometimes to see the best way to get a leg up. Or before crossing a stream on scattered rocks, it may take you a while to see a good way to cross without wetting your feet. This sort of trying to see a way seems entirely different from practical inference as inference is usually described. Surely the squirrel was not thinking in propositional form, doing syllogisms in its head. But if rationality is the capacity to make trials and errors in one's head, it certainly was exhibiting a form of rationality. This form of rationality seems to be common to humans and, I imagine, many other animals as well.[2]

Moving to the other extreme, I want to discuss a second kind of rationality that is characteristic of humans but that it seems unlikely other animals are capable of. This is the ability to recognize contradictions among representations of facts lying beyond immediate perception and to make corrections in thought accordingly.

My arguments—perhaps I had better say speculations—will unfold as follows.

First, non-human animals generally have no interest in facts that don't pertain directly to practical activity. They do not represent or remember dead facts.

Second, an important criterion of correct recognition of objects and properties needed for guidance of practical activity is success in these practical activities. Perhaps, then, it is not just that non-human animals have no interest in representing dead facts. Perhaps they have no means of developing and testing abilities to recognize objective states of affairs other than through the consequences of practical activities that depend on these recognitional abilities.

Third, I suggest that to learn to recognize dead facts requires the ability to use another sort of test of correct recognition of objective objects and properties, hence of objective states of affairs, namely, triangulation, or the convergence or agreement of a variety of different methods of recognizing the same objective state of affairs. But recognizing agreement implies the ability also to recognize disagreement, hence the ability to recognize contradictions in thought.

Fourth, I suggest that the emergence of useful contradiction in thought depends on the development of representations in thought (and perhaps in language) that have subject–predicate structure and that are sensitive to internal negation. This structure allows thought openly to display the ways it is representing the world as coherent or incoherent prior to possible uses of those representations in helping to govern behaviour. It offers a criterion of correct reidentification of objects and properties that does not depend on success or failure in practical activity.

Fifth, this development produces a more sophisticated Popperian animal, one that can make trials and errors *in its attempts to represent the world* prior to risky employment

---

2 Editors' note: for other discussions of the relationship between simulation and rationality in this volume, see chapters by Hurley, Proust, and Currie.

of those representations as guides to action. I suggest that it is unlikely that non-human animals are capable of this properly propositional form of mental representation. I will take up these various points in turn.

(1) It is true that non-human animals may learn or remember, may systematically store away, knowledge of the layout and of many significant features of the geography of the locales in which they live, knowledge of conditional probabilities among events significant for the animal, knowledge of hundreds of places in which they have cached food, knowledge of the social hierarchy of the group in which they live, and so forth, and that all of these things may be learned prior to use of this information to govern rewarding behaviour. But these kinds of knowledge seem all to have been determined in advance by the experience of the species as useful in guiding practical activities of importance for survival. Moreover, this knowledge is typically called on only in contexts in which, according to the experience of the individual or the species, it has immediate uses of predetermined kinds. Non-human animals may also collect many skills out of context of immediate use but, again, these skills seem always to be of a kind their evolutionary pasts have shown to be significant. I am thinking here of play, which seems to occur in all mammals at least, but in non-human animals seems always to be practice for well-defined species-typical adult activities. Similarly, through rigorous and careful step-by-step training by humans, individuals of many higher species can laboriously be brought to recognize perceptual affordances of kinds quite remote from any they were specifically designed to learn. Recognizing these affordances may involve recognizing properties and kinds of objects with no history of relevance to the animal's species. But they seem to be recognized only as things that have proved useful in the individual's previous experience and only as affording those known uses.

What this suggests is that what non-human animals actively represent is, primarily or exclusively, affordances. Their active inner representations are primarily pushmi-pullyu representations, rather than representations of dead facts. The other side of this coin would naturally be a failure of the animal to represent what to do dissociated from perceived possibilities of implementation. Pure goals would not be represented torn apart from the perception of affordances directing the animal towards those goals, or at the very least, perception of what will support searching behaviours, designed to raise the probability of encounter with more rewarding affordances. Thus the hungry animal perceives aspects of its environment as for traversing, or for sniffing, or for searching with its eyes, these behaviours being designed to bring it into contact with more direct food-tracking affordances, and so forth. Motivation would always be directly grounded in perception, including perception of the animal's interior, of course, of its current needs as well as its current opportunities. Accordingly, Merlin Donald says of the behaviour of apes:

> complex as it is, [it] seems unreflective, concrete, and situation-bound. Even their uses of signing and their social behaviour are immediate, short-term responses to the

environment...Their lives are lived entirely in the present, as a series of concrete episodes...(Donald 1991, p. 149)

and

...the use of signing by apes is restricted to situations in which the eliciting stimulus, and the reward, are clearly specified and present, or at least very close to the ape at the time of signing (1991, p. 152).

The pushmi-pullyu animal solves only problems posed by immediate perception. It does so by deciding from among possibilities currently presented in perception, or as known extensions from current perception, as in knowingly moving from a known place toward another place known to afford what the animal currently needs.

Human beings, on the other hand, spend a great deal of time collecting both skills and pure facts that no experience, either of the individual or the species, has yet shown to be of any relevance to practical activity. Children practice hula hoops, Rubik's cubes, wiggling their ears, cracking their knuckles, standing on their hands, and turning around to make themselves dizzy without falling down. People memorize baseball scores and batting averages and, some of them, timetables for railroads all over the country. They are capable of learning thousands of facts about what has occurred or is occurring at times and in places to which they have no potential access, let alone past or present practical acquaintance. They are curious about what will cause what and why, wholly apart from any envisioned practical applications for this knowledge. They may be curious about how things work, where they came from, what properties and dispositions they have, in a completely disinterested way. And, of course, many of these interests may eventually bear unforeseen fruit. The adage is that if you keep a thing for 7 years, it will eventually find a use. Having stored enough tools and materials in the attic over the years, some of it is eventually bound to come in handy, granted one is an inventive enough tinker with ideas. Dead facts can come alive. The disposition to collect skills and facts that have no foreseen uses is a disposition that has foreseen uses, foreseen through a history of natural selection.

(2) My task now is to make it plausible that collecting dead facts requires a skill that is different from those needed for collecting merely live facts, and that this skill plausibly rests on the development of propositional structure in thought and the use of internal or predicate negation. The central idea here is that the ability to recognize or reidentify the *same* distal object, the *same* significant kind of distal object, or the *same* objective distal property again when these are encountered in different perspectives relative to the observer, or evidenced through different intervening media, or evidenced to different senses, is an extremely difficult, far from trivial skill. Learning any kind of practical technique for interacting productively with the environment requires, of course, that one be able recognize the objective features of the situation of action that previously were relevant to success on new occasions. Having discovered a technique for opening hazelnuts or for escaping from foxes is of no use to the red squirrel or its species unless it is possible to recognize on new occasions when a hazel nut has been found or when a

fox is encountered. But the variety among proximal stimuli that may indicate the same relevant kind of distal situation is enormous. The capacity to represent, unequivocally, that a hazelnut is present and where, or that a fox is present and where, will typically rest on the capacity to translate a wide diversity of proximal stimuli that may proximally manifest presences of these things into mental representations of univocal affordances.

How does the animal, or the species, learn to select out just those members of the set of proximal stimuli that indicate hazelnut or fox? My suggestion is that when perception is used in the guidance of immediate practical activity, the criterion of correct recognition of affording objects or properties lies in practical successes. Roughly, you are right that this is the same affording object, or kind, or property again if you can successfully deal with it in the same way again. The proof of the pudding is in the eating. But if this answer is correct, it opens wide a second question. How does a human learn to recognize, through diverse manifestations, new objects, kinds, and properties that have, as yet, no practical significance for it?

To find this an interesting question, of course, you have to make some realist assumptions. You have to assume that nature has some say in what can count as the same property again or the same object or the same theoretical kind. Humans are not free to determine randomly what will count as objectively the same thing again when encountered on other occasions. This is evident, of course, when reidentifying is for the sake of practical learning. It has been considered more problematic in theoretical as distinguished from practical contexts. In Millikan (1984, Chapters 15–17; Millikan 2000, Chapters 1, 2, and 7) I argued for a realist interpretation of basic objects, kinds, and properties, and I cannot repeat those arguments here. Rather, I will run on the common sense assumption that the question whether the same individual object has been reencountered usually has an objective answer, and that whether the same property, or at least a property within a close range, has been reencountered generally has an objective answer. And I will assume that there are real kinds. These are not kinds that divide the world of individual objects into mutually exclusive categories. But they are not defined by arbitrary sets of necessary and sufficient properties either. A real kind covers instances that have numerous properties in common for the same underlying reason, for example, because the inner constitution of these instances is the same and numerous superficial properties causally depend on these underlying properties, or because the members of the kind are historically connected, perhaps by having been reproduced or copied from one another, (Millikan 2000, Chapter 2).

Thus the question whether the same real kind has been reencountered often has an objective answer. Our question then is how an organism learns to recognize the same objective individual, property, or real kind through the wide diversity of its manifestations to the various senses, through a variety of intervening media, as encountered in a wide variety of different orientations to the organism. The question becomes particularly acute when we notice that humans, unlike non-human animals, apparently manage to identify novel kinds of events and states of affairs at huge spatial and temporal distances from them, as in collecting information about historically remote events, or events on the other side of the world or in outer space, or inside atoms.

(3) The correct answer, I suggest, is the traditional one, that the test of truth in theoretical as distinguished from practical knowledge is non-contradiction. Our ability to collect merely theoretical knowledge depends on the disposition to adjust beliefs and methods of identifying used in forming beliefs until these are consistent or, putting this more perspicuously perhaps, until they are stable. This is done by employing as many methods as possible of triangulation, attempting to arrive at the same belief by many methods. If the same belief is confirmed by sight, by touch, by hearing, by testimony, by various inductions one has made and by theories about what ought to be so or at least might be so, this is a good test not only for the objectivity of the belief but for each of the general methods one has employed in identifying and reidentifying the objects, properties, relations, and so forth that the belief concerns. The same object that is square as perceived from here should be square as perceived from there and square by feel and square by checking with a carpenter's square and square by measuring its diagonals. If a person knows French when I find him today, that same person should know French when I find him tomorrow and as inferred from such facts as that he buys *Le Monde* every Saturday. If Sadie arrived in Germany on March 22 as determined by noting the plane on which she left Boston, she should have been in Germany on March 22 as determined, for example, from what she says later and from what those who went with her say, and from what the immigration records and credit card bills show. That beliefs are reconfirmed by use of a variety of methods for checking their truth is evidence both for the objectivity of their subject matters and for the reliability of the general methods we use in reidentifying the sorts of objects, kinds, and properties they concern.

More important than confirmation, however, is that the beliefs formed by these methods tend not be *contradicted* by further experience. Failure to reconfirm a belief is not evidence against it. If I look again from another angle and fail this time to see that the object is square, this is not evidence against its being square. For perhaps I can't see the object at all, or although I see it, perhaps I can't make out its shape from here or against the light. To have evidence against its being square, first I must see it, and then I must see its shape and then, further, I must see that this shape is some contrary of square, such as round or oblong, and I must understand that round or oblong is incompatible with square. For a judgement to be said to remain stable, it must be possible that it should have been rendered unstable. And this requires that the thinker be capable of recognizing evidence for the truth of its contraries, and grasp that these contraries are incompatible with it.[3]

(4) The upshot, I am suggesting, is that the capacity to learn, out of the context of practical activity, to recognize what is objectively the same object, kind, or property again rests

---

[3] For much more detail on this subject, see Millikan 1984, Chapters 14–17.

on the capacity to form representations with subject–predicate structure, where certain predicates are understood as contrary to one another[4] so that contradiction is possible. Notice that simple representational systems do not contain contrary representations. Signals used to alert conspecifics to danger, for example, do not have contraries. A dozen danger signals at a dozen times and places do not contradict one another. Perhaps there really is that much danger around. Nor does one bee dance contradict another. There may well be nectar both those places. Similarly, perceptual representations telling of a variety of different affordances at different places do not contradict one another. The animal may not be able to avail itself of all those affordances at once. It may have to make choices. But the perceptual representation of one affordance doesn't contradict the perceptual representation of another.

Perceptual representations don't contradict one another because what is perceived is relations that affording objects and situations have to the perceiver *as from here and now*. Simply as perceived, what has a particular relation to the perceiver as from here and now is intrinsically unstable. Perceptual representations need to be updated continually. But updating one's perceptions is not changing one's mind. That certain objects and properties are here and there relative to me now does not conflict with there being different objects and properties here and there relative to me at other times. A representation of an affording situation as from here and now is not a representation with an articulate subject term ready to be stored away for potential use or re-evaluation on other occasions. Not that a particular time, place, and perspective couldn't in principle be a subject of judgement. But to represent a particular time, place, and perspective as a subject of judgement would require that one grasp possible ways of reidentifying that subject, for example, through evidence from the testimony of other people, or through theories by which one attempts to reconstruct past events. This kind of grasp of objective reidentifiable locations in space-time is not given merely in perception.

(5) My suggestion, then, is that the capacity to adjust beliefs until they are consistent, hence the capacity to think in subject–predicate form where the predicate is sensitive to negation, is needed primarily by an animal that reconstructs in thought large portions of its world that it has not yet dealt with in practice. This is a form of rationality that it seems less likely that non-human animals achieve.

References

Dennett, D. C. (1996). *Kinds of Minds*. New York: Basic Books.

Donald, M. (1991). *Origins of the Modern Mind*. Cambridge, MA: Harvard University Press.

Jeannerod, M. (1997). *The Cognitive Neuroscience of Action*. Oxford: Blackwells.

Millikan, R. G. (1984). *Language, Thought and Other Biological Categories*. Cambridge MA: MIT Press.

---

[4] On certain grounds. See Millikan 1984, Chapter 16.

Millikan, R. G. (1996). Pushmi-pullyu representations. In: J. Tomberlin, ed. *Philosophical Perspectives*, vol. 9, pp. 185–200. Atascadero CA: Ridgeview Publishing. [Reprinted in L. May and M. Friedman, eds. (1996). *Mind and Morals*, pp. 145–161. Cambridge, MA: MIT Press.]

Millikan, R. G. (2000). *On Clear and Confused Ideas*, Cambridge: Cambridge University Press.

Millikan, R. G. (2004). *Varieties of Meaning; the Jean Nicod Lectures 2002*, Mit Press.

Norman, J. (2002). Two visual systems and two theories of perception: An attempt to reconcile the constructivist and ecological approaches. *Behavioral and Brain Sciences*, **25**: 73–144.

Chapter 5

Animal reasoning and proto-logic

José Luis Bermúdez

Abstract

This chapter addresses a theoretical problem that arises when we treat non-linguistic animals as thinkers in order to explain their behavior in psychological terms. Psychological explanations work because they identify beliefs and desires that show why the action in question *made sense* from the agent's perspective. To say that an action makes sense in the light of an agent's beliefs and desires is to say that it is the rational thing to do (or, at least, *a* rational thing to do) given those beliefs and desires. And that in turn means that, in at least some cases, an agent might *reason* her way from those beliefs and desires to acting in the relevant way. Most models of reasoning, however, treat it in terms of logical operations defined over linguistic structures, which makes it difficult to see how it might be extended to non-linguistic creatures. This paper develops a framework for thinking about the types of reasoning engaged in by non-linguistic creatures. It explores non-linguistic analogs of basic patterns of inference that can be understood at the linguistic level in terms of rules of inference involving elementary logical concepts. The three schemas discussed (reasoning from an excluded alternative and two types of conditional reasoning) are highly relevant to animal practical reasoning, and I show how animals might apply them without deploying any logical concepts.

We find ourselves committed to providing an account of animal reasoning as soon as we grant that forms of animal behavior require psychological explanation—as soon as we grant that in certain situations animals behave the way they do because of their beliefs about their environment and about how best to achieve their goals.[1] Psychological

....

[1] The practice of applying belief–desire psychology (Kacelnik's 'PP-rationality') to animals (and prelinguistic infants) is well established in cognitive ethology, comparative psychology, and developmental psychology. In this paper I will be assuming the legitimacy of this practice. I have explored it and defended it against various philosophical objections in Bermúdez 2003. See also and compare Dickinson and Balleine 1993; Heyes and Dickinson 1995; and Kacelnik; Dretske; Millikan; and Hurley, this volume.

explanations work because they identify beliefs and desires that show why the action in question *made sense* from the agent's perspective—just as psychological predictions work by showing the course of action that *makes sense* in the light of the agent's beliefs and desires. To say that an action makes sense in the light of an agent's beliefs and desires is to say that it is the rational thing to do (or, at least, *a* rational thing to do) given those beliefs and desires. And that in turn means that, in at least some cases, an agent might *reason* her way from those beliefs and desires to acting in the relevant way. Reasoning and rationality are correlative notions, at least where rational behavior is understood to mean more than simply adaptive behavior.[2]

Studying animal reasoning is important both for the light it sheds upon animal cognition and animal behavior, and for the way it helps us to understand what is distinctive about the types of reasoning that are made available by language mastery (henceforth: linguistic reasoning)—and hence, by extension, for how we understand the relation between psychological explanations as applied to the behavior of linguistic and non-linguistic creatures respectively. It is clear that an account of animal reasoning will have to strike a delicate balance. On the one hand, there must be sufficient parallels between animal reasoning and linguistic reasoning for comparable models of belief–desire explanation to be applicable in both cases. On the other hand, however, such an account must be sensitive to the significant differences between linguistic and non-linguistic cognition. The acquisition of language makes available types of reasoning that are impossible in the absence of language.

I have argued elsewhere that logic requires language (Bermúdez 2003, Ch. 9). Language offers the possibility of *intentional ascent*—of thinking about thoughts. A thought can only be 'held in mind' in such a way that it can be the object of a further thought if it has a linguistic vehicle. Only when it has a linguistic vehicle can the internal structure of a thought be manifest. If this is right, then those types of reasoning that exploit the internal structure of a thought are only available to language-using creatures. Paradigmatic here is the quantificational reasoning typically formalized in the *predicate calculus*. This is the reasoning that allows us to conclude, for example, that *this F* must be *G* because all *F*s are *G*—or that, since *a* is *F*, at least one thing is *F*. But the thesis of language-dependence holds also for types of reasoning that do not exploit the internal structure of thoughts. Consider the types of inference that involve logical concepts such as *disjunction* (...or...) and *material implication* (if...then...). The validity of these inferences (typically formalized in the *propositional calculus*) is a function of the truth-values of the

---

2 Editors' note: Bermúdez is here primarily concerned to make out the possibility of proto-reasoning in non-linguistic animals, rather than with evidence that it is actually present. Compare the skepticism of, for example, Kacelnik (this volume) about the empirical tractability of something like the concept of PP rationality/belief–desire rationality applied to animals. An interesting question is whether evidence about proto-reasoning in Bermúdez' sense may be more forthcoming than evidence about full-fledged PP-rationality.

thoughts featuring in them (as opposed to the internal structure of those thoughts). The inference-schema of *modus ponens* (that allows one to conclude q from $p{\rightarrow}q$, and p) is valid just as long as it is not the case that p is true and q false. Making these truth-functional inferences requires understanding the relation between the truth-values of thoughts. But thinking about the truth-value of a thought is a form of intentional ascent. It requires holding the thought in mind and ascribing to it a higher-order property.

The challenge, therefore, in developing an account of animal reasoning is to identify forms of reasoning at the non-linguistic level and then explain them without assuming that the animal is deploying elementary logical concepts or exploiting the internal structure of a thought. I shall discuss three such basic types of reasoning that it would be natural (when thinking about language-using creatures) to characterize in terms of mastery of certain primitive logical concepts. As we shall see, there is an alternative way of understanding them at the non-linguistic level—in terms of what we might term *proto-logic*.

The first type of reasoning can be described as reasoning from an excluded alternative. This is the type of inference that takes a creature from recognition that one of an incompatible pair of states of affairs holds to the recognition that the other does not hold. Here is an example. Imagine a creature that has learnt that the lion and the gazelle will not be at the watering hole at the same time and, moreover, is in a position to see that the gazelle is drinking happily at the watering hole. The creature can conclude with confidence that the lion is not in the vicinity. This type of reasoning is one of the ways in which a creature can learn about what is not immediately perceptible. One can see easily, for example, how this sort of inference could be life-preserving for a creature that is just as threatened by the lion as the gazelle is. It is natural to formalize it in the propositional calculus as an instance of disjunctive syllogism (the transition from 'A or B' and 'not-A' to 'B', where 'A' stands for 'The gazelle is not at the water-hole' and 'B' for 'The lion is not at the water-hole').

A second such way of moving beyond the here-and-now comes with straightforward conditional reasoning (typically formalized as *modus ponens*). This is the reasoning that takes one from recognition that there is a conditional dependence between two states of affairs (the second will be the case if the first is the case) and recognition that the first state of affairs is indeed the case to the conclusion that the second state of affairs is the case. Conditional reasoning of this type is deeply implicated in a range of different activities. The detection of patterns of behavior seems closely bound up with the possibility of conditional reasoning. A creature that knows that if the gazelles see the lion they will run away and that recognizes (perhaps on the basis of its understanding of the gazelles' visual perspective) that the lion will shortly be detected by the gazelles, is in a position to predict that the gazelles will soon take flight.

The third fundamental type of inference is also based upon recognition of a conditional dependence between two states of affairs—but in this case (formalized in terms of *modus tollens*) the reasoning proceeds from recognition that the second state of affairs does not hold to recognition that the first state of affairs is not the case. So, to stick with the gazelles, an observer (perhaps a fellow predator) who is too far away to have a view about the visual perspective of the gazelles can infer from the fact that they are happily feeding where they are that they have not yet seen the lion.

In standard propositional logic these three fundamental forms of inference are understood in terms of the three propositional operators of disjunction, negation, and the material conditional. All three operators are functions from propositions to propositions that take complete thoughts as both arguments and values. Clearly, if we are to find analogs of these three types of reasoning at the non-linguistic level then we will need to find ways of understanding them so that they do not involve propositional operators.

We can begin by simplifying the problem. Reasoning from an excluded alternative can be understood as a form of conditional reasoning. Let 'A' stand for the sentence 'The gazelle is not at the watering hole' and 'B' for the sentence 'The lion is not at the watering hole'. The disjunction 'A or B' is truth-functionally equivalent to the conditional 'If not-A, then B' (that is to say, 'If the gazelle is at the watering hole, then the lion is not at the watering hole'). Both sentences will be true just if it is not the case that the gazelle and the lion are both at the watering hole.[3] The process of reasoning from an excluded alternative will involve a grasp, on the one hand, of the conditional 'If not-A, then B' and, on the other, of the antecedent of that conditional (where that antecedent is in some sense negative). All we need, therefore, is to find a way of understanding analogs for negation and the conditional that do not operate upon complete thoughts.

Let us start with negation. Modern, that is to say post-Fregean, logic is founded on the idea that, as far as the fundamental logical form of sentences is concerned, the linguistic act of negation applies essentially to sentences—and correlatively, at the level of thought, that negation is a logical operation upon propositions. It may seem that, in the sentence 'Socrates is not wise' (and still more so in the sentence 'Socrates is unwise'), a particular property, the property of wisdom, is being held not to apply to Socrates. However, the surface form of natural language sentences is deceptive. The negation operator actually applies at the level of the sentence 'Socrates is wise' rather than at the level of the predicate '—is wise'.[4] The sentence 'Socrates is unwise' is a sentence that is true just if the sentence 'Socrates is wise' is false.

Many philosophers have thought that this claim obscures an important distinction. The sentence 'Socrates is wise' can be false in circumstances in which Socrates does not exist—such as now, for example. Yet these are not, many have thought, circumstances in

---

[3] I am taking 'or' in its inclusive sense, according to which a disjunction remains true if both its disjuncts are true. The exclusive sense of 'or' can be defined by adding to the inclusive sense the further requirement that the two disjuncts not both be true. This requirement can itself be given a conditional reading. 'Not-(P and Q)' is equivalent to 'If P, then not-Q'.

[4] For an extended discussion of this view of negation see Frege 1918–1919.

which it would be appropriate to say that Socrates is unwise. One way of putting the point would be to say that, whereas the two sentences 'Socrates is wise' and 'Socrates is unwise' are contraries (that is, they cannot both simultaneously be true), the two sentences 'Socrates is wise' and 'It is not the case that Socrates is wise' are contradictories (that is, one or other of them must be true). This, in fact, is how the distinction between predicate negation and sentential negation was originally put by Aristotle in the *Prior Analytics* (I.46). Aristotle insisted, and in this he was followed by almost all logicians until Frege, that there is a fundamental logical difference between negating a sentence and negating a predicate.

There is no need to go into the question of whether the distinction between predicate negation and sentential negation is a genuine logical distinction (as opposed, for example, to a distinction in the pragmatics of ordinary language best accommodated at the level of conversational implicature)—or the related question of whether a Fregean or an Aristotelian account of negation is a better way to understand how negation operates in ordinary language.[5] For present purposes the important point is that the distinction between predicate negation and sentential negation gives us a way of understanding negation (or rather, *proto-negation*) at the non-linguistic level as involving a thought with a negative predicate (subject to the qualifications to be noted in the next paragraph)—as opposed to the truth-functional construction of a complex thought. In terms of understanding animal reasoning, the problem of understanding how a creature without language can be capable of negation becomes the problem of how a creature without language can think thoughts in which the predicate component is one rather than the other of a pair of contraries. The task becomes one of understanding how the non-linguistic creature can grasp pairs of concepts that are contraries—the concepts of presence and absence, for example, or of safety and danger, or of visibility and invisibility.[6] To return to our observer at the water-hole, we should understand the thought that the gazelle is not at the water-hole as the thought that the gazelle is absent from the water-hole—rather than as the denial of the thought that the gazelle is at the water-hole. Such a thought would be the contrary of the thought that the gazelle is at the water-hole—but it would not be constructed from that thought in a truth-functional manner.

This line of thought opens up a possibility that is not recognized in Millikan's contribution to this volume. Millikan is exploring the thesis that that the representations of non-linguistic creatures are all what she terms *pushmi-pullyu* representations (that is, representations exclusively from the animal's point of view of the affordances that

---

[5] For extended discussion of these and related matters see Sommers 1982, and Grice 1989. Bochvar 1981 provides a formal development of the idea that there are two fundamentally different types of negation.

[6] I am assuming, for the sake of simplicity, that the mental representations of animals should be described as concepts. I have argued elsewhere that concept possession requires language (Bermúdez 1998). Those who agree can translate the claims in the text about concepts into their chosen vocabulary.

the distal environment presents for action and reaction). She draws a sharp distinction between such pushmi-pullyu representations and those types of thinking that move beyond the practical sphere into the realm of the theoretical. Theoretical reasoning, according to Millikan, is thought about objectively existing, independent, and reidentifiable particulars and properties. Thinking of this type requires subject–predicate structure and a negation operator. Millikan comments:

> ... the capacity to learn, out of the context of practical activity, to recognize what is objectively the same object, kind, or property again rests on the capacity to form representations with subject–predicate structure, where certain predicates are understood as contrary to one another[7] so that contradiction is possible. Notice that simple representational systems do not contain contrary representations. Signals used to alert conspecifics to danger, for example, do not have contraries. A dozen danger signals at a dozen times and places do not contradict one another. (This volume, pp. 124–5)

According to Millikan, then, contrariety cannot be perceived—and nor does it manifest itself in successive perceptual representations (even if these are incompatible with each other, this is evidence only of change not of contrariety).

It seems to me, however, that Millikan does not recognize the possibility of exploiting and acting upon contrariety without explicitly understanding and representing contrariety. I am fully in agreement with Millikan's claim that non-linguistic animals cannot understand the notion of contrariety.[8] An understanding of contrariety is a highly complex cognitive achievement, simply because understanding that two propositions are contraries involves understanding that it is not possible for them both to be true at the same time, and hence requires not simply being able to think about truth-values but also about time and modality. However, a creature can master pairs of contrary concepts (such as the concepts of presence and absence) and deploy those concepts in inferences using proto-negation without a full understanding of the notion of contrariety. It is no more plausible to think that the effective deployment of contrary concepts requires a theoretical grasp of the concept of contrariety than it is to demand that the effective deployment of number concepts requires a theoretical grasp of the concept of number.

Proto-negation, understood in terms of contrariety, permits primitive versions of the two basic types of inference involving negation that I identified earlier. The first type of inference involves reasoning from an excluded alternative. Let us consider the earlier example. We are trying to characterize how a creature might reason from the thought that the gazelle is at the watering-hole to the thought that the lion is not at the watering hole.

---

7 On certain grounds. See Millikan 1984, Chapter 16.

8 Our grounds, however, are not the same. Chapter 3 of Bermúdez 2003 argues against what I term the minimalist conception of non-linguistic thought (roughly, the thesis that the thoughts of non-linguistic creatures should be understood in perceptual terms). It is true that much of animal cognition can be understood in terms of Millikan's 'pushmi-pullyu' representations, but there are very significant forms of animal cognition that cannot be understood in those terms.

The reasoning here can be assimilated to standard conditional reasoning by treating the central premise as a conditional—namely, the conditional that if the gazelle is at the watering-hole then the lion is not at the watering-hole. The notion of proto-negation shows how this can be understood without deploying propositional negation. The conditional in question becomes 'If the gazelle is present (at the watering-hole) then the lion is absent (at the watering-hole)'. Deploying this thought (apart from the need, to be explored further below, to develop a non-linguistic analog of the truth-functional conditional operator) is a matter of reasoning practically in accordance with the fact that presence and absence are contrary concepts. Any creature that reasons in this way will also be able (again subject to a satisfactory account being given at the non-linguistic level of conditional reasoning) to undertake reasoning approximating to *modus tollens*. Starting with the conditional 'If the gazelle is present (at the watering-hole), then the lion is absent (at the watering hole)' such a creature will be able to proto-negate the consequent by forming the thought that the lion is present and hence to arrive at the proto-negation of the antecedent (namely, 'The gazelle is absent').

This inference is not valid in virtue of its form in the way that an instance of *modus tollens* is valid in virtue of its form. There is no formal rule that will take one from the premises of the argument to the conclusion—since the transition from premises to conclusion works only because of the particular pair of contrary concepts involved. It is, moreover, a highly specialized form of practical reasoning, applicable only where the creature in question has the appropriate pairs of contrary concepts. Nonetheless, instances of this inference-schema are of course valid in the semantic sense—that is to say, their premises cannot be true and their conclusion false.

This discussion of proto-negation still leaves us with an important challenge. We need an analogous way of understanding how some precursor of the conditional operator can operate at the non-linguistic level. The conditional operator is a truth-functional propositional operator forming a complex thought from two thoughts in such a way that the complex thought is true in all circumstances except those in which the first component thought is true and the second component thought is false. This requires intentional ascent and hence is unavailable at the non-linguistic level. Conditional thought clearly links two different things. But if those things cannot be complete thoughts, then what can they be?

The proposal I would like to explore is that we look for the sources of conditional reasoning in a primitive form of causal reasoning. Whereas conditional reasoning involves a propositional operator establishing a truth-functional relation between complete thoughts, causal reasoning works on the basis of a causal condition holding between one state of affairs and another. Since causal relationships do not hold between complete thoughts, an understanding of causality presupposes no intentional ascent, and hence does not require language. Causal reasoning, in the sense in which I understand it, should be distinguished from the type of cognition involved in instrumental conditioning. Instrumental conditioning (as discussed in this volume, for example, in

Chapter 7 by Allen and Chapter 8 by Papineau and Heyes) depends upon a creature's 'registering' a contingency between its own behavior and changes in the environment (the contingency between pecking and food delivery, for example). In cases where this contingency is causal (as it is in most cases of instrumental conditioning), exploiting the contingency involves a form of causal reasoning. The converse does not hold, however. The causal relations exploited in causal reasoning need not be in any sense related to the agent's own causal powers (which is why the representations involved in causal reasoning will not always count as pushmi-pullyu representations in Millikan's sense).

Philosophers frequently reflect upon the relations between causation and conditionals, and it is often suggested that a proper understanding of conditionals will be an ingredient of an adequate account of causation.[9] At the very least, a causal explanation of a particular event entails certain conditional predictions about what would happen in suitably similar situations.[10] It is clear that, in the order of analysis, thought and talk about conditionals is more fundamental than thought and talk about causation. No one has ever proposed that we understand conditionals in terms of causation. Conditional sentences and conditional thoughts assert the existence of dependence relations and causation is just one of a range of possible dependence relations. Nonetheless, the order of acquisition frequently fails to duplicate the order of analysis. It is highly plausible on experimental and observational grounds that the capacity for causal cognition is very widespread in the animal kingdom and available at a very early stage in human development (Leslie 1982, and the essays in Sperber 1995)—which is exactly what one would predict on evolutionary grounds. The ability to detect certain types of causal regularity and to distinguish genuine causal relations from accidental conjunctions has obvious survival value. Causal dependence relations (which may hold between the agent's behavior and changes in the environment, or between agent-independent states of affairs in the environment) are directly observable, highly salient, and pragmatically significant in a way that no other dependence relations are. It seems plausible both that causal relations should be more primitive than conditional relations and that a creature arrives at an understanding of conditional dependence by abstracting away from the more familiar and everyday relation of causal dependence. Perhaps the child's first step towards an understanding of conditional dependence is observing that a certain relation holds between the truth-values of two separate thoughts when a causal relation holds between the states of affairs that those thoughts characterize. Once this first step (which of course involves intentional ascent and hence requires language) has been taken, it is a relatively straightforward matter to

---

[9] The most radical proposal in this area is that singular causal claims of the form 'event c caused even e' can be analyzed in terms of counterfactual conditional claims of the type 'had event c not occurred, even e would not have occurred'. A counterfactual theory of causation is proposed in Lewis 1973.

[10] These predictions are most frequently viewed as involving (non-truth-functional) subjunctive conditionals (about what would or would not happen), rather than truth-functional indicative conditionals. There is a useful discussion of causation and conditionals in the first chapter of Mellor 1995.

notice that there are other types of situation (and correlatively other types of dependence relation) that share that same feature. And thus the abstract concept of conditional dependence is grasped.

How might causality be understood by non-linguistic animals? Certain aspects of the full-fledged concept of causation are clearly unavailable at the non-linguistic level. The full-fledged understanding of causation has a modal dimension that comes with the thought that a cause is sufficient for the effect it brings about, and this is, in effect, the thought that it is not possible for the cause to occur without the effect occurring. On the assumption that modal thinking involves a type of intentional ascent and hence requires semantic ascent,[11] the notion of sufficiency is not available at the non-linguistic level. Different theorists will view this with different degrees of concern. Some analyses of causation take the idea that a cause is sufficient for its effect as central (e.g. Mackie 1965). Other accounts do not. If we follow Mellor (Mellor 1995) in holding that what makes it the case that one fact, c, causes another fact, e, is that the conditional probability of e given c is greater than the conditional probability of e given not-c, then cases in which causes really are sufficient for their effects cease to be central to the understanding of causation. But there is no suggestion that non-linguistic creatures can have a full understanding of causation. The proposal is simply that (at least some) non-linguistic creatures have a basic capacity to track causal relationships holding between events or facts and that this basic capacity allows them to engage in a primitive form of conditional reasoning.

All accounts of causation, from David Hume's pioneering account onwards, are agreed that certain forms of regularity are at the heart of the notion. And it seems overwhelmingly plausible that the core of the understanding of causation at the non-linguistic level is based on registering regularities in the distal environment. It is easy to see where this type of understanding might originate. On the one hand, it seems plausible to take a sensitivity to environmental regularities to be a basic part of the innate endowment of any creature capable of learning about the environment. On the other, one might expect any creature to be peculiarly sensitive to regularities between its own actions and ensuing changes in its immediate environment. Of course, as regularity theories of causation have been forced to acknowledge, there are many regularities that are not causal, and it is in the capacity to distinguish genuinely causal regularities from accidental regularities that one might expect differences between different species of non-linguistic creature and, for that matter, different stages of development within any given species.[12] We have evidence from dishabituation studies of infant perceptual expectations about object behavior that, even at only a few months old, they show surprise at examples of 'action at a distance' and many developmental psychologists have suggested that their understanding of the physical

---

[11] For an argument for this assumption see section 9.4 of Bermúdez 2003.

[12] See Call's contribution to this volume for empirical evidence on how non-arbitrary causal relations enhance learning in great apes.

world is governed by the principle that objects can only interact causally when they are in physical contact (Spelke 1990). If correct, this suggestion indicates that the simple association of two events cannot be sufficient for registering causal dependence. At a minimum, the associated events must be spatio-temporally contiguous.

The regularities to which non-linguistic creatures are sensitive (unlike those usually stressed in regularity analyses of causation) need not be exceptionless. Indeed, there are very good evolutionary reasons why one would expect causal cognition to be sensitive to probabilistic regularities (Brunswik 1943). Decision-making in the wild is decision-making under uncertainty and no creature that waited for an exceptionless regularity would fare well in evading predators and obtaining food. This is another reason not to be concerned about the unavailability at the non-linguistic level of the idea that causes are sufficient for their effects. It may well be the case that every apparent example of probabilistic (or indeterministic) causation at the macro-level can be explained in terms of hidden variables, so that what looks like a probabilistic regularity is really a manifestation of a deeper underlying exceptionless regularity. If this were the case then the only reasons for adopting a probabilistic understanding of causation would be apparent examples of indeterministic causation at the micro-physical level. But the emphasis is very different when it comes to non-linguistic creatures' understanding of causation. As far as the practicalities of foraging and reproducing are concerned, the apparent probabilistic regularities are paramount.

Proto-causal understanding tracks relationships, which can be either deterministic or probabilistic, between states of affairs. This is why an understanding of causation is available at the non-linguistic level. It also explains why primitive versions of certain fundamental inference forms are available at the non-linguistic level. We can term this proto-conditional reasoning. Let us return to our three basic inference forms. We are looking for analogs at the non-linguistic level of the basic inference forms of disjunctive syllogism, *modus ponens* and *modus tollens*. The basic logical operations involved here are negation and the conditional. If negation at the non-linguistic level is understood in terms of the mastery of pairs of contrary predicates as suggested in the earlier discussion of proto-negation, and if we view the relevant conditionals as proto-conditionals tracking the causal relations between states of affairs, then we have all we need for non-linguistic analogs of our three basic forms of inference. *Modus ponens* can be understood straightforwardly in terms of a proto-conditional together with an understanding, which may take the form of a perception or a memory, that the antecedent holds. The consequent will straightforwardly be detached. We can view *modus tollens* in terms of the combination of a proto-conditional with the proto-negation of the consequent of that conditional resulting in the detachment of the proto-negation of the antecedent. As we saw earlier, the disjunctive syllogism 'A or B, not-A, therefore B' can be understood in terms of a causal conditional with not-A as its antecedent and B as its consequent.

We began with two thoughts. The first is that, since psychological explanations of the behavior of non-linguistic creatures are rationalizing explanations and since the notions of rationality and reasoning are correlative notions, the application of belief-desire psychology to animals (in the manner widespread in cognitive ethology and comparative psychology) stands or falls with the possibility of explaining how animals might engage in practical reasoning. The second is that, although animal reasoning must have sufficient commonalities with linguistic reasoning for it to be plausible to apply similar explanatory models of belief–desire explanation in the linguistic and non-linguistic cases, we cannot view non-linguistic reasoning as involving logical concepts and logical rules of inference. The suggestions about proto-logic sketched out in this paper try to do justice to both these thoughts by proposing analogs at the non-linguistic level for schemes of practical inference that can be understood at the linguistic level in terms of rules of inference involving elementary logical concepts. The three schemas discussed (reasoning from an excluded alternative and the two types of conditional reasoning) are highly relevant to animal practical reasoning, and I have shown how animals might apply them without deploying any logical concepts. The ability to deploy pairs of contrary concepts (without an explicit understanding of the notion of contrariety) and to be sensitive to causal regularities in the distal environment can provide animals with the tools for relatively complex forms of practical reasoning—and certainly for forms of practical reasoning that are sufficiently complex to underwrite the extension of belief-desire psychology to non-linguistic creatures.

References

Bermúdez, J. L. (1998). *The Paradox of Self-Consciousness*. Cambridge MA: MIT Press.

Bermúdez, J. L. (2003). *Thinking Without Words*. New York: Oxford University Press.

Bochvar, D. A. (1981). On a three-valued logical calculus and its application to the analysis of the paradoxes of the classical extended functional calculus. *History and Philosophy of Logic*, 2: 87–112.

Brunswik, E. (1943). Organismic achievement and environmental probability. *Psychological Review*, 50: 255–272.

Dickinson, A. and Balleine, B. (1993). Actions and responses: The dual psychology of behaviour. In: N. Eilan, B. Brewer, and R. McCarthy, eds. *Spatial Representation*, pp. 277–293. Oxford: Blackwell.

Frege, G. (1918–1919). Negation. [Translated by P. Geach and R. H. Stoothoff (1984). In: B. McGuiness, ed. Collected Papers on Mathematics, Logic, and Philosophy, pp. 373–389. Oxford: Basil Blackwell.]

Grice, H. P. (1989). *Studies in the Ways of Words*. Cambridge, MA: Harvard University Press.

Heyes, C. and Dickinson, A. (1995). Folk psychology won't go away: Response to Allen and Bekoff. *Mind and Language*, 10: 329–332.

Leslie, A. M. (1982). The perception of causality in infants. *Perception*, 11: 173–186.

Lewis, D. (1973). Causation. *Journal of Philosophy*, 70: 556–567.

Mackie, J. L. (1965). Causes and conditions. *American Philosophical Quarterly*, 2: 245–264.

Mellor, D. H. (1995). *The Facts of Causation*. London: Routledge.

Millikan, R. G. (1984). *Language, Thought, and Other Biological Categories*. Cambridge MA: MIT Press.

Sommers, F. (1982). *The Logic of Natural Language*. Oxford: Clarendon Press.

Spelke, E. S. (1990). Principles of object perception. *Cognitive Science,* 14: 29–56.

Sperber, D., ed. (1995). *Causal Cognition*. New York. Oxford University Press.

Chapter 6

Making sense of animals

Susan Hurley

Abstract

We shouldn't overintellectualize the mind. Non-human animals can occupy islands of practical rationality: they can have domain-specific reasons for action even though they lack full conceptual abilities. Holism and the possibility of mistake are required for such reasons to be the agent's reasons, but these requirements can be met in the absence of inferential promiscuity. Empirical work with animals is used to illustrate the possibility that reasons for action could be specific to symbolic or social contexts, and connections are made to simulationist accounts of cognitive skills.

6.1 Reasons without conceptual abilities: belief vs. action

Does having reasons require conceptual abilities? If we focus on perception and belief and theoretical reasons, as opposed to intention and action and practical reasons, the answer may seem to be *yes*. But I'll argue that acting for reasons does not require conceptual abilities—not, at least, the full-fledged domain-general conceptual abilities associated with theoretical rationality and inferential promiscuity. I will appeal to practical rationaliy to argue that the space of reasons is the space of action, not the space of conceptualized inference or theorizing, and to show how non-human animals can act for reasons.

The most powerful motivation for allowing that having reasons does not require full conceptual abilities derives from practical reason rather than epistemology. An intentional agent who lacks domain-general conceptual and inferential abilities and does not conceptualize her reasons can still act for reasons that are her own, reasons from her perspective. Her point of view may provide islands of practical rationality rather than a continuous space of reasons. Reasons for action can be domain-specific and lack conceptual generality. I'll illustrate this possibility by reference to empirical work.

This introductory section makes five preliminary points about the character of my argument, and then sketches its general structure.

First, a contrast. Some may argue that non-conceptual content is needed to provide an epistemological grounding for perceptual and/or theoretical beliefs. Others may deny this, and claim that non-conceptual content could not do this work in any case.

My argument is not in this territory.[1] Rather, it locates the possibility of having non-conceptual reasons in the practical sphere.

Suppose for the sake of argument that non-conceptual content is not needed for, or able to serve, epistemological purposes, and that conceptual content is needed. This supposition would not settle the question whether reasons must be conceptualized. It might seem to do so, if we have already overintellectualized the mind by giving epistemology priority over practical reason, by treating reasons as fundamentally reasons for belief held by epistemic subjects rather than reasons for action held by intentional agents.[2] But if either have priority, reasons for action are primary and reasons for belief less fundamental. And at any rate, reasons for action are not reasons for belief about what should be done. If they were, it would be very difficult to understand how there could be truths about conflicting reasons for action. Practical rationality is not theoretical rationality with practical content.[3] So, even if reasons for belief must be conceptualized, it would not follow that reasons for action must be.

Second, my argument is framed not in terms of conceptual *content* but rather in terms of conceptual *abilities*, which are less abstract and contentious and more operational than conceptual content.[4] Whatever conceptual content is, it provides

---

[1] The position taken here in no way allies me with the idea of non-conceptual scenario content. See Noë 2002 on why this is problematic on both phenomenological and empirical grounds.

[2] For example, Bennett's argument (1964) for denying rationality to non-linguistic animals turns on their lack of a kind of epistemic sophistication, and assimilates reasons to evidence relevant to theoretical knowledge.

More recently, Brewer (1999) has argued that reasons must be conceptual. His basic argument has two steps. First, giving reasons requires identifying propositions as premises and conclusions of the relevant inferences. Second, for reasons to be the subject's own reasons, at the personal level and from his point of view, they must consist in some mental state of his that's directly related to the pro-positional premise of the relevant inference: the premise proposition must be the content of the mental state in a sense that requires the subject to have all the constituent concepts of the proposition. Otherwise, the mental state will not be the subject's own reason (Brewer 1999, 150–152).

But Brewer's discussion is also one-sidedly oriented toward perception and belief, as opposed to intention and action. He typically speaks of reasons for judgments or beliefs, adding parenthetically: '(or action)', to keep practical reasons in play (e.g. 150, 151, 168), but these gestures toward action don't do the work needed. See my discussion in Hurley 2001, in which an earlier version of some of the present arguments is made.

[3] As I've argued elsewhere, this claim is fully compatible with the claim that there are truths about reasons for action; it does not depend on non-cognitivism about such reasons. Williams (1973) appears to make the mistake of assuming that for there to be truths about reasons for action they must be understood as reasons for belief about what should be done. For discussion see Hurley 1989, Chs 7–9.

[4] See also Tomasello and Call 1997, 351: 'What is needed are more articulated views of primate social cognition that specify different types of social knowledge and skills, without an excessive concern for their computational aspects' or for whether species 'have' certain representations. Also, cf. MacIntyre 1999.

conceptual abilities. If information that an object has a property is conceptualized, it satisfies a *generality constraint*: it has a fine-grained intrapropositional structure that enables the subject to decompose and recombine its elements promiscuously and transfer them to other contexts, and to generalize and make quantificationally structured inferences that depend on such context-free decompositional structure (for a useful discussion see Segal 1996, 148). My concern here is with the fine-grained and domain-general character of conceptual abilities rather than with the structure of whatever internal subpersonal processes may support such abilities. Conceptual reasoning abilities are governed by correspondingly fine-grained normative constraints, and are not bound to specific domains but transfer systematically to states of affairs removed from the subject's immediate environment and needs.

Note that being bound to a specific context or domain (such as foraging, or the detection of cheaters, or competition for food within a social dominance hierarchy—see the examples that follow) should not be confused with the use of demonstratives or indexicality, which philosophers sometimes describe in terms of the 'context-dependence' of claims employing 'here', 'now', 'this', 'I', and so on.[5] Of course conceptual content can be demonstrative; but it does not follow that conceptual abilities can be domain-specific. Nor does it follow from the domain-generality of conceptual abilities that they do not involve indexicality. My argument in no way depends on a rejection of demonstrative concepts. Arguably, practical reason must involve indexicality; but the specificity of reasons for action to a particular domain would not follow from that. For present purposes, indexicality and domain-specificity are independent issues.

Third, when I argue for the possibility of having non-conceptual reasons for action, I intend to be arguing that such reasons can correctly be attributed to the intentional agents in question. They can be the agent's own reasons, reasons at the personal or animal level. The personal level is distinguished from the subpersonal level by the holism and normativity of the personal level. We can interpret or make sense of action at the personal level, folk psychologically, in terms of holistically related and normatively constrained mental states. Or we can describe and explain an action at the subpersonal level, in purely causal, structural or functional, or neural-mechanism terms. When we do the former, we can correctly attribute reasons to persons as their reasons for acting. (See *Philosophical Issues* 3(1), 2000, an issue devoted to essays on the personal/subpersonal distinction.) I suggest that we can similarly distinguish, in terms of holism and normativity, an animal from a subanimal level, even if animals lack conceptual abilities—the way animals can correctly be made sense of from the way animals are made. Often we can make patchy sense of animal action at the animal level and correctly attribute reasons for acting to animals, even if these reasons are specific to certain domains. The animal level emerges as an archipelago from the sea of causes.

----

[5] Michael Luntley fails to distinguish them; see his 2003, at p. 410.

It may seem plausible to give the having of reasons links in two directions (see e.g. Brewer 1999, 49, 54, 56, 77, 82, 150–52). First, having reasons can be linked with the personal or animal level and the agent's perspective: reasons make whatever they are reasons for appropriate from the perspective of the agent in question. Second, having reasons can be linked with making general inferences from conceptually structured premises to conceptually structured conclusions, hence with conceptual abilities. But these two links pull against one another; we should thus distinguish thinner and richer senses of 'having reasons'. For example, one way of making these links explicit would be to claim that having a perspective as an agent requires having reasons, and that having reasons requires conceptual abilities and inferential promiscuity. But this would equivocate between thinner and richer senses of 'having reasons', since having a perspective as an agent does not require conceptual abilities and inferential promiscuity (even if the reverse requirement does hold).

It could still be objected that while there may *be* reasons to act that an agent has not conceptualized, these cannot be the agent's *own* reasons, reasons for the agent, at the personal or animal level (see and cf. Dennett 1996, Ch. 5, 6). But why not? Reasons for action at the personal or animal level are understood in terms of the requirements of holism and of normativity. Perceptual information leads to no invariant response, but explains action only in the context set by intentions and the constraints of at least primitive forms of practical rationality. Perceptions and intentions combine to make certain actions reasonable and appropriate from the animal's perspective. Means and end can decouple: an intentional agent can try, err, and try again, can try various different means to achieve the same end. The holism of intentional agency provides a minimal, coarse kind of recombinant structure: an intentional agent has the ability to combine a given intention with different perceptions, given ends with different means. Moreover, intentional agency is not merely a complex pattern of dispositions. It essentially involves normative constraints, relative to which mistakes make sense; actions can be inconsistent or instrumentally irrational.

These ideas of holism and normativity are familiar from the writings of Davidson, Dennett, and others, though they are here applied primarily to perceptions and motor intentions rather than beliefs and desires. Moreover, they are here detached from the requirement that the creatures in question have conceptual abilities or are themselves interpreters of other minds (cf. Davidson 2001, 105). Full-fledged conceptual abilities are not required for the conditions of holism and normativity to be met. The coarse recombinant structure provided by the holism of intentional agency falls well short of the fine-grained structure and context-free inferential promiscuity of conceptual abilities, as I explain further below. Holism and normativity characterize the personal or animal level, at which it is correct to regard an agent as acting for reasons that are his own. Of course, acting *for* a reason, rather than merely in the presence of or in agreement with a reason, requires the reason to cause the action 'in the right way', not deviantly. But this demand must be also met in the presence of fine-grained, context-free, inferentially promiscuous conceptual abilities.

So, the weight of my claim that non-conceptual reasons can be reasons for animals, at the animal level, rests on the claim that the holism and normativity of intentional agency can be present in animals who lack full-fledged conceptual abilities.

Fourth, consciousness of reasons is not a requirement of having reasons at the animal or personal level; I set issues of consciousness aside in this discussion. An unconscious reason for acting can nevertheless be the agent's own reason, in virtue of holism and normativity. I hold that conscious information must be available to the agent as a reason for acting, but not that information that serves as an agent's reason for acting must be conscious (see Hurley 1998, Ch. 4). For example in Freudian cases, or cases of self-deception, the partitioning of an agent into subsystems can be driven by normative constraints of consistency. Even if some such subsystems are subconscious, they can still count as at the personal level. By contrast, the subpersonal level is a level of causal/functional description to which normative constraints and reasons are not relevant. Why require consciousness of reasons at the animal level if not at the personal level?

Nevertheless, points closely related to those made here I elsewhere apply to *perspectival self-consciousness*.[6] Part of what it is to be in conscious states, including perceptual states, is to have a unified perspective, from which what you perceive depends systematically on what you do and *vice versa*, and such that you keep track, at the personal level, of this interdependence of perception and action. When I intentionally turn my head to the right, I expect the stationary object in front of me to swing toward the left of my visual field. If I intentionally turn my head and the object remains in the same place in my visual field, I perceive the object as moving. If my eye muscles are paralyzed and I try to move them but fail, the world around me, surprisingly, appears to move. Such perspectival self-consciousness is an essential aspect of perception, and it essentially involves ordinary motor agency as well as perception. Motor intentions are as important as perceptual experience to making a self-world distinction and a mind-independent world available to an agent. Many current views of perception as active emphasize the dynamic role of the agent's motor intentions and actions, as well as attention, in controlling his experience. Our understanding of the *world* as distinct from, and independent of, the *self* is most deeply grounded in environmental recalcitrance in the face of our rational efforts at motor control: control just is the maintenance of a target value by *endogenous* adjustments for *exogenous* disturbances.

Perspectival self-consciousness can be conceptual, but does not have to be. As an animal moves through its environment, its intentional motor actions dynamically control its perceptual experience against exogenous environmental disturbances,

---

6 See Hurley 1998, Ch. 4; see and cf. Van Gulick 1988; Bermúdez 1998.

simultaneously with its perceptions providing reasons for action. It can keep track of contingencies between its perceptions and motor intentions, in a practically if not theoretically rational way. In doing so it can use information about itself and its environment more or less intelligently, to meet its needs. Such a perspective is correctly described at the animal level, though of course there may be interesting things to say about how it is enabled at the subanimal level.[7] But it doesn't follow that the animal has a general concept of itself or its conscious states or of the self, or the ability to reason theoretically or systematically about aspects of self and environment in a variety of ways detachable from specific contexts and needs. It may not be able to generalize from self to others. Its perspectival uses of information about itself may be context-bound.

Fifth, following and generalizing views of Millikan (1991, 1993) and Dennett (1991), I think we should be wary of projecting properties or structures between the subpersonal (subanimal) and the personal (animal) levels of description (see Hurley 1998). For personal level descriptions *really* to apply does not require that an inter-level isomorphism requirement be met; the subpersonal architecture that explains and enables thought need not correspond to the structure of thought (see McDowell 1994). From failure of such isomorphism we can't infer the absence of true thought, or a defeat for realism about thought.[8] I don't accept an inter-level isomorphism requirement either for persons or for animals, either in the presence or absence of conceptual abilities. Personal or animal level descriptions may not map tidily onto descriptions of subpersonal functions or neural mechanisms; different vocabularies may be needed. Interpretation is not neuroscience, and the scientific experimental study of animal behavior can, but need not, involve neuroscience. We can simply be interested in both personal level folk psychology and subpersonal level functions and mechanisms. I'm deeply compatibilist here. There's no need to see these levels of description as in competition rather than as peacefully co-existing, going about their different business. That's how we normally do regard them, and so we should.

Moreover, the scientific study of animal minds can and does proceed at both the animal and subanimal levels. Good science doesn't require an isomorphism between correct interpretation of what the animal is doing and subanimal processes or architecture. This is an application of a more general point: the embodied, situated brain is nothing if not a complex dynamic system, and such systems give rise to patterns of macrobehavior that are not isomorphic with underlying microchanges of variables and parameters: there can be discontinuities at one level and continuities at

---

[7] Hurley 1998 develops the idea that a dynamic singularity at the subpersonal level enables perspectival self-consciousness at the personal level, within a two-level view of the unity of consciousness and the interdependence of perception and action.

[8] I don't argue against an isomorphism requirement here, but it may help to avoid misunderstanding to make explicit my rejection of it.

the other (Elman *et al.* 1996; Kelso 1995; Brooks 1999). Animal behavior and its psychological interpretation can be studied scientifically and experimentally without necessarily becoming the study of the architecture of animal brains (see Tomasello and Call 1997, 351, quoted above, note 4)—though of course it often is illuminating to bring neuroscience and behavioral results together. However, if the architecture of real neural networks turns out to be that of a reliable but interpretatively inscrutable bowl of spaghetti, that in itself would be no threat to rationality. Look-up table worries are overwrought, biologically irrelevant, and anyway don't generalize to inscrutable spaghetti.

This means that we can endorse realism and externalism at both the personal and subpersonal levels. We needn't accept the alignment of concerns about internal architecture with realism about minds, on the one hand, and interpretivism with pragmatism, on the other hand.[9] To be real, reasons needn't be internally contained or independent of external or practical or social affairs. Nor does the interpreter-dependence of concepts *per se* unfit them for application to animals. Perhaps interpretation says as much about us as it does about the world—those whose behavior we interpret; but if so, this holds for the interpretation of people as much as animals. Similarly, while interpretative facts are facts about how our framework applies to their behavioral features, so are lots of creditable facts. I suspect it may be difficult to say things about the world without also saying things about ourselves or our framework of concepts. But that doesn't mean we're not saying things about the world.

Thoroughgoing externalism at both levels, about intentional content, qualities of consciousness, agents' reasons for action, and vehicles of content, makes it easier to combine realism and interpretivism. If minds aren't necessarily confined to the head, then investigating minds externally as well as internally implies no lack of realism about minds; facts about minds can essentially include external facts. Connections to natural and social environments, as well as internal wiring, can be essential to the causal powers of minds themselves. This is true at both the personal and subpersonal levels, at both the level of content and the level of vehicles of content; minds can leak into the world at both levels of description. Though they do different jobs, neither level is privileged with respect to reality. A slogan to sum this up might be: realism at the personal level as well as the subpersonal; externalism at the subpersonal level as well as the personal.[10]

The isomorphism requirement suggests that the constraints identifying rationality are to be found at the level of architecture. By contrast, my arguments below focus on interpretation rather than architecture, the animal rather than the subanimal level.

---

[9] This view is neither eccentric nor original: Dennett and Davidson often encourage realist readings of their positions; consider also the practical realism of Lynne Rudder Baker (1995).

[10] For details, see my 1989 and 1998.

I consider questions of rationality to be about capacities and their correct interpretation at the animal level, rather than about internal subanimal-level architecture, and am concerned with flexibility at the level of behavioral capacities rather than at the level of architecture.[11] I try, below, to characterize rational capacities at the animal level rather than whatever subanimal-level processes or mechanisms may reliably enable those capacities, though I make some comments about architecture in passing. Thinking about mechanisms and processes can help to understand how it is possible for certain capacities to become dissociated, in particular the capacity for intentional agency from conceptual abilities. That is the point of the discussion below of feedback, simulation, and other useful architectural tricks. Different architectures can enable rationality in human beings and other animals; rationality can be variably realized, both functionally and neurally. However, focusing on interpretation rather than architecture implies no dismissal of the philosophical interest of issues about the architecture of minds at a subinterpretative level or interlevel relations. There are philosophically interesting things to say about interlevel relations even if an interlevel isomorphism requirement is rejected.

The argument that follows falls into five main parts. First, the holism of intentional agency and its relationship to conceptual abilities will be characterized. Second, the normativity of intentional agency will be considered. Third, empirical work will be examined to illustrate the possibility that intentional agents can have context-bound reasons for acting. Fourth, I'll explain how my argument shares ground with recent work in simulation theory, in locating the possibility of having non-conceptualized reasons in the practical sphere. Finally, I'll ask: what is the *point* of making sense of animals?

6.2 **Degrees of generality: holism vs. inferential promiscuity**

Intentional agency—something that many animals have and plants lack—occupies a normative middle ground between a mere stimulus-response system and full context-free conceptual abilities (Hurley 1998, Ch. 4). An intentional agent can act intentionally, for reasons that are her own reasons. Intentional agency involves a familiar kind of holism: relations between stimuli and responses are not invariant. Rather, actions depend holistically on normatively constrained relationships between intentions and perceptions, between ends and means. A given intention or end will yield different actions given different perceptions about means/ends contingencies, and *vice versa*.

In what follows, 'holism' is used in a familiar Davidsonian sense; beliefs and desires combine and recombine flexibly in making sense of actions. Such holism is a specific

---

[11] My situated approach to rationality does not fit tidily into either side of Kacelnik's distinction (this volume) between PP-rationality and E-rationality: while my approach is psychological and philosophical, it is concerned with the flexible structure of the behavioral interactions of persons and animals with their environments, rather than with the structure of the internal processes that contribute to explaining these interactions. In this respect it may have more in common with an ecological conception of rationality (see Gigerenzer *et al.* 1999).

kind or aspect of flexible generality, as are domain-freedom and flexibility that extends across ends and means. 'Flexible generality' is used here as a generic term admitting differences of degree and of specific kind. It can be illustrated in various ways. There are degrees of flexible generality, in abilities to use information, abilities to use behavioral techniques, and so on. Moreover, since flexible generality is multi-dimensional, rationality may be best viewed for certain purposes in disaggregated terms rather than as unified. In particular, it is useful to focus on specific capacities and the degree to which we share them with other animals; in this way rationality can be disaggregated into island-like capacities, bound to specific contexts. An interest in whether animals are rational agents does not require that rationality has a deep unity or that all its aspects can be compared on some one spectrum; it is an interest in various specific ways in which the capacities of animals may be continuous as well as discontinuous with our own.[12]

The holism of intentional agency is located in a space that can be further articulated, into various types and degrees of flexibility and generality. A recurring theme of work on animal cognition has been the importance of escaping from a crude dichotomy between an inflexible, rigidly context-bound stimulus–response system, on the one hand, and full-fledged conceptual, inferential, and mind-reading abilities, on the other. Various finer distinctions between locations in the space of flexible generality can usefully be drawn.[13] In this section I'll sketch parts of this space, and focus in particular on the difference in level of flexible generality between the holism of intentional agency and the inferential promiscuity of conceptual abilities.

Start with a classically conditioned response, such as salivating in, or approaching, the place where food is normally delivered. Food induces salivation; a certain place is associated with food; so that place comes to induce salivation. An animal doesn't salivate in that place in order to get food; salivation is not a means to the end of getting food. Classically conditioned responses are insensitive to 'omission schedules': if salivating causes food to be omitted, the animal will salivate anyway; if approaching the place where food has normally been delivered causes food to be omitted, the animal with nevertheless go on approaching. Thus classical responses, such as salivation and food approach, are not counterfactually sensitive to changes in information about the ends/means contingencies the animal faces. In this respect they are inflexible. Nevertheless, such responses may be sensitive to changes in ends, or in the values of

---

[12] Here I am especially indebted to commentary by and discussion with Kim Sterelny (see his 2003). Cf. my earlier arguments against centralism about reasons for action, in Hurley 1989, Part I.

[13] See and cf. Call and Tomasello 1996; Tomasello and Call 1997, 11, 229, 272, 382, etc., on the three-fold distinction between understanding another organism as animate and directed vs. as an intentional agent vs. as a mental/epistemic subject; Premack 1978, on how the ability to transfer or generalize skills across domains is a matter of degree; Harris 1996 on understanding another as an intentional agent vs. as an epistemic subject.

outcomes: if food is devalued by association with nausea or by satiation, food approach in rats will decline.[14] Flexibility is not all or nothing.

According to Heyes and Dickinson (1993), responses that are not counterfactually sensitive to changes in information about ends/means contingencies are not intentional actions. Intentional action requires counterfactual sensitivity both to ends or goals (or desires) and to instrumental information (or beliefs) about whether the action conduces to the end in question, and requires that these interact rationally to produce the intention to act. While approach to a food delivery area does not meet these conditions in rats, lever pressing to obtain food does, they argue. Rats reduce lever pressing when contingencies change so that it is no longer instrumental for obtaining food, even though they do not reduce approach behavior. So lever pressing is sensitive to changes in information about means—ends contingencies.

However, lever pressing is, strangely, less immediately sensitive to changes in ends than is approach behavior. When sugar is devalued by association with nausea, approach behavior declines at once, but lever pressing does not decline until the rats have had been re-exposed to the devalued food, after the devaluation but before testing. Only then does lever pressing decline in accord with the devaluation of sugar. So-called 'incentive learning' is needed before instrumental action, lever pressing, is affected by the devaluation of a previous goal, even though as measured by the immediate decline in approach behavior, the rats already knew about the devaluation. They have to be reminded that they no longer like sugar, before they stop pressing, even though they already knew it, in some sense.[15] Classical and instrumental behavior can thus appear to access different knowledge bases.[16] Approach behavior responds at once to the devaluation of an end, but does not respond to a change in means/end contingency, while lever pressing does not respond to the devaluation of an end until after further incentive learning, and does respond to a change in means/end contingency. Again, flexibility is not all or nothing.[17]

The middle ground of intentional agency can be characterized in terms of flexible, holistic relations between ends and means.[18] As Tomasello and Call explain, an intentional agent distinguishes ends from means, recognizes that there can be different means to the same end, that the same behavior can be a means to different ends, that the same behavior can be an end or a means. Ends and means articulate, come apart, recombine; intentional action results from the relations between them. Understanding

---

[14] See Balleine and Dickinson 1991, Balleine 1992; Heyes and Dickinson 1993.

[15] See Balleine and Dickinson 1991; Balleine 1992; Heyes and Dickinson 1993.

[16] Thanks to Nicholas Rawlins for discussion of these issues.

[17] See also Russell *et al.* 1994, 303; Tomasello and Call 1997, 11.

[18] See Tomasello and Call 1997, 318, 361, etc.; Call and Tomasello 1996; Tomasello 1999; see also and cf. Sterelny 2001, Chs 11, 12.

an action as intentional involves understanding the more or less rational interaction between the agent's ends and his perceptions of means/ends contingencies in producing his action, so that given movements are not always followed by the same results. Intentional action is flexibly adjusted in various circumstances in the ways needed to bring about the agent's goal (Tomasello and Call 1997, 10, 318, etc.).

The holistic flexibility of intentional agency contributes a degree of generality to the agent's skills: a given means can be transferred to a novel end, or a novel means adopted toward a given end. The end or goal functions as an intervening variable that organizes varying inputs and outputs and allows a degree of transfer across contexts.[19] As a result, understanding another organism as an intentional agent permits transfer or generalization from a specific circumstance/behavior contingency to others: if she has ends that call for deception, she may be expected not only to give leopard alarm calls when there are no leopards, but also to give eagle alarm calls when there are no eagles.[20] However, Tomasello argued (in 1999) that being an intentional agent, whose ends and means are related in holistically flexible and transferable ways, does not entail that you can understand others as intentional agents: the additional momentous step of identifying with others is also needed for this. He then regarded chimps as intentional agents who cannot—except possibly for enculturated chimps—take this further step of identification with others (but cf. his later view, in Chapter 17 by Tomasello and Call in this volume; see also the earlier view in Tomasello and Call 1997, Ch. 11–13; cf. Davidson 2001, 105).

But again, flexibility and generality are not all or nothing. Tomasello and Call emphasize that flexibility and generality are inherent in cognitive adaptations, but can vary in nature and degree and can be confined to a specific domain or function.[21] This point is illustrated by their claim that being an intentional agent, which entails a degree of flexible generality and transferability, of ends and means across contexts, does not entail being able to understand others as intentional agents, which entails an additional degree of flexible generality and transferability, across perspectives (Tomasello and Call 1997, 190, Ch. 12–13). Similarly, being an intentional agent does not entail being an epistemic subject, with the further abilities to distance oneself from one's current perceptions and entertain the possibility that they are misleading.[22] An agent could understand that a certain behavior is not a good means to her ends in light of her perceptions without understanding more specifically that her perceptions are not a

---

[19] See, for example, Whiten 1996b, 282–288; Smith 1996, 350–351; Dennett 1996, 124–125; Taylor Parker and Russon 1996; Whiten 1996a.

[20] See Cheney and Seyfarth 1990, 182, 307; see also Tomasello and Call 1997, 383; Tomasello 1999, 25.

[21] Tomasello and Call 1997, 10–11, 179; 380, 383, 384; 417, etc.; see also Cheney and Seyfarth 1990, 262, etc.; Premack 1978, 424, etc.

[22] See Bennett 1964 on the demanding kind of generalization away from the present and particular required for the ability to assess evidence.

good guide to the way things are. Putting these two points together, it may be possible to understand others as intentional agents but not as epistemic subjects, whose beliefs can differ from one's own or be false.[23]

For present purposes, the most telling application of the point that flexible generality is not all or nothing does not contrast intentional agency with epistemic or mind reading abilities. Rather, it focuses on the way the flexible generality involved in intentional agency itself can be limited to specific domains or functions (see Tomasello and Call 1997, 179). An intentional agent's reasons for action can be bound to specific contexts and not generalize; there can be islands of practical rationality. For example, a primate could have reasons in social contexts that she cannot generalize to non-social but logically similar contexts. Suppose a monkey observes that conspecific A is dominant over B and that B is dominant over C and, never having observed A and C together, registers that A is dominant over C, and is able to use this information in instrumentally appropriate ways in relation to various goals. Nevertheless, she might be unable to generalize the ability to make transitive inferences to foraging contexts, such as: tree A has more fruit than tree B, which has more than tree C, so tree A has more fruit than tree C.[24] Evolution might have conferred the ability to make transitive inferences in the social context, if it was most valuable there, without conferring the conceptual abilities needed to transfer it readily to other contexts.[25] This empirical possibility (I do not need to claim it is more than a possibility) illustrates how holistic means/ends flexibility might obtain without conceptual flexibility.

The relatively coarse ends/means structure of intentional agency contrasts with the finer structure of conceptual abilities. Practical knowledge of how to do things can be implicit in procedures that are tied to a specific practical context or function, even if they show a degree of flexibility within that context.[26] By contrast, a creature with conceptual abilities can decompose, transfer, and recombine the conceptualized intrapropositional elements of information, and thus can recognize fine-grained inferential structures, such as that involved in transitive inferences, that are common to quite different contexts. The exercise of such recombinant conceptual abilities liberates reasons to operate across contexts; conceptual abilities underwrite inferential promiscuity.

This section has contrasted the degree and kind of flexible generality involved in the holism of intentional agency and in the inferential promiscuity of conceptual abilities.

---

[23] See Harris 1996; Tomasello and Call 1997, 189.

[24] See and cf. Cheney and Seyfarth 1990, 83 ff, 257–258; Tomasello and Call 1997, 184. Intensive training is needed to get primates to make transitive inferences in non-social contexts in the lab, while they make them from a young age in social contexts in the wild. See also Tomasello 1999, 18.

[25] On the compatibility of rationality with horizontal modularity, see Hurley 1998, Ch. 10. Cf. Allen and Bekoff 1997, 96, on how, given teleological views, intentionality can lack compositionality for evolutionary reasons. The latter refers to intentionality in the technical sense of *aboutness*, not in the sense of intentional agency that I am concerned with, but the points are nevertheless closely related.

[26] See and cf. Cheney and Seyfarth 1990, 83 ff, 250; Gomez 1996; Gordon 1995a, 108, 112.

Holism is required for correct attribution of reasons for action to an intentional agent, as his own reasons. But this condition can be met even if the agent lacks conceptual abilities.

6.3 A quibble: are conceptual abilities themselves a matter of degree?

The previous section employed a familiar conception of conceptual abilities as involving context-free recombinant skills and inferential promiscuity. But it may be objected that since flexibility and generality can differ in kind and degree, so can conceptual abilities. As a result, there would be no sharp distinction between non-conceptual and conceptual abilities, and conceptual abilities may themselves be domain-specific to some degree.[27] The holism of intentional agency involves a degree of flexible generality, even if of a relatively coarse kind that can be domain-specific. So, if an animal is an intentional agent and can act for reasons, it has at least rudimentary conceptual abilities.

But this quibble is not really an objection to the substance of my position so much as a notational preference. There is not a precise and agreed definition of conceptual abilities across philosophy and psychology, or even within them, that rules this notational preference in or out. I could adopt this notational variant, and recast the substance of what I've been saying in these terms. My notion has been as follows: Flexibility and generality come in different kinds and degrees, and can be present in intentional agency even when they are domain-specific. By contrast, conceptual abilities are domain-general. Thus a creature without conceptual abilities can have reasons for action. An alternative notation would be as follows: Conceptual abilities come in different kinds and degrees, and can be present in intentional agency even when even to some extent domain-specific. By contrast, full conceptual abilities are domain-general. Thus a creature without full conceptual abilities can have reasons for action.

On the one hand, acting for a reason does not require the ability to infer *everything* that follows from it. Not even we human beings can do that, and we are paradigmatic possessors of conceptual abilities and conceptualized reasons. On the other hand, it is not plausible to claim that a creature can act for a reason in a particular case, without being able to generalize the reason in *any* way.[28] The non-human agents discussed below appear to operate on islands of practical rationality, rather than in a continuous space of reasons. Thus, a middle ground is attractive, according to which acting for a reason requires *some* degree of ability to generalize one's reason, even if it is limited to certain contexts. The quibble describes this middle ground in terms of rudimentary

---

[27] Again, keep in mind that domain-specificity is not indexicality; my notation is consistent with indexical concepts. See also and cf.: Noë 2004; Premack 1978, 424; Segal 1996, 148, 151; Astington 1996, 192–193. Note that treating conceptual abilities as a matter of degree contrasts with much of the discussion of conceptual vs. non-conceptual content, which gives the distinction an all or nothing character.

[28] I am grateful here for comments from Ram Neta.

rather than full conceptual abilities, while I describe it in terms of intentional agency without conceptual abilities. My substantive position has no deep stake in one notation as opposed to the other.

But doesn't my notation presuppose too exalted a conception of conceptual abilities (see Noë 2004)? After all, human beings, those paradigmatic possessors of conceptual abilities, arguably display context-bound rationality. For example it has been claimed human beings are able to detect violations of conditional rules far more reliably when these constitute cheating on social contracts than in many other contexts, even if the rule violations are formally equivalent.[29] Wouldn't my notation force us to say that human beings therefore lack conceptual abilities?

No. We can instead say that human conceptual abilities are not always exercised. In the case of human beings, unlike other animals, we can have linguistic evidence that conceptual abilities are present, even if not exercised in a particular case. Indeed, the point may be stronger than this: linguistic abilities may be the basis of conceptual abilities,[30] and so provide a basis for saying that conceptual abilities are present even when not exercised.

So the argument will proceed using my preferred notation, but with the understanding that its substance could be recast in the alternative notation. The substantive point is that acting for a reason requires the flexible generality implied by the holism of intentional agency, but not the domain-general recombinant skill and inferential promiscuity that my notation associates with conceptual abilities and the alternative notation associates with full conceptual abilities. In my notation, the result is that animals who lack conceptual abilities can act for reasons; in the alternative notation, the result is that animals who lack full conceptual abilities can act for reasons.

6.4 Normativity and the possibility of mistake

I've argued that an intentional agent who lacks conceptual abilities can nevertheless have reasons for acting that are his own reasons, if the conditions of holism and normativity are met. I'll have somewhat less to say here about normativity than about holism.

In the sense relevant here, the sense required for intentional agency, the normativity of reasons for acting at the personal or animal level requires that the possibility of mistake be established. And this requires that there is a difference between the agent's making a mistake in acting for one reason and his acting for a different reason, a difference between his making a mistake in following one rule and his following a different rule. Ultimately, what is required is a solution to, or dissolution of, Kripke's (1982) version of

---

[29] Cosmides 1989; Cosmides and Tooby 1992, 166, etc.; see discussion in section 6.6 below.

[30] See the discussion, in section 6.5 below, of: Boysen and Bernston 1995; Boysen *et al.* 1996; Boysen 1996; Boysen *et al.* 1999; Thompson *et al.* 1997; see also and cf. Bennett 1964; McDowell 1994; Dennett 1996, Chs 5, 6; Smith 1996 in Smith and Carruthers.

the rule-following problem, applied to intentional agency. Rule-following issues are deep issues for everyone, and arise for creatures with full-fledged conceptual abilities as well as for mere intentional agents. The correct solution to, or dissolution of, the problem is controversial (though I believe that externalism is required in response). At any rate, it is not obvious that the issue is any more tractable for subjects with full-fledged conceptual and theoretical abilities than for mere intentional agents with practical abilities.[31]

I'll explain why I think that intentional agents who lack conceptual abilities can nonetheless make mistakes, thus that the normativity condition for attributing reasons for actions to such agents can be met. But I admit that the case for this is less than conclusive, pending a fully satisfactory resolution of rule-following issues.

What is the relationship between holism and normativity? It is arguable that normativity and the possibility of mistake require a kind of generality, an ability to go on in the same way, to follow a general rule flexibly in various cases (see and cf. Bennett 1964). Reasons are inherently general; a mistake in a particular case is a mistake against a background provided by the general rule one is trying to follow. However, as argued above, both the holism of intentional agency and the inferential promiscuity of conceptual abilities involve flexible generality, though to different degrees. I don't see why normativity should require the more flexible, finer grained generality of conceptual abilities and inferential promiscuity, as opposed to the lesser degree of flexible generality involved in the mere holism of intentional agency.

However, even if holism can satisfy normativity's requirement of generality, holism does not entail normativity. Consider control systems. Tomasello and Call invoke a control system view of intentional action, as involving a goal or end, perception of the current situation and its relation to the goal, and flexible choice of behavioral means to the goal that is sensitive to differences in the current situation and its relation to the goal (Tomasello and Call 1997, 10, 318, etc.). But a thermostat is a simple negative-feedback control system, which adapts its output by reference to comparisons between a variable target or goal and variable current input, where current input is the joint result of independent environmental changes and feedback from its own output. It exercises control, which involves a simple holism: in different circumstances different outputs may be instrumental to achieving the same target, and the same output may be instrumental to achieving different targets. A thermostat exercises such control in virtue of having certain holistically characterizable, dynamic dispositions. But such dispositions alone, without further context, surely do not underwrite normativity. They do not give a sense to the claim that the thermostat has made a mistake: that it is mistaken in its pursuit of one target, as opposed to that it is simply pursuing a different target.

---

[31] See and cf. Blackburn 1995; Hurley 1998, Ch. 6; Proust, this volume. Note that if mind reading is a practical rather than a conceptual skill, as some simulationists hold, one could agree with Davidson that 'rationality is a social trait' (2001, 105) and that the possibility of mistake requires an intersubjective world, without agreeing about the conceptual abilities required.

So the holism of simple control does not suffice for the possibility of mistake or for normativity—or for the thermostat to be acting for reasons. But consider next how simple control can be enhanced, both internally and externally. It can be made more complex in various ways, and it can be embedded in teleological contexts that provide functions. Both kinds of enhancement help to understand how the normativity of intentional agency can emerge from simple control, even in the absence of conceptual abilities.

Consider some ways in which simple control can be made more complex, increasing further the flexible generality of the system, yet without going all the way to inferential promiscuity. Real time feedback is slow; a 'forward model' that predicts feedback can be trained up using real time feedback and then used to provide a control system with simulated feedback to speed up controlled performance. In effect, a copy of the system's output can be used on-line to predict the resulting input, so as more smoothly to prepare any needed adjustments in output. Predictions can also be made off-line, by simulating the results of different possible outputs, which could enable assessment of the instrumental value of alternatives, deliberation, and planning. Such simulations of control could also be projected to the perspectives of other agents, enabling mind reading and more flexible predictions and control of the behavior of others, across a wider range of circumstances than is provided by circumstance/behavior contingencies that must be learned in each specific context.[32] A control system's target can be multimodal or otherwise complex, and may require trade-offs between the achievement of different elements of the target.[33] Such trade-offs can be controlled by another linked control system with its own target. So one control system can be used to determine the goal of another control system, giving rise to a hierarchy of control. But the reins of control may also be tangled and heterarchical, with no consistent hierarchy. There need be no transfer of tricks from one specific practical domain to others, as in the example of transitive inference above; even sophisticated forms of control can be domain-specific.[34] Nevertheless, as control becomes more complex and less transparent, properties of self-organization may emerge in the overall system (see Kelso 1995). Complex dynamic systems may be theoretically and nomologically opaque in ways that favor process-driven simulation over conceptualized theorizing as a mode of prediction and control of such systems.

Notice that forward models used on-line to simulate and predict can make erroneous predictions—can make 'mistakes' that should prompt the system to default to real feedback for control functions and in order to retune its forward models. However,

---

[32] See Cheney and Seyfarth 1990, 250, 307; on the distinctive benefits of simulating intentional agency, see Tomasello and Call 1997, 417 ff; Tomasello 1999, 25, 70 ff; see also and cf. Whiten 1996b, in Carruthers and Smith.

[33] See and cf. Sterelny 2001, on how multitracking opens up the possibility of error, 245–247.

[34] This sketch is closely related to what I call 'horizontal modularity', in Hurley 1998, Ch. 10.

such 'mistakes' are local, technical glitches, not of global significance for the whole system. That is, to function effectively, the system doesn't need to monitor whether it is using simulated or real feedback for on-line control functions, so long as its local switching and tuning devices operate smoothly. Such low level simulation supports a local sense of 'mistake', but not normativity and the possibility of mistake at the level of the whole system, whether animal or person.[35] (But see also section 6.7 below.)

Does system-level normativity emerge as forms of control and simulation become more complex and flexible, or do we just get more complex and inscrutable patterns of dispositions? Perhaps such increases in a system's internal complexity and flexibility are necessary for the possibility of animal-level mistake, but not in themselves sufficient. But note that such complex forms of control and simulation are likely to be found in living bodies that are embedded in environments and histories, which provide a teleological context. The way teleological context can underwrite the possibility of mistake is familiar from teleological accounts of content; evolutionary processes provide functions that allow a gap to open between what a system actually or usually does, and what it is supposed to do (Millikan 1984).

Cultural teleological context supplies a richer kind of normativity than does biological teleological context. The onset of culture and cultural evolution marks an important discontinuity between human and animal minds, a discontinuity probably related to the onset of conceptual and mind reading abilities. But here I'm looking for continuities that reach across these differences.

I suggest that teleological external resources, as well as internal resources, are needed to understand how animals can make mistakes. On an externalist view of reasons (which I also apply to human persons), for reasons to be an agent's reasons requires the possibility of mistake, and the possibility of mistake requires not just a system with some degree of internal complexity and flexibility, but also exogenous constraints.[36] What makes an act the wrong means to one end, rather than the right means to a different end, is not solely internal to the agent whose act it, but is also a matter of the agent's relations to his environment. Nevertheless, the reasons for which the agent acts can be his own reasons. An agent's own reasons are in part externally constituted.

Is the combination of relatively complex and flexible forms of control with teleological context enough to make the kind of mistake required for intentional agency possible—enough for there to be a difference between an act's being the wrong means to one end and its being the right means to a different end? I'm inclined to say

---

[35] See Hurley 2005a, 2006 for more on the relationship between simulation and mistake, and the distinction between local and system-level simulation and mistake. Cf. Proust, this volume.

[36] On how the possibility of mistake requires constitutive exogenous constraints, substantive as well as formal, see Hurley 1989, Part 1; Introduction to Bacharach and Hurley 1991; see and cf. Hurley 1998, Ch. 6, on the difference between making a mistake in following one rule and following a different rule. See also and cf. Greenberg, in progress.

yes. Normativity admits of different kinds and levels and degrees. But the kind of mistake possible for a relatively complex and flexible, teleologically embedded system seems to me adequate to meet the normativity condition for correctly attributing practical reasons to an intentional agent, even in the absence of conceptual abilities.

This combination may not suffice for the agent consciously to act for one reason rather than another, but consciousness is not what I'm talking about here. As indicated earlier, a reason can be the agent's reason for acting, a reason at the personal or animal level, even if it is not his conscious reason for acting. And appealing to qualities of consciousness does not resolve issues about the normativity of rule-following in any case, as Kripke famously explained.

If someone wants to deny that the possibility of mistake for intentional agents can be established in this way, I don't have a knockdown argument. But I also don't see how conceptual abilities and inferential promiscuity do any more to show how mistakes are possible (as opposed to presupposing this). After all, the most compelling statement we have of the difficulty of solving the normativity issue, in Kripke's account of the rule-following problem, applies to persons with conceptual and inferential and even linguistic abilities. If an agent with conceptual abilities makes a transitive inference in a social context but fails to do so in a non-social context, has he made a mistake in following the transitivity rule, or is he just following a different rule—the 'stransitivity rule', which requires transitive inferences in social contexts but not in others? Despite the greater flexible generality of conceptual abilities and inferential promiscuity, resolving such issues requires constitutive external constraints, substantive as well as formal, on the eligibility of contents and patterns of content. If externalism about reasons is a threat to the sense in which reasons are the agent's own reasons, it is no less a threat when the agent has conceptual abilities.

So, with or without conceptual abilities, normativity and the possibility of mistake are required for an agent to act for reasons that are his own reasons. According to externalism about reasons, an intentional agent's reasons can be his own reasons, even though the possibility of mistake this requires depends on exogenous constraints provided by relations to the agent's environment, as well as a degree of internal complexity and flexibility. Such external constraints are part of what it is for an agent's reasons to be his own. I suggest that such internal and external enhancements to simple control make mistake in the sense required for intentional agency possible, even in the absence of conceptual abilities. And at least they leave the possibility of mistake no less problematic than it is for creatures with conceptual abilities.

6.5 Illustrations of context-bound reasons for action: symbolic context

So far I have been defending, at a fairly abstract level, the claim that the conditions required for an agent's reasons for action to be his own reasons, holism and normativity, can be met for creatures who lack conceptual abilities. Consider now some concrete illustrations of how an intentional agent's reasons for action might be bound to specific contexts, beginning with symbolic contexts.

Sarah Boysen's chimp Sheba arguably displays an island of instrumental rationality that does not generalize. Sheba was allowed to indicate either of two dishes of candies, one containing more than the other. The rule was: the candies in whichever dish Sheba indicated went to another chimp, and Sheba got the candies in the other dish. Sheba persisted in indicating the dish containing more candies at a rate well above chance, even though this resulted in her getting fewer candies. Boysen next substituted numerals in the dishes for actual candies. She had previously taught Sheba to recognize and use the numerals '1' through '4'. 'Without further training, Sheba immediately invoked the optimal selection rule', that is, she began to choose the smaller numeral at a rate well above chance, thereby acquiring the correspondingly larger number of candies for herself (Boysen et al. 1999, 229). The substitution of numerals seemed to make instrumental reasons for action available to her, as they seemed not to be when she was faced directly by the candies. When the numerals were again replaced by candies, Sheba reverted to choosing the larger number.[37]

On one interpretation, the candy task is similar to an omission schedule, to which classical performance such as food approach is insensitive. Perhaps indicating food is like approaching food, so that even if indicating food causes the animal not to get it, or to get less of it, she would still do it. However, when numerals are substituted, instrumental rationality, which is sensitive to omission schedules, becomes operative. However, the difference here is not in the bodily action, as in food approach vs. level pressing, but in the context of action: gesturing toward food vs. toward symbols, numerals. The symbolic context appears to have provided a scaffolding that made instrumental reasons available to Sheba, but these reasons were bound to the context the symbols provided.

Boysen comments on how these results hint at the 'explosive impact' that symbols may have had in overcoming the behavioral limitations of our hominid ancestors (in Russon et al. 1996, 187). Comparison of other work by her group with earlier work by Premack also suggests the role of a symbolic environment in overcoming cognitive limitations (Thompson et al. 1997). With 'dogged training' of language-naïve chimps, Premack had eventually been able to find some evidence of ability to perform match-to-sample tasks for identity and difference relations. Shown AA, Premack's chimps were rewarded for choosing BB rather than EF; shown CD, they were rewarded for choosing EF rather than BB, and so on. By contrast, Boysen's group got immediate success with chimps who had been trained to use tokens for *same* and *different* (though they'd had no experience with linguistic strings or language training *per se*). Shown AA, Boysen's chimps were rewarded for choosing a heart token (for same); shown CD,

---

[37] Boysen and Bernston 1995. A similar pattern of results was obtained in a design with no passive partner chimp, so conspecific social context was not critical; see Boysen et al. 1996.

they were rewarded for choosing a diagonal token (for different), and so on. When these chimps were transferred to the match-to-sample task for identity and difference, with no explicit training or differential reinforcement of correct responses, they were immediately well above chance.

However, Boysen also compares the candy/numeral results to results with children in Russell's windows task (Boysen 1996, 184–186). This comparison suggests what may seem a rival interpretation of the candy/numeral results, in terms of an executive function deficit, or inability to inhibit certain responses, as opposed to context-bound rationality.[38]

In the training phase of the windows task, the child points to either of two closed boxes, one of which contains chocolate. The experimenter opens the box pointed to, and if there is chocolate in the box, the experimenter gets it. If there is no chocolate in the box pointed to, then the child gets chocolate from the other of the boxes. In the testing phase, the boxes are given clear windows, so that only the child can see which box contains chocolate. Will the child then employ strategic deception and point to the empty box, in order to get the chocolate from the other box?

Russell *et al.* (1991) found that children under 3 years and autistic children continued pointing to the baited box throughout the testing period. They suggest that executive rather than mind-reading limitations are at the root of the difficulties 3 year olds and autistic children have with the windows task (and possibly with false belief tasks more generally). Success at the windows task would require them to inhibit a gesture toward a salient object, and this is what they have trouble with. They fail even at a modified version of the windows task in which there is no opponent to deceive and hence no mind-reading demand, only an inhibition demand (Russell *et al.* 1994).[39] More recently, Hala and Russell (2001) found that giving 3-year-olds an ally, or asking them to indicate a box using a pointer arrow on a dial rather than their finger, improved their performance on the windows task, though not from the first trial. They suggest that both an ally[40] and the artificial pointing device provide symbolic distance between ends and means, and they cite Boysen's work on the way symbols can enable cognitive performance (Hala and Russell 2001, 134–136).[41]

Two questions are relevant here: (1) Is the difference between failure (in the original version) and success (in the ally and artificial pointer conditions) on the windows task

---

[38] Thanks here to comments from Greg Currie.

[39] Note that Samuels *et al.* (1996) failed to replicate these results, and found that 3-year-olds could perform the windows task and variation on it.

[40] For a Vygotskian perspective, see also and cf. Astington 1996, 193, on how 3-year-olds can pass false belief tests given the 'scaffolding' or context provided by a conspiratorial ally; Call and Tomasello on the possibility that mind reading in chimps is scaffolded by enculturation.

[41] Recall also and cf. the greater availability of instrumental reasons to rats in lever pressing than in food approach.

better interpreted in terms of inability to inhibit certain non-instrumental responses as opposed to scaffolding of rationality by symbolic context? (2) Is the difference between failure with candies and success with numerals better interpreted in terms of inability to inhibit or scaffolding by symbolic context?

The answer to (1) is unclear, as the work done does not clearly discriminate between the inhibition and scaffolding accounts. Russell *et al.* originally contrasted a failure-of-executive-function account with a mind reading account, but in the more recent work the executive function account blurs into the symbolic distance account. In theory, inhibition is negative and removes interference, thereby revealing underlying instrumental reasons, while scaffolding is positive, and makes instrumental reasons available. But in practice the difference may not be clear. It is not clear whether the ally and pointer contexts that provide symbolic distance make instrumental reasons available to the 3-year-olds, or whether these contexts permit the children to inhibit a conflicting non-instrumental response. They do not 'inhibit' from the first trial in the ally and pointer conditions, at any rate, and children seemed not to have learned the rule from previous training (Hala and Russell, 2001). Unfortunately, children who failed in the original windows task were not asked to explain their responses (Russell *et al.* 1991, 339).

On (2), there is additional relevant evidence. In a later variant of the candy experiment (Boysen *et al.* 1999), mixed candy/numeral choice pairs were used in addition to candy/candy and numeral/numeral pairs, with a group of chimps. Again, with unmixed choice pairs, the chimps did not learn to choose the smaller group of candies in order to get the larger group, but numerals got immediate success. However, in a feasibility study, three of the chimps were as successful with mixed candy/numeral pairs as with pure numeral pairs. The other two chimps were at chance with mixed pairs: once was biased toward choosing only candies, regardless of number, while the other was biased toward choosing only numerals.

I suggest that success with the mixed candy/numeral pairs is better accounted for in terms of symbolic context making instrumental reasons available than in terms of symbols enabling inhibition, to the extent these are indeed different hypotheses. As Boysen *et al.* point out (1999, 233), numerals have no intrinsic incentive features, so one might expect an exaggerated response bias toward the candies in mixed candy/numeral pairs. But this was not generally found.[42] It would seem to require greater inhibition to counter a response to something sweet as opposed to nothing, than to more of something sweet as opposed to less. But overall performance on mixed

---

[42] But see Boysen *et al.* 1999 on individual differences in responses. See also Boysen's chapter, this volume, for a survey and update of this line of research. The results in the feasibility study cited here for mixed candy/numeral pairs differ from those obtained in later work, discussed in Boysen's chapter, with mixed pairs.

pairs was comparable with performance on pure numeral pairs, which is what would be expected if symbolic context makes instrumental reasons available.[43]

Further work may tease these two accounts apart. Animals may differ, and neither account may fit all chimps tested. My point here is not tied to the actual empirical outcome in these cases or in follow-up work.[44] It is rather to use these experiments to illustrate how an intentional agent *could* have reasons for action that are specific to symbolic contexts and fail to generalize, and what would count as evidence for this.

6.6 **Illustrations of context-bound reasons for action: social contexts**

Consider next how an intentional agent could have reasons for action that are bound to social contexts and fail to generalize. We've already considered one possible illustration: transitive inferences could be made for dominance hierarchies but not quantities of food.

Tomasello suggests that non-human primates have a special ability to understand the social relations of conspecifics that hold among third parties, such as the mother/child relation. They are also unusual in their ability to learn relations among objects: for example to choose a pair of objects that display the same relation as a sample pair. However, mastering relations among objects is a difficult task for non-human primates, taking hundreds or thousands of trials, whereas understanding of third party relations among conspecifics is seemingly effortless.[45] Their skill with relations does not readily generalize from the social to the non-social domain.

Cosmides' and Tooby's work on the Wason effect suggests that even for human primates certain reasons for action are bound to specific social contexts and fail to generalize readily.[46] Wason (1996) asked people to test a simple instance of 'if p then q': if a card has 'D' on one side, it has '3' on the other side. Subjects observed four cards,

---

[43] I suggest that the fact that substituting rocks for candies did not improve performance significantly, while substituting numerals for candies did, is also better explained in terms of the scaffolding of rationality by symbols than in terms of the enabling of inhibition (see and cf. Boysen *et al.* 1996, 80, 84). However, my preference for the scaffolding over the inhibition account is also in part a reflection of a big picture preference, for a horizontally modular rather than a vertically modular conception of rationality; see Hurley 1998, Ch. 10.

[44] Hence it is not undermined by different results in later work with mixed candy/numeral pairs; again, see Boysen, this volume.

[45] Tomasello 1999, 18; see also and cf. Tomasello and Call 1997; Cheney and Seyfarth 1990; note 20 above.

[46] Cosmides and Tooby 1992; Cosmides 1989; for refinements and qualifications see Gigerenzer and Hug 1992. In considering how reasons can be context-bound, issues about domain-specificity should be distinguished from issues about innateness; Cosmides and Tooby (1992) tend to run these issues together. Domain-specificity does not entail innateness.

showing on their upturned sides: D, F, 3, 7. They were asked which cards they should turn over to determine whether the rule was correct. The right answer is: the D card and the 7 card. Most people (90–95%, including those trained in logic) choose either just the D card or the D card and the 3 card.

But Cosmides and Tooby show that people *do* get the right answer when they are asked to test instances of 'if p then q' that express a rule of the form: if you take a benefit, you must meet a requirement. One interpretation of this result is that people are very good at detecting cheaters (see and cf. Chater and Oaksford 1996; Oaksford and Chater 1994; Gigerenzer and Hug 1992); they can readily perceive reasons to act in order to flush out cheaters. But their reasons are highly domain-specific, and do not generalize readily, even to other social contexts. People do not get the right answer even for the converse type of rule: if you meet a requirement, you get a benefit. When an agent acts on her perceptions so as to flush out a cheater, she can be acting on her own reasons, available from her own perspective, even though they are not inferentially promiscuous. She may have domain-general conceptual abilities—so that she could be brought to see how her reasons generalize—but fail to exercise them.

An interesting though speculative recent twist on Cosmides' and Tooby's demonstration of the domain-specificity of reasons is suggested by early results of non-verbal false belief tests given to chimps and dolphins. These non-verbal false belief tests are being developed and applied by Hare, Call, Tomasello, Tschudin, and others, in a series of recent articles and work in press and in progress. The interpretation below is one possible interpretation of early results, and is subject to further empirical work. But for present purposes it serves to illustrate a relevant possibility.

One version of a non-verbal mind-reading test uses a hider/communicator paradigm, and has been applied to children and to chimps.[47] In the training phase of this paradigm, the subject perceives two opaque boxes, which are then hidden from her view. The hider then proceeds to hide a reward in one of the two boxes, while the communicator watches. The subject cannot see which box the hider puts the reward in. But the subject can see that the communicator can see which box the hider puts the reward in. The barrier is removed, and the subject is then allowed to choose between the two opaque boxes, still unable to see which contains the reward. The communicator truthfully indicates to the subject which box the reward is in. The subject learns to choose the box the communicator has indicated in order to obtain the reward.

In the critical trials, the procedure is altered as follows. After the hider has put the reward in one of the boxes, the communicator leaves the scene. The barrier is removed so that the subject can see the boxes, though not which box contains the reward. While the subject watches, the hider switches the positions of the boxes. The communicator then returns, and indicates the box not containing the reward to the subject, since this box is

[47] Call and Tomasello 1999; Call *et al* 2000; work is in progress with bottlenose dolphins; see Tschudin, this volume.

now in the position of the box into which the communicator saw the reward placed. The subject can choose either the box indicated by the communicator, or the other box. The correct response is to choose the other box, since the communicator did not see that the boxes were switched and thus has a false belief about which contains the reward.

When this non-verbal test of mind-reading ability is applied to children of varying ages, the results are strongly correlated with the results of verbal false belief tests. In general, children under 4 years old make the wrong response, and select the box the communicator indicates in the false belief trials as well as the control trials. Children over 5 years old make the correct response, and select the box not indicated by the communicator when the boxes are switched, though they select the box indicated by the communicator in the control trials. Chimps readily learn to choose the box indicated by the communicator in the training phase and control trials, but fail profoundly in the critical false belief trials, and continue to choose the box indicated by the 'deluded' communicator.

If this test is accepted as an indication of the ability to reason about the mental states of others, chimps appear to lack such ability. However, these results contrast with results for chimps of a different non-verbal mind reading test, suggesting that chimps may have a domain-specific ability to act for reasons that depend on perceptions of the mental states of others.

This second version of a non-verbal mind reading test uses a dominant/subordinate paradigm.[48] The dominant and subordinate chimps compete for food. In some conditions the dominants had not seen the food hidden, or food they had seen hidden was moved to a different location when they were not watching (whereas in control conditions they saw the food being hidden or moved). At the same time, subordinates always saw the entire baiting procedure and could monitor the visual access of their dominant competitor as well. The results are that subordinates more often approach and obtain the food that dominants have not seen hidden or moved. Similarly, the subordinate gets more food when a new dominant chimp is substituted for the one who saw the baiting. This suggests a kind of mind reading ability on the part of chimps: that subordinates are rationally sensitive to what dominants did or did not see during baiting.

How should the apparent ability of chimps to act rationally in light of the probable mental states of others in the dominant/subordinate paradigm be reconciled with the apparent lack of this ability in the hider/communicator paradigm? It is too soon to say with any confidence.[49] But an empirical speculation of particular interest here is that

----

[48] Hare *et al.* 2000, 2001; Tomasello and Call, this volume.

[49] It can be argued that there is a confound present, and that the subordinate may have learned either 'no point trying when that guy could see where the food went' or 'no point trying when that guy's face/head/body was present during baiting'. See and cf. Heyes 1998; Tomasello and Call, this volume; Povinelli and Vonk, this volume; thanks here to Cecilia Heyes for discussion.

the hider/communicator paradigm provides a context in which there is co-operation over finding food, while the dominant/subordinate paradigm provides a context in which there is competition over finding food.[50] It may be natural for chimps to compete over food; their ability to act rationally in light of the mental states of others may be evolutionarily tuned to competitive practical contexts rather than co-operative ones. This provides another possible illustration of how practical reasons might be domain-specific, and so fail to have full conceptual generality.

6.7 **Simulation: context-bound reasons vs. conceptual abilities**

I have located the possibility of having non-conceptualized reasons in the practical sphere. In this respect, my position has common ground with recent work on simulation as a way of understanding mind reading and the enhanced social rationality it enables.

Recall from Section 6.4 above that simulation can be used in various ways to enhance control. It is a generally useful trick, which evolution may stumble on for various purposes other than (or as well as) mind reading. Some uses of simulation to enhance control do not require the simulator to distinguish simulations from non-simulations, while others do. When a forward model is used online to automatize and speed up a practical skill, to free it from the constraints of real time feedback, the agent need not know whether he is using a simulation or not. But more sophisticated uses of simulation, to consider counter-factual alternatives in order to assess their instrumental value and plan, or to mind read, do require the simulator to keep track of whether he is simulating or not. A possible/actual distinction, or a self/other distinction, thus arises for the agent.[51] These contribute both to the flexible generality of the agent's skills and to the sense in which he can make mistakes and hence to satisfying the normativity requirement.

I suggest that such sophisticated forms of simulation could provide reasons for action, even though they do not by themselves provide the inferential promiscuity of conceptual abilities (see also and cf. Millikan, this volume; Proust, this volume; Bennett 1964, 116–117). Simulationist accounts of cognitive skills view them as rooted in practical rather than theoretical or conceptual abilities.[52] As a result, such skills should be liable to the domain-specificity of practical skills. That is, we should expect the reasons for action provided by process-driven[53] simulation to be relatively domain-specific, since what one

---

[50] This was raised as one possibility by Josep Call, in discussion. See Tomasello and Call, and cf. Povinelli and Vonk, this volume; see also Povinelli 1996, 322–3 on the possibility that chimps only deploy their representations of mental states in certain ecological contexts.

[51] This line of thought is further developed in Hurley 2005a, 2006.

[52] On this contrast, see for example Freeman 1995; Gordon 1995a, 108, 113–114; Gopnik and Wellman 1995, 232; see also Gordon 1995b, 1996.

[53] In Goldman's sense (1995).

can simulate is more or less bound to what one can do, and know-how itself tends to be relatively domain-specific.

For example a simulationist account of mind reading explains how simulation can provide an agent with reasons for action in social contexts even if the agent cannot conceptualize and theorize agency.[54] An agent may be able to do certain things, such as compete for food, and not others, such as co-operate over food, for evolutionary reasons. As a result, she may be able to simulate doing certain things and not others, and hence to act for reasons that depend on mind reading in competitive but not co-operative contexts, even though similar inferences are available in both contexts.

Of course, it has proved extremely difficult to demonstrate unequivocal mind reading abilities in non-human animals. But my point here is not that animals can mind read, whether by simulation or otherwise. It is that simulationist accounts of cognitive skills provide general-purpose insight into how control can be enhanced to provide reasons for action that are nevertheless context-bound. This general point holds, even if non-human animals cannot mind read.

6.8 What is the point of making sense of animals?

I've argued that the functions and interest of interpretation are not wholly discontinuous at the human/non-human boundary. We are certainly interested in causally explaining human behavior and mental capacities in evolutionary and ecological and neuroscientific terms. But we are also interested in making sense of human beings, including those with whom we have no practical interactions. Why should only the former interests be valid when we turn to non-human animals? Why shouldn't we also allow ourselves to be interested in animals at both levels (as in fact we are)?

Interpreting animals and interpreting human beings can both have practical functions. Animals appear not to be interpreters themselves, so interpreting animals will be unilateral. But unilateral interpretation can still have practical functions, if not the same ones that mutual interpretation has.

We should be wary of projecting onto humanity at large the perspective of cosmopolitan city dwellers unused to living and working with animals but used to interacting with many different human cultures. This is not the perspective of most practitioners of folk psychology throughout human history, or even today. Almost all human beings co-ordinate behaviors with fellow humans in their local community, and no doubt interpretation has important functions in this context. But beyond that, I reckon that many human beings have actually done more in the way of regular co-ordination of behavior with non-human animals than with human beings from

---

[54] See Gordon 1995a, 113–114; 1996, 14; but compare collapse worries concerned with tacit theory; Stone and Davies 1996, 132. Compare also Heal (1996), who emphasizes the relevance of simulation to dealing with the relations between the thoughts of another.

strange and distant cultures. Many human beings normally do interact in meaningful ways with animals and have found it natural to interpret animals, albeit unilaterally, in these contexts. We often overdo it, but it's not obvious that interpreting the animals people live and work with has no practical function.

Even if it is right to suspect that making sense of animals has often had practical functions for human beings, there are important differences between the functions of interpreting animals and people (whether familiar or strange). It is not just that our power over animals or our distance from them may attenuate the practical benefits of interpreting them (that could be true of people also). Even if other animals have minds for us to interpret, most current evidence suggests that they are not mind readers themselves. Asking what is rational for a creature to do when it plays 'against nature' is very different from asking what is rational for a creature to do when it plays against another rational agent whom it is trying to interpret and who is also trying to interpret it. If non-human animals are not mind readers, then game-theoretic problems of mutual interpretation and prediction do not arise in the same way for our relations with them, and strategic rationality does not really get a grip on animals. However, it may still be useful to make sense of non-human animals playing against nature as rational.[55] Moreover, playing against nature may include playing against the behavior of other animals, and it can be hard to say just where sophisticated behavior-reading ends and mind reading starts.

However, I doubt that our interest in interpreting either human or other animals can be fully explained in pragmatic terms. We often are interested in making sense of creatures very different from ourselves, even when doing so transparently has no practical point. When we encounter people from strange or distant cultures, our interest in interpreting them is often compelling, regardless of any practical reason to do so or whether we will ever encounter them again. We are extremely interested in making sense of people we will only ever read about, whether real or fictional, past or present. The organizing function of interpretation in human communities cannot fully account for the natural, immediate, and profound human interest in making sense even of other human beings.

Here's a just-so story. At some point in the infancy of humanity our ancestors cracked the mind-reading trick and rather went wild with it. They projected minds all over the world, not just to non-human animals—finding spirits in trees and stones, gods in stars and clouds. Presumably, for all the confusion this involved, our ancestors reaped some net benefits from the ability to identify with others, and some truths about other minds were mingled with the many falsehoods. Having started out doing intentional explanation promiscuously, we eventually grew up and out of our animism;

---

[55] Indeed, given the interdeterminacies of mutual prediction that plague game theory, non-strategic rationality may be more predictive than strategic rationality.

we worked our way pure to causal explanation. We then started to retrench on seeing intentionality everywhere: anthropomorphizing the world was to be avoided as an embarrassing sign of our former collective intellectual immaturity. But our hands have been relatively clean for some time; we can now afford to think again. There may be some overlooked truths mingled with the discarded falsehoods. There may be something after all to make sense of in the strange non-human minds around us, without anthropomorphizing them: continuities as well as discontinuities to be described between human and non-human minds.

The alienness of non-human animals is part of what makes it so interesting to consider whether they can sometimes be understood as rational agents, without simply projecting substandard humanity onto them. In particular, it is interesting to try to understand rational agency in the absence of its distinctively human concomitants such as conceptual and mind-reading abilities.

Our explanatory interests in animals are, like rationality itself, multidimensional. An interest in whether animals are agents does not *follow* from an interest in their evolution and ecology; but neither does it follow that this is *not* an interesting question. Our interest in agency needn't derive from our interests in evolution and ecology, even if the latter are relevant to understanding whether animals are agents. We human beings are keen interpreters, interested in making sense of others as rational, in various ways and to various degrees, whether or not they do so in return. Interpreting has practical functions, but our interest in interpreting also extends beyond these. It is a self-standing explanatory interest, and doesn't need underwriting in terms of other explanatory interests. If someone from Alpha Centauri doesn't share that interest, I have no argument that he should. But many down here do, and I see no argument that they should not either. Why subject our interest in making sense of others to the Alpha Centauri test?

Perhaps the worry is that interpretation is just a human hobby. But we could have that worry about many of our worthwhile scientific and other activities. We might not be able to persuade someone from Alpha Centauri to share them either, but that in itself wouldn't be a reason to give them up.

6.9 **Summary and concluding remarks**

I've considered examples of how an intentional agent's own reasons to act could be bound to symbolic or social contexts. For present purposes, these cases only need to illustrate an empirical possibility; further empirical work may well be needed to determine if this is the right interpretation to give in particular cases. Cases such as these should motivate us to admit the possibility that non-human animals who lack conceptual abilities can nonetheless be intentional agents and act for reasons. We should admit this possibility in order to be in a position to characterize the practical abilities and perspectives of non-human animals correctly, as neither too rich nor too impoverished. It is possible for such creatures to act for reasons while doing very little in the way of conceptually structured inference or theorizing.[56] They can be intentional agents even if the normativity of their non-conceptual intentional agency plays no role in an epistemological project.[57] Animals can occupy islands of instrumental rationality, without being in the business of trying to justify their beliefs, or of trying to understand others as engaged in justifying their beliefs.

I have also argued that reasons for action in such cases need not be regarded as 'sub-animal' level phenomena, but can properly be attributed to the intentional agents in question—even if they lack full conceptual abilities. Why? For the old familiar reasons: holism and normativity, which characterize the personal or animal level. Perceptual information may not lead to an invariant response, but explain action only in the context set by intentions and the constraints of at least primitive forms of practical rationality. Perceptions and intentions may combine to make certain actions reasonable and appropriate from the animal's perspective, and mistakes are possible. As explained, the holism and normativity of intentional agency bring with them a kind of coarse recombinant structure, but this falls well short of enabling the context-free inferential promiscuity of conceptual abilities. An animal's various goals could nevertheless give him reasons to act in one way rather than another in particular circumstances—his own reasons, reasons from his own perspective. The animal may not *conceptualize these as reasons*—but to require that would be to beg the question at issue.

We shouldn't overintellectualize what is to have a mind. We don't have to choose between conceptualized, inferentially promiscuous reasons and the fine, rich kind of justification they provide, on the one hand, and the absence of reasons that are reasons for the agent, on the other: this dichotomy is spurious. The space of reasons is not coextensive with the space of conceptualized inference and theory, but rather with the space of intentional actions at large. This is a space in which non-human animals can and do act, a space largely occupied by instrumental motor actions embedded in specific natural domains. Perhaps the metaphor of a space of reasons should be replaced with the metaphor of islands of reasons emerging from a sea of causes. For us human animals, language provides bridges that finally link these islands together.[58]

---

[56] This can be true even for human agents with conceptual abilities. Indeed, inferential mediation can be self-defeating in relation to some virtuous reasons for action: when a virtuous person jumps into a river to save a drowning child, she does so because that child is drowning. If someone jumps in because that is what virtue demands in this case and she wants to be virtuous, her reason for acting is quite different and the virtue of her action is compromised.

[57] Moreover, some animals, as well as very young children, may be able to understand others as intentional agents in relation to goals even if they cannot understand them as epistemic subjects. See Harris 1996, 213; Povinelli 1996, 329; Carruthers 1996, 265; Gomez 1996, 341–2.

[58] However, see also Tomasello 1999, 138 ff, on failures to generalize patterns across 'verb islands'.

Acknowledgements

For discussion and comments, thanks to Greg Currie, Martin Davies, Julia Driver, Mark Greenberg, Celia Heyes, Ram Neta, Alva Noë, Matthew Nudds, Hanna Pickard, Joëlle Proust, Nicholas Rawlins, Helen Steward, and members of audiences to whom I've presented related material on various occasions. Many thanks to *Mind and Language* for permission to reuse material from Hurley 2003a and 2003b, and to Peter Godfrey-Smith and Kim Sterelny for their thoughtful (2003) commentaries on Hurley 2003a. My replies to them (2003b) are incorporated into this version of the material.

References

Allen, C., and Bekoff, M. (1997). *Species of Mind*. Cambridge: MIT Press.

Astington, J. (1996). What is theoretical about a child's theory of mind? In: P. Carruthers and P. K. Smith, eds. *Theories of Theories of Mind*, pp. 184–199. Cambridge: Cambridge University Press.

Bacharach, M., and Hurley, S. (1991). *Foundations of Decision Theory*. Oxford: Blackwell.

Balleine, B. W. (1992). The role of incentive learning in instrumental performance following shifts in primary motivation. Journal of Experimental Psychology, Animal Behavior Processes, **18**: 236–150.

Balleine, B. W., and Dickinson, A. (1991). Instrumental performance following reinforcer devaluation depends upon incentive learning. *Quarterly Journal of Experimental Psychology*, **43**(B): 279–296.

Bennett, J. (1964/1989). *Rationality*. Indianapolis: Hackett.

Bermúdez, J. L. (1998). *The Paradox of Self-Consciousness*. Cambridge MA: MIT Press.

Blackburn, S. (1995). Theory, observation and drama. In: M. Davies and T. Stone, eds. *Folk Psychology*, pp. 274–290. Oxford: Blackwell.

Boysen, S. T. (1996). More is less: the elicitation of rule-governed resource distribution in Chimpanzees. In: A. Russon, K. Bard and S. Taylor Parker, eds. *Reaching into Thought: The Minds of the Great Apes*, pp. 176–189. Cambridge, UK: Cambridge University Press.

Boysen, S. T., and Bernston, G. (1995). Responses to quantity: perceptual vs. cognitive mechanisms in chimpanzees (*Pan troglodytes*). *Journal of Experimental Psychology and Animal Behavior Processes*, **21**: 82–86.

Boysen, S. T., Bernston, G., Hannan, M., and Cacioppo, J. (1996). Quantity-based inference and symbolic representation in chimpanzees (*Pan troglodytes*). *Journal of Experimental Psychology and Animal Behavior Processes*, **22**: 76–86.

Boysen, S. T., Mukobi, K. L., and Berntson, G. G. (1999). Overcoming response bias using symbolic representations of number by chimpanzees (*Pan troglodytes*). *Animal Learning and Behavior*, **27**: 229–235.

Brewer, W. (1999). Perception and Reason. Oxford: Oxford University Press.

Brooks, R. (1999). *Cambrian Intelligence*. Cambridge: MIT Press.

Call, J., Agnetta, B., and Tomasello, M. (2000). Social cues that chimpanzees do and do not use to find hidden objects. *Animal Cognition*, 3: 23–34.

Call, J., and Tomasello, M. (1996). The effect of humans on the cognitive development of apes. In: A. Russon, K. Bard and S. Taylor Parker, eds. *Reaching into Thought: The Minds of the Great Apes*, pp. 371–403. Cambridge, UK: Cambridge University Press.

Call, J., and Tomasello, M. (1999). A non-verbal theory of mind test: the performance of children and apes. *Child Development*, **70**: 381–395.

Carruthers, P. (1996). Autism as mind-blindness. In: P. Carruthers and P. K. Smith, eds. *Theories of Theories of Mind*, pp. 257–273. Cambridge: Cambridge University Press.

Carruthers, P., and Smith, P. K., eds. (1996). *Theories of Theories of Mind*. Cambridge: Cambridge University Press.

Chater, N., and Heyes, C. (1994). Animal concepts: content and discontent. *Mind and Language*, **9**: 209–246.

Chater, N., and Oaksford, M. (1996). Deontic reasoning, modules, and innateness: A second look. *Mind and Language*, **11**: 191–202.

Cheney, D. L., and Seyfarth, R. M. (1990). *How Monkeys See the World*. Chicago: University of Chicago Press.

Cosmides, L. (1989). The logic of social exchange: has natural selection shaped how humans reason? Studies with the Wason Selection Task. *Cognition*, **31**: 187–276.

Cosmides, L., and Tooby, J. (1992). Cognitive adaptations for social exchange. In: J. Barkow, L. Cosmides and J. Tooby, eds. *The Adapted Mind*, pp. 163–228. New York: Oxford University Press.

Davidson, D. (2001). Rational animals. In: D. Davidson, ed. *Subjective, Intersubjective, Objective*, pp. 95–105. Oxford: Clarendon Press.

Davies, M., and Stone, T., eds. (1995a). *Folk Psychology*. Oxford: Blackwell.

Davies, M., and Stone, T., eds. (1995b). *Mental Simulation*. Oxford: Blackwell.

Dennett, D. C. (1991). *Consciousness Explained*. Boston: Little Brown.

Dennett, D. C. (1996). *Kinds of Minds*. London: Weiden and Nicholson.

Elman, J., Bates, E., Johnson, M., Karmiloff-Smith, A., Parisi, D., and Plunkett, K. (1996). *Rethinking Innateness*. Cambridge: MIT Press.

Freeman, N. H. (1995). Theories of the mind in collision: plausibility and authority. In: M. Davies and T. Stone, eds. *Mental Simulation*, pp. 68–86. Oxford: Blackwell.

Gigerenzer, G., and Hug, K. (1992). Domain-specific reasoning: social contracts, cheating, and perspective change. *Cognition*, **43**: 127–171.

Gigerenzer, G., Todd, P., and the ABC Research Group (1999). *Simple Heuristics that Make us Smart*. New York: Oxford University Press.

Godfrey-Smith, P. (2003). Folk Psychology under Stress: Comments on Susan Hurley's Animal action in the space of reasons, *Mind and Language*, **18**: 266–272.

Goldman, A. (1995). Indefense of the Simulation Theory. In: M. Davies and T. Stone, eds. *Folk Psychology*. Oxford: Blackwell. pp. 191–206.

Gomez, J.-C. (1996). Non-human primate theories of (non-human primate) minds. In: P. Carruthers and P. K. Smith, eds. *Theories of Theories of Mind*, pp. 330–343. Cambridge: Cambridge University Press.

Gopnik, A., and Wellman, H. M. (1995). Why the child's theory of mind really *is* a theory. In: M. Davies and T. Stone, eds. *Folk Psychology*, pp. 232–258. Oxford: Blackwell.

Gordon, R. (1995a). The simulation theory: objections and misconceptions. In: M. Davies and T. Stone, eds. *Folk Psychology*, pp. 100–122. Oxford: Blackwell.

Gordon, R. (1995b). Simulation without introspection or inference from me to you. In: M. Davies and T. Stone, eds, *Mental Simulation*. Oxford: Blackwell, pp. 53–67.

Gordon, R. (1996). Radical simulationism. In: P. Carruthers and P. K. Smith, eds. *Theories of Theories of Mind*, pp. 11–21. Cambridge: Cambridge University Press.

Greenberg, M. (in progress). *Thoughts Without Masters*.

Hala, S., and Russell, J. (2001). Executive control within strategic deception: a window on early cognitive development? *Journal of Experimental Child Psychology*, **80**: 112–141.

Hare, B., Call, J., Agnetta, B., and Tomasello, M. (2000). Chimpanzees know what conspecifics do and do not see. *Animal Behaviour*, **59**: 771–785.

Hare, B., Call, J., and Tomasello, M. (2001). Do chimpanzees know what conspecifics know? *Animal Behaviour*, **61**: 139–151.

Harris, P. (1996). Desires, beliefs, and language. In: P. Carruthers and P. K. Smith, eds. *Theories of Theories of Mind*, pp. 200–220. Cambridge: Cambridge University Press.

Heal, J. (1996). Simulation, theory, and content. In: P. Carruthers and P. K. Smith, eds., *Theories of Theories of Mind*, pp. 75–89. Cambridge: Cambridge University Press.

Heyes, C. (1998). Theory of mind in nonhuman primates. *Behavioral and Brain Sciences*, **21**: 101–148.

Heyes, C., and Dickinson, A. (1993). The intentionality of animal action. In: M. Davies and G. Humphreys, eds. *Consciousness*, pp. 105–120. Oxford: Blackwell.

Hurley, S. (1989). *Natural Reasons*. New York: Oxford University Press.

Hurley, S. (1998). *Consciousness in Action*. Cambridge: Harvard University Press.

Hurley, S. (2001). Overintellectualizing the mind. *Philosophy and Phenomenological Research*, **58**: 423–431.

Hurley, S. L. (2003a). Animal action in the space of reasons. *Mind and Language*, **18**: 231–256.

Hurley, S. L. (2003b). Making sense of animals: Interpretation vs. architecture. *Mind and Language*, **18**: 273–281.

Hurley, S. (2005a). The shared circuits model: How control, mirroring, and simulation can enable imitation and mind reading. Interdisciplines webforum on mirror neurons, archived at: www.interdisciplines/mirror/papers/5.

Hurley, S. (2005b). Social heunztics that make us smarter. *Philosophical Psychology*, **18**: 585–611.

Hurley, S. (2006). Active perception and perceiving action: the shared circuits model. In: T. Gendler and J. Hawthorne, eds. *Perceptual Experience*. New York: Oxford University Press.

Kelso, S. J. A. (1995). *Dynamic Patterns: The Self-Organization of Brain and Behavior*. Cambridge, MA: MIT Press.

Kripke, S. A. (1982). *Wittgenstein on Rules and Private Language*. Cambridge: Harvard University Press.

Luntley, M. (2003). Non-conceptual content and the sound of music. *Mind and Language*, **18**: 402–426.

MacIntyre, A. (1999). *Rational Dependent Animals*. London: Duckworth.

McDowell, J. (1994). The content of perceptual experience. *Philosophical Quarterly*, **44**(175): 190–205.

Millikan, R. G. (1984). *Language, Thought and Other Biological Categories*. Cambridge MA: MIT Press.

Millikan, R. (1991). Perceptual content and Fregean myth. *Mind*, **100**: 439–459.

Millikan, R. (1993). Content and vehicle. In: N. Eilan, R. McCarthy, and B. Brewer, eds. *Spatial Representation*, pp. 256–268. Oxford: Blackwell.

Noë, A. (2002). Is perspectival self-consciousness nonconceptual? *Philosophical Quarterly*, **52**(207): 185–194.

Noë, A. (2004). *Action in Perception*. Cambridge, Mass: MIT Press.

Oaksford, M., and Chater, N. (1994). A rational analysis of the selection task as optimal data selection. *Psychological Review*, **101**: 608–631.

Povinelli, D. J. (1996). Chimpanzee theory of mind? The long road to strong inference. In: P. Carruthers and P. Smith (eds). *Theories of Theories of Mind*, pp. 293–329. Cambridge: Cambridge University Press.

Premack, D. (1978). On the abstractness of human concepts: why it would be difficulty to talk to a pigeon. In: S. Hulse, H. Fowler, and W. Honig, eds. *Cognitive Processes in Animal Behavior*, pp. 423–451. Hillsdale, NJ: Erlbaum.

Rudder Baker, L. (1995). *Explaining Attitudes*. Cambridge: Cambridge University Press.

Russell, J., Jarrold, C., and Potel, D. (1994). What makes strategic deception difficult for children— the deception or the strategy? *British Journal of Developmental Psychology*, **12**: 301–314.

Russell, J., Mauthner, N., Sharpe, S., and Tidswell, T. (1991). The 'windows task' as a measure of strategic deception in preschoolers and autistic subjects. *British Journal of Developmental Psychology*, **9**: 331–349.

Russon, A., Bard, K., and Taylor Parker, S., eds. (1996). *Reaching into Thought: The Minds of the Great Apes*. Cambridge, UK: Cambridge University Press.

Samuels, M. C., Brooks, P. J., and Frye, D. (1996). Strategic game playing in children through the windows task. *British Journal of Developmental Psychology*, **14**: 159–172.

Segal, G. (1996). The modularity of theory of mind. In: P. Carruthers and P. K. Smith, *Theories of Theories of Mind*, pp. 141–157. Cambridge: Cambridge University Press.

Smith, P. K. (1996). Language and the evolution of mind-reading. In: P. Carruthers and P. K. Smith, eds. *Theories of Theories of Mind*, pp. 344–354. Cambridge: Cambridge University Press.

Sterelny, K. (2001). *The Evolution of Agency*. Cambridge, UK: Cambridge University Press.

Sterelny, K. (2003). Charting Control-Space: A Commentary on Susan Hurley's 'Animal Action in the Space of Reasons'. *Mind and Language*, **18**: 257–266.

Stone, T., and Davies, M. (1996). The mental simulation debate: a progress report. In: P. Carruthers and P. K. Smith, eds. *Theories of Theories of Mind*, pp. 119–137. Cambridge: Cambridge University Press.

Taylor Parker, S., and Russon, A. (1996). On the wild side of culture and cognition in the great apes. In: A. Russon, K. Bard, and S. Taylor Parker, eds. *Reaching into Thought: The Minds of the Great Apes*, pp. 430–450. Cambridge, UK: Cambridge University Press.

Thompson, R., Boysen, S., and Oden, D. (1997). Language-naïve chimpanzees (*Pan troglodytes*) judge relations between relations in a conceptual matching-to sample task. *Journal of Experimental Psychology: Animal Behavior Processes*, **23**: 31–43.

Tomasello, M. (1999). *The Cultural Origins of Human Cognition*. Cambridge, MA: Harvard University Press.

Tomasello, M., and Call, J. (1997). *Primate Cognition*. New York: Oxford University Press.

Van Gulick, R. (1988). A functionalist plea for self-consciousness. *Philosophical Review*, **97**: 149–181.

Wason, P. (1966). Reasoning. In: B. Foss, ed. *New Horizons in Psychology*, pp. 135–151. London: Penguin.

Whiten, A. (1996a). Imitation, pretense, and mind-reading. In: A. Russon, K. Bard, and S. Taylor Parker, eds. *Reaching into Thought: The Minds of the Great Apes*, pp. 300–324. Cambridge, UK: Cambridge University Press.

Whiten, A. (1996b). When does smart behaviour-reading become mind-reading?. In: P. Carruthers and P. K. Smith, *Theories of Theories of Mind*, pp. 277–292. Cambridge: Cambridge University Press.

Williams, B. (1973). Ethical consistency. In: B. Williams (ed.), *Problems of the Self*, pp. 166–186. Cambridge, UK: Cambridge University Press.

Part II

Rational versus associative processes

Chapter 7

Transitive inference in animals: Reasoning or conditioned associations?

Colin Allen

Abstract

It is widely accepted that many species of non-human animals appear to engage in transitive inference, producing appropriate responses to novel pairings of non-adjacent members of an ordered series without previous experience of these pairings. Some researchers have taken this capability as providing direct evidence that these animals reason. Others resist such declarations, favouring instead explanations in terms of associative conditioning. Associative accounts of transitive inference have been refined in application to a simple five-element learning task that is the main paradigm for laboratory investigations of the phenomenon, but it remains unclear how well those accounts generalize to more information-rich environments such as primate social hierarchies, which may contain scores of individuals. The case of transitive inference is an example of a more general dispute between proponents of associative accounts and advocates of more cognitive accounts of animal behaviour. Examination of the specific details of transitive inference suggests some lessons for the wider debate.

7.1 Transitive inference

Transitive relationships are frequently important to animals, especially those living in social groups. Some of these relationships are manifest in perception: if A is larger than B, and B is larger than C, then simple inspection of A next to C will reveal that A is larger; no reasoning is required. But it is also possible to draw the inference that A is larger than C without having to see A and C side by side. Most adult humans have the capacity for such reasoning, demonstrating their understanding of the transitivity of the larger-than relationship. Other transitive relationships are not directly manifest in perception. If A is a faster runner than B and B is a faster runner than C, it will not always be able to tell just by looking at them. But once these relationships are known, the judgement that A is a faster runner than C is an inference that can be based on the transitivity of the *faster-than* relationship.

Social dominance relationships are also typically not manifest in perception (Martin *et al.* 1997), providing a domain in which a capacity for transitive inference would seem to be very useful. If animal A dominates B, and B dominates C, there need be no common perceptual marker of this dominance, and no direct comparison of A to C will reveal the relationship between them. Being able to infer A's dominance of C in a linear dominance hierarchy could be very advantageous in an environment in which losing a struggle could result in injury, and reduced access to food and other resources, with potentially serious consequences for fitness (see, for example, Beaugrand *et al.* 1997).

Transitive relationships define orderings: $A > B > C$, etc. Cognitive approaches to transitive inference postulate that animals explicitly represent such orderings and use these representations to infer relationships between pairs of non-neighbouring elements (Zentall 2001).[1] Behaviouristically-trained psychologists favour explaining apparent transitive inferences in terms of the past reinforcement history of the individual elements without invoking any explicit representation of the entire series. This characterization of the distinction between cognitive and associative approaches to transitive inference will suffice for my present purposes; I will have more to say about the distinction below.

7.2 Explaining (the appearance of) transitive inference

The simplest experiment that generates the appearance of transitive inference consists of training an animal with two pairs of (arbitrarily-labelled) stimuli, A+B− and B+C−. (Here '+' means that selection of this item is rewarded and '−' means this item is not rewarded. The letters A through C are our labels for the stimuli, not the actual stimuli themselves, which may be arbitrary shapes, smells, etc.) When trained in this way using standard operant conditioning procedures, pigeons and rats presented with the novel pair AC are highly likely to select A. However, this particular result admits of a very simple associative explanation. In training, A was always rewarded and C never rewarded. Hence the preference for A over C can be explained entirely in terms of the past reinforcement history for the individual elements; the animal is simply picking the one that has been rewarded in the past.[2]

---

[1] Editors' note: see also Pepperberg (this volume) on transitive inference.

[2] Editors' note: Allen is concerned here with the character of the processes supposedly required by PP-rationality for genuine transitive inference. Compare this with Kacelnik's worry (this volume) that very little behaviour is ruled out by transitivity of revealed preference in E-rationality, given the possibility of state- or context-dependent preferences. Given behaviour that apparently expresses transitive inference, Allen is concerned with the distinction between rational or associative explanations of the processes that generate such behaviour. Kacelnik's concern is rather with how such behaviour can be identified in the first place.

This result leads to a slightly more sophisticated experiment, that has become the industry standard for laboratory investigations of transitive inference in animals. In the five-element procedure, the animals are trained with four pairs of stimuli: A+B−, B+C−, C+D−, and D+E−. Once they have reached a certain criterion level of correct performance on these pairs, the subjects are then tested with the novel pair BD. Many kinds of animal (e.g. rats, pigeons, monkeys) tested in this way reliably select B. In the training set, B is rewarded exactly as frequently as D (on average, 50 per cent of the time—that is always when paired with C and E respectively, and never when paired with A and C). Consequently there is no explanation of the preference for B over D simply in terms of the past history of direct reinforcement of selections of each of these individual elements. Successful transfer on this test has been taken by many to provide strong evidence for inference by animals. Hence, for example, Dusek and Eichenbaum (1997, p. 7109) write: 'An appropriate choice between the two non-adjacent and non-end elements, B and D, provides unambiguous evidence for transitive inference.'

Despite this claim, however, ambiguity between associative and cognitive accounts remains. Fersen *et al.* (1991) had already suggested that differential conditioning effects could account for the selection of B over D through a mechanism of 'value transfer'. According to the theory of value transfer, in any simultaneous discrimination task, some of the value associated with the S+ is transferred to the accompanying S−. According to the value transfer theory, in the A+B−, B+C−, C+D−, D+E− training set, even though both B and D are individually rewarded at the same rate, B is seen in association with A, which is always a winner. This is hypothesized to give B a positive boost in comparison to D.

A test of positive transfer theory is described by Zentall (2001; see also Zentall and Sherburne 1994). Pigeons were trained on just two pairs: AB and CD. B and D were never rewarded when selected. Selection of A was rewarded on 100 per cent of the occasions it was selected (represented by $A_{100}B_0$), whereas selection of C was rewarded on just 50 per cent of the occasions it was selected (represented by $C_{50}D_0$). Once the pigeons had reached a criterion level of performance—reliably selecting A over B and C over D—they were tested with the novel pair BD. The training set does not justify a transitive inference, but pigeons trained in this fashion nevertheless tend to select B over D, hence confirming the value transfer hypothesis that B gains simply by being paired with the reliably rewarded A.

Despite these results, Zentall (2001) does not entirely discount the possibility that a cognitive explanation of the pigeons' behaviour might be correct. At the very least, he accepts the utility of a cognitive perspective as a heuristic for devising novel experiments, with the ultimate goal of constructing better associative models. Describing ongoing work that might support an account in terms of explicit representation of the ABCDE series, he writes, 'Whether this line of research will provide evidence in support of a cognitive account is less important than the fact that the investigation of the transitive inference effect led to a series of experiments that clarified the interaction between the S+ and S− in a simultaneous discrimination.' In a similar vein, DeLillo *et al.* (2001) show that the BD generalization is modelled by a standard associative learning mechanism: backward error propagation in an artificial neural network (Rumelhart and McClelland 1986). DeLillo and colleagues conclude that 'a simple error-correcting rule can generate transitive behaviour similar to the choice pattern of children and animals in the binary form of the five-term series task without requiring high-order logical or paralogical abilities.' The implication here is that the burden of proof lies with those who would argue that animals are engaged in reasoning with an explicitly represented ordering.

In defence of animal reasoning, McGonigle and Chalmers (1992) exclaim in their title 'Monkeys are rational!' This follows up an earlier paper that, in its title, asked 'Are monkeys logical?' (McGonigle and Chalmers 1977). Building on the standard five-element test, McGonigle and Chalmers don't base their case solely on BD generalization, but also on various effects of series position on performance, such as end anchoring, where performance is better on comparisons where the pair includes an item from the start or finish of the series, and the symbolic distance effect (SDE), where increased separation in the series between the pair of elements tested leads to degraded performance. To tease apart different components of the SDE, McGonigle and Chalmers (1992) also used a triadic version of the five-element task. In this variant, animals were trained with triples of stimuli—AAB, ABB, BBC, BCC, CCD, CDD, DDE, DEE—and tested with BCD. Monkeys do generalize a correct response to the novel BCD stimulus, but curiously, their performance is somewhat worse on this task than in generalizing to BD in the standard five-element task, despite the explicit presence of C between B and D. McGonigle and Chalmers (1992, p. 224) conclude that 'the SDE has been over-interpreted as a ranking phenomenon'. Nevertheless, they write of their monkeys that 'some sort of explicit seriation ability may be within their scope'.

De Lillo *et al.* undermine the significance of arguments based on serial-positioning effects by claiming that their artificial neural network model displays SDE, end-anchor, and other effects reported in the literature on transitive inference on animals and children. They concede that certain features of adult human learning are not captured by their model—for instance the fact that young children do equally well whether the training pairs are presented randomly or in serial order, whereas adult performance is seriously degraded by random presentation (De Boysson-Bardies and O'Regan 1973). Referring to the work by McGonigle and Chalmers, De Lillo *et al.* also admit that they know of no 'connectionist implementations of the task robust enough to deal with both binary and triadic versions' (p. 67). In their final remarks, they take the proven capacity of a simple, randomized neural network to perform the basic BD generalization task to suggest that 'the binary, non-verbal, five-term-series task might not be suitable for detecting ontogenetic or phylogenetic trends in the development of the cognitive skills underlying inferential abilities. In order to find behavioural differences of potential comparative significance, it might prove a more fruitful exercise to manipulate the training procedures...and the structure of the task itself (such as the triadic testing introduced by McGonigle and Chalmers 1977, 1992), instead of extending the same binary version of the paradigm to yet more non-human species or younger children.' (p. 68).

A muffled 'Hurrah!' might be heard at this point coming from ethologists who have never been particularly impressed with the weight that comparative psychologists have placed on the five-element task. Muffled, because replacing one simplistic paradigm with another would be only limited progress and the DiLillo recommendation might not seem to go far enough in overthrowing the grip that simplistic lab experiments have on thinking about the mechanisms underlying animal behaviour.

Although not directly targeting the work I have described above, Seyfarth and Cheney (2002) offer a naturalist's critique of those who would extend the results of similar associative learning experiments to explaining the capacities of animals living in complex natural societies. They point out that social dominance hierarchies can be quite large—a typical troop of baboons might consist of '80 or more individuals drawn from eight or nine matrilineal families arranged in a linear dominance rank order' (p. 379). In the five-element paradigm, the number of possible pairs (treating AB the same as BA, etc.) is ten. The five-element task artificially limits the number of pairs actually encountered during training to four, whereas baboons living in an 80-member troop may confront any of the 3160 different possible dyads. Furthermore, because of alliances where two animals may combine forces against another, dominance interactions may often take place among triads, of which there are 82 160 possible combinations (Seyfarth and Cheney 2002; see also Connor, this volume, who conveys the further complexity introduced by second-order alliances among bottlenose dolphins). Also, dominance hierarchies have to be rapidly relearned whenever there is a reversal in dominance. (For a description of this phenomenon in coyotes, see Bekoff 1977.) Whereas pigeons may take months to learn just four dyads, monkeys rapidly learn a much bigger hierarchy and flexibly adjust to changes in the hierarchy.

The laborious way in which lab animals learn the key associations in the five-element task could be due to the lack of any natural ordering among the stimuli. Typical visual stimuli used with pigeons include, for example, square symbols distinguished by arbitrary shading patterns in black and white. Unlike dominance hierarchies, the experimenter-imposed ordering on these stimuli has no intrinsic biological signi-ficance to the animals, nor any connection to any naturally transitive relationship. In this way, they are like the conventional ordering of the letters of the alphabet, which must be laboriously learned by young children; imagine trying to teach the alphabet by giving the elements only in adjacent pairs, possibly randomized: WX, DE, JK, BC, etc. (and definitely no singing!).

These issues become extremely significant when trying to think of ways in which one might like to extend the existing experiments. For instance, to address the positive transfer hypothesis put forward by Fersen and tested by Zentall, an obvious suggestion is to train the animal on the pairs that define a seven-element series, ABCDEFG and then test on the middle span CE. Because the nearest neighbours of C and E have equivalent reinforcement histories, successful generalization of the correct response to CE could not be accounted for by the positive transfer theory (although there could be a secondary transfer theory that accorded positive transfer to C on the basis of its association with B, which gets direct positive transfer from A; this move can of course be iterated to the point of implausibility, although the location of that point might be the subject of considerable differences among commentators depending on their commitment to the hegemony of associationist principles). Some psychologists with whom I have discussed the seven-element experiment have remarked only somewhat facetiously that the experiment might well be impracticable with pigeons as one could approach the pigeon's life span in trying to train it to reach criterion performance on all six pairs simultaneously. But Bond *et al.* (2003) have recently trained five representatives each of two different corvid species, pinyon jays and the less intensely social scrub jays, on the seven-element task. Although the pinyon jays reached criterion performance levels much more quickly than scrub jays, a three-stage training process involving hundreds of exposures to the training set was required for members of both species to reach the criterion level of performance.

Lest it be thought that these relatively slow learning rates reflect some sort of inherent limitation of the avian brain with respect to transitive inferences, it is worth pointing out that Beaugrand *et al.* (1997) demonstrate the occurrence in domestic chickens of rapid observational learning of dominance relations, apparently involving a transitive inference, as the result of watching a single aggressive interaction between a familiar dominant and an unfamiliar conspecific. When the stranger dominates the familiar dominant, observers never initiate an attack on the unfamiliar animal, and submit immediately if attacked. But when the familiar dominant is observed to defeat the stranger, observers are much more likely to attack and defeat the stranger than in circumstances where no information is available about the stranger.

The relatively slow learning rate in the standard operant conditioning paradigm may, then, have much to do with the nature of the task. In the original Piagetian version of the five-element paradigm, designed to test transitive inference in young human children, arbitrary symbols or colours are paired with different length rods. During training and testing the length of the rods is obscured and the children are rewarded for choosing the longer rod from pairs that are distinguishable only by the associated symbol or colour. Unlike the typical task that pigeons are confronted with, here the arbitrary relationship among the visual cues is mapped onto a naturally transitive relationship in the length of the rods. It is quite possible, then, that the arbitrary markers acquire significance as proxies for the properties (specific lengths) underlying the transitive ordering. I am unaware of any similar attempts to pair stimuli with real transitive relationships within a traditional, behaviouristic animal learning paradigm.

7.3 **Reasoning or conditioned associations?**

It appears to be a shared presupposition among many comparative psychologists and cognitive ethologists that behaviourism and cognitivism are exclusive alternatives. This is echoed by Dennett's (1983) early suggestion that behaviourism is the null hypothesis against which cognitive accounts are tested. But is this a correct representation of the dialectical situation? Why should the availability of an associative model trump the adequacy of a serial-representation model? Are associationist explanations *always* to be preferred? A certain Cartesian residue might be detected here: if we have an associative 'mechanism' we can avoid attributing (unsubstantiated) rational thought. That residue seems apparent when Clayton *et al.* (Ch. 9, this volume) write: 'the issue of whether an animal is psychologically rational turns on the nature of the processes causing its behaviour; specifically on whether this behaviour is caused by psychological mechanisms or by intentional processes.' In these days of computers, we know perfectly well how to build mechanisms to represent arbitrary sequences of elements, so characterization of a mechanist/mentalist dichotomy does not apply to the current debate about transitive inference. I agree with Papineau and Heyes (Ch. 8, this volume) that contemporary psychology is premised on materialism. This means that rational (or intentional, or cognitive) processes, whether in humans or other animals, have to supervene on psychological mechanisms. (For reasons I'll explain shortly, we can embrace the slippage between 'cognitive', 'rational', and 'intentional' that is present in other contributions to this volume.)

The question, then, is whether anything worthwhile can be made of the distinction between different kinds of psychological process or mechanism. Heyes and Papineau rightly ask this question about the distinction between 'associative' and 'rational', and, after critiquing two suggestions for making the distinction in very general terms, they declare themselves sceptical about its utility. Earlier in this paper I adopted a rough and ready distinction between associative and cognitive explanations of transitive inference. Cognitive approaches to transitive inference, I wrote, postulate that animals explicitly represent such orderings and use these representations to infer relationships between pairs of non-neighbouring elements. This did not, and was not intended to, supply a general account of the distinction between cognitive and associative mechanisms; instead it merely reflected the categories employed by the psychologists themselves (especially Zentall) who draw distinctions between putative explanations of the behavioural phenomena.

We can certainly 'put on our philosophical hats' and ask whether the distinction between 'associative' and 'cognitive' is generally defensible. But, in my view, the resulting discussion is likely to be too abstract to be instructive. Cognitive approaches to a variety of psychological phenomena are likely to bear family resemblances rather than being analytically definable. Elements of reasoning (hence rationality) and meaning or semantic content (hence intentionality) appear in specific cognitive explanations (hence the acceptable slippage among these terms) but it would be a mistake to say that cognitive approaches stand or fall with the practical syllogism or a commitment to unlimited inferential power. Even if a generalized distinction between cognitive (or rational, or intentional) and associative mechanisms is not tenable, it is nevertheless the case that the distinction between the different kinds of explanation of transitive inference is specifically defensible. Cognitive approaches to transitive inference appeal to a retrievable unified representation of an entire ordered series which functions as a template for specific inferences about the relative ordering of arbitrary pairs of elements drawn from the series. Whatever efficiencies might be provided by such representations (Cheney and Seyfarth, for example, suggest that cognitive accounts but not standard associative accounts can deal adequately with the fact that individuals may be simultaneously classified in multiple ways), a cognitive mechanism that exploits such a representation of the series might, nevertheless, be quite limited. It may not have a completely general transitive inference capability for its ability to store sequences will almost certainly be limited by length of the series, and may also be limited in the content domains to which it can be applied. But this does not detract from the fact that, from an engineering/programming perspective, the distinction is clear enough between, on the one hand, how one would go about building the kind of system that builds and uses a unified representation of a transitively governed series and, on the other hand, how one would build a system that is limited to associating stimuli with actions to various degrees of strength.

DeLillo *et al.* (2001) demonstrate that the latter approach is capable of something akin to transitive inference. Their model has limitations, but associative mechanisms are clearly a very important part of the psychological make up of all animals, including humans. So it is not a stretch to imagine that associative mechanisms play a role in many behaviours that appear to conform to norms of transitive inference. Rather than this excluding a role for cognitive mechanisms, it is quite possible that multiple mechanisms may coexist in a single organism. Indeed there is direct evidence from Dusek and Eichenbaum (1997) that the training used in the standard five-element transitive inference paradigm engages two different systems in rats. Comparing rats with hippocampal lesions to controls, Dusek and Eichenbaum found that lesioned rats retained the ability to learn pairwise discriminations between (olfactory) stimuli, but were impaired in their ability to make some of the transitive generalizations about these pairs made by intact rats, including the BD generalization. Given the importance of the hippocampus for declarative memory, Dusek and Eichenbaum suggest that it plays a role in 'representation of orderly relations among stimulus items' (p. 7113). (See also Wu and Levy 1998, for a computational model of transitive inference inspired by the hippocampus.)

Given the existence of parallel systems, it is also likely that these would be engaged somewhat differently depending on the context of learning. In the standard five-element task there is nothing in the stimuli or the training situation to suggest that the elements belong to a transitive ordering (the experimenter could, after all, follow the three pairs A+B−, B+C−, C+D− with the transitivity destroying D+A−). The animals in the experimental situation must discover an arbitrary transitive ordering of the stimuli in a situation where they would not be primed to expect such an ordering. In more natural contexts, where the cues to be learned are markers of real transitive relationships, it would be unsurprising to discover that learning of the series is facilitated by priming of the relevant brain areas. Nonetheless, if Dusek and Eichenbaum are correct, the five-element paradigm does lead eventually to explicit serial representation in the hippocampus, albeit through a more laborious process than might be triggered by more natural contexts.

7.4 **Reflections on animal cognition**

'Animal cognition' is nowadays a term embraced by scientists with a variety of methodological backgrounds, but it is clear that many of the old methodological divisions remain. Many laboratory-based psychologists are sceptical of the claims that field-based ethologists make about the cognitive abilities of their research subjects. Conversely, many ethologists remain deeply sceptical about the significance of results derived from laboratory studies of animal learning. The debate on transitive inference provides a microcosm of this dispute. From the behaviourists' perspective, the reinforcement history of animals studied under field conditions cannot be known in sufficient detail to permit reliable inferences about the animals' cognitive abilities. The ethologist doesn't know what experiences have shaped the behaviour of the 80 or so baboons in a troop. From the naturalists' perspective, laboratory studies are insufficiently rich to provide reliable knowledge of the mechanisms underlying the cognitive capacities of animals in the field. The comparative psychologist doesn't know how to train a rat or a pigeon to produce (or appear to produce) transitive inferences about an 80-element series. How to bridge this methodological divide is, in my view, one of the most important outstanding problems blocking the development of a unified scientific approach to animal cognition. Even though I have doubts about the abstract distinctions they draw, the empirical contribution of Clayton and her colleagues in this volume is important because it brings laboratory methods to bear on cognitive questions involving genuinely naturalistic problems.

Turning to neuroscience, it is appropriate to think (as Konrad Lorenz did) that it has an integral part to play in the development of a complete science of animal behaviour. The work of Dusek and Eichenbaum described above provides intriguing insights. But neuroscientists, being lab scientists, have so far tended to adopt the learning paradigms of comparative psychology, with scant regard for ecological validity (see also Chemero and Heyser 2003). Neuroscientists, after all, typically want to get home in time for the 6 o'clock news (or perhaps the 9 o'clock news if they are particularly driven). Not for them the rigours of 6 months in the bush following monkeys from tree to tree. Perhaps one day it will be possible to study the brain functions of animals non-invasively and in real time under field conditions, and at that point we can match complex social behaviours to events in the nervous system. But we should probably not hold our breath waiting for that moment, which depends on the rate of technological progress exceeding the rate of habitat loss.

Lorenz imagined a grand synthesis of functional, evolutionary, developmental, and neurological approaches to animal behaviour (Lorenz 1981). Such a synthesis has not yet been achieved, and it remains questionable whether it ever will be so long as lab and field scientists keep attacking each other on methodological grounds rather than finding ways to work on common problems. If the answer to the question 'Transitive inference in animals: reasoning or conditioned associations?' is perhaps 'Both', then we should be asking how to study these processes in all the contexts in which animals naturally employ them using a variety of approaches. But the hope for a fully synthetic approach to animal behaviour may be forlorn given the cultural divide that exists between behaviouristic psychologists and neuroscientists on the one hand, and ethologists on the other. This divide goes much deeper than the methodological problems indicated so far. It extends to issues about the proper treatment of animals for scientific purposes—to whether keeping animals in cages or cutting into their brains is an acceptable corollary of our desire to understand how real animals think, reason, and act.

Acknowledgements

I thank Marc Bekoff and Matt Nudds for their comments. Questions from participants at the meeting the editors organized in Oxford in 2002 were also helpful, and I'm also grateful for feedback from audiences at Texas A & M University and the Texas ARMADILLO conference in San Antonio in 2002.

References

Beaugrand, J. P., Hogue, M. E., and Lague, P. C. (1997). Utilisation cohérente de l'information obtenue par des poules domestiques assistant à la défaite de leur dominante contre une étrangère: s'agit-il d'inférence transitive? *Processus cognitifs et ajustement écologique*, pp. 131–137. Société Française pour l'Étude du Comportement Animal, Toulouse: Presses de l'Université Paul Sabatier.

Bekoff, M. (1977). Quantitative studies of three areas of classical ethology: social dominance, behavioral taxonomy, and behavioral variability. In: B. A. Hazlett, ed. *Quantitative Methods in the Study of Animal Behavior*, pp. 1–46. New York: Academic Press.

Bond, A. B., Kamil, A. C., and Balda, R. P. (2003). Social complexity and transitive inference in corvids. *Animal Behaviour,* **65**: 479–487.

Chemero, A., and Heyser, C. (2003). *What mice can do: affordances in neuroscience.* Presented at Society for Philosophy and Psychology, Pasadena, CA.

De Boysson-Bardies, B., and O'Regan, K. (1973). What children do in spite of adults. *Nature,* **246**: 531–534.

De Lillo, C., Floreano, D., and Antinucci, F. (2001). Transitive choices by a simple, fully connected, backpropagation neural network: implications for the comparative study of transitive inference. *Animal Cognition,* **4**: 61–68.

Dennett, D. C. (1983). Intentional systems in cognitive ethology: The 'Panglossian paradigm' defended. *Behavioral and Brain Sciences,* **6**: 343–390.

Dusek, A., and Eichenbaum, H. (1997). The hippocampus and memory for orderly stimulus relations. *Proceedings of the National Academy of the Sciences USA,* **94**: 7109–7114.

Fersen, L. von, Wynne, C. D. L., Delius, J. D., and Staddon, J. E. R. (1991). Transitive inference formation in pigeons. *Journal of Experimental Psychology: Animal Behavior Processes,* **17**: 334–341.

Lorenz, K. (1981). *The Foundations of Ethology.* New York: Springer-Verlag.

McGonigle, B., and Chalmers, M. (1977). Are monkeys logical? *Nature,* **267**: 694–696.

McGonigle, B., and Chalmers, M. (1992). Monkeys are rational! *Quarterly Journal of Experimental Psychology,* **45B**: 198–228.

Martin, F., Beaugrand, J. P., and Lagüe, P. C. (1997). The role of hen's weight and recent experience on dyadic conflict outcome. *Behavioural Processes,* **41**: 139–150.

Rumelhart, D. E., and McClelland, J. L. (1986). *Parallel Distributed Processing.* Cambridge, MA: MIT Press.

Seyfarth, R. M., and Cheney, D. L. (2002). The structure of social knowledge in monkeys. In: M. Bekoff, C. Allen, and G. M. Burghardt, eds. *The Cognitive Animal: Empirical and Theoretical Perspectives on Animal Cognition,* pp. 379–384. Cambridge, MA: MIT Press.

Wu, X., and Levy, W. B. (1998). A hippocampal-like neural network solves the transitive inference problem. In: J. M. Bower, ed. *Computational Neuroscience: Trends in Research,* pp. 567–572. New York: Plenum Press.

Zentall, T. R. (2001). The case for a cognitive approach to animal learning and behavior. *Behavioural Processes,* **54**: 65–78.

Zentall, T. R., and Sherburne, L. M. (1994). Transfer of value from S+ to S− in a simultaneous discrimination. *Journal of Experimental Psychology: Animal Behavior Processes,* **20**: 176–183.

Chapter 8

Rational or associative? Imitation in Japanese quail

David Papineau and Cecilia Heyes

Abstract

Much contemporary psychology assumes a fundamental distinction between associative explanations of animal behaviour, in term of unthinking 'conditioned responses', and rational explanations, which credit animals with relevant 'knowledge' or 'understanding' or 'concepts'. This paper argues that this dichotomy is both unclear and methodologically unhelpful, serving only to distract attention from serious questions about which cognitive abilities are present in which animals. We illustrate the issues by considering recent experimental work on imitation in Japanese quail.

8.1 Introduction

According to René Descartes, a wide gulf separates animals from humans:

> '... [A]fter the error of those who deny the existence of God ... there is none that is more power-
> ful in leading feeble minds astray from the straight paths of virtue than the supposition that the
> soul of the brutes is of the same nature with our own' (Descartes 1637)

For Descartes, the gulf between animals and humans had a metaphysical basis. The universe was composed of two fundamentally different substances, mind and matter, and animals lacked minds. Because of this, animals were incapable of rational thought and action. Where human activities were governed by the operations of the soul acting through the pineal gland, animals were mere machines, automata governed by nothing but the laws of physics.

When we ask 'Are animals rational?' the question can no longer have anything like the same significance it had for Descartes. Contemporary psychology is premised on materialism. Humans, no less than animals, are complex machines, whose operations depend on scientifically describable psychological and neurological processes. If there is still a live issue as to whether animals are rational, it cannot depend on whether or not animals are blessed with the same special mind-stuff as humans.

Perhaps nowadays we understand the question differently, in terms of two styles of cognitive explanation. This certainly seems to be the way that many psychologists understand the issue. On the one hand lie 'rational' explanations of behaviour, explanations that advert to norm-governed reasoning involving belief-like representations. On the other side lie non-rational explanations, in terms of 'behaviourist' or (more accurately) associative psychological processes.

Ideally, this contrast would give rise to fruitful experimental work. Given some potentially 'rational' behaviour, we try to get the alternative norm-governed and associative accounts to make contrasting predictions about how the behaviour would change under various experimental manipulations, and then test these predictions empirically.

However, we are doubtful that this methodological strategy is as cogent as it initially seems.[1] Of course, we are all in favour of heuristic attitudes that lead to the experimental exploration of animal behaviours and the processes that mediate them. But we doubt that a simple dichotomy between 'rational' and 'associative' is the best way to motivate such experimental work. We suspect that the rational–associative dichotomy is just Descartes dressed up in modern garb. In place of Descartes' immaterial mind we have the accolade of 'folk psychology', and in place of his brute matter we have 'associative machines'. Psychologists are asked to decide whether: (a) animals 'know about', or 'understand', or 'have a concept of' such-and-such (causality, space and time, others' minds); or whether (b) they are driven by unthinking associative mechanisms.

This way of setting things up may only obscure the real issues. We know from basic Darwinian principles that there are no unbridgeable gulfs in cognition: each cognitive innovation must proceed by adding to, or tinkering with, pre-existing cognitive systems. In line with this, we want to know precisely which cognitive mechanisms have evolved in which animals. To be told that certain animals can 'understand' such-and-such—or alternatively that they are simply displaying 'conditioned responses'—is not helpful. What we want to know is how they work, not whether they can cross the Rubicon of folk psychology.

8.2 Imitation in Japanese quail

We would like to back up the remarks above by considering a recent experiment by Akins and Zentall on imitation in Japanese quail.

---

[1] Editors' note: Consider how these methodological concerns may apply to empirical work described in various chapters in this volume, including those by Allen, Clayton *et al.*, Call, Shettleworth and Sutton, and Addessi and Visalberghi. Compare these concerns about the rational–associative distinction with Povinelli and Vonks' methodological concerns (this volume) about the mind reading–behaviour reading distinction.

Akins and Zentall (1998) allowed naïve observer quail to watch a trained 'model' or 'demonstrator' quail depressing a lever using either its beak or its feet. Half of the observers saw the demonstrator rewarded with food for this behaviour, while the other half saw the demonstrator responding repeatedly 'in extinction', that is without reward.

When the observer birds were subsequently given access to the lever themselves, those that had seen their demonstrator's behaviour being rewarded tended to imitate the model—to peck the lever if their model had used its beak, and to step on the lever if their model had used its feet. However, when the model's behaviour was unrewarded, there was no evidence of imitation—the proportion of pecking and stepping responses was the same for observers of beak use and foot use. Thus, the observer quail in this experiment showed sensitivity to demonstrator reward.[2]

8.3 **The significance of sensitivity to demonstrator reward**

Akins and Zentall's result is a good test case for the heuristic utility of the 'rationality–associative' dichotomy for two reasons. First, it is easy, perhaps irresistible, to interpret sensitivity to demonstrator reward in folk-psychological terms. We naturally assume that the birds who imitated did so because they *wanted* food and *believed* that performing the same action as the demonstrator would enable them to get it. Second, standard associative learning theory is apparently unable to explain sensitivity to demonstrator reward. This kind of imitation would thus seem to be a good candidate for 'rational' behaviour, as opposed to mere associative response.

It will be worth going slowly for a moment in order to explain exactly why sensitivity to demonstrator reward is difficult to explain in standard associative terms. An initial stumbling block is known as the 'correspondence problem' (Heyes 2005; Nehaniv and Dautenhahn 2002). Given that a quail sees something very different when he watches another quail pecking a lever (him-pecking) and when he pecks the lever himself (me-pecking), how does the observer know that it is pecking behaviour, rather than his stepping behaviour, which is 'the same' as that of the pecking demonstrator? After all, neither me-stepping nor me-pecking looks or feels much like him-pecking. It appears that associative learning theory is too flimsy to deal with this conundrum, and that heavy cognitive machinery—such as 'supramodal' mental representation (Meltzoff and Moore 1997) or 'perspective-taking' (Bruner 1972; Tomasello *et al.* 1993)—is necessary to solve the correspondence problem.

However, let us suppose that the correspondence problem can somehow be overcome, by an associative mechanism (Heyes 2001) or otherwise. (When the imitating animal is a bird, rather than a person or other ape, the correspondence problem is typically evaded by calling the phenomenon 'mimicry', and assuming that it is the product of

---

[2] Editors' note: see also Pepperberg's use of model-rival training with African grey parrots, this volume.

some innate tendency.) Still, even if we assume that there is some automatic tendency for animals blindly to reproduce the behaviour of conspecifics, this does not immediately clear the way for an associative explanation of sensitivity to demonstrator reward—that is, of a tendency to reproduce behaviour *only* when it is observed to be followed by some reward.

It might seem as if it would just be a short step from blind mimicry to full sensitivity to demonstrator reward. Don't the animals just have to observe what result generally follows the demonstrator's behaviour, and draw a straightforward inference? However, if we think about the basic associative learning mechanisms by which animals might acquire information about general patterns in their environment, we can see that the inference in question is not straightforward at all.

Suppose, for the sake of the argument, that the basic associative learning mechanisms are classical and instrumental conditioning. Classical conditioning allows animals to take one stimulus, F, as a sign of another, G, whenever Fs have previously been followed by Gs in the animals' environment. Instrumental conditioning allows animals to perform some behaviour B in pursuit of reward R, whenever Bs have previously led to Rs in the animals' environment. In combination, these two kinds of mechanisms can lead to a high degree of flexibility in gearing behaviour to local environmental regularities.

Neither of these mechanisms, however, will give rise to the kind of sensitivity to demonstrator reward found in Japanese quails. It may not be immediately obvious why. But consider things from the perspective of the bird observing the demonstrator. The observer quail sees the demonstrator peck at the lever, say, and subsequently sees the demonstrator being rewarded with food. Note, however, that in this scenario the *observer* hasn't yet performed any behaviour, nor has *it* received any subsequent reward. So there is nothing here for instrumental conditioning to get a grip on, nothing to make the *observer* peck in pursuit of food.

True, if the observer quail were prone to blind mimicry, then this alone would dispose it to peck when it observes the demonstrator pecking. But this won't explain why the observer quail only pecks when it sees the demonstrator being rewarded.

What about classical conditioning? Won't this lead the observer to associate the pecking with the reward? Yes—but here the association will be between *observing* the pecking and *observing* the demonstrator receiving the reward, not between *doing* the pecking and *receiving* the reward. This classical association might lead the observer quail to respond to the sight of *another* pecking bird in the way it previously responded to the appearance of food, for instance by approaching. But this too will do nothing to explain why the observer itself learns to peck only when the demonstrator receives the food.

In short, there is nothing obviously rewarding to the observer in seeing some other bird being fed. So, even assuming some tendency to blind mimicry, there seems nothing in standard associative learning theory to explain why the observers' pecking tendencies should get reinforced specifically when the demonstrator quail gets fed.[3]

8.4 Rational versus associative explanations of the Japanese quail

According to the simple dichotomy outlined earlier, the alternative to associationism is rationalism. If the behaviour of the Japanese quail cannot be explained in associative terms, then we should account for it in terms of rational cognition. Thus we should credit the observer bird with a desire for food, and a belief about how to get it, and view its behaviour as the rational product of these propositional attitudes. For instance, we might reconstruct its reasoning as follows:

The demonstrator's pecking produced food.

So my pecking will produce food.

I want food.

———

So I'll peck.

No doubt those who advocate some explanation along these lines will not wish to credit Japanese quail with unlimited cognitive powers. There are unquestionably some constraints on the range of subjects that the quail can think about—they can't think about electricity, for instance, or genetics. In line with this, it might be claimed that their rationality is only displayed in the operation of certain 'domain-specific modules', such as their navigation module and their social reasoning module, along with their imitation module (see and cf. Hurley, this volume, on context-bound reasons for action). Still, despite the restriction of rationality to such specific domains, the idea would remain that, within these domains, the quail are capable of forming contentful beliefs and desires, and of using reason to figure out the consequences thereof.

So, validating the rational–associative distinction, we seem to have found that sensitivity to demonstrator reward is subject to rational but not to associative explanation. But there are two twists in this tale, and here is the first: with a little imagination, old fashioned, stimulus–response (S–R) associative learning theory *can* explain sensitivity to demonstrator reward. Suppose that in the past the observer bird has fed while seeing other birds feeding (quails are highly social animals); this could well have

---

[3] Interestingly, this is a special case of a more general limitation. As well as not being able to absorb means–end information from observation of other animals, creatures limited to standard associative mechanisms won't be able to absorb means–end information from any kind of non-participatory observation, and for the same reason. Observing that cause A produces effect B might make you anticipate B when you see A, via classical conditioning, but it won't get you *doing* A in pursuit of B, when there is nothing rewarding about merely observing B. (Cf. Papineau, in press.)

led to the sight of another bird feeding becoming a 'secondary reinforcer' for the observer, with the result that sight of another feeding itself becomes rewarding. Now put this together with a general tendency for mimicry. This incipient tendency to peck blindly, say, when observing another bird pecking, would then become paired with a rewarding event for the observer, *if* the observer sees the demonstrator feeding. This model would then explain why only the birds that saw the demonstrator being rewarded would acquire a disposition to peck—only they would have been (secondarily) rewarded by the observation of another bird feeding (see and cf. Addessi and Visalberghi, this volume, on socially facilitated feeding in capuchin monkeys).

A variety of experiments would be needed to test this associative hypothesis against the rationality alternative, but here's a simple one that would provide a good start. Consider a variant experiment in which hungry observers see demonstrators stepping and pecking, but where the two demonstrator behaviours provide access to two visually distinct but equally palatable foods. Then, before the observers are subsequently tested, they are allowed to feed to satiety on *one* of these foods. The prefed food is, therefore, devalued. If the S–R associative account is correct, one would not expect this prefeeding to have any impact on the proportion of pecking and stepping responses made by the observer quail on test. This is because the (secondary) reinforcement during observation would merely have forged a simple S–R association between, say, the sight of the lever and the urge to peck it, detached from any expectation of reward. On the other hand, if the rationality hypothesis were correct, and the animals were forming explicit beliefs about the effects of different actions, we would expect the birds to perform less of the response that produced the now devalued food during observation. (See Saggerson, George & Honey 2005 for a similar experiment published after this was written.)

This experiment has not yet been performed, but let us suppose that, when it is done, the quail perform less of the response that had produced devalued food. Would this outcome show that quail are rational, and thereby demonstrate the heuristic value of the rational–associative dichotomy? Unfortunately not, because (and here's the second twist in the tale) the 'rational' behaviour could also be explained in associative terms. It would be incompatible with S–R associative learning, but not with associative learning of a relationship between the response and its outcome. On this account, the quail learn by demonstrator observation that, for example, pecking is associated with red food and stepping is associated with green food. However, devaluation of, say, the red food, prevents the experimental context from activating the stepping response by establishing a negative feedback loop between the animal's incentive system and its representation of the red food outcome (see Dickinson's 1994 associative-cybernetic model of instrumental action for details).[4]

---

[4] No similarly obvious associationist explanations offer themselves for some of the very impressive cognitive feats performed by the scrub jays in Clayton *et al.* (this volume). But this does not undermine the point, stressed in our final section below, that a bald assertion that some piece of cognition is 'rational' rather than 'associative' is no substitute for a specific hypothesis about the mechanisms responsible.

8.5 **Morals**

At one level, our doubts are simply about the clarity of the distinction between 'rational' and 'associative'. The standard way of drawing this distinction is in terms of *representation*: where rational thought involves the processing of genuine representational states, associative responses rest on nothing but blind mechanical response.[5] However, it is not clear that this contrast stands up to philosophical scrutiny, especially when we reflect on the more complex kinds of associative hypotheses considered in the last section. After all, animals who have learned by conditioning must be internally modified in certain ways, and one natural way of characterizing these modifications is in terms of their content. For example an animal who has learned a response–outcome association as a result of being rewarded by D after doing B in conditions C is quite naturally described as embodying the information that B in C will lead to D. In this connection, note that it is certainly the biological purpose of such instrumental conditioning to produce such modifications just in those cases when B in C will indeed lead to D. Note also how this causal information will interact with a current perception of C and a drive for D to yield the behaviour B (in a way quite analogous to the familiar practical inference 'I know that B in C will lead to D; I can see I am in circumstances C; I want D; so, I do B').[6,7]

Perhaps we shouldn't be trying to draw the rational–associative distinction in terms of representation *per se*, but in terms of the kind of processes to which the animals can subject their representations. It is certainly true that associative mechanisms are limited in terms of the kinds inferences that they can perform. Consider an animal who has formed a classical association between the sight of another animal doing B in C and its receiving a reward D. In a sense, this animal embodies the information that

[5] Sometimes the distinction is also drawn in terms of normativity: only rational processes are guided by rules of correct and incorrect reasoning. But this strikes us as an obscure idea. Surely the first concern of cognitive science is to figure out how animals *do* think, not how they *ought* to think. Moreover, one of us (Papineau 1999) would argue that any normative considerations that do apply to cognitive states are *derivative* from their representational nature (you ought to believe what is true, and you ought to reason so as to preserve truth); in which case distinguishing rational from associative processes in terms of normativity collapses into distinguishing them in terms of representation.

[6] For a more detailed discussion of representationalist readings of associative mechanisms, see Papineau, in press.

[7] Dretske (this volume) argues that behaviour arising from associatively learned cognitive representations is minimally rational, since it is guided by the content of those representations, but that this is not true of behaviour guided by hard-wired cognitive structures. We agree that associatively learned behaviour is guided by contentful representations, and to this extent can be counted as 'rational' (though note our caveats about inferential limitations immediately below). However, we are not convinced by Dretske's contention that, if just the same cognitive structures were designed by genetic selection rather than learning, this would somehow yield less rationality.

doing B in C will be followed by D. But this information is embodied in the wrong place to influence behaviour. Since it is embodied in a classical association, rather than in an instrumentally conditioned response–outcome disposition, the information can't combine with the animal's own drive for D and its perception that it is currently in circumstances C to generate behaviour B, as would an instrumentally conditioned response–outcome disposition.[8]

So perhaps we should characterize rational cognition as cognition that is not limited in the inferences it can perform. Note how this was implicit in the rationalist explanation offered above for the Japanese quail's learning abilities: this simply presupposed that the information about a pecking–food association derived from observation could then inform the birds' own choice of action. (The demonstrator's pecking produced food/So my pecking will produce food/I want food//So I'll peck.)

However, if *this* is what rationalist explanation means, then we have substantial doubts whether it is of any serious utility for psychology. To explain behaviour by positing unlimited inferential powers is little short of offering no explanation at all. Such unlimited intellectual powers may be a kind of cognitive ideal, and maybe humans using language (or other artificial aids) come close, but cognition in non-linguistic animals must be realized by specific architectures, and such architectures will inevitably place limitations on the kinds of inferences animals can draw, and consequently on which behaviours they will display.

It runs against every principle of evolutionary thinking to suppose that some animals, in some domains, have magically jumped across the chasm dividing associative mechanisms from ideal cognition. Evolution builds adaptations by tinkering and gradual additions, not via miraculous saltations, and this applies to cognitive evolution as much as elsewhere (Heyes, 2003). Many different animals can perform many varieties of flexibly intelligent behaviour. But it is not helpful to try to explain such behaviour by suggesting that these animals have somehow achieved 'rationality', if this means that they are somehow unconstrained in the inferences they can draw from their information. What we need to know is precisely which extra cognitive mechanisms evolution has endowed them with, and which extra bits of reasoning they are thereby able to perform.

Some contemporary students of animal behaviour happily characterize themselves as children of the 'cognitive revolution'. If the cognitive revolution means that other possibilities beyond strict behaviourism should be explored, then it is clearly a good thing. But if it means that psychologists should stop worrying about cognitive

---

8 Information embodied in classical associations can influence behaviour in a different way: if an animal is already disposed to perform behaviour B in condition F in pursuit of D, and then forms a classical association between G and F, this will lead it to do B in condition G too. (We can here think of it as putting together the information that B in F will lead to D, with the classical information that all Gs are Fs, to derive the conclusion that B in G will lead to D.)

mechanisms, then it is arguably a step backwards. There is little point in asking whether animals 'know about', or 'understand', or 'have a concept of' such-and-such, if these questions are not accompanied by an enquiry into what animals can *do* with their information.

So, to sum up, it doesn't seem as if the rational–associative dichotomy is satisfactory, whichever way we cut it. If we try to define 'rational' in terms of representation, then there seems no good reason to rule out associative mechanisms as rational. And if we try to define 'rational' in terms of unlimited inferential abilities, then this idealizes rationality to a degree that renders it useless as a research tool.

Maybe, when we fully understand the structure of animal cognition, we will be better placed to decide what kinds of animal achievement are worth dignifying as rational. But by then we probably won't care.

References

Akins, C. K., and Zentall, T. R. (1998). Imitation in Japanese quail: The role of reinforcement of demonstrator responding. *Psychonomic Bulletin and Review*, **5**: 694–697.

Bruner, J. S. (1972). Nature and uses of immaturity. *American Psychologist*, **27**: 687–708.

Descartes, R. (1637). *A Discourse on Method*, p. 46. Translated by John Veitch (1912). London: J. M. Dent and Sons, Everyman's Library Edition.

Dickinson, A. (1994). Instrumental conditioning. In: N. J. Mackintosh, ed. *Animal Learning and Cognition*, pp. 45–79. San Diego: Academic Press.

Heyes, C. M. (2001). Causes and consequences of imitation. *Trends in Cognitive Sciences*, **5**: 253–261.

Heyes, C. M. (2003). Four routes of cognitive evolution. *Psychological Review*, **110**: 713–7.

Heyes, C. M. (2005). Imitation by association. In: S. Hurley and N. Chater, eds. *Perspectives on Imitation*. Cambridge, MA: MIT Press.

Meltzoff, A. N., and Moore, M. K. (1997). Explaining facial imitation: a theoretical approach. *Early Development and Parenting*, **6**: 179–192.

Nehaniv, C. L., and Dautenhahn, K. (2002). The correspondence problem. In: K. Dautenhahn and C. L. Nehaniv, eds. *Imitation in Animals and Artifacts*, pp. 41–61. Cambridge, MA: MIT Press.

Papineau, D. (1999). The normativity of judgement. *Aristotelian Society Supplementary Volume*, **73**: 17–43.

Papineau, D. (in press). Human minds. In: A. O'Hear, ed. *Minds and Persons*. Cambridge University Press.

Saggerson, A. L., George, D. N. & Honey, R. C. (2005). Initative learning of stimulus-response and response-outcome association in pigeons. *Journal of Experimental Psychology: Animal Behaviour Processes*, **31**: 289–300.

Tomasello, M., Kruger, A. C., and Ratner, H. H. (1993). Cultural learning. *Behavioral and Brain Sciences*, **16**: 495–552.

Chapter 9

The rationality of animal memory: Complex caching strategies of western scrub jays

Nicola Clayton, Nathan Emery, and
Anthony Dickinson

Abstract

Scrub jays cache perishable and non-perishable foods, and their caches
may be pilfered by conspecifics. Caching and recovery by scrub jays is
psychologically rational in the sense that these behaviours responded
appropriately to conditions that should have changed the birds' beliefs
and desires. For example scrub jays were allowed to cache worms and
peanuts in a visuospatially distinct tray. At recovery, birds search initially for
worms after a short retention interval because they believe that the worms
are still edible, but switch to searching for peanuts at a long retention
interval because they believe that worms are now degraded. If jays acquire
new information after caching, such that worms are no longer edible
when recovered at the short interval, this should affect their belief about
the state of their caches. Jays update their cache memory, and on
subsequent trials of the short interval, search selectively in peanut sites.
In a second example, scrub jays cached either in private (when another
bird's view was obscured) or while a conspecific was watching, and then
recovered their caches in private. Scrub jays with prior experience of
stealing another bird's caches subsequently recached food in new sites
during recovery trials, but only when they had been observed caching.
Naïve birds did not. We suggest that experienced pilferers had formed a
belief that observers will pilfer caches they have seen, and recache food in
new sites to fulfil their desire to protect their caches. Since recaching is
not dependent on the presence of the potential thief, the jays must recall
the previous social context during caching, and flexibly use this
information to implement an appropriate cache protection strategy,
namely recache the food in locations unbeknownst to the pilferer.

9.1 **Introduction: intentional and mechanistic psychology**

In common with other scatter-hoarding animals, western scrub jays *(Aphelocoma californica)* hide surplus food in discrete locations within their territories, which they recover in times of need (Vander Wall 1990). Psychology offers two classes of explanation for such behaviour: the mechanistic and the intentional. Mechanistic accounts appeal to psychological processes that gain their explanatory power by analogy to physical processes. A classic example is associative learning theory. According to this theory, learning about the relationship between events, whether they are stimuli or responses, consists of the formation of excitatory (or inhibitory) connections between nodes activated by these events. The mechanism by which such associative structures control behaviour is the transmission of excitation (or inhibition) from one node to another until the activation of the terminal node of the associative chain is sufficient to generate the observed behaviour.

Associative theories have had an enduring influence on the study of animal learning and cognition ever since Thorndike (Thorndike 1911) formulated his Law of Effect on the basis of the first controlled, comparative studies of learning in animals. The development of associative theory continued throughout the last century to reach the complexity and sophistication of multilayered networks with distributed representations (Rumelhart and McClelland 1986). We do not need to appeal to such complex structures, however, to explain cache recovery by jays in terms of associative mechanisms. All that we need to assume is that associative nodes activated by the visual cues around the cache become connected with those excited by the food stored at that site, perhaps a peanut, at the time of caching. Consequently, re-exposure to the cache site stimuli activates the food nodes that, in turn, are associated with nodes controlling an approach response either innately or through prior learning.

We suspect, however, that we should have little success in persuading our proverbial grandmother of the merits of this account. By analogy with remembering where she hid her cache of chocolate when overtaken by a craving, our granny would probably explain the jay's recovery behaviour in terms of beliefs and desires. Her explanation might run something like this: the jay, being hungry, has a desire for food and, as a result of caching the peanut, has a belief that searching in the cache site will yield food. If pressed about why having this belief and desire causes the bird to search the cache site, granny would probably tell us to stop being obtuse—given this belief and desire, the only sensible or rational thing to do is to search the cache site. Moreover, if she would tolerate further Socratic enquiry, we should find that what she means is that cache searching is rational because, of necessity, it must fulfil the jay's desire for food if the belief that searching the cache site will yield food is true.

It is the rationality of this practical inference process that distinguishes the intentional explanation from the mechanistic one. Our account of rational behaviour accords with that outlined by Dretske (Chapter 3, this volume). The processes by which associative structures control behaviour are constrained only by their mechanistic-like properties. The level of activation of the food nodes, and hence the probability and vigour of searching, is simply a function of the strength of the associative connections, the input activation, the nodes' thresholds for activation, whether or not there is concurrent inhibitory input, etc. But in and of themselves, none of these processes yield behaviour that necessarily conforms to any cannons or principles of rationality. Of course, evolution will have ensured that behaviour governed by such mechanistic processes is adaptive in that it contributes to the reproductive fitness of the jay by maintaining its nutritional state, and in this sense associatively-controlled cache searching can be regarded as *biologically rational* (see Kacelnik, Chapter 2, this volume). But such biological rationality must be distinguished from the *psychological rationality* of the practical inference process that generates cache searching from a belief about the consequences of this behaviour and the desire for these consequences.

According to this analysis, the issue of whether an animal is psychologically rational turns on the nature of the processes causing its behaviour; specifically on whether this behaviour is caused by psychological mechanisms or by intentional processes. The jay's behaviour is psychologically rational to the extent that it is caused by the interaction of a belief and desire in such a way that performance of the behaviour in question fulfils the desire if the belief is true (and fails to do so if the belief is false).[1] Such an account is intentional because it requires that the antecedent mental states, the belief and the desire, have intentional properties, such as truth and fulfilment, because their content represents current or desired states of affairs.

There are a number of points to note about this analysis. First, it is not sufficient for an intentional explanation that the behaviour is simply caused by a belief and desire; rather that it has to be caused by the right process, namely the rational process of practical inference. It may well be that the sight of a cache site in which a food-desiring jay believes that searching will yield food causes an increase in heart rate as a component of general autonomic arousal. But such activation is not caused intentionally. Unlike cache searching, whether or not the jay's autonomic nervous system is activated has no necessary relation to whether its desire will be fulfilled and is, therefore, to be explained in terms of psychological mechanisms rather than by intentional processes. Perhaps in the past being in these belief and desire states has been associated with autonomic arousal so that the reinstatement of these states once again triggers an increase in heart rate.

---

[1] Editors' note: For related distinctions, see Kacelnik on PP rationality and Dretske on minimal rationality; cf. Allen, Papineau and Heyes for methodological reflections and scepticism about related distinctions (all in this volume). Dretske distinguishes behaviour that is (merely) caused by a state with intentional properties, and the *content* of a state with intentional properties being explanatorily relevant to some behaviour; on his view, the latter is a necessary condition for rational behaviour.

Second, according to this analysis, rationality is a property not of an animal, but of the processes causing its behaviour. So, to characterize a jay as rational does not imply that all of its behaviour is intentional, but rather that it is capable of at least some intentional action. We do, of course, take humans as the canonical case of a rational animal because the concordance between our actions and the expression of our beliefs and desire through language provide evidence of intentional causation. This is not to say, however, that all of our behaviour is rational. In fact, we may well be surprised to discover how little of our daily life is in fact under intentional control! The complex sequences of behaviour by which we drive, walk, or cycle to work may well appear to be purposive and goal-directed but, on further investigation, to consist of a chain of mechanistically elicited habits, albeit complex and highly structured ones, triggered by the stimuli along our route. So even we are creatures of a dual psychology; the mechanistic and the intentional. Therefore the issue of whether an animal is psychologically rational is really the issue of whether any of its behaviour warrants an intentional account.

Finally, it is important to note that behaviour does not carry its rationality on its sleeve. There may be nothing obvious about the manifest behaviour during our journey to work that marks whether it is under intentional or habitual control. Indeed, it is most likely that it started out under intentional control and only became habitual with repetition (Dickinson 1985, 1989). And, for the same reason, the processes of behavioural control cannot be determined by simple observation of the adaptive nature of the behaviour (Heyes and Dickinson 1990). As we shall see in our discussion of the psychological processes controlling food caching and recovery by western scrub jays, determining whether these processes are *psychologically* mechanistic or intentional is a complex enterprise which requires converging lines of evidence.

9.2 **The content of desires**

For a variety of reasons, there is a reluctance to accept that non-linguistic creatures can be endowed with an intentional and therefore a rational psychology. One issue relates to the vehicles that carry the content of beliefs and desires and, specifically, to whether only explicit and therefore potentially public languages, be they natural or artificial, are capable of providing the requisite vehicles for the content of beliefs and desires. This is not the place to debate this issue other than to note that certain philosophers of mind, most notably Fodor (1977), have vigorously argued that explicit languages are grounded on a language of thought that is shared, at least in some rudimentary form, by some non-linguistic creatures. By contrast, others have disputed the claim that a language, natural or purely mental, provides the only psychological vehicles for representational or intentional content (see Bermúdez, Chapter 5, this volume).

If one accepts, at least in principle, that animals other than humans can possess mental states with intentional or representational content, there still remains the problem of determining their content in the absence of an explicit, communicable language. As we have already noted, unlike speech acts, behaviour does not manifest its intentionality explicitly. However, it is possible to make some empirical progress on this issue in specific cases.

Consider once again our jay recovering a peanut from its cache site. Is this behaviour motivated by a desire for food or by a desire for a peanut? In other words, is the bird motivated by a general desire or by a specific one? We have attempted to answer this question by using the technique of devaluing the food immediately prior to recovery. In this study (Clayton and Dickinson 1999), jays cached peanuts in one cache site and dog food kibbles in a second cache site before being allowed them to search for these foods after a retention interval. The important feature of this study was that immediately prior to recovery we attempted to remove the desire for one of the foods, while maintaining the desire for the other. It is well known that we and other animals show food-specific satiety—a surfeit of one type of food, however delicious initially, rapidly looses its pleasure while maintaining, or even enhancing, the attraction of other foods. Consequently, the birds were prefed one of the two foods, either peanuts or kibbles, immediately prior to giving them a choice between searching in the two cache sites. To the extent that cache searching was motivated by a general desire for food, this prefeeding should have produced just a general decrease in their desire for food and hence an equivalent reduction in searching in both sites. By contrast, motivation by a specific desire for peanuts should have reduced searching selectively in the site in which the birds had cached the kibbles, and the opposite searching preference should be shown by jays that had been prefed peanuts.

The results favoured the selective rather than the general content of desires. The birds searched preferentially in the cache sites in which they had cached the non-prefed food; in other words, the birds that had been prefed peanuts searched in the kibbles site and those that had been prefed kibbles searched in the peanut site. This finding corresponds with what we know about motivation in other animals for which food-related desires are both specific and learned (Dickinson and Balleine 1994, 2002). An important feature of the experimental design is that we pilfered the food from the cache sites before allowing the birds to search for their caches on test trials. If we had not done so, as soon as the birds had recovered a cache of the prefed type, they would have discovered that this food was no longer attractive and thereby extinguished searching in this site, and, as a consequence, their pattern of searching would not have needed to reflect the interaction of their relative desires for the two foods with beliefs about where they were cached. By pilfering the caches we ensured that on test trials the relative preference must have been mediated by memory for the location of these caches.

What this procedure did not ensure, however, is that this memory took the form of a belief rather than of an association between the cues of the cache site and the type of food stored there (or, more strictly speaking, between the nodes activated by the cues and food). An associative account of the preference only requires that the repetitive activation of the food nodes during pre-feeding induced a temporary refractory state in these nodes. Consequently, the activation of the prefed food nodes at the time of recovery by the cache-site cues would have been reduced by their refractory states, thereby producing a preference for searching in the cache site of the non-prefed food. In an attempt to differentiate between the associative and the intentional accounts, we repeated the specific-satiety procedure using the following, more complex, design.

To understand this design, it is necessary to describe our procedures in more detail. The cache sites used in our studies are ice-cube trays, each consisting of two parallel rows of ice-cube moulds filled with a substrate such as sand. The trays are made spatially and visually distinct by surrounding each of them with a structure built of toy building blocks of various colours and shapes so that they are topographically unique. In this way, a given bird caches in different trays on different days, without having to reuse the trays. The birds must attend to and learn about the trays' cues because they cache different foods in the two sides of a particular tray and have to use these cues to remember where in a tray they have cached a particular type of food.

So it was in the experiment outlined in Fig. 9.1. Each bird cached three peanuts in one side of two trays, the Same and Different Trays, on separate caching episodes, and three kibbles in the other side of each tray on two further caching episodes. For example the jay, whose design is illustrated in Fig. 9.1, cached the peanuts in the left side of the trays during caching episodes 1 and 2, and the kibbles in the right sides of the trays during caching episodes 3 and 4. On the assumption that each pairing of the

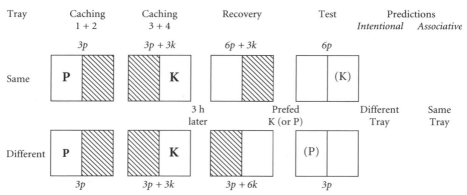

Fig. 9.1 The experimental design for differentiating between the associative and the intentional account of cache recovery. Numbers in italics represent the increase in associative strength of the connection between the tray cue and peanut (*p*) and kibble (*k*) nodes. When restricted to caching or recovering a particular food type in just one side of the caching tray, the other side of the tray is covered with a Plexiglas cover. The Plexiglas strip is removed during test so that birds can search both sides of the caching (indicated by hatching) tray. P: peanuts; K: kibbles. The brackets indicate the caches that would have remained in the tray at recovery if they had not been previously removed on test.

tray cues with a food item enhanced the associative strength of the connection between the tray cue nodes and the food nodes, the strength of the connection between the tray cue nodes and the peanut nodes should have been $3p$ where p is the increment in associative strength produced by a single peanut. Similarly, tray cue-kibble associative strength should have been $3k$ where k is the increment in strength produced by pairing a single kibble with the trays' cues. Therefore, after these caching episodes both the same and the different trays cues should have a total associative strength with food nodes of $3p + 3k$ (Fig. 9.1). The trays were then removed from the birds' homecages for a retention interval of 3 hours.

After the retention interval, the trays were returned to the birds who were then allowed to recover the three peanuts from the Same Tray and the three kibbles from the Different Tray. These tray cue–food pairings should have further enhanced the total associative strengths to $6p + 3k$ for the Same Tray but to $3p + 6k$ for the Different Tray. Finally, the birds were prefed one of the foods to satiety before being given the opportunity to search again in the two trays, but this time with both sides of each tray available. At issue is whether they directed most searches to the Same Tray that should have still contained the food of the *same* type as that which had been prefed, or to the Different Tray that should have still contained the food that was *different* to the prefed type. Of course, we had pilfered all the remaining food items prior to this final recovery test.

The prediction of the associative theory is clear: the birds should have searched preferentially in the Same Tray. Recall that prior to prefeeding the Same Tray had a total associative strength of $6p + 3k$ whereas the Different Tray had a strength of $3p + 6k$. On the assumption that prefeeding renders the nodes for the same food refractory, the total activation of the food nodes by the tray cues should have been $6p$ for the Same Tray but only $3p$ for the Different Tray after prefeeding on kibbles. If searching in a tray is determined by the strength of activation of food nodes by the tray cues, then the jays should have searched in the same tray rather than the different one.

But clearly, such a search pattern would not be the rational choice. After the regime of caching, recovery, and prefeeding, the Same Tray should have contained only the less desirable food items, the kibbles in the example illustrated in Fig. 9.1, whereas the desirable food, namely peanuts, should have been in the Different Tray. Our birds behaved rationally in this test by showing a marked preference for searching for what should have been intact caches in the Different Tray. This finding accords with an intentional account that assumes that searching is controlled by specific desires interacting with beliefs about the content of the cache sites, beliefs that were acquired during caching but were also updated by subsequent experience.

There are two general points to be drawn from these studies. The first is that, given one adopts an intentional account of animal behaviour, the problem of determining the content of intentional representations is not an entirely intractable one. Second, these studies illustrate an empirical strategy for evaluating intentional, and therefore psychologically rational, accounts of behaviour. There is never going to be a behaviourally decisive test of the intentionality of animal behaviour. All one can do is to derive predictions from an intentional account of a specific behaviour, in this case cache recovery, and test these predictions within a procedure that discriminates this explanation from a specific, mechanistic alternative such as the associative account (see and cf. Papineau and Heyes, Chapter 8, this volume; see also Kacelnik, Chapter 2, this volume, for scepticism about the empirical tractability of 'PP rationality' in studies of animals). Having illustrated how this strategy can be implemented in the case of desires, we now turn to the investigation of cache beliefs.

9.3 The structure and content of cache beliefs

Human cognitive psychology classifies beliefs as declarative memories to distinguish them from procedural memories, which encompass various forms of acquired motor and cognitive skills, responses, and habits. Declarative memory can be further subdivided into two forms: semantic or general knowledge and episodic recall (Tulving 1972, 1983). Semantic memories are general beliefs or knowledge, whereas episodic memories are recollections of specific, particular life events. Thus, for a jay the knowledge about the location and properties of the reliable sources of food in its territory would be an example of general knowledge, whereas the recollection of a particular caching episode at the time of recovery would be an example of episodic memory. Contemporary accounts of human memory (i.e. Tulving and Markowitch 1998) view episodic memory as being embedded within a more general declarative framework in such a way that specific episodic information can interact with general declarative knowledge. This interaction ensures that action based upon information represented in a recollection of specific episode is informed by the agent's corpus of general knowledge.

For a number of years, we have been investigating whether cache recovery by western scrub jays is mediated by at least a declarative-like memory (Clayton and Griffiths 2002; Clayton et al. 2000; Clayton et al. 2001; Griffiths et al. 1999). Our studies capitalized on the fact that these jays are omnivorous, eating and caching a variety of foods such as insects, larvae, and nuts. One problem the jays face, living as they do in the Californian Central Valley, is that some of their most preferred foods, such as invertebrates, decay if left too long in the cache before recovery. In contrast, other foods, such as nuts, are relatively durable. Consequently, as a result of experience with caching and recovering various foods after different cache-recovery (retention) intervals, the jays may acquire general knowledge about the rates at which different foods perish which they can then deploy in conjunction with episodic-like memory for specific caching events to determine their choices at recovery.

As a concrete illustration of this interaction between general knowledge and episodic-like recall, consider the case in which a jay caches perishable crickets in one site and non-perishable peanuts in another. As crickets are one of the bird's preferred foods, it should choose to search for cricket caches in preference to peanut caches. The problem is, however, that crickets perish if left in the cache for too long, so the choice at recovery should depend upon the length of the retention interval. When recovering caches after a short retention interval, the jay should search preferentially for crickets, but this preference should reverse after longer retention intervals so that the bird searches for peanuts if it believes that the crickets will have perished.

The fact that our colony of scrub jays are hand-raised, and therefore have no prior experience with decaying foods, allowed us (Clayton *et al.* 2001) to investigate whether the birds are capable of learning this reversal in recovery preference. As in the previous study, on each trial the birds cached one food in one side of a trial-unique caching tray and the second food in the other side of the same tray, but in this case the foods were crickets and peanuts. On some trials they were allowed to recover both peanuts and fresh crickets after one day, whereas on other trials the opportunity for recovery was delayed for 4 days by which time the crickets had decayed.[2]

The jays rapidly learned to search for crickets when fresh and to search for peanuts when the time interval between caching and recovery was such that the crickets should have degraded. On the first two trials the majority of birds directed their first search to the cricket side of the tray after the 4-day retention interval, but by the third trial all birds switched their preference and searched in the peanut side first. However, this preference switch was under temporal control because the majority of birds continued to direct their first search to the cricket side on all training with the 1-day retention interval. The reason why we recorded the first direction of the first search on these training trials was because once a bird had found a food cache it did not have to rely on memory for its caches to determine its preference—all it need do was avoid the side on which it had just found a less preferred food item, be it a peanut or decayed cricket.

It remains possible, however, that the birds could have detected the type of food buried in each side of the caching trays before making even their first search, and so we conducted a series of probe trials in which we pilfered all the food caches prior to the recovery test. In addition, we also tested searching at recovery after untrained retention intervals of 2, 3, and 5 days to see how the birds interpolated and extrapolated from the trained retention intervals of 1 and 4 days (Clayton *et al.* 2003). The profile of searching shows that the jays had inferred from their training experience that crickets remained palatable for up to 3 days but then perished. The majority of birds searched the cricket side first after retention intervals of up to 3 days and before switching their preference to the peanut side after longer intervals.

---

[2] Clayton *et al.* (2001) included trials with a third food type, meal worms, and included a shorter 4-hour retention interval, but the procedural description has been simplified for exposition in a way that does not vitiate the interpretation.

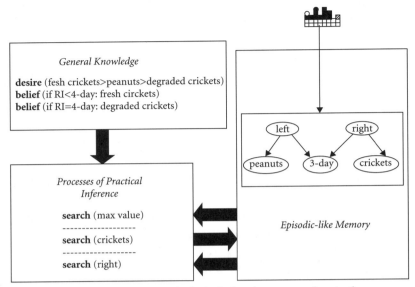

Fig. 9.2 The architecture and contents of a jay's declarative memory (see text).

The ability to learn about the degradation of crickets and to apply this knowledge to new cache sites is readily analysable in terms of declarative memory. A possible architecture for the jays' declarative memory in this task is illustrated in Fig. 9.2. According to this model, the jays acquired two forms of general knowledge during their training experience. First, they learned about the relative desirability of the various foods they experienced: fresh crickets, degraded crickets and peanuts. Second, they acquire beliefs about the temporal decay of the crickets: specifically that crickets are fresh at retention intervals (RIs) of less than 4 days, but degraded after longer intervals.

The second component of the declarative system is an episodic-like memory which encodes in a bound and integrated representation the content of a specific life event, which in this case is the caching of peanuts in the left side and crickets in the right side of a particular tray 3 days ago. We refer to such a representation as a 'what–where–when' memory because it is important that all three features of the experience are represented if this memory is to resemble episodic memory. Recent discussions of human episodic memory have emphasized the phenomenological characteristics of recollection, such as 'autonoetic awareness' (Tulving and Markowitsch 1998). However, such awareness cannot be assessed in non-linguistic animals because there are no agreed behavioural markers of consciousness in the absence of language. It is for this reason that we have referred to the memories mediating cache recovery by western scrub jays as episodic-like (Clayton *et al.* 2000, 2001; Griffiths *et al.* 1999) by reference to Tulving's (1972) original characterization of episodic memory as a form of memory that 'receives and stores information about temporally dated episodes or events, and temporal–spatial relations among those events' (p. 385). By this 'what–where–when' criterion, the bird's memory of the caching episodes is episodic-like because it involves recall of the content (what) and location (where) of their cache, and a temporal component (when).

This model allows us to offer a sketch of the interaction between general knowledge and episodic-like memories through the processes of practical inference. Our assumption is that the birds have an enduring intention to search for the most desired food items when in the motivational state that promotes cache recovery. Encountering a cache site, in this case a particular tray, retrieves the what–where–when episodic-like memory of caching in that tray. Integrating this memory with general knowledge about its relative food desires and the degradation profiles of these foods allows the bird to derive the more specific intention to search for crickets. This intention, when taken in conjunction with the episodic-like memory that crickets were cached in right-hand side of the tray, leads to the derivation of an intention that can be directly expressed in behaviour, specifically the intention to search in the right-hand side.

9.3.1 The flexibility of cache memories

We do not intend that the details of this model should be taken at face value. Rather its function is to illustrate a cardinal feature of declarative memory systems that is central to the issue of animal rationality, namely that declarative memories or beliefs have a representational form that allows their content to be deployed flexibly. It was this issue of the flexibility of knowledge representations that motivated Winograd's (1975) classic analysis of the relative merits of declarative and procedural representations for artificial intelligence. In that analysis, he pointed out that the flexibility of declarative representations is bought at the computational cost of having inference processes that can operate on their content. Indeed Tulving (2001) explicitly raised the issue of mnemonic flexibility with reference to the role of declarative memory in cache recovery when he asked '...could Clayton and her colleagues (or someone else) get their scrub jays, who remember what kind of food is where, to do something other with that information than act on it 'inflexibly'...?' (p. 1513).

Although the concept of 'flexibility' is not well defined, our model of declarative memory allows for the flexible deployment of episodic-like information in relation to a bird's general, semantic-like, knowledge; a flexibility that arises from their interaction through practical inference processes. Recall that our jays experienced the variable palatability of crickets only after 1 and 4 days during training, so their belief that crickets degrade between 3 and 4 days after caching is a generalization from this training experience. But what would happen if we falsified this belief during the retention interval for a caching episode? At issue is whether the birds could use this new general knowledge about degradation profiles to alter their recovery preference, even though this new information was not available at the time when the cache memory was encoded.

To address this issue, following the test of recovery after different retention intervals we gave our jays interleaved caching and recovery trials using the design illustrated in Fig. 9.3

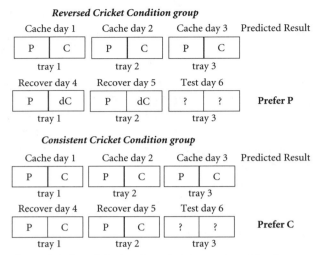

Fig. 9.3 The experimental design of the interleaved trials experiment for the reversed and consistent groups. P: peanuts; C: fresh crickets; dC: degraded crickets; ?: on test, all caches from both sides of the tray are removed to establish where the birds will search in the absence of any cues directly emanating from the food.

(Clayton *et al*. 2003). Although the design looks complex, the basic idea was to give the jays in the reverse condition information that the crickets did in fact decay during the retention interval. So these birds were allowed to cache crickets and peanuts in each of three different trays on successive days. For example, a jay might cache the two foods in tray 1 on Monday (day 1), and then cache more crickets and peanuts in a different tray 2 on Tuesday (day 2) and in a third tray 3 on Wednesday (day 3). Remember that none of the birds had been given any information about whether crickets are fresh or degraded after 3 days because they had been trained only with retention intervals of 1 day and 4 days. The birds were then provided for the first time with direct information about the fate of their cricket caches after a 3-day retention interval so that this is information was available only after they had finished caching in all three trays. Therefore, on Thursday (day 4) the reverse group recovered peanuts and crickets from Monday's tray 1 and, against their expectation, discovered the crickets had in fact perished after the 3- day retention interval. This new information was reinforced by recovery from Tuesday's caching tray 2 on Friday (day 5) before these jays were given a critical test in which they searched in Wednesday's tray 3 on the Saturday (day 6). Of course, being a test trial no food was present at recovery. The recovery preferences of the birds in this reverse condition on this test trial was contrasted with that of another set of birds from the degrade group. The jays in this consistent condition had their expectation that the cricket would be fresh after 3 days confirmed. Searching in tray 1 on Thursday (day 4) and tray 2 on Friday (day 5) yielded fresh and palatable crickets.

Predicting the search preference for the consistent group is straightforward. The experience of recovering palatable crickets from trays 1 and 2 should have confirmed the jays' generalized belief(s) about the decay profile for crickets and therefore these birds should have searched for crickets in the left side of tray 3 on the test recovery period on day 6 (Fig. 9.3). And this is what all four of these jays did. For the reversed condition, however, the experience of recovering perished crickets from trays 1 and 2 on days 4 and 5, respectively, should have disconfirmed the jays' beliefs that crickets were fresh after 3 days and replaced this general knowledge belief with one representing the crickets as degraded after this retention interval. Consequently, when the episodic-like memory of caching in tray 3 on day 3 was retrieved by the presentation of this tray on day 6, the interaction of this memory with the birds' general knowledge through the practical inference processes should have derived an intention to search for peanuts. A further interaction with the episodic-like memory of where peanuts were cached in tray 3 should have lead to an intention to search in the left side of tray.

Indeed, all four birds in the reversed condition directed their first search during the test on day 6 to the peanut side of tray 3. Importantly, this switch in preference did not reflect a general change in their beliefs about whether or not to search for crickets because when the birds were tested with a 1-day retention interval they reverted to their prior preference for crickets. We therefore interpret this reversal of the search preference as evidence that our jays can integrate information about the caching episode with new information presented during the retention interval in a rational manner.

9.4 **The rationality of caching strategies**

As we have emphasized, the evaluation of the psychological rationality of animal action is a matter of bringing converging evidence to bear of the issue because intentionality is not necessarily manifest in behaviour. It is often claimed that the demands of social interactions, and especially competition from conspecifics, is a major factor in the evolution of cognition (Humphrey 1976; Jolly 1966). In this respect, it is notable that for some species, food caching and recovery are activities that occur within a social context, not least because caches are susceptible to pilfering by other individuals (Van der Wall 1990). Many species are known to pilfer the caches of other birds, and several species of corvid, including our scrub jays, use observational memory to locate the caches of their competitors and pilfer them when the food-storer has left the scene (see Clayton *et al.* 2001). But these food-caching corvids also engage in a number of strategies that serve to reduce the probability that their caches will be stolen by competitors. For example ravens will delay caching if other ravens are in the vicinity, waiting until the potential pilferers are distracted and have left the scene (Heinrich and Pepper 1998; Bugnyar and Kotrschal 2002). In other instances they will preferentially store food behind obstacles so that other ravens cannot see where the caches are being made (Bugnyar and Kotrschal 2002). But when food is readily available and many other individuals are present it might be better to cache while you have the opportunity to do so, in the hope that your competitors are distracted! There are a number of reports of corvids, including ravens and jays, caching rapidly when other individuals are present, only to return alone when the competitors have left the scene, moving or recaching the caches they had hidden in the presence of conspecifics to new cache sites (e.g. Bugnyar and Kotrschal 2002; Emery and Clayton 2001; Goodwin 1956; Heinrich 1999).

As noted earlier, the apparent complexity of a particular behaviour, in this case caching, does not speak to the issue of whether or not the behaviour is rational. So is there any evidence that our scrub jays can make rational decisions about the future consequences of caching and stealing? Consider the apparent observation of birds moving the food they had hidden in the presence of other individuals, and recaching the food items in new places when those observers were no longer present. Of course in the wild, one might explain the presence or absence of another bird as purely coincidental to the caching and recaching events. So first, one needs to test whether it is the presence of an observer at caching, and absence of one at recovery, that elicits the food-cacher's recaching behaviour. This is what we set out do in the first test (Emery and Clayton 2001) illustrated in Fig. 9.4.

We allowed our scrub jays to cache food under two conditions, namely in one uniquely-configured tray when watched by another jay (the observer), or in another

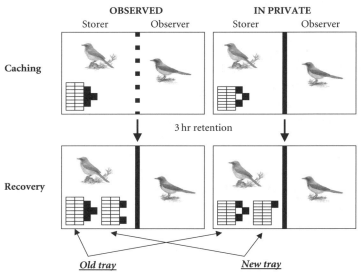

Fig. 9.4 The experimental design of the first social context of caching experiment, where storer jays cached worms in unique trays either when observed by another jay (observer) or in private (with the view of the observer occluded. After a 3-hour period, the storers recovered their caches in private, independent of the social context during caching. An additional, second unique tray provided opportunities for re-caching.

uniquely-configured tray when a cover was placed over the other bird's cage so that the storer could hide its food in private. Subsequently, the storer was presented with the tray in which it had cached (old tray) as well as a new tray (also with a unique arrangement of toy blocks) in which it could recache, and allowed to recover and recache the food in private so that at the time of recovery the conditions were identical. In other words, the only difference between the two treatments was in whether the jay had been observed during the caching phase or had been allowed to hide its food in private. If recaching really depends on the presence of an observer at the time of caching, then we should only expect to see recaching if the bird had been observed, and not when it had been allowed to cache in private. And so it was in this experiment. The jays continued to cache food, irrespective of whether another bird was watching them or not. The key result was that birds recached significantly more items during recovery when they had been observed during caching. Moreover, they specifically placed these recovered caches in new cache sites the observer had not seen, but only when they had been observed caching. When they had cached in private, they were just as likely to place the few recaches that they did make in original cache sites as in the new ones. Because the two conditions were identical at the time of recovery, the difference between treatments meant that the birds had to remember whether or not they had been watched during the previous caching condition in order to know when to recache during recovery, and whether to recache in new sites.

Before considering an intentional account of recaching, we attempted to rule out some more obvious mechanistic explanations. For example generalized recaching may simply be triggered automatically by the memory of being in the presence of another bird recently. Consequently in the next study, we investigated whether recaching is mediated by an episodic-like memory of the specific caching episode during which the jay was observed. To do so, we used an interleaved trials design similar to that described for the reversed and consistent conditions of the cricket degradation study, except that this time we gave the birds 'observed' and 'in private' caching trials (Fig. 9.5). Thus some birds cached in the observed tray while watched by a conspecific and then cached in the other, in-private tray immediately afterwards. The remaining jays received these two caching episodes in the reverse order. Subsequently the storers were allowed to recover from both trays in the presence of a novel tray in which they could recache. The birds mainly recached from the observed tray, moving them to new sites. By contrast, the birds showed little tendency to recache items from the in-private tray, nor did they discriminate between old and new sites for those items that they did recache. So this result suggests that jays remember not only whether or not they were being observed, but can also recall the specific tray in which they cached while being watched, rather than relying on a default mechanism to recache if they have been observed recently.

The obvious intentional account of recaching assumes that the birds have general beliefs that being observed during caching causes the loss of the cached food items and

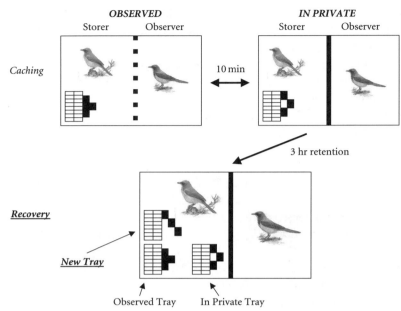

Fig. 9.5 The experimental design of the interleaved trials experiment on the social context of caching. Storer jays either cached in one unique tray when observed (observed tray) then 10 min later in a second unique tray in private (in private tray) or *vice versa*. After a 3-hour period, the storers were allowed to recover their caches from the observed and in private trays outside of the view of the observer (in private). An additional, third unique tray was provided for re-caching.

that this loss can be prevented by recaching. When taken in conjunction with a low desire for pilfered caches relative to intact caches, the process of practical inference would rationalize recaching with respect to the content of these beliefs and desires. However, this account immediately raises the issue of where these beliefs come from; unless one is prepared to countenance the idea of innate beliefs with intentional content (as opposed to innate associative structures), the answer must lie with the past experience of the storer. Therefore, in a final study, we compared recaching by the jays described above, all of whom had experience of caching and of observing and subsequently stealing other birds' caches, with two other groups of jays (Emery and Clayton 2001). The observer group had experience of watching other birds cache, but had never been given the opportunity to steal those caches. By contrast, the pilferer group had experience of stealing other birds' caches, but were not tested explicitly for their ability to remember the location of those caches within an experimental context. (We cannot say that these jays had no experience of observing other birds caching outside of the experiments.) We found that experience did matter. Indeed, jays that had prior experience of pilfering another bird's caches subsequently recached food in new cache sites during recovery trials, but only when they had been observed caching. Jays without this pilfering experience did not, even though they had observed other jays caching. These results therefore suggest that jays relate information about their previous experience as a thief to the possibility of future stealing by another bird, and modify their caching strategy accordingly.

We draw two conclusions from this study. The first is that recaching is psychologically rational in that, of the explanations currently available, only an intentional account explains the differences in recaching behaviour between those with and without experience of having been a pilferer in the past. Second, it is important to note that recaching appears to be based on mental attribution or 'mind reading' (see this volume, Part V).[3] The inference that jays with prior pilfering experience appear to make in this situation is that a conspecific with similar prior experience would share the same beliefs as they have, namely that caches can be stolen. It is this inference that both rationalizes and causes recaching. This conclusion, of course, goes beyond the issue of the simple behavioural rationality and clearly requires further examination.

9.5 Summary and conclusions

In this chapter we have reviewed some of our studies of cache recovery and recaching by western scrub jays with respect to the issue of whether these behaviours are psychologically rational. We have assessed psychological rationality by investigating whether the jays' behaviour warrants an intentional belief–desire explanation. With respect to desires, motivational manipulations that should have yielded differential desires for two types of cached foods caused the birds to search preferentially for the most desired food in a task in which a simple associative account predicted the opposite preference. Moreover, the response to these motivational manipulations provided a procedure for investigating the content of the birds' desires.

Our analysis of the beliefs underlying cache recovery was set within the framework of a declarative memory system that distinguished between general beliefs or knowledge and episodic memories. According to this analysis, the information about a specific caching event is represented in a what–where–when memory with episodic-like properties, whereas general knowledge about the relative desirability of difference food types and the way in which desirability changes with the time in the cache, is represented by semantic-like memories. Searching for caches at recovery is then generated by the interaction of these two types of memory through processes of practical inference.

---

[3] Editors' note: It is interesting to assess this evidence for mental attribution in light of the methodological constraints urged by Povinelli and Vonk, this volume. Does it satisfy their concerns? It can be argued that this is some of the best evidence to date precisely because of the levels of control: (a) recovery conditions are always conducted in private and therefore any difference between treatments must result from differences at the time of caching, hours earlier and (b) only some of the jays—and those with specific prior experience of being pilferers—engage in this behaviour. The editors are grateful to the authors of this chapter for discussion of this point.

It is the nature of this interaction that gives cache recovery its rational character and endows this behaviour with a degree of flexibility. This flexibility was demonstrated in procedures in which we gave the jays new general information during the retention interval that was not available to the birds at the time of caching. The reversed search preference at recovery demonstrates that they were capable of integrating this new information with their memory of the specific caching episode in a way predicted by the rational–declarative model.

Further evidence for mnemonic integration comes from studies demonstrating that the jays can integrate information from their own experience of pilfering other birds' caches with memories of caching episodes in which they themselves were observed caching food. This integration produced recaching of food items into new sites, a rational defensive action on the part of the storer. Although none of these examples of the flexibility of caching and recovery behaviour is decisive on its own, taken together they provide strong, converging evidence for the intentional and rational control of behaviour especially when contrasted with the failure of standard associative accounts.

Many, of course, have claimed that mammals, and especially primates (see Call, Chapter 10; Tomasello and Call, Chapter17; Boysen, Chapter 22, this volume) and cetaceans (see Herman, Chapter 20; Tschudin, Chapter 19, this volume) are capable of rational cognition, and indeed one of us (AD) has argued that even the goal-directed behaviour of rodents is mediated by intentional processes (Heyes and Dickinson 1990; Dickinson and Balleine 2000). Our research on caching and recovery by western scrub jays contributes to the growing body of evidence that behavioural markers of rational cognition are as compelling for at least some birds (see Pepperberg, Chapter 21; Papineau and Heyes, Chapter 8; Kacelnik, Chapter 2, this volume) as they are for mammals in spite of their radically divergent neurobiology and evolutionary trajectory (see Emery and Clayton 2004, for a detailed discussion).

Acknowledgments

This research was supported by grants from National Institute of Health (NS35465 and MH62602), and the Grindley Trust to Nicola Clayton, a BBSRC Grant (S16565) to Nicola Clayton and Anthony Dickinson, and conducted within the UK Medical Research Council Co-operative Grant (G9805862). Nathan Emery was supported by a Royal Society University Research Fellowship.

References

Bugnyar, T., and Kotrschal, K. (2002). Observational learning and the raiding of food caches in ravens, *Corus corvax*: Is it 'tactical' deception? *Animal Behaviour*, **64**: 185–195.

Clayton, N. S., and Dickinson, A. (1999). Memory for the contents of caches by scrub jays (*Aphelocoma coerulescens*). *Journal of Experimental Psychology: Animal Behavior Processes*, **25**: 82–91.

Clayton, N. S., and Griffiths, D. P. (2002). Testing episodic-like memory in animals. In: L. Squire and D. Schacter, eds. *The Neuropsychology of Memory*, 3rd edn, pp. 492–507. New York: Guilford Publications.

Clayton, N. S., Griffiths, D. P., and Dickinson, A. (2000). Declarative and episodic-like memory in animals: Personal musing of a scrub-jay. In: C. Heyes and L. Huber, eds. *The Evolution of Cognition*. Cambridge, pp. 273–288. MA: MIT Press.

Clayton, N. S., Griffiths, D. P., Emery, N. J., and Dickinson, A. (2001). Elements of episodic-like memory in animals. *Philosophical Transactions of the Royal Society (London), B,* **356**: 1483–1491.

Clayton, N. S., Yu, K. S., and Dickinson, A. (2001). Scrub-jays (*Aphelocoma coerulescens*) form integrated memories of the multiple features of caching episodes. *Journal of Experimental Psychology: Animal Behavior Processes,* **27**: 17–29.

Clayton, N. S., Yu, K. S., and Dickinson, A. (2003). Interacting cache memories: evidence of flexible memory use by scrub jays. *Journal of Experimental Psychology: Animal Behavior Processes,* **29**: 14–22.

Dickinson, A. (1985). Actions and habits: the development of behavioural autonomy. *Philosophical Transactions of the Royal Society (London), B,* **308**: 67–78.

Dickinson, A. (1989). Expectancy theory in animal conditioning. In: S. B. Klein and R. R. Mowrer, eds. *Contemporary Learning Theories: Pavlovian Conditioning and the Status of Traditional Learning Theories,* pp. 279–308. Hillsdale, NJ: Lawrence Erlbaum Associates.

Dickinson, A., and Balleine, B. (1994). Motivational control of goal-directed action. *Animal Learning and Behavior,* **22**: 1–18.

Dickinson, A., and Balleine, B. W. (2000). Causal cognition and goal-directed action. In: C. M. Heyes and L. Huber, eds. *The Evolution of Cognition,* pp. 185–204. Cambridge, MA: MIT Press.

Dickinson, A., and Balleine, B. (2002). The role of learning in the operation of motivational systems. In: H. Pashler and R. Gallistel, eds. *Stevens' handbook of Experimental Psychology, vol. 3, Learning, motivation, and emotion,* 3rd edn, pp. 497–533. New York: John Wiley and Sons.

Emery, N. J., and Clayton, N. S. (2001). Effects of experience and social context on prospective caching strategies in scrub jays. *Nature,* **414**: 443–446. [See also *Nature,* **447**: 349.]

Emery, N. J., and Clayton, N. S. (2004). Comparing the complex cognition of birds and primates. In: L. J. Rogers and G. S. Kaplan, eds. *Comparative Vertebrate Cognition: Are Primates Superior to Non-Primates?,* pp. 3–55. The Hague: Kluwer Academic Press.

Fodor, J. A. (1977). *The Language of Thought.* Hassocks, UK: Harvester Press.

Goodwin, D. (1956). Further observations on the behaviour of the jay. *Ibis,* **98**: 186–219.

Griffiths, D., Dickinson, A., and Clayton, N. S. (1999). Episodic memory: what can animals remember about their past? *Trends in Cognitive Science,* **3**: 74–80.

Heinrich, B. (1999). *Mind of the Raven.* New York: Harper Collins.

Heinrich, B., and Pepper, J. (1998). Influence of competitors on caching behaviour in common ravens, *Corvus corax. Animal Behaviour,* **56**: 1083–1090.

Heyes, C., and Dickinson, A. (1990). The intentionality of animal action. *Mind and Language,* **5**: 87–104.

Humphrey, N. K. (1976). The social function of intellect. In: P. P. G. Bateson and R. A. Hinde, eds. *Growing Points in Ethology,* pp. 303–317. Cambridge: Cambridge University Press.

Jolly, A. (1966). Lemur social behaviour and primate intelligence. *Science,* **153**: 501–506.

Rumelhart, D. E., and McClelland, J. L. (1986). *Parallel Distributed Processing.* Cambridge, MA: MIT Press.

Thorndike, E. L. (1911). *Animal Intelligence: Experimental Studies.* New York: Macmillan.

Tulving, E. (1972). Episodic and semantic memory. In: E. Tulving and W. Donaldson, eds. *Organization of Memory,* pp. 381–403. New York: Academic Press.

Tulving, E. (1983). *Elements of Episodic Memory.* New York: Oxford University Press.

Tulving, E. (2001). Episodic memory and common sense: how far apart? *Philosophical Transactions of the Royal Society (London),* **B356**: 1505–1515.

Tulving, E., and Markowitch, H. J. (1998). Episodic and declarative memory: role of the hippocampus. *Hippocampus,* **8**: 198–204.

Van der Wall, S. B. (1990). *Food Hoarding in Animals.* Chicago: University of Chicago Press.

Winograd, T. (1975). Frame representations and the declarative-procedural controversy. In: D. Bobrow and A. Collins, eds. *Representation and Understanding: Studies in Cognitive Science,* pp. 185–210. San Diego, CA: Academic Press.

Part III

Metacognition

Chapter 10

Descartes' two errors: Reason and reflection in the great apes

Josep Call

Abstract

Reasoning and reflection have traditionally been considered uniquely human attributes. Many animals, including the great apes, are often regarded as masters at making associations between arbitrary stimuli while at the same time they are rarely considered capable of reasoning and understanding the causality behind even simple phenomena. In this chapter, I defend a view opposite to this predominant position. Apes (and possibly other animals) are actually quite good at understanding and reasoning about certain physical properties of their world while at the same time they are quite bad at associating arbitrary stimuli and responses. In other words, if two stimuli have a causal connection (as when food inside a shaken cup makes noise) apes perform better than if stimuli hold an arbitrary relation (as when an unrelated noise indicates food), even if the contingencies of reinforcement are the same. Neither a history of reinforcement based on traditional associationism or a biological predisposition to respond differentially to certain stimuli combinations explains these results. Instead, I postulate that subjects reason and use logical operations based on inference by exclusion to locate the hidden food. In addition to the ability to reason about physical phenomena, I argue that apes (and other animals) also have some access to their understanding of the problem. More precisely, they have metacognitive abilities that allow them to know what they know or do not know. Thus reasoning and reflection may not be the bastions of human uniqueness, as Descartes once thought. Rather, these skills may have evolved (or coevolved) in other animals as well because they allowed them to solve problems in the world more efficiently.

10.1 **Introduction**

When I was growing up, my elementary school teacher taught us that there were two kinds of animals: rational and irrational. Rational animals were those that are able to reason and think, capable of moral judgments, and possessing a conscience. Irrational animals not only lacked such high qualities, they also lacked feelings, a sense of the future, language, and culture. Put another way, we were taught that human beings are rational animals, and that animals in the rest of the animal kingdom are irrational or non-rational.

This distinction between humans and non-human animals (animals henceforth) has its roots in some of the greatest Western thinkers. Descartes in particular made a sharp distinction between animals and humans. Perhaps partly influenced by the automata activated by the laws of hydraulics that populated the gardens of 16th century France, Descartes maintained that animal behavior could be reduced to a series of automatic reflexes. Human behavior, in contrast, is not purely reflexive but is controlled by rational thought. His famous 'cogito ergo sum' ('I think, therefore I am') sums up the gist of his message. According to Descartes, humans are rational and conscious whereas animals are non-rational and unconscious; humans can reason and reflect whereas animals can do neither.

This idea of a clear dichotomy between humans and animals has had a pervasive influence, not just on elementary school teachers and philosophers, but also on much of contemporary psychology. Beginning with such illustrious figures as James (1890), Loeb (1918), or Thorndike (1898), the idea that animals can learn to associate stimuli and responses, but possess no understanding and are incapable of reasoning about even simple events in the world is widespread. Animals are often branded as association champions: capable of associating arbitrary stimuli and responses, yet incapable of understanding the causality behind even simple phenomena.

In this paper, I will defend a position that is exactly the opposite of this predominant view. I will argue that apes (and possibly other animals) are actually quite good at understanding and reasoning about certain physical properties of their world, but that they are quite bad at associating arbitrary stimuli and responses. In other words, if two stimuli have a causal connection (for example, shaken food inside a cup makes noise) apes perform better than if stimuli hold an arbitrary relation (for example, noise indicates food), even if the contingencies of reinforcement are maintained equal. Furthermore, I will argue that in addition to being able to reason about their physical world, apes can reflect on the problems they try to solve, have some access to their own understanding of the problems, and have the capacity to recognize that they make mistakes. These claims go against the positions of some of the most prominent cognitive scientists of our time. For instance, Tulving and Madigan (1970) maintain that 'one of the truly unique characteristics of human memory [is]: knowing about knowing' (p. 477).

This chapter is organized into three parts. First, I will review some recent evidence supporting the idea that apes engage in causal reasoning about physical problems related to finding food. Second, I will review some evidence that suggests that apes have access to their own mental processes. Finally, I will conclude with some reflections on the evolution (perhaps it would be more appropriate to say coevolution) of reasoning and reflection in animals.

10.2 **On reasoning**

It is widely believed that many animals are extremely good at associating arbitrary cues with responses while at the same time being rather limited in their ability to understand the causal relations between certain physical events. For instance, an animal may learn that a noise predicts the presence a food, but may not understand that it is the food that causes the noise. Associating such cues is seen as common, while having a causal understanding is seen as exceptional, in many cases unproven. It is true that there are many cases that suggest cue learning, but there are other cases that, although at first sight may appear as cases of cue learning, on closer inspection cannot be so easily explained by conditional associations.

Premack (1995) contrasted what he called natural causes and arbitrary causes. Natural causation is based on an underlying theory that explains the relation between two events; for instance, gravitational theory states that an unsupported object will fall to the ground due to the gravitational force of the earth. Arbitrary causation is based on pairing the two events in time and space; for instance, the color green denotes right of way. Premack argued that chimpanzees were quite skilful at solving problems depicting natural causes. In one study, Premack and Premack (1994) presented chimpanzees with two boxes and two types of fruit (for example banana and apple). Chimpanzees were allowed to witness the experimenter deposit one kind of fruit in each of the boxes so that both boxes were baited. Later subjects saw the experimenter eating one of the fruits (for example, banana) and the question was whether, given the opportunity to select one of the boxes, they would select the box in which the experimenter had deposited the food that he was not currently eating (that is, apple), presumably because it still contained the fruit. Chimpanzees solved this problem quickly without trial-and-error, showing that they were able to infer that if the experimenter was eating the banana, the box where the banana was deposited would be empty. This is called *inferential reasoning by exclusion* (see also Bermúdez' discussion, Chapter 5, this volume). The alternative to this inferential strategy is a discriminative learning strategy in which subjects learn through multiple trials that the presence of the banana in the experimenter's mouth is a discriminative sign for choosing the other alternative. Premack (1995), however, ruled out this explanation on the grounds that chimpanzees did not learn gradually to select the correct container, but selected it from the beginning. Next I will present some additional studies on this topic in which I further explored great apes' understanding of natural causes, and directly contrasted problems depicting natural and arbitrary causes.

10.2.1 **Shaken food inside a cup makes noise**

Call (2004) extended this previous work on inferences to include the use of auditory stimuli and to probe more deeply into the understanding that apes (not just chimpanzees) have about such events. The basic idea of this study was to contrast two types of problem that present similar superficial cues and reward contingencies but that differ in their causal structure. In one such study, we presented two opaque cups and baited one of them. In the causal problem (a natural cause, in Premack's terminology), we shook both cups, which resulted in the baited one making a noise. In the arbitrary problem (an arbitrary cause, in Premack's terminology), we tapped both cups but we only produced a noisy tapping for the baited cup. Thus in both problems a sound indicated the presence of the food, but the problems differed in their causal structure: in one case the noise was caused by the food, in the other it wasn't. The results showed that apes selected the baited cup above chance in the causal, but not the arbitrary problem. This is even more striking since we ran the arbitrary test after the apes had succeeded in the causal test; moreover, in the arbitrary test they were actually rewarded every time they selected the cup associated with the tapping noise, something which occurred on approximately 50 per cent (that is, 12) of the trials. Additional tests indicated that substituting the tapping noise with a recorded shaking noise made no difference, and control tests showed that animals were not using cues given inadvertently by the experimenter, or the smell of the reward, to locate the food.

In another test, we further probed the apes' understanding of the relation between movement and noise; in particular, we investigated whether they were able to make inferences regarding the location of food. Again, we presented subjects with two cups. This time, however, we shook the empty cup and lifted the baited one so that no sound was produced from either cup. If subjects know that shaken food produces a noise, they should select the lifted cup because if a shaken cup makes no noise, then the food must be in the other cup. Results indicated that subjects selected the baited cup more than in a control condition in which we lifted both cups. Additional experiments showed that this result could not be explained as a learned avoidance response to the silent shaken cup since they did not avoid the shaken silent cup when it was paired with a cup, which we rotated silently. In a further test, three subjects were able to appreciate that certain movements but not others can produce a noise. We presented a silent shaken cup (moved laterally) paired with a silent stirred cup (that is, moved in a circle). Here, both cups were moved, but food inside the cups would only produce a sound if the cup was shaken and not if the cup was stirred. The subjects who succeeded at this task know that certain movements, but not others, are likely to produce a sound.

In summary, these results show that the apes are better at solving causal as opposed to than arbitrary problems, and that they can make inferences regarding the location of food based on the differential movements of cups in the absence of any sound cue. One interpretation of these results is that apes do not simply associate a sound with the presence of food, but attribute the sound to the food itself, they understand that the food is the cause of the noise.

10.2.2 Solid food under a board occupies space

One possible criticism of the previous paradigm is that it is based on auditory information, and this may be special for some reason. Therefore, it is unclear whether these results represent a general understanding of causality about the physical world. We decided to investigate this possibility by using tests that do not involve auditory information but visual information. Some studies with human infants have shown that they are sensitive to certain properties of objects such as solidity. In a classic study, Baillargeon *et al.* (1985) presented infants with an object and a screen located in front of the object, which flipped towards the object. In one condition, the screen flipped over the object until it occluded it and then stopped. In another condition, the screen described the same movement as before, but once it reached the object it continued to move so that it apparently passed through the object.

Using this inclined barrier idea, Call (submitted) investigated whether bonobos, gorillas, and orangutans can understand that a piece of food hidden underneath a screen occupies space. Again, we contrasted a causal problem with a matched arbitrary problem. In the causal problem, we presented apes with two small wooden boards (25 × 9 cm) on a platform and hid a reward under one of them in such a way that the orientation of the boards differed. The empty board remained flat on the surface of the platform while the baited board acquired an inclined orientation due to the presence of the reward under it (Fig. 10.1). Given a choice between the two boards, the apes preferentially selected the inclined board in 80 per cent of the trials. There were no significant differences between species. It's possible that the apes had previously learned that objects with an inclined orientation are often rewarded, or that they preferred inclined objects because they look larger: a response acquired before the experiment is often very hard to rule out, particularly with species that have to live in complex and stimulating environments to develop properly. Fortunately, the arbitrary test in this study allowed us to rule out the possibility of a previous association. In this arbitrary test we replaced one of the boards with a solid wooden wedge that produced the same visual effect as a baited board when placed on the platform. Prior to the test, we showed the wedge to the subjects so that they could inspect it. We then paired it with a flat board and measured the subjects' preferences. Results indicated that there was no preference for the wooden wedge over the flat board. Subjects selected the wedge and the flat board in 53 per cent and 47 per cent of the trials, respectively. This result is even more striking if one notes that, as in the noise paradigm, we ran this arbitrary test after the apes had succeeded in the causal test.

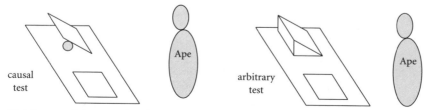

Fig. 10.1 Set-up of the causal and arbitrary tests in the board paradigm. Note that the perceptual information received by the subject is very similar in both tests.

Moreover, they were rewarded every time they selected the wedge in this test, which on average they did 50 per cent of the time (that is, on 12 trials).

In summary, subjects only preferred the wooden inclined object in the causal test, not the arbitrary one, presumably because, as in the previous experiment, they understood that the inclined orientation was caused by the presence of food.

10.2.3 Arbitrary cues are hard to learn

One could argue that the arbitrary tests that we used are too hard for the apes and that is why they scored higher in the causal tests than the arbitrary ones. On this suggestion, differences are detected not because offering a causal structure helps, but because the specific arbitrary tests that we used to compare with causal tests are particularly difficult. This argument, however, is disproved by the results of various color and shape discrimination learning tests. As part of other studies, we tested discrimination learning in the same apes that we used for the causality tests and found that these results resembled those obtained in the arbitrary tests presented here, not those in the causal tests. Figure 10.2 presents a comparison between causal, arbitrary, and simple discrimination tests after 24 trials. Only causal tests were above chance ($p < 0.001$ in both cases). Moreover, a much higher proportion of subjects obtained individual scores above chance in the causal than in the arbitrary or discrimination tests; and none was above chance after the first 12 trials in the arbitrary or discrimination tests, whereas several apes were above chance in the causal tests.

In summary, subjects performed better in causal than in arbitrary tests, both in auditory and visual modalities. Moreover, there was no evidence of learning during the testing sessions: subjects either performed well from the beginning of the session

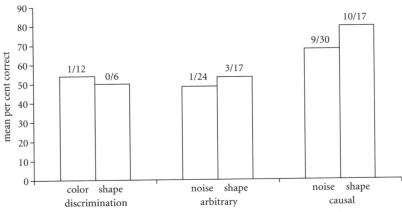

Fig. 10.2 A comparison of the mean percentage of correct trials in the discrimination, arbitrary, and causal tests. Also shown is the number of subjects that obtained individual scores above chance in each of the tests.

or poorly throughout. Finally, the performance on arbitrary tests closely resembled that found in basic discrimination tests, which suggests that the arbitrary tests that we used were not particularly difficult. Taken together, these results suggest that apes do not simply associate a cue with the presence of food, but rather understand that the food is the cause of the cue.

10.3 Rationality by association, biological predisposition, or logic

A common response to evidence such as that presented here for rationality and causal understanding in apes is to seek alternative mechanisms that could also explain the evidence but which do not attribute rationality or causal understanding to apes. Two mechanisms, in particular, have historically played a prominent role in comparative research. Some authors have appealed to associative learning and a history of reinforcement; others have appealed to innate preprogrammed dispositions to behave in certain ways in response to certain stimuli. I next consider whether either of these two mechanisms can account for the different results, described in this chapter, between causal and arbitrary tests. After rejecting each of these explanations, I offer a third explanation based on logical thinking and knowledge about physical phenomena.

10.3.1 History of reinforcement

Learning theorists like to think that previous associations or a generalizations of previous associations are all that is needed to explain phenomena like those presented in this chapter. Thus acquired aversions or preferences for certain stimuli may be postulated to explain such results, as opposed to an understanding of the causal structure of the problem. However, such accounts do not explain some of the results in the literature, including those in this chapter.

For example, it might be argued, that the preference for the cup associated with a noise in the shaken food experiment is the result of a previously learned history of reinforcement. However, leaving aside for the moment those cases in which subjects solved the problem satisfactorily when there was no noise, an explanation in terms of association cannot explain why subjects responded to shaken food and noise, but not to the tapping noise or to the tape recording of the noise.

It could be argued that the noise is not exactly the same across these cases. However, this causal test provided the subjects' first experience of this particular noise associated with the particular cup and particular movements. So if a specific noise is critical, they should not have passed the causal test; but they did. If one abandons the idea of noise specificity and argues instead that all that is shown is that animals can generalize from past experience, one then faces the difficulty of explaining why they failed the arbitrary tests whose perceptual structure closely mirrored that of the causal test. Moreover, when given the opportunity within our testing session to learn to use the noise as a cue for the location of the reward, subjects failed to do so, even though they were rewarded on an average of 50 per cent of the trials due to their chance responding.

Now let's go back to the test in which subjects selected above chance the baited cup in the absence of auditory cues. Recall that subjects preferred to choose a noiseless lifted cup over a noiseless shaken cup. This finding cannot be explained by invoking the perception of auditory cues because there were none. Yet, due to the fact that this experiment took place after subjects had experienced the pairing of a noiseless shaken cup with a noisy shaken cup, one could argue that subjects had developed an aversion to the noiseless shaken cup. However, the follow-up test revealed that subjects did not avoid the noiseless shaken cup when paired with a noiseless rotated cup. This means that learned cup aversion cannot satisfactorily explain these results.

Another example of the limitations of a history of reinforcement to account for these results is the arbitrary task with the wedge. In study described above, the apes performed well in the causal task in which the food under a board produced an inclined orientation but did poorly in the arbitrary task in which the objects presented closely resembled those of the causal task. No doubt other *post hoc* explanations in terms of a hypothetical history of reinforcement could be put forward to account for the differences between the results on the causal task and the arbitrary task. Unfortunately, these hypotheses are rarely tested, and so far the ones that have been suggested to us and tested failed to account for our results. Therefore, an explanation of the differences between causal and arbitrary tests presented in this chapter in terms of some learned association does not seem plausible.[1]

10.3.2 Biological predispositions

The second alternative explains differences in learning as a result of innate aversions or preferences. The phenomenon of differences in learning is not new. Learning theorists have traditionally appealed to some biological predisposition of certain species to explain differences in learning across tasks. Seligman (1972) coined the term 'preparedness' to indicate that animals may be prepared to learn some things faster and better than others, particularly those with great biological significance or importance for their survival. This would typically apply to things like taste aversion and to predator detection and avoidance, both of which may channel learning in very constrained ways. Bolles (1970) referred to species-specific defense reactions to explain way rats and pigeons learn different behaviors with different degrees of ease. Another case of specialized learning can be seen in bees that show a greater facility for learning certain colors and odors over others (Menzel 1985), something which has obvious relevance for bees' survival.

---

[1] Editors' note: Cf. Papineau and Heyes, this volume, for skepticism about the empirical tractability of issues between associative and rational explanations of animal behavior.

It is hard to see, however, how that same biological urgency would apply to the problems presented in this chapter. Clearly, solving these problems does not have the same urgency as defensive reactions to predators or the avoidance of poisonous food. Besides, apes do not experience the artifacts used in the tests in their natural environment and, more importantly, detecting food items by the shape of objects over them or by the noise they make does not form part of the natural feeding habits of the great apes. These animals find their fruit in trees or as part of the vegetation that surrounds them. Moreover, if subjects are predisposed to select the inclined board or objects that produce auditory cues, we would still have to explain why they did not select those alternatives in the arbitrary tasks whose reinforcement contingencies and perceptual arrangements closely resembled those of the causal tasks. Therefore, a biological predisposition does not seem a plausible explanation for the differences between causal and arbitrary tasks presented in this chapter.

10.3.3 Knowledge, logic, and inference

If the difference between causal and arbitrary problems cannot be satisfactorily accounted for by appealing to a biological predisposition or to a learned association, one must find another alternative. I propose that the differential success in these two kinds of problems resides in the ability of individuals to grasp the causal structure of the problem. This ability exists because subjects have developed a system of knowledge and inference about observable phenomena. It is important to emphasize that this system of knowledge is not totally incompatible with the other two mechanisms. It is very likely that experience or a core knowledge system may be important in developing the knowledge about the world that allows individuals to make inferences and solve novel problems. However, I argue that this knowledge system cannot be simply equated with biological predispositions or learned associations. Rather, it is an emergent system that goes beyond perceptual generalizations or biological predispositions (Call 2001). It represents knowledge about the causality of certain physical events, and supports the use of logical inference in choosing among alternatives so as to maximize food intake.[2]

10.4 On reflection

The ability of animals to reflect on their own knowledge states has, until now, received little attention. While we know a great deal about cognition in animals, we know much less about metacognition. In other words, although it is clear that many animals know how to solve problems, it is unclear whether they also know that they possess this knowledge and, more importantly, whether they know when they do not know. If animals are capable of metacognition, this would open up the possibility that they

---

[2] Editors' note: See also Bermúdez' hypothesis (this volume) that understanding of causal relations may underwrite proto-conditional inference in non-linguistic animals.

understand that they can make mistakes when they lack appropriate knowledge (see also Shettleworth and Sutton, Chapter 11, this volume).

10.4.1 Monkeys and dolphins know when they are uncertain

Recently, several authors have investigated this area with two basic paradigms. In one paradigm, investigators produce uncertainty in their subjects by presenting stimuli that are hard to discriminate; subjects' escape responses are then measured. The question addressed is whether subjects tend to escape hard-to-discriminate trials when given the opportunity to press an escape key. In one example, Smith *et al.* (1995) presented a discrimination task in which a dolphin and human subjects had to decide whether the stimulus was high or low pitch. (In another version of this task (Smith *et al.* 1997), rhesus monkeys had to distinguish between stimuli with low and high pixel densities.) If the stimulus had the highest pitch (2100 Hz) subjects had to press the HIGH key whereas if the stimulus fell between 1200 and 2099 Hz, they had to press the LOW key. The closer the stimuli drew to the 2100 Hz pitch, the harder it became to decide whether it was either low or high pitch. Subjects were only rewarded for correct responses and incurred a time-out period without reward for incorrect choices. In addition to the LOW or HIGH key choice, subjects had a third option: the escape key. Subjects could press this key to skip the current trial and go to the next one; pressing escape was always reinforced with the limitation that its excessive use gradually delayed the presentation of the next trial, thus becoming aversive. The results showed that dolphins and humans (and rhesus monkeys in the visual version of this task) increased the use of the escape key as the discrimination became increasingly difficult. They rarely used the escape key for low pitch stimuli and the percentage of escape responses also decreased for the highest pitch stimulus.

10.4.2 Monkeys know when they have forgotten

Hampton (2001) also used the uncertainty/escape paradigm and tested two rhesus monkeys with a delayed matching to sample task presented on a computer screen. In a third of the trials, the subjects experienced the presentation of a sample stimulus followed by four alternative stimuli after a delay of approximately 30 seconds. In the remaining two-thirds of the trials, subjects were offered the possibility to decline or take a trial. Declining a trial invariably resulted in the delivery of a not-so-preferred reward (monkey chow). Subjects could attempt to get a highly preferred reward (a peanut) by choosing to take the test (and answering correctly). Incorrect choices resulted in a rewardless time-out period. Hampton (2001) found that subjects performed better in those trials in which they were free to decline a test compared with the compulsory test. This suggests that the monkeys may have known when they had forgotten the correct answer.

Two additional tests reinforced this conclusion. In a second experiment, Hampton (2001) increased the delay interval between the presentation of the sample and the alternatives in an attempt to foster forgetting in the monkeys. As expected, results showed that monkeys' performance declined proportionally to the amount of delay between the sample and the alternative stimuli. More importantly, monkeys chose to decline a higher proportion of trials in those trials with longer delays between the sample and the alternatives. Hampton (2001) also presented some trials without a sample. As expected, one of the monkeys declined tests more often when no sample had been offered. Hampton (2001) concluded that rhesus monkeys know when they have forgotten and they can respond to this lack of information by escaping the situation, that is, declining to take the test.[3]

10.4.3 **Apes know what they have not seen**

A second paradigm has investigated the metacognitive abilities of animals by investigating whether subjects seek additional information before making a choice between two alternatives when presented with incomplete information. Call and Carpenter (2001) presented 2-year-old children, orangutans, and chimpanzees with the following situation. Food was placed inside one of two hollow tubes perpendicularly oriented toward the subjects. In order to get a reward, subjects had to touch the baited tube. There were two baiting conditions. In the visible condition, the experimenter placed the food inside the tube in full view of the subject. In the hidden condition, the experimenter baited one of the tubes but prevented the subjects from witnessing the baiting so although they knew there was food in one tube they did not know which. The question was whether subjects would preferentially look inside the tubes before choosing in those tests in which they had not witnessed the baiting.

The results showed that subjects looked into the tubes before choosing more often when they had not seen the baiting. In addition, in 20–30 per cent of the trials subjects made a selection immediately after encountering an empty tube, indicating that they could succeed at the task without seeing the food. Subjects were not, in other words, simply searching until finding the food. Interestingly, this result combines the metacognitive and cognitive dimensions in the same problem. After being prevented from witnessing the baiting, subjects sought information to resolve their lack of knowledge (metacognitive dimension) and upon finding the empty tube they inferred that the food was in the other tube (cognitive dimension). Recall that this type of inference is analogous to the ones described in the previous section when, upon finding an empty cup (or a noiseless shaken cup), subjects selected the other cup. Also note that, because the shaken silent tube is not aversive, it is likely that the empty tube is not aversive either. Therefore it is likely that, when they choose the baited tube after finding the empty one, they are not simply avoiding the empty tube, but are actually inferring that the food is in the non-inspected tube.

---

[3] Editors' note: See and cf. Shettleworth and Sutton, this volume, for further details and a methodological assessment of these tests of metacognition.

Recently, we replicated our original results with a different group of orangutans and chimpanzees, and extended them to gorillas and bonobos (Call, in press). The positive results with the gorillas were particularly interesting because, unlike other ape species, gorillas seem to have great difficulty with the mirror-self-recognition test (Suarez and Gallup 1981). In contrast with the positive results with apes, dogs failed this test despite being able to retrieve food with high accuracy when they saw where the food was located (Bräuer *et al.*, 2004). Thus the main finding was that, when presented with a situation with incomplete information about the location of food, apes, including gorillas (but not dogs), seek additional information to make a correct choice.

These results indicate that subjects have access to what they have seen or not seen, they know when they are uncertain, and whether or not they remember a past event. In other words, subjects know what they know and do not know, and can respond appropriately in order to remedy their lack of knowledge, either by escaping the situation or seeking additional information.

10.5 Arbitrary, causal, and symbolic connections

Throughout this chapter a distinction has been made between tasks involving causal connections and tasks involving arbitrary associations. The evidence suggests that apes find the former easier to solve than the latter; I have argued that this is because they can understand the underlying causal structure of the former tasks. Of course, this does not mean that apes cannot learn to solve problems that involve arbitrary associations; with enough trials, they certainly can. However, it is not clear whether the capacities that result from acquiring such arbitrary associations the have the same properties as those that result from acquiring information about causal connections. Although there is no data directly on this point, I venture that once data is available, it will still show that the capacities enabled by acquiring information about causal connections differ from those enabled by acquiring information about arbitrary associations. One may predict that extinction or contingency reversals may be harder for causal connections than for arbitrary connections. Likewise, the acquisition curves may also differ. Information about arbitrary connections may be acquired gradually, in accord with traditional associative accounts, information about causal connections may be acquired more suddenly, as in insightful learning. These questions and predictions await further research.

The nature of the capacities provided by learning arbitrary associations has important implications for debates on symbol acquisition in animals, since symbols, by definition, are arbitrarily connected with the referent they represent. Although it is clear that apes can acquire and use symbols effectively, it remains unclear whether those symbols have the same connotations that they have for humans.[4] Unlike apes,

---

4 Editors' note: see and cf. Boysen; Savage-Rumbaugh *et al*, this volume, on the way acquiring (arbitrary) symbolic associations can enhance the capacities of apes.

human beings have an uncanny ability for quickly making sense of and learning arbitrary connections. It is tempting to think that the special meaning of symbols for human beings and the ease with which we learn and use symbols is facilitated by our capacities to reason and reflect.

10.6 **The evolution of reasoning and reflection**

The evolution of reasoning and reflection has allowed some species and individuals to become more efficient foragers when faced with certain problems. This cognitive adaptation is akin to other behavioral and morphological adaptations. I will use an example from feeding adaptations to illustrate this point. Some animals have evolved digestive systems that allow them to exploit low-quality resources that other animals cannot use effectively. For example howler monkeys feed on leaves and have a larger gut than spider monkeys of equal body-size that feed on fruits. It is this morphological adaptation that allows howler monkeys to feed almost exclusively on leaves. Similarly, cognitive adaptations such as the ability to reason and reflect have evolved to take advantage of properties of the external world. This cognitive adaptation allows individuals to perceive the causal structure of the world, not just statistical regularities.

Acquiring the capacities to reason and reflect does not mean that subjects would loose their capacity to use associative processes to detect statistical regularities. Just as howler monkeys can still feed on fruits despite being folivores, apes can still solve problems in ways that depend on association and reinforcement, despite having developed a degree of causal understanding and some capacity for reason and reflection. Associative mechanisms probably represent a safe fall-back when the reason and reflection system cannot be engaged. Likewise, certain animals may be restricted to traditional learning or have evolved fixed strategies such as preference for certain colors or odors. These animals may not engage in the same inferential processes as apes. Recently we tested dogs in some of the inferential problems presented in this chapter and found that although they can use sound to detect food, they do not infer that the food must be in the other container when the shaken one remains noiseless (Bräuer *et al.*, submitted). Similarly, Watson *et al.* (2001) found that dogs, unlike children, seem to use association-based system rather than a logic-based system in an object permanence task. In particular, upon finding empty containers in their search, dogs gradually decelerated their search responses, which is consistent with an extinction mechanism. In contrast children accelerated their searches or remained unaltered, which is consistent with the expectation of finding the target object in the next search.

The evolution of cognitive and metacognitive processes (which includes reflection and consciousness) is a topic that has recently received considerable attention. Some view metacognitive processes as an epiphenomenon consequent on other changes in cognitive processes, such as the development of language (for example, Macphail 2000), while others view metacognition as an evolved mechanism with a specific function. Humphrey (1993) has championed a version of this latter view and coined the term 'inner eye' to refer to a system that evolved to read the minds and predict the behavior of other people. Dickinson and Balleine (2000) suggest that non-reflexive consciousness represents the interface between the motivational and cognitive systems and it is probably present in several vertebrates such as corvids, rats, and primates. Finally, the research reviewed in this chapter (see also Smith *et al.*, in press; Terrace and Metcalfe, in press) suggests that dolphins, apes, and rhesus macaques are capable of monitoring their own internal representations and perceptions and suggests that reflective abilities may be more widely distributed among animals than previously thought (Macphail 2000; Humphrey 1986). If so, the ability to reflect may be better characterized as a continuum rather than a discrete variable. It is conceivable that different species have evolved different versions of reflective abilities tailored to the cognitive mechanisms that they support or enable.

What view should be taken of the relationship between reasoning and reflection? Is it possible to have 'clever brains and blank minds', as Humphrey (1986) put it, or are clever brains associated with non-blank minds? It is clear that some cognitive processes can occur without metacognitive monitoring; in blindsight and related phenomena there is a clear dissociation between cognitive and metacognitive processing. On the other hand, certain forms of reasoning may have arisen only with the participation of reflective processes. Frye *et al.* (1998) have postulated that different levels of metacognitive processing result in different levels of knowledge in children. They use a stepwise approach to explain the changes in both metacognition and knowledge level that occur during children's cognitive development. It is conceivable that a similar approach can be successfully applied to different animals that have evolved different levels of cognitive and metacognitive processes. For example, in contrast with the great apes, there is no evidence that dogs are capable of engaging on certain inferential operations (Watson *et al.* 2001; Brauer *et al.* submitted) or that dogs know what they have or have not seen (Bräuer *et al.*, 2004; Call, in press). Similarly, pigeons, in contrast with rhesus monkeys, seem to solve transitivity and ordinality problems without a scalar representation of stimuli and show little evidence of internal monitoring (Inman and Shettleworth 1999).

It is tempting to speculate that the ability to engage in certain inferential abilities is tied to the ability to monitor internal states of knowledge. Future studies should be devoted to investigating the relationship between cognitive and metacognitive processes. It would be particularly interesting to investigate metacognition in those species that have shown certain inferential abilities such as parrots and corvids (Pepperberg 1999; Bond *et al.* 2003). Contrary to what some authors have argued (Macphail 2000), reflection may not be a consequence of the evolution of cognitive skills such as language; rather, it may be the other way around. Indeed, a certain level of reflection may have enabled the development of certain reasoning and cognitive skills.

10.7 **Conclusion**

Reasoning and reflection have traditionally been considered uniquely human attributes. Many animals, including the great apes, are often regarded as expert at associating arbitrary stimuli while at the same time they are rarely thought capable of reasoning and understanding the causality behind even simple phenomena. In this chapter, I have defended the opposite view. Apes (and possibly other animals) are actually quite good at understanding and reasoning about certain physical properties of their world while at the same time they are quite bad at associating arbitrary stimuli and responses. In other words, if two stimuli have a causal connection (as when food inside a shaken cup makes noise) apes perform better than if stimuli hold an arbitrary relation (as when unrelated noise indicates food), even if the contingencies of reinforcement are the same. In addition to the ability to reason about physical phenomena, I have argued that apes (and other animals) can also reflect on a problem at hand and have some access to their own understanding of a problem. Thus, reasoning and reflection may not be as unique to human beings as Descartes thought; they may have evolved (or coevolved) in other animals also because they allowed them to solve problems in the world more efficiently.

References

Baillargeon, R., Spelke, E. S., and Wasserman, S. (1985). Object permanence in five-month-old infants. *Cognition,* **20**: 191–208.

Bolles, R. (1970). Species-specific defense reactions and avoidance learning. *Psychological Review,* **77**: 32–48.

Bond, A. B., Kamil, A. C., and Balda, R. P. (2003). Social complexity and transitive inference in corvids. *Animal Behaviour,* **65**: 479–487.

Brauer, J., Call, J., and Tomasello, M. (2004). Dogs use an allocentric perspective taking without evidence of access to their own perception. *Animal Behaviour Applied Science,* **88**: 299–317.

Bräuer, J., Kaminski, J., Riedel, J., Call, J. and Tomasello, M. (submitted). Making inferences about the location of hidden food: Social dog—causal ape. *Journal of Comparative Psychology.*

Call, J. (2001). Chimpanzee social cognition. *Trends in Cognitive Sciences,* **5**: 369–405.

Call, J. (2004). Inferences about the location of food in the great apes. *Journal of Comparative Psychology,* **118**: 232–241.

Call, J. (2005). The self and the other: a missing link in comparative social cognition. In: H. Terrace and J. Metcalfe, eds. *The Evolution of Consciousness in Animals and Humans.* New York: Oxford University Press, 321–341.

Call, J. (submitted). Apes know that food takes up space. *Cognition.*

Call, J., and Carpenter, M. (2001). Do apes and children know what they have seen? *Animal Cognition,* **4**: 207–220.

Dickinson, A., and Balleine, B. W. (2000). Causal cognition and goal-directed action. In: C. M. Heyes and L. Huber, eds. *The Evolution of Cognition,* pp. 185–204. Cambridge, MA: MIT Press.

Frye, D., Zelazo, P. D., and Burack, J. A. (1998). Cognitive complexity and control: I. Theory of mind in typical and atypical development. *Current Directions in Psychological Science,* **7**: 116–121.

Hampton, R. R. (2001). Rhesus monkeys know when they remember. *Proceedings of the National Academy of Sciences,* **98**: 5359–5362.

Humphrey, N. (1993/1986). *The Inner Eye*. New York: Vintage Books.

Inman, A., and Shettleworth, S. J. (1999). Detecting metamemory in nonverbal subjects: A test with pigeons. *Journal of Experimental Psychology: Animal Behavior Processes,* **25**: 389–395.

James, W. (1890). *The Principles of Psychology*, vol. 2. New York: Dover Publications.

Loeb, J. (1918). *Forced Movements, Tropisms, and Animal Conduct*. Philadelphia: J. P. Lippincott.

Macphail, E. M. (2000). The search for a mental rubicon. In: C. M. Heyes and L. Huber, eds. *The Evolution of Cognition*, pp. 253–271. Cambridge, MA: MIT Press.

Menzel, R. (1985). Learning in honey bees in an ecological and behavioral context. *Fortschritte der Zoologie*, **31**: 55–74.

Pepperberg, I. M. (1999). *The Alex Studies*. Cambridge, MA: Harvard University Press.

Premack, D. (1995). Cause/induced motion: intention/spontaneous motion. In: J. P. Changeux and J. Chavaillon, eds. *Origins of the Human Brain*, pp. 286–308. Oxford: Oxford University Press.

Premack, D., and Premack, A. J. (1994). Levels of causal understanding in chimpanzees and children. *Cognition*, **50**: 347–362.

Riedel, J., Bräuer, J., Kaminski, J., Call, J., and Tomasello, M. (in preparation). The use of communicative and causal cues in dogs and chimpanzees.

Seligman, M. E. P. (1972). *Biological Boundaries of Learning*. New York: Appleton-Century-Crofts.

Smith, J. D., Schull, J., Strote, J., McGee, K., Egnor, R., and Erb, L. (1995). The uncertain response in the bottlenosed dolphin (*Tursiops truncatus*). *Journal of Experimental Psychology: General*, **124**: 391–408.

Smith, J. D., Shields, W. E., Schull, J., and Washburn, D. A. (1997). The uncertain response in humans and animals. *Cognition*, **62**: 75–97.

Smith, J. D., Shields, W. E., and Washburn, D. A. (in press). The comparative psychology of uncertainty monitoring and metacognition. *Behavioral and Brain Sciences*.

Suarez, S. D., and Gallup, G. G. Jr. (1981). Self-recognition in chimpanzees and orangutans, but not gorillas. *Journal of Human Evolution*, **10**: 175–188.

Terrace, H., and Metcalfe, J. (2005). *The Evolution of Consciousness in Animals and Humans*. New York: Oxford University Press.

Thorndike, E. L. (1898). Animal intelligence. An experimental study of the associative processes in animals. *Psychological Review, Monograph supplement*, **2**: 551–553.

Tulving, E., and Madigan, S. A. (1970). Memory and verbal learning. In: P. H. Mussen and M. R. Rosenzweig, eds. *Annual Review of Psychology*, vol. 21, pp. 437–484. Palo Alto, CA: Annual Reviews Inc.

Watson, J. S., Gergely, G., Csanyi, V., Topal, J., Gacsi, M., and Sarkozi, Z. (2001). Distinguishing logic from association in the solution of an invisible displacement task by children (*Homo sapiens*) and dogs (*Canis familiaris*): Using negation of disjunction. *Journal of Comparative Psychology*, **115**: 219–226.

Chapter 11

Do animals know what they know?

Sara J. Shettleworth and Jennifer E. Sutton

Abstract

Using well-established paradigms for studying animal perception and memory, researchers have begun to ask whether animals can monitor the status of their knowledge in a behavioural task—whether they know what they know. Generally, such metacognitive ability is tested by giving animals the opportunity to avoid (or 'escape') a test of memory or perceptual discrimination. The pattern of escapes can then be analyzed in a number of ways, including whether the subject escapes more often from difficult tests, where a correct answer is less likely, than from easy tests. A number of non-metacognitive strategies can be used by animals in these experiments, however, and it is important to control carefully for alternative explanations. Moreover, only rigorous, controlled tests will determine whether current suggestions of species differences in metacognitive abilities are correct.

11.1 Introduction

Some of the most interesting and controversial work in animal behaviour involves the search for relatively complex mental processes in non-human animals. Investigations of episodic memory (Clayton *et al.*, Chapter 9, this volume), mind reading (Tomasello and Call, Ch 17; Povinelli and Vonk, Chapter 18, this volume), intentional deception (Kummer *et al.* 1996), and metacognition (Hampton 2001; Inman and Shettleworth 1999; Smith *et al.* 2003; Sole *et al.* 2003) have established experimental situations in which animals may be tested for behaviour functionally similar to behaviour that is accompanied by distinctive mental states in humans.

The term *functional similarity* (Hampton 2001) captures the idea that the best we can do in such investigations is to define rigorously the behaviour accompanied by a given mental process and see if the animals show it. If the process is one usually assessed by verbal report, we will never be able to have the same kind of evidence for it in other species as in humans. For example episodic memory in adult humans is an integrated memory for the what, where, and when of an event that comes with an awareness of re-experiencing that event (Tulving 1972). As discussed by Clayton *et al.* (this volume), it may be possible to devise ways to probe an animal's memory to see if it remembers the 'what', 'where', and 'when' of an event in an integrated way, but this is not the same as asking about its private experiences while recalling the event. People may never agree on the degree to which similar non-verbal behaviour implies similar mental events in very different species (c.f. Griffin 2001), but when the phenomena under study are well enough defined, agreement on functional similarity may be possible.[1] The fact that researchers interested in episodic memory in non-human animals refer to what they study as 'episodic-like' memory (Griffiths *et al.* 1999; Morris 2001) acknowledges this basic methodological and philosophical truth.

In this chapter we discuss whether animals know what they know. That is, do any animals have metacognitive abilities? There are well-established paradigms for assessing cognitive processes such as memory, perceptual processing, categorization, timing, and numerical competence in non-human animals (henceforth simply 'animals'), and we know a great deal about how they perform in these tasks (Roberts 1998; Shettleworth 1998). But in addition to knowing which stimulus was presented most recently or how long it lasted, do animals know *that* they know such things? That is to say, in addition to reporting what it had just seen or heard, could any animal additionally report on the status of its knowledge? Adult humans can. In a variety of laboratory paradigms, reports of a 'feeling of knowing' in memory tests or perceptual certainty in psychophysical tasks are positively associated with accurate performance on these tasks (Nelson 1996). Thus people not only feel that they have metacognitive abilities, but metacognition is generally accurate.

In everyday human life, metacognition clearly contributes to rational, efficient, behaviour. For example, before setting out on a shopping expedition, a person will be aware of whether or not he needs to consult a map. Similarly, one usually knows whether or not it's necessary to look in the phone book before making a call. We also behave as if we assume accurate metacognition in others. For example, lost in a new city, one approaches a stranger not by asking 'Where is the train station?' but more likely by saying, 'Can you tell me the way to the train station?' If the stranger replies 'Sorry, I can't', we ask someone else. It is precisely the dissociation between knowing and awareness of knowing in some brain-damaged patients that makes them so fascinating (e.g. Weiskrantz 1986), but on the whole we can safely treat others as if they can tell us whether or not they know something, or how sure they are that they know it. Can members of any other species do the same thing, or are they like patients who have knowledge without awareness?

11.2 **How can we test animals' metacognition?**

As the preceding examples indicate, the challenge to researchers is simultaneously to test animals' ability to perform a cognitive task (to answer a direct question like

---

[1] Editors' note: see and cf. the methodological points about distinguishing rational and associative processes made by Papineau and Heyes, this volume.

'What is the way to the station?') and to test their ability to report on the state of their knowledge (to answer a question like 'Do you know the way to the station?'). Consequently, metacognition experiments include both a test of memory or perceptual discrimination that can vary in difficulty, and an additional response that allows the animal to accept or decline these tests. To encourage the animals to use it appropriately, this 'don't know' or 'uncertain' response receives a small or delayed reward, more than the non-reward typically given for an incorrect response in the cognitive test but less than could be obtained by responding correctly in that test. Thus this opportunity to 'opt out' or escape from an impending test should be taken only if the animal knows it is likely to fail the test.

Any method for studying the functional similarity of animal behaviour and complex human mental processes must carefully control for alternative interpretations. Therefore, one further step is needed to be sure such choices are based on metacognition rather than on learning about distinctive external stimuli that predict the animal's performance (and therefore reward rate) in the primary cognitive task. For example suppose we are using a test of memory for shapes in which we vary memory strength by varying the interval between showing the animal the shape and testing its memory, as illustrated in Fig. 11.1. The longer the interval between this stimulus (the 'sample') and

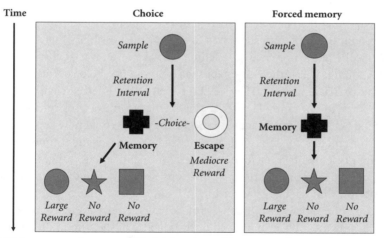

Fig. 11.1 Trial sequences from a modified delayed matching to sample procedure used to test metamemory in animals. On 'choice' trials (left panel), the animal is given an opportunity to decline or accept a test of memory after the retention interval and before the comparisons are presented. On 'forced' trials (right panel), only the memory option is offered after the retention interval, and the animal is required to take the memory test. Some experiments referred to in the text which used this procedure also included trials when only the escape option followed the retention interval. Whether the square, circle, or star serves as the sample (and hence, as the rewarded choice in the memory test) varies randomly from trial to trial, as does the type of event (choice, forced memory, or forced escape) following the retention interval.

the test, the worse performance will be. Our subject therefore can maximize its reward rate by choosing to complete the memory test after a short interval, when it is likely to receive a large or immediate reward, and by opting out after a long interval, when the relatively small reward for escaping exceeds the reward expected from performing at chance in the memory test, that is, without the aid of memory. That is to say, in this case behaviour functionally similar in one respect to behaviour consistent with metamemory could result from discriminating a long from a short retention interval and choosing accordingly. We know that animals are very good at discriminating among time intervals (Shettleworth 1998), so this is a very real possibility. Obviously, then, functional similarity must be defined more richly, so that it specifies a constellation of behaviours that together are uniquely consistent with metacognition.

There are two ways in which this has been done, and in a handful of experiments both have been used. One is to include randomly presented trials without the escape option, the 'forced trials' in Fig. 11.1. On such trials, the animal is forced to perform the primary memory task whether it knows the answer or not. An animal that is escaping selectively on trials when it 'knows it does not know' will do worse on these forced trials than on trials that it freely chose when it could have escaped. Notice, however, that because many animals tend to do much worse than normal when conditions change, performance may also be impaired if trials without the escape option are relatively novel. Therefore, it is important to mix in such trials throughout the experiment rather than presenting them only occasionally.

The second, and very important, way to address the possibility that subjects simply learn to escape in the presence of specific external stimuli is to include transfer tests with new stimuli. This is the well-established policy in psychology of using convergent operations or 'triangulation' (Heyes 1993) to test for an inferred cognitive process in multiple ways. In our example, after testing memory with varying retention intervals and finding behaviour consistent with metamemory, we could now keep the retention interval constant at some intermediate value and vary the length of time for which the shapes are presented to vary memory strength in a new way. Longer presentations should improve memory, shorter presentations weaken it, and the animals' use of the escape option should change accordingly. It should also change immediately upon introduction of this new test. Other ways of varying memory strength can be imagined, such as varying the inter-trial interval or the salience of the stimuli. In one clever test with monkeys (Hampton 2001), the to-be-remembered stimuli were occasionally omitted altogether. An animal using metamemory must report 'I don't remember' on such trials, which is what the monkeys did.

Not all tests of metacognitive abilities in animals have provided such clear cut results as did the one by Hampton just mentioned. In the rest of this chapter, we briefly review the research in this area, identify some gaps in knowledge, and sketch possible future directions. Smith *et al.* (2003) have recently reviewed research on animal metacognition in more detail. We emphasize more than they do the methodological issues sketched above, and we are accordingly less inclined to interpret the results as evidence for human-like awareness or metacognition.

11.3 **Uncertainty in perceptual discriminations**

The first published studies of metacognition in animals asked whether animals could report on their level of perceptual certainty when performing a difficult discrimination (Smith *et al.* 1995; Shields *et al.* 1997; Smith *et al.* 1997; Sole *et al.* 2003). In the first such study, Smith *et al.* (1995) tested a dolphin in an auditory psychophysical task. The dolphin was presented with a single tone from an underwater speaker on each trial. Two underwater response paddles were also available, with a response to one resulting in reinforcement after a high-pitched tone and the other resulting in reinforcement after a low-pitched tone. After a series of training phases, the final testing phase contained probe trials where the pitch of the relatively lower tone was raised, making it more similar to the high tone. Along with this increase in difficulty, the dolphin also had a third response paddle available, which allowed it to escape from categorizing the tone and endure a delay followed by an easily discriminated tone instead. The important prediction was that the dolphin should preferentially choose the escape option and the less-preferred delayed reward on difficult trials but attempt to categorize the tone on easier trials for the immediate payoff.

At first, the dolphin chose the escape option whenever it was presented, so an additional cost was added by increasing the delay to the easy trial if the escape option had been chosen frequently in the previous trials. Eventually, the dolphin escaped more when the tone was hard to classify than when the tone was easily classified as high or low. In a comparable experiment (Smith *et al.* 1997), monkeys were trained to classify the density of illuminated pixels on a video monitor as either low or high. Like the dolphin, the monkeys escaped most on difficult trials, with stimuli in the middle of the continuum.

A nice feature of both of these studies is that people were tested on almost identical tasks so their performance could be directly compared to the animals'. It was indeed similar, and the data from humans, monkeys, and dolphin could all be accounted for by a signal detection model in which responses are based on dividing the stimulus continuum into 'low', 'uncertain', and 'high' regions. The human participants reported that they chose the escape option when they experienced a feeling of uncertainty about how to classify the stimulus. Explaining human and animal data using a single model makes it tempting to conclude the same underlying, conscious process is employed by all the participants. However, although signal detection models refer to setting criteria and making decisions, there is no necessary implication that these processes are conscious (Macmillan and Creelman 1991). Moreover, because the studies with the monkeys and the dolphin did not include regular, randomly presented, tests of the animals' discrimination performance in the absence of the escape option, it is unclear whether or not they were actually using the escape response selectively when they 'knew they did not know' how to classify the stimulus. In addition, because no transfer tests were included, it seems possible that the animals had simply learned that for stimuli in a certain part of the tone frequency or pixel density continuum, the escape response had the shortest average delay to reward. Thus, although these studies are ground-breaking and the results are consistent with those of some of the more stringent tests reviewed below, they lacked all the controls required for a conclusive demonstration of metacognition.

Pigeons have been widely tested in psychophysical procedures and the results successfully accounted for by signal detection models (Blough and Blough 1977; Commons *et al.* 1991). This suggests that pigeons should behave similarly to the subjects in the studies of Smith *et al.* if they were given a similar test. This was the idea behind a study by Sole *et al.* (2003). Like the monkeys in Smith *et al.* (1997), the pigeons were required to classify the number of pixels on a computer monitor as sparse or dense. They were also provided with an escape option that could be chosen instead of classifying the image. However, this experiment also included two control procedures lacking in the earlier studies. First, rather than a delay to an easy trial, the reward for escaping was three pigeon pellets, a mediocre reward compared to the six pellets that could be gained if the stimulus was correctly classified. This reward was constant for each pigeon rather than changing dynamically throughout the experiment, which helped to ensure that the birds knew the outcome of each available option and facilitated modelling the results. Second, after experience with the escape option, the birds received a transfer phase where some stimuli were red instead of green. They needed time to acquire the density discrimination with these new stimuli and performed poorly at first. Therefore, in the first session of transfer when performance was poor, the birds should choose to escape if uncertainty was truly governing the use of the escape option. Finally, classification accuracy was compared on forced and freely chosen trials throughout the experiment.

Like the monkeys and the dolphin in the experiments by Smith and colleagues, the pigeons escaped more difficult trials than easy trials. However, accuracy on forced versus chosen tests did not differ. Moreover, when presented with the new transfer stimuli that were difficult to classify, the birds performed poorly but did not opt for the escape option any more than they did once they were classifying the new stimuli accurately. Thus the pigeons did not selectively choose to take the test when they 'knew that they knew' how the pixels should be classified. However, their performance could be accounted for very well by a signal detection model that assumed they were maximizing perceived reward for correct responses in a consistent way across all the experimental conditions. That is to say, they behaved as if they had learned which was the most profitable response to each region of the continuum of internal responses evoked by the displays. Fitting the model did not require postulating some extra reward such as 'reduction of uncertainty' on the difficult trials.

Is it the case then, that pigeons do not monitor uncertainty while monkeys, dolphins, and humans do? Because there were procedural differences between experiments, it is still difficult to form a firm conclusion. The tests with pigeons may be viewed as more rigorous given the multiple predictions afforded by the design, but it is still not known how monkeys, dolphins, or people would behave with identical tests.

11.4 Monitoring the status of a memory

Like psychophysical investigations, the study of memory processes in animals is well established (Roberts 1998; Shettleworth 1998) and provides a strong basis for tests of metacognition. A substantial series of experiments on monkeys and one on pigeons has sought to determine whether animals can assess the status of a memory. These tests of metamemory have employed two different paradigms: serial probe recognition and matching-to-sample.

Smith *et al.* (1998) used the well-established serial probe recognition task with monkeys. In a serial probe recognition task (Wright 1989), a series of images is displayed (a 'list' of to-be-remembered items), and then a probe image is presented. The subject's task is to determine whether the probe image was or was not displayed in the preceding list. Under the conditions used by Smith *et al.* (1998), monkeys typically show a recency effect, performing best when the probe item was presented late in the list. Longer lists are also more difficult than short ones. In Smith *et al.*'s (1998) study, images in the list were chosen randomly from a bank of over 100 digital files, enabling a different combination of images on each trial (so-called trial-unique items).

After the list of two to six images was presented, a probe image and three stimuli indicating different response options appeared on the screen. Responses to one stimulus were reinforced if the probe image was presented in the previous list, one was reinforced if the image was not presented, and one stimulus served as the 'escape' option and was followed by a delay and then the correct answer. Only one item from each list was tested, and the position of the tested item was varied randomly. For example on one trial the probe might be the first item from the most recent list. Then a new list would be presented and the last item from that list presented as the probe, and so on. Importantly, the use of this probe procedure together with trial-unique stimuli eliminated the possibility that the monkeys could learn specific stimulus–outcome pairings. Instead, choosing the escape option on difficult trials could be confidently attributed to monitoring the status of memory for the probed item. The monkeys chose the escape option more often on tests of earlier items that were more difficult to remember. That is, accuracy showed a significant recency effect, and escaping mirrored this effect, with fewer escapes for most recently presented items. Generally, accuracy was higher on trials when the monkeys chose to take the test than when the escape option was not available and they were 'forced' to answer, although they did not receive both types of trial in a single session.

Hampton (2001) also tested rhesus monkeys' metamemory, but he used a delayed matching-to-sample procedure similar to that depicted in Fig. 11.1. On each trial, the monkeys were shown an image to study (the 'sample'), followed by a delay with no stimulus present. After this retention interval, the monkeys chose between two stimuli that led to different conclusions of the trial. A response to one stimulus resulted in a memory test, where the sample was presented along with three distractor stimuli, and a correct match was rewarded with a highly-valued peanut. Responses to the other stimulus led to one final stimulus instead of a memory test, and a primate pellet reward, a treat considered rather mediocre by the monkeys. Tests at the end of each day confirmed that the monkeys always preferred peanuts to pellets.

Hampton's (2001) study was a powerful test for a number of reasons. First, the monkeys were required to make a decision about the status of their memories *before* the test was presented. That way, retrieval cues from the presence of the correct stimulus or interference from the incorrect stimuli could not influence the memory judgement. This procedure is demanding because the animal must access the strength of its memory for the most recent sample stimulus in the absence of that sample. Second, the use of two different kinds of rewards produced a natural cost for opting out of the memory test and eliminated the need to punish the monkeys for overusing that option. Finally, Hampton manipulated the strength of monkeys' memories for the sample in more than one way, providing an opportunity for the analysis of transfer performance. Two monkeys were first trained on the task just described with a single retention interval, chosen so they made errors on about 20 per cent of trials. Consistent with the use of metamemory, both animals performed better on freely chosen than on forced memory tests. To test whether use of the escape option was equivalent to reporting 'I don't remember', they were then given probe trials where no sample was presented. Finally, their memory strength was manipulated by increasing the retention interval.

The results of all three phases of the study were consistent with memory monitoring. The monkeys were more accurate when they chose to take a test than when they were forced to, they declined memory tests much more often when no sample had been presented, and they performed more poorly and declined more tests after longer delays. In addition, performance on freely chosen memory tests declined less at longer delays than did performance on forced tests, consistent with the monkeys selectively choosing the memory test when they knew they remembered the sample. Taken together, the serial probe recognition test of Smith *et al.* (1997) and the matching-to-sample experiments by Hampton (2001) provide strong evidence that monkeys can monitor the status of a memory.

In the first published investigation of any metacognitive process in pigeons, Inman and Shettleworth (1999) used a procedure nearly identical to Hampton's (2001) and found that procedural factors were crucial to determining whether pigeons' behaviour was consistent with memory monitoring. They used a matching-to-sample design with several retention intervals, and as in Sole *et al.*'s (2003) study, escaping the test resulted in fewer pellets than a correct answer. As in Hampton's (2001) transfer test, at the longer delays memory should be worse, escape should be chosen more often, and there should be a bigger difference between forced and chosen memory performance. Unlike the case in Hampton's experiment, however, but as in the study of Shields *et al.*, in Inman and Shettleworth's Experiment 1 the memory test and the escape option were presented at the same time on choice trials, rather than each one following an explicit choice as depicted in Fig. 11.1. In this experiment, three of the four birds chose the escape option more at the longer delays, where memory should be poor, than at the short delays. The birds were also less accurate on forced than on freely chosen memory tests.

The pattern of results was different in a second experiment with the same birds. Here, the escape option was presented before the opportunity to complete the memory test, as in Hampton's experiments and as shown in Fig. 11.1. On those trials, delay length had no effect on the probability of escaping the test, and the birds were no more accurate on chosen than forced tests. Thus, when the birds had to consult their memory strength before choosing whether or not to have their memory tested, they failed the test of metamemory, whereas if they were confronted with the items in the memory test and the escape option at the same time, they passed. One might want to conclude that the pigeons can pass an easy but not a difficult test of metamemory. However, one reason why this interpretation might not be warranted is that in Inman and Shettleworth's second experiment, most birds were performing so well in the primary test of memory that there was little to be gained by escaping from it at even the longest retention interval. Another potential problem is the lack of evidence that the pigeons actually had any preference for the mixture of six and zero pellets determined by their proportion of correct responses over the three sure pellets from the escape option. Thus, Inman and Shettleworth had to conclude that while their results were not incompatible with pigeons being able to use metamemory, they were not conclusive one way or the other.

Given that Hampton's (2001) study using very similar procedures to Inman and Shettleworth's subsequently provided strong evidence for metamemory in monkeys, we (Sutton and Shettleworth, in preparation) took up the study of pigeon metamemory again in an experiment very similar to Hampton's. One important change from the earlier studies with pigeons is that we verified pigeons' preferences among the rewards we used: all birds in an independent study preferred a constant 12 pellets every time to a constant five pellets, but they preferred a constant five pellets to receiving 12 pellets on 25 per cent of trials and zero pellets otherwise. In the main experiment, memory tests had four alternatives with 12 pellets' reward for a correct choice, and escaping was rewarded with five pellets. Pigeons should therefore escape the memory tests when they know that they do not know and therefore can expect only a 25 per cent chance of getting 12 pellets.

Like Hampton's monkeys, the birds were first trained with a single retention interval in a procedure like that depicted in Fig. 11.1. Unlike the monkeys, at this stage they performed no better on freely chosen than on forced tests of memory. This same pattern of results was maintained when testing continued with a mixture of a long and a short retention interval. However, most birds apparently learned that the value of taking the memory test declined relative to the value of escaping at the longer delay, since they escaped more at this delay. Finally, the birds did not consistently reject the test of memory after no sample had occurred. Like Inman and Shettleworth's Experiment 2, these results indicate that pigeons cannot report on their metacognitions under conditions where monkeys can.

11.5 Summary and future directions

Consistent with the literature on human subjects (e.g. Nelson 1996), we and other researchers using animals have tended to lump together tests of perceptual certainty with tests of metamemory as if they assay a single cognitive process. This assumption may be encouraged by the fact that nearly all the tests of either one to date consist of offering the animal the option of escaping from the test of cognition (the uncertain response, Smith *et al.* 1997). But perceptual certainty ('Do I know what I am seeing?') and memory monitoring ('Do I remember what I saw?') are not necessarily the same process, even though people as well as animals might report on it by 'saying' they are uncertain. Some species might show one and not the other. For example whereas pigeons passed one easy test of metamemory (Inman and Shettleworth 1999, Experiment 1), they failed a parallel test of perceptual certainty (Sole *et al.* 2003). In the absence of further studies with pigeons, the existing data are not conclusive, but they do suggest that future studies might more carefully distinguish between perceptual certainty and memory monitoring abilities. Consistent with this suggestion, in humans 'blindsight', or vision without awareness (Weiskrantz 1986) and specific loss of episodic but not semantic memory are associated with damage to different brain areas.[2]

A further important distinction is that between showing the animal the item(s) to be classified or recognized at the same time as offering it an alternative to completing the primary task and offering this choice before presenting the primary task (as in Fig. 11.1). To our knowledge, the latter procedure has not been used in a perceptual task with animals. In a memory task it is surely the more difficult procedure because it requires the animal to call to mind an absent stimulus. Intuitively, displaying the test stimuli and asking 'Do you recognize one of these or would you rather not answer?' tests memory strength in a more direct way than asking 'Do you think you will recognize the sample when you see it?' As pointed out in the introduction to this chapter, tests which do the latter, are most closely analogous to tests of human metamemory. So far,

---

[2] Editors' note: It would be interesting to relate the paradigms described here for studying metacognition in animals to the paradigms used to study blindsight in animals: would blindsighted monkeys opt to escape no-stimulus classification tasks? See Stoerig and Cowey 1999, Stoerig *et al.* 2002.

however, they have been used only in one of the studies with monkeys (Hampton 2001) and two with pigeons (Inman and Shettleworth 1999, Experiment 2; Sutton and Shettleworth, in preparation).

All the studies we have reviewed so far involve training animals very extensively to give an explicit report on memory strength or perceptual certainty. In contrast, the examples from everyday life sketched at the beginning of this chapter illustrate how people make spontaneous implicit use of metacognition. Might these suggest new ways to look for metacognitive abilities in animals, related to ways they might spontaneously use these abilities in biologically relevant contexts? Call and Carpenter (2001) have recently devised such a test (see also Call, Chapter 10, this volume). Chimpanzees, orang-utans, and children saw food (for the apes) or a sticker (for the children) hidden in one of two or three horizontal tubes and were allowed to retrieve it. Sometimes it was difficult to know which tube was baited either because the subject did not see the baiting or because there was a delay before a choice was permitted. These conditions increased the chance that subjects would bend down and peer into the tubes before choosing, as if they knew they did not know and were looking for more information. Not all the apes' data were consistent with looking into tubes being information-seeking, however. For instance, subjects often continued looking after they had found the food, and they did not always look when the baiting had been concealed. Hampton *et al.* (2004) adapted this task for rhesus macaques and most monkeys looked down the tubes more often when they were ignorant of which one held the bait. This finding is particularly important because it means that memory awareness in this species is supported by convergent evidence from two quite different kinds of task.

So far, tests of whether animals know what they know suggest that despite the fact that both monkeys and pigeons can perform difficult tests of perceptual discrimination or memory, only the monkeys, and perhaps other primates, can report on their cognitive states while they are doing these tasks. Thus only primates and perhaps a dolphin have been proven rational in a strong sense that requires having access to the reasons for their behaviour. But because their choices in similar tasks optimize reward under the constraint of imperfect discrimination, pigeons are at least biologically rational (see Kacelnik, Chapter 2, this volume). One challenge for the future might be to seek naturalistic situations in which a biologically rational animal must be rational in the stronger sense as well.

References

Blough, D., and Blough, P. (1977). Animal psychophysics. In: W. K. Honig and J. E. R. Staddon, eds. *Handbook of Operant Behavior*, pp. 514–539. Englewood Cliffs, NJ: Prentice-Hall.

Call, J., and Carpenter, M. (2001). Do apes and children know what they have seen? *Animal Cognition*, **4**: 207–220.

Commons, M. L., Nevin, J. A., and Davison, M. C., eds (1991). *Signal Detection: Mechanisms, Models, and Applications*. Hillsdale, NJ: Lawrence Erlbaum Associates.

Griffin, D. R. (2001). Animals know more than we used to think. *Proceedings of the National Academy of Sciences (USA),* **98**: 4833–4834.

Griffiths, D., Dickinson, A., and Clayton, N. S. (1999). Episodic memory: what can animals remember about their past? *Trends in Cognitive Science,* **3**: 74–80.

Hampton, R. R. (2001). Rhesus monkeys know when they remember. *Proceedings of the National Academy of Sciences (USA),* **98**: 5359–5362.

Hampton, R. R., Zivin, A., and Murray, E. A. (2004). Rhesus monkeys (*Macaca mulatta*) discriminate between knowing and not knowing and collect information as needed before acting. *Animal Cognition,* **7**. Published online 23 April, 2004. DOI: 10.1007/s10071–004–0215–1.

Heyes, C. M. (1993). Anecdotes, training, trapping and triangulating: do animals attribute mental states? *Animal Behaviour,* **46**: 177–188.

Inman, A., and Shettleworth, S. J. (1999). Detecting metamemory in nonverbal subjects: A test with pigeons. *Journal of Experimental Psychology: Animal Behavior Processes,* **25**: 389–395.

Kummer, H., Anzenberger, G., and Hemelrijk, C. K. (1996). Hiding and perspective taking in long-tailed macaques (*Macaca fascicularis*). *Journal of Comparative Psychology,* **110**: 97–102.

Macmillan, N. A., and Creelman, C. G. (1991). *Detection theory: A user's guide.* Cambridge: Cambridge University Press.

Morris, R. G. M. (2001). Episodic-like memory in animals: psychological criteria, neural mechanisms and the value of episodic-like tasks to investigate animal models on neurodegenerative disease. *Philosophical Transactions of the Royal Society (London) B,* **356**: 1453–1465.

Nelson, T. O. (1996). Metacognition and consciousness. *American Psychologist,* **51**: 102–116.

Roberts, W. A. (1998). *Principles of Animal Cognition.* New York: McGraw Hill.

Shettleworth, S. J. (1998). *Cognition, Evolution, and Behavior.* Oxford: Oxford University Press.

Shields, W. E., Smith, J. D., and Washburn, D. A. (1997). Uncertain responses by humans and rhesus monkeys (*Macaca mulatta*) in a psychophysical same-different task. *Journal of Experimental Psychology: General,* **126**: 147–164.

Smith, J. D., Schull, J., Strote, J., McGee, K., Egnor, R., and Erb, L. (1995). The uncertain response in the bottlenosed dolphin (*Tursiops truncatus*). *Journal of Experimental Psychology: General,* **124**: 391–408.

Smith, J. D., Shields, W. E., Allendoerfer, K. R., and Washburn, D. A. (1998). Memory monitoring by animals and humans. *Journal of Experimental Psychology: General,* **127**: 227–250.

Smith, J. D., Shields, W. E., Schull, J., and Washburn, D. A. (1997). The uncertain response in humans and animals. *Cognition,* **62**: 75–97.

Smith, J. D., Shields, W. E., and Washburn, D. A. (2003). The comparative psychology of uncertainty monitoring and metacognition. *Behavioral and Brain Sciences,* **26**: 317–339.

Sole, L. M., Shettleworth, S. J., and Bennett, P. J. (2003). Uncertainty in pigeons. *Psychonomic Bulletin and Review,* **10**: 738–745.

Stoerig, P. and Cowey, A. (1997). Blindsight in man and monkey. *Brain,* **129**: 535–559.

Stoerig, P., Zontanou, A., and Cowey, A. (2002). Aware or unaware: assessment of cortical blindness in four men and monkeys. *Cerebral Cortex,* **12**: 565–574.

Sutton, J. E., and Shettleworth, S. J. (in preparation). Further tests of metamemory in the pigeon.

Tulving, E. (1972). Episodic and semantic memory. In: E. Tulving and W. Donaldson, eds. *Organization of Memory,* pp. 381–403. New York: Academic Press.

Weiskrantz, L. (1986). *Blindsight.* Oxford: Clarendon Press.

Wright, A. A. (1989). Memory processing by pigeons, monkeys, and people. *The Psychology of Learning and Motivation,* vol. 24, pp. 25–70. San Diego: Academic Press.

Chapter 12

Rationality and metacognition in non-human animals

Joëlle Proust

Abstract

The present chapter approaches the subject of animal rationality on the basis of dynamic-evolutionary considerations. Rationality is defined as a disposition that *tends to be realized* by a control system that can adapt to changing circumstances and that relies on cognition to do so. The specific selective pressures exerted on agents endowed with information-processing capacities are analysed. Rationality reflects the characteristics of these pressures (in particular the contingent fact that the relevant environment is variable and competitive). It is hypothesized that a primary form of rationality consists in a set of metacognitive skills. They offer an evolutionary stable response to the various demands of the internal and external flows of information in a variable and competitive environment. Metacognition provides a form of procedural reflexivity that can, but does not have to be redeployed through metarepresentations. Finally, the claim that this early form of rationality based on metacognition involves normativity is discussed.

12.1 **Rationality**

There are currently many different ways of understanding the concept of rationality. Alex Kacelnik (Chapter 2, this volume) distinguishes three of them: PP-rationality, as discussed by philosophers and psychologists, focuses on justification and requires that beliefs or actions be based on (adequate) reasoning; E-rationality, as economists conceive it, focuses on utility-maximizing behaviour; and, finally, B-rationality, the biologists' conception, involves maximization of inclusive fitness across circumstances, though it does not specify the processes through which such maximization is reached. One can only agree with Kacelnik that the three approaches respond to disciplinary concerns, but that illuminating the connections between them is a daunting task.

The difficulty is reflected in current efforts to capture a definition of rationality sufficiently general to apply to non-human as well as human forms of behaviour.

Some philosophers[1] have tried to use the PP approach in a top–down manner, beginning with a conception of human rationality which they then adapt and apply to non-humans. Such an generalization, however, raises a number of questions for which no fully satisfactory answer is available: what might be the closest equivalent in non-linguistic animals of reporting one's own reasons? How might formal inference schemas, such as *modus ponens*, proceed in the absence of a linguistic vehicle? Why should this form of rationality represent a maximization of fitness? Does it connect with utility maximization?

Others[2] work from the bottom up: they rely on evolutionary considerations to offer suggestions about how to understand rationality, in all animals (linguistic or not). In their view, rationality emerges from a set of representational and control devices that have evolved due to specific evolutionary pressures. According to this approach, rationality belongs in the first instance to biology rather than to psychology and philosophy.

The present chapter belongs to the bottom–up construction of rationality. The method chosen is to study the causal-informational constraints on cognitively operated control systems in order to understand how rationality develops. This method might help generate connections between the three senses of rationality distinguished by Kacelnik. Herbert Simon's concept of 'bounded rationality' (Simon, 1982) had already introduced important functional links between the minds we actually have and the economists' conception of rationality. In his view, rationality should not be conceived as the maximization of expected utility, but as a more modest 'satisficing' of this utility; a rational agent does not aim to get optimal results, but results that are sufficient to fulfil her needs. Indeed, rationality involves the effort of an entity to reach its goals while having to use cognitive resources that, in a physically realized mind, are necessarily limited. Put in a dynamic evolutionary perspective, Simon's idea of rationality leads to investigating how the informational resources used by simple cognitive systems, being limited, bring with them new selective pressures.

I will here sketch a possible evolutionary path that leads from fitness maximization to mental processes that are precursors of verbal rational thinking. I will defend the claim that the first site of rationality is metacognition, a capacity allowing an animal to monitor the informational quality of its inner environment (i.e. to predict what it can do given a signal with a specific discriminability level). Metacognition and metarepresentation will be contrasted, as will different forms of reflexivity inherent to control systems. Finally, the claim that rationality involves normativity will be re-examined from the standpoint of an evolutionary approach to rationality. This part of the discussion will hopefully help in establishing a few important links between B- and PP-rationality.

....

[1] See in particular Bermúdez (2003, and this volume), and Hurley (2003, and this volume), for two different attempts at top—down theorizing on animal minds; though Hurley also engages in bottom—up theorizing (this volume, and 2005) and regards them as compatible.

[2] See in particular Sober 1994, Godfrey-Smith 1996, 2003, Sterelny 2003b, and this volume, Sousa 2004.

12.2 **Rationality as adaptive control by cognitive means**

12.2.1 **Bounded rationality in a dynamic evolutionary perspective**

To understand rationality in a broad biological sense, we have to grasp the selective pressures from which it results, but also to identify which mental processes can respond to these pressures. An interesting proposal has been made in the literature on the kind of selective pressures that are relevant to the emergence of cognitive systems. The Environmental Complexity Hypothesis, developed by Peter Godfrey-Smith (1996) and Kim Sterelny (2003b) states that behavioural flexibility has been crucial in the variable environments in which organisms had to survive. As a result, a major selective pressure was exerted that favoured extracting information and creating alternative control procedures. I endorse this hypothesis, and here explore its implications for the evolution of rationality. The claim this chapter develops is that there is an architectural solution to the problem of adjusting to changing circumstances and needs: adaptive control, effected by cognitive means. Before introducing this concept, we first need to understand what a control structure is.

12.2.1.1 **Control structures**

Control implies the capacity to determine which motor command to a system will be associated with some desired result, such as grasping a ball. Control systems involve a loop in which information has a double flow (Fig. 12.1). One is the top–down flow: a command is selected and sent to an effector. The other is the bottom–up flow; reafferences (i.e. feedback generated by the former command) inform the control level of the adequacy of the activated command. What is crucial in any control system is the fact that observed feedback can be *compared* with expected feedback.

There are many forms of control that regulate the vital functions in living creatures (cell growth, digestion, blood circulation, sensorimotor and neuroendocrine reflexes, etc.) as well as in machines. In the simpler forms of regulators, such as thermostats or Watt's flyball governors, the very physical organization of a mechanical device allows

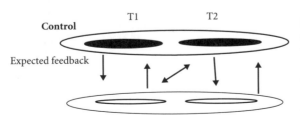

Fig. 12.1 A simple control system: a command is sent at time t1 to an object level. Feedback allows comparing the observed with the desired response and correcting the command at t2 if necessary.

unwanted perturbations to be neutralized, and brings the system back to a desired state. Because the causal structure of the physical interactions is designed so as to map the informational structure in the comparator, information plays no causal role in these simple control systems; they are called 'slave control systems' because the range of their 'responses', given a particular input, is strictly and inflexibly determined by the machine design; these systems cannot learn and cannot change their goals.

12.2.1.2 Adaptive control

Devices that use information as a specific causal medium between regulating and regulated systems constitute a new step in the evolution of control systems. They allow a different form of control to emerge, called 'adaptive control'.

Adaptive control operates on partially unknown systems. It requires the additional ability to determine input–output couplings in varying and uncertain situations. A particularly important class of adaptive control systems uses feedback from previous performance to select a course of action. In these systems, learning can influence command selection. Two forms of regularities must be present for an organism to adaptively interact with its environment. First, regulation laws determine which affordances are associated with specific motor commands in specific environments. Second, feedback laws determine what portion of the regulation space is accessible to an organism with a given learning history. A key role in adaptive control accrues to memory, that is to representation. These representations are constructed, retained, and retrieved in what is called 'an internal model' of the events to be controlled.

Let us take an example in the realm of action. Skilled motor behaviour[3] involves using two types of internal models, or of sets of representations in the brain, that can predict the sensory consequences of action commands: *forward models* store the causal relationships from motor commands to sensory consequences, enabling 'prediction' of sensory results of motor commands (Fig. 12.2); *inverse models* conversely transform a desired sensory consequence into the motor command that would achieve it, enabling selection of appropriate means to desired results.

When the environment changes rapidly, and many times over a life time, it becomes very advantageous to be able to retain the control parameters that worked well in similar cases. But it also becomes crucial to be able to select the *correct* setting for a new situation; that is, to behave flexibly. The evolutionary response to this dual demand seems to have taken two successive steps: simple representations have been used, then cognitive ones.

---

[3] Because of their importance in motor learning, control systems have been particularly studied in the context of motor behaviour. Cf. Wolpert *et al.* 1995, 1998, 2001.

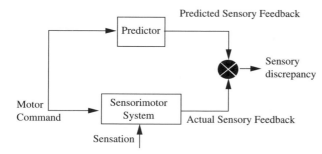

Fig. 12.2 A feed-forward model of action. A motor command is both sent to the sensorimotor system and to a predictor; a comparison is effected between predicted and actual feedback; anticipated or observed discrepancies will trigger a motor command revision.

Internal models can be formed and maintained in a very specialized way, in systems that do not refer to facts about the world. The earliest kinds of 'internal models' involved in adaptive control were most plausibly of this kind. Examples of such non-semantic representation are found in very simple entities that store information in order to control their motor behaviour more flexibly, like molluscs or insects.[4] Other examples are provided by the various feedback loops that regulate posture.

Cognitive representations, on the other hand, are of an entirely different kind. They are informational states selected and functioning to refer to a specific event or property.[5] A crucial feature of this form of reference is the capacity to reidentify objects over time—whether or not currently observed—as having various, possibly changing, properties. A control structure including cognitive representations is equipped to carry out a semantic function. As frequently emphasized by philosophers,[6] cognitive representations allow control to become much more powerful, by exploiting their disposition to combine in indefinitely new ways—a disposition that non-cognitive representations, being 'modular', lack. Internal forward models using cognitive representations *ipso facto* gain a much wider range, in predicting, imagining, and planning behaviour.

Let us summarize. Information plays a major role in the two types of dynamic laws described above, regulation and feedback laws. Internal models, which have a crucial role in adaptive control, develop from local and specialized to multipurpose forms of data storage. The ability of a system to represent 'cognitively' the dynamics of external or endogenous events and to expand and revise learned associations becomes the key to flexible regulation. Recombining acquired representations allows the predictive capacities of a representational system to deal with new contexts. Cognition as an adaptive control system thus evolved with the function of acquiring, storing, revising, and recombining representations.

---

[4] See Proust 1997, 1999, 2003.

[5] See Dretske, 1988 and Proust 1997.

[6] See for example Evans' 'generality principle' in Evans 1982.

12.2.2 Flexible control via cognition: competition, and the control of information

We can now offer a first attempt to define rationality in very general terms. Rationality is the capacity of certain control systems to adapt to changing circumstances by means of cognition. That is, it seems natural to define rationality as the property of having cognitively-operated adaptive strategies. This definition is not tautological, since there are systems, such as bacteria and bodily cells, that do not rely on cognition to adjust to changing circumstances. This definition is too weak, however. For some cognitive systems may 'believe' that a strategy they selected is adaptive when it is not. But a specific strategy that is ill-devised, that does not take into account relevant and well-known constraints, cannot be regarded as rational. Moreover, our attempted definition is equivocal about the entity that is evaluated; is it a specific strategy, an individual organism, or a generic cognitive system? A way out is to define rationality as a disposition that tends to be realized by a control system that can adapt to changing circumstances and that relies on cognition to do so.

This definition emphasizes the observation made in Section 12.2.1 that the kind of cognitive disposition that will end up being rational cannot be decided independently of the way the world is. Whether a rational disposition will emerge depends essentially on the dynamics of the environments in which the organisms of a given species have to develop and reproduce themselves. A plausible hypothesis is that cognitive representation is crucial for survival only in environments that require flexible and efficient types of control. In these environments, information itself—as part of the many adaptive control structures required in everyday activity—becomes a good for which to compete.

As a result, the main pressure for increased rationality will apply to beings that can manipulate the informational quality of their own and others' internal and external environments. This capacity presupposes that informational access and discriminability can, in part, be modulated deliberately. This is what we now have to understand.

12.3 Flexibility and informational environments

12.3.1 The costs and benefits of flexibility

Godfrey-Smith (1996) and Sterelny (2003b) have shown that each type of activity—flexible or rigid—has a metabolic cost. Maintaining a control system does not come for free; the investment it represents has to be matched by the returns it provides. Although a rigid system is usually cheaper, it may sometimes also have better returns. Let us sketch a simplified case developed in Godfrey-Smith (1996), in which an organism has to cope with two possible states of the world. This example will show that being flexible involves a form of uncertainty that brings with it different pay-offs in different contexts. Suppose that there are only two kinds of prey of unequal value present in the environment, making a particular behaviour more adequate than the other. The probabilities that each kind of prey is present are P and $1-P$, respectively. The organism may either have a fixed response to the first type of prey, or to the second, or have a flexible response depending on the case identified, which may turn out to be incorrect. Now the payoff of each kind of behaviour, when successful, may be different. *Ceteris paribus*, a flexible response should be favoured over a rigid one only when the ratio of the probabilities of producing the correct rather than the incorrect response by using a specific cue and behaviour outweighs the ratio between the expected importance of each of the two possible states of the world.[7] (Expected importance is defined as the importance of a state of the world multiplied by its probability.) For example a hard-wired, rigid type of behaviour, such as swallowing every moving object in a predetermined perimeter, may bring only a small but regular reward. Even so, it may well yield more resources in a given environment than a flexible behaviour with a higher probability of error and more variance in its returns.

This reasoning illustrates the fact, already mentioned, that the potential of a strategy depends on the environment in which the organism has to survive. Some environments can be particularly tolerant of rigid responses. For example sea snails (*Littorina*) manage very well with a very simple device that simply sums the intensity of light and gravity in order to move up or down according to tide level. Every biologist knows, however, that most environments are not of this kind. When the control parameters have to be frequently updated, information becomes an essential instrumental resource. In this case, flexible responses cannot be prewired. Cognitive flexibility becomes valuable when variability affects the species in the long run. It permits the generation of more control opportunities; but this presupposes a capacity for detecting new regularities and forming updated mappings from arbitrarily new contexts to possible actions; these two capacities, completed by a motivational system for ordering preferences, are constitutive of cognitive systems.[8]

In sum, cognitive flexibility emerges not for its own sake, but as an evolutionary response to an environment complex enough (dangerous and sparse in resources) to force the organism to represent alternative courses of action. Cognitive flexibility requires specific resources and incurs particular costs, which have been explored in the recent literature.

12.3.2 The costs and benefits of cognitive flexibility

In the case of cognition, the costs are not simply those incurred by any kind of adaptation, of maintaining a dedicated system, such as a brain or some other functional system. Cognition brings with it a particular type of cost, linked to the kind of resource that

---

[7] I here follow Godfrey-Smith 1996, p. 209 ff. See also Moran 1992, Sober 1994.

[8] Cf. Boyd and Richerson 1995.

information is. This section examines the specific costs attached to using non-cognitive or cognitive representational means to control behaviour. Signal detection theory (SDT) will help to present the general argument. All organisms able to control their behaviour more or less flexibly—that is using 'simple' or 'cognitive' representations—are subject to a difficult trade-off between responding or not responding, as is shown in SDT. Cognitive organisms, however, have to face simultaneously detection *and* categorization costs. Let us see how.

Signal detection theory teaches us that, when an agent has to detect, over the course of many trials, the presence of a signal in a noisy context,[9] there exists an overlap between two distributions: the probability distribution that a signal was detected given that no signal was produced but only noise, and the probability distribution that a signal was detected given that a signal was produced as well as noise (Fig. 12.3). This overlap obliges the agent to choose a decision criterion. The position of the decision criterion determines the probability of each response. There are two decision strategies that can be chosen: security in signal (strict criterion) or security in response (lenient criterion). For example a small mollusc, *Aplysia*, is able to form 'simple' representations of affordances and learn from experience; it can move its detection criterion according to whether the stimulation on its tail is strong and rare (in this case it tends to withdraw its siphons: it applies a strict criterion) or weak and frequent (in this case no detection occurs: a lenient criterion is selected).

Bayes' theorem predicts where an ideal observer, all things being equal, should locate the criterion for each kind of probability distribution. In fact, as already noted, all things are rarely equal. As the example developed in Section 12.3.1 suggests, it is a very different situation, in terms of consequences, to miss an opportunity (false negative), or to expend energy when the world does not warrant it (false positive). The decision criterion can thus be moved up and down according to the payoffs, that is the cost of a miss versus the benefit of a hit.[10]

Flexible control thus brings with it new puzzles; not only, as in non-flexible strategies, does the system incur *objective* costs when the probabilities of response no longer match probabilities of actual events. It also incurs costs *linked to the process of information acquisition and evaluation*. A representational control system must decide whether it is worth it to go on acquiring additional information or whether it should exploit the evidence already gathered. It can react too quickly or too slowly; it can mislocate its decision criterion. It can thus ruin even favourable opportunities.

....

[9] Information is here assumed to be constant, and the *a priori* probabilities of signal and noise to be equal.

[10] This problem has been studied closely in Godfrey-Smith 1996.

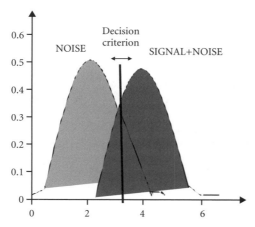

Fig. 12.3 Receiver operating characteristic (ROC) curve used in signal detection theory. The area of overlap indicates where the perceptual system cannot distinguish signal from noise. The receiver may move her decision criterion in a strict way (to the right) or in a lenient way (to the left) depending on the respective consequences of a false positive or of a false negative.

The same point holds again for identification, a specialty of cognitive-representation control systems (see Section 12.2.1). In this case, not only must an agent find out that some relevant signal is present in the perceptual flow; it must also categorize it—an operation that involves the (expensive and revisable) construction of new categories.[11] An organism may detect a signal correctly, but miscategorize it. Therefore an additional choice of response strategy has to be made, opening up a new range of costs and benefits to cognitive organisms. In the context of predation, a statistically rare exemplar may involve sure death; in that case, it is better to produce false positives than false negatives. Conversely, in hunting, for example, it may be better to pursue prey that are clearly identified than to run after everything that moves.[12]

Clearly, having more degrees of freedom in decision-making creates specific selective pressures on how to decide correctly (i.e. how to learn quickly from past encounters the most beneficial strategy). But SDT suggests that an additional feature of information processing must have influenced the evolution of rational behaviour. Adequate detection and identification depend not only on the agent correctly deciding how to move the decision criterion, but also on the signal being *discriminable*. Discriminability[13] depends on the strength of a signal and on the amount of noise (noise is reduced when there is less 'spread' or overlap between signal and noise). If the signal is very strong, and if there is no possibility of mistaking it with another, then it is

---

[11] Stimuli can vary on one or on several dimensions, in which case a multidimensional version of SDT is applied Ashby 1992.

[12] Not necessarily always, as we saw earlier. An example of an inflexible strategy for hunting is the insect capture in toads, and for mating, the strategy of the hoverfly; both strategies launch a response to a stimulus defined in proximal terms as moving dots.

[13] A dimension called d' in Signal Detection Theory.

highly discriminable, and can be extracted more easily. Success in signal detection tasks of all kinds thus depends crucially on the capacity of an observer to cope with noise. This universal constraint explains why cognition-related selection exerts the most severe pressures on information processes that deal with noise. As we shall see below, there are two varieties of noise that play a crucial role in the evolution of rationality: noise that threatens to infect one's own cognitive system and noise that can be manipulated to infect others'.

12.3.3 **Transparent vs. translucent informational environments**

Let us first borrow a distinction from Kim Sterelny (Chapter 14, this volume and 2003). An environment, or rather a relevant dimension of an environment, is 'informationally transparent' when the cues available for crucial resources (or dangers) are reliable and discriminable. In such cases, the costs of information are low, because the cues can be easily coded in the genes or learned and reliably exploited. An example offered by Sterelny is the migratory cues used by shore birds. Day length is an invariant cue that objectively predicts the best moment to migrate.

In a translucent environment, however, the cues are less reliable or can be manipulated by predators. In such environments, there is a cost in mining information (because of the risks incurred in exploring the presence or value of the cues) and/or in acting on it (when the cues are not reliable, the action becomes ineffective). An important consequence is that, in such environments, it becomes important to devise strategies not only for reaching external goals, but also for extracting and using information. Ground squirrels, for example, use 'interactive' methods to find out whether a rattlesnake is large (that is: dangerously venomous) and warm (that is: quick and accurate).[14] They confront the snake (approach it, 'flag' their tails, and jump back repeatedly) in order to provoke a defensive response, since such a response includes an auditory signal that cannot be faked or hidden from view by the vegetation. The sound of the rattle tells them whether the animal is large (higher amplitude) and/or quick (faster click rate).

Given the uncontroversial fact that access to information is crucial for survival in cognitive organisms, and given the constraints imposed on discriminating signal from noise, it becomes obvious that, in order to survive and outsmart their predators and competitors, cognitive systems will have to rely on ways of manipulating the informational properties of environments. This is how a whole range of activity having to do with noise evaluation, reduction, or manipulation came to be selected in phylogeny. If this is correct, and given also that information can be accessed more or less easily, then rationality is essentially related to the capacity to assess informational quality and to restore transparency whenever it is possible and useful to do so (by changing either the internal or external environments). This is how 'epistemic actions' emerged and became a central part of animals' repertoire.

---

[14] See Owings 2002.

12.3.4 **Epistemic action: action for informational ends, by physical or informational means**

The term 'epistemic action'[15] refers to an action whose goal is to manipulate the informational quality of a cognitive environment: either to increase transparency for oneself (and for offspring) or to decrease it for others. A physical action aims at reaching a certain physical goal state by producing a specific bodily movement. Similarly, an epistemic action aims at reaching (or causing in others) a certain epistemic state, disposition, or property, by relying on the agent's own (or the other's) perceptual and cognitive apparatus, as well as on certain physically relevant properties.

Although epistemic actions are aimed at bringing about changes in mental states (one's own or others'), they can be performed in ways that are either physical–ecological or cognitive-representational: there are two varieties of epistemic action, physical and doxastic, corresponding to the means by which they are implemented.

(a) The means used to an epistemic end can be physical; a niche[16] can be organized in a way that makes it informationally transparent in some important dimensions. Building one's home on top of a mountain has the advantage of allowing a full view of the environment. On the other hand, squirting ink or throwing dust in the eyes of a predator, or introducing random variation into one's movements, are physical ways of preventing a hostile agent from using spatial tracking information.

(b) Doxastic actions are epistemic actions performed through cognitive-representational means; they include speech acts, for example, and all communicative ways of conveying information to other minds. Doxastic actions are in general less resource-consuming than physical ones (they typically use internal resources). There are various types of epistemic actions relying on doxastic means. Some involve signalling and other communicative devices. Misrepresenting facts (by issuing deceptive signals) is a standard way of manipulating belief and motivation in other cognitive agents. Agents with no reflexive understanding of mental concepts can perform doxastic actions. Examples are offered by tactical deception in primates, false signalling in birds, etc. Doxastic actions backed up by mind-reading capacities however reach a much larger range of manipulative and evaluative opportunities. (More on this in Section 12.4.1)

12.3.5 **Metacognition: action for informational ends by informational means**

There is a class of doxastic actions particularly relevant for rationality: those that use informational means to achieve informational ends that concern the system itself.

---

[15] I use Sterelny's term 2003b, and this volume. I refer to this notion in previous work under the term of 'mental act' (see Proust 2001). In this chapter the word 'epistemic' designates not the means used but the end pursued.

[16] On the importance of niche construction in evolution, see Odling-Smee *et al.* 2003 and Sterelny 2003b.

A prominent end, as we saw above, consists of improving the signal-to-noise ratio in the system itself. This is exactly what metacognition is about; it is cognition *about* one's own cognitions. Its function is to monitor current mental activity (perception, memory, thought, action, emotion) and to control it (by sending appropriate commands). What makes it metacognitive, rather than simply cognitive, is that its goal is to improve flexibility and informational quality in the cognitive processes, rather than transform the external world or exploit affordances. It transforms phylogenetically older, automatic, forms of cognition into adjustable, multivalued controlled processes.

Metacognitive organisms are apt to monitor and evaluate their performances in the various functional capacities that constitute a mind. These include: judging the adequacy of a particular response and correcting it when necessary (*retrospective monitoring*), evaluating such matters as one's ability to carry out a new task (*prospective monitoring*), the difficulty of items to be learnt (*ease of learning* judgements), or the past acquisition of a given memory (*feeling of knowing* judgements). Attending and planning respectively aim to form a clearer picture of some object, property, or context, and to make better use of its potential resources. Other facets of metacognition are the control of motivation, of emotion, and of the social impact of informational and motivational states.[17]

Clearly, the capacity to muddle the epistemic environment of competitors and predators, on the one hand, and the need to achieve or restore transparency for one's own use, on the other, constitute fundamental evolutionary pressures from which metacognitive processes have finally emerged.[18] To remain viable in a competitive world where information becomes a good, an individual organism has to secure a low noise/information ratio for itself, while finding ways of increasing it in its competitors. Other adaptations are steps in the same direction, toward safer informational procedures. Sterelny mentions multicue tracking—that is, multimodal perception and categorization—and the decoupling of perception from action (see next paragraph); there are other procedures one might add: for example detecting goals from other's movements (a way of predicting what is the invariant end point of movement in spite of various realizations), of which gaze orientation is a crucial element; joint attention; and tactical deception. In human beings, these various adaptations combine to make mind reading possible.[19]

These new informational techniques do not come for free, however. Multimodality involves expensive attentional mechanisms for binding the various modal-specific features and a calibration device for maintaining coherence across the various modal-specific spatial maps.[20] The decoupling of perception from action allows alternative action models to be maintained for the same type of situation, only one of

[17] See Proust 2001, 2003, in press.

[18] See for example Sober 1994, Ch. 4, Proust 2003a and Sterelny, this volume.

[19] Cf. Baron-Cohen 1995, Whiten 1997, Proust 2003b.

[20] On the important role of calibration in higher cognition, see Proust 1999.

them is the preferred or 'correct' one in a given case (see Sterelny 2003b). To prevent confusion between the various models, specific inhibitory mechanisms are needed; these mechanisms, however, turn out to be brittle and slow to depotentiate.[21]

12.3.6 From the external to the internal environment

To understand how metacognition was selected—a covert way of evaluating one's own plans and mental processing—it may be useful to understand how selective pressures have tended to favour covert informational techniques and, in particular, covert simulations.

Simulation indeed is a central feature in any adaptive control system. Classic control theory explicitly argues that internal models have to be simulations of the dynamic facts they control. According to Roger Conant and W. Ross Ashby, the most accurate and flexible way of controlling a system consists in taking the system itself as a representational medium. In an optimal control system, therefore, the regulator's actions are 'merely the system's actions as seen through a specific mapping'.[22]

Now simulation can be performed in a cognitive system either openly or covertly. As we saw above, some epistemic actions include an overt physical process of testing.[23] Reducing the costs should tend to favour covert ways of executing them (thus reducing predictability of one's own endeavours by others). For covertly simulating a process—that is, representing it dynamically in its connections with a specific context—generates 'internal feedback', detailed anticipations of changes brought about by an action in the absence of any actual engagement in the world. As Millikan observes (Chapter 4, this volume), covert simulation allows the agent to perform actions in its head. Representations are tried and revised in safety.

There are several varieties of covert simulation in nature. As we saw in Section 12.2.1, internal models used in adaptive control can be either highly specialized (as in posture control) or cognitively operated (as in planning). Covert simulation can therefore be either cognitive or non-cognitive, or can integrate several embedded models (including, for example, strategical and motor simulations) as in sport competition.

What evidence do we have that the various metacognitive capacities reviewed in Section 12.2.3 involve simulation as a necessary component ? Neural imagery of action-related thinking indeed suggests that simulation is a pervasive process through which the brain categorizes and predicts the effects of types of action. Planning an action (imagining it, rehearsing it, watching others perform it) uses, in part, the motor

---

[21] Cf. Proust 2003b.

[22] See Conant and Ashby 1970.

[23] Kim Sterelny (2003b) discusses the function of mutual probing in male–male mate competition. The stag needs to compare its bodily strength with a competitor; their parallel walks might be a ritualized way of appreciating physical superiority while saving the cost of an actual fight. Another case of testing was offered above, through the ground squirrel's 'tail flagging-to-rattlesnake' behaviour. These examples illustrate the vital importance of securing knowledge in a translucent environment.

areas involved in executing the action.[24] It is also quite plausible that self-simulation (within the context of a task) is what allows an agent to perform metacognition: to know what it knows, to evaluate its own uncertainty; to assess what it can or cannot retrieve, etc. Various forms of this kind of metacognitive control are available in many species.[25] Although still little-known, they are clearly fundamental to our appreciations of the extent to which non-linguistic animals are rational.[26] As we will see later, they constitute an implicit precursor to a disposition that is crucial in human rationality, which enables an agent to report on her reasons to act.

Let us summarize the discussion so far. We first suggested that the biological form of rationality that is suited for a wide application to non-human as well as to human animals is captured by the notion of cognitively operated adaptive control. We saw that flexibility in behaviour presupposes multivalued regulation. We then argued that this form of regulation has been selected as a result of the dynamic properties of the environments relevant to survival; it created in turn new selective pressures of a 'Machiavellian' type. To remain viable, each organism must work at maintaining the informational quality of its own environment, both internal and external, while 'modulating' (selectively restricting) the other organisms' access to it.[27] This is how metacognition finally emerged, relying on covert simulatory processes.

12.4 **From metacognition to rationality**

12.4.1 **Metacognition, metarepresentation, and mind reading**

If the function of metacognition is to modulate the way in which information is used in a representational system, the question arises of a possible overlap between metacognition and mind-reading capacities. Whereas metarepresentation is a theoretical (concept-based) capacity for reporting (explicitly) on the contents of mental representations, metacognition is a practical (implicit) capacity for guiding mental activity. In this section, I defend the view that metacognition does not depend on metarepresentation, while the latter presupposes metacognitive abilities.

Metacognition comprises the various skills involved in controlling one's own and others' mental states for their informational adequacy. Control, when understood in these terms, has sometimes been taken to presuppose a metarepresentational capacity, which is also often taken to constitute the essential core of the ability to read minds (to attribute mental states to other agents).[28] A metarepresentation is a representation

---

[24] See Decety *et al.* 1994, Jeannerod 1999.

[25] For a presentation of the theoretical framework of metacognition, see Nelson and Narens 1992.

[26] Editors' note: see also Millikan, this volume; Hurley, this volume, Ch. 6.4, 6.7.

[27] Communication with conspecifics is also modulated by a tension between trustworthiness and manipulation, as predicted by game theory. See Sober 1994, Hauser 1997, Miller 1997, and Proust 2003b.

[28] For example Frith 1992.

that contains, and is about, another representation; metarepresentations used in mind reading include attitude markers (such as 'believes that', 'desires that'), which categorize the mental status of the first-order representation (for example as in a metarepresentation such as 'she believes that the apple is ripe'). Many psychologists and philosophers of different persuasions (theory-theorists, modularists, and even some simulation theorists) argue that you cannot understand what a mind is if you cannot think about some mental content as constituting a belief or desire.[29] But if this claim is true, the objection continues, then any form of metacognitive control over one's own mind or others' involves reasoning with mental concepts. In this view, for example, an organism cannot plan to act if it cannot metarepresent itself as having intentions—if it cannot form the representation: 'I believe that I intend to do A'. Similarly, a perceiver cannot attend if it does not metarepresent itself as having perception, and so on. One might thus argue that meta*cognition* must depend on meta*representation*, and hence on folk psychology (if one takes folk psychology to be a system of mental concepts allowing an agent to reason in mental terms about actions). This assumption obviously bears on the validity of any attempt to articulate the notion of rationality independently of a reflexive belief-desire or folk-psychological framework.

The view that metacognition must necessarily rely on either metarepresentation or on mind reading, however, is now rejected by experimental findings in cognitive ethology. It has been shown that animals can develop a metacognitive sensitivity to the quality of the information available to others or to themselves. Such a sensitivity does not seem to involve a metarepresentation, that is a conceptual representation of the relevant informational links. Let us turn to the evidence.

Concerning the capacity to manipulate others: there is now ample evidence that non-human animals can form reliable strategies for masking their intentions from others, and for adjusting their actions to the presence of competitors, without any understanding of what it means to believe something (in the sense of forming true or false representations). Animals can arrive at appropriate control routines by relying exclusively on behavioural cues.[30]

Concerning the capacity to evaluate one's own mental capacity and to adjust one's actions to what one knows about the internal quality of the information, new findings[31] suggest that animals who have not been shown to have mind-reading capacities, such

---

[29] The present argument does not need to assume that mind reading requires metarepresentation. However, I would defend the view that mind reading involves embedded simulations that are a semantic equivalent of metarepresentation.

[30] See Clayton *et al.* this volume; for a discussion of the extent to which chimpanzees understand certain mental states (like seeing) as opposed to behaviours (like looking), see Call and Tomasello 1999, and this volume, Tomasello *et al.* 2003, and Povinelli and Vonk 2003, and this volume.

[31] Cf. Hampton 2001, Inman and Shettleworth 1999, Smith *et al.* 1997, 1998, 2003, Shettleworth and Sutton this volume, Proust 2003a.

as monkeys and dolphins,[32] have metacognitive capacities allowing them to assess a situation relative to their present capacity—to form a judgement of competence—and rationally to revise their strategies on the basis of this judgement.[33] Smith and his colleagues have studied the comparative performance of human subjects and rhesus monkeys in a visual density discrimination task.[34] Participants had the option to decline the test when the discrimination task was sensed to be too difficult, and get access to an easier, less rewarded task. The decisions of rhesus monkeys closely paralleled those of their human counterparts. These capacities may, *prima facie*, look modest—to be able to choose not to complete the task and turn to another, less rewarding but more promising one—but they are far from trivial, and turn out to be relatively rare in non-human animals. A visual density discrimination task such as this requires the evaluation of the quality of the information offered given the epistemic requirements of a successful response.[35]

Other, more limited forms of metacognitive skills, have been found in birds. For example when subjected to metamemory tasks, pigeons seem to lack the capacity of monkeys to form a judgement of competence without having current perceptual access to the stimuli.[36] Jays seem, however, to have a form of metamemory. They are able to remember in which social circumstances they stored a particular item, and whether or not a conspecific was present.[37] Ravens can evaluate their own specific capacities; they are good at appreciating—without any antecedent task-specific learning—whether they can or cannot perform a physical task prior to executing it (for example raising an object of a given mass attached to a string).[38] However no evidence in favour of mind-reading capacities has yet been collected involving these animals.[39]

---

[32] Editors' note: cf. Tschudin, this volume, for recent work on whether dolphins can mind read.

[33] For a careful presentation of the methodological issues involved in this type of experimentation, see Shettleworth and Sutton, this volume.

[34] Smith *et al.* 2003.

[35] Editors' note: See Shettleworth and Sutton's comments (in Ch. 11) on the lack of controls in tests of metacognition for perceptual discrimination, as opposed to in the tests for metamemory they describe.

[36] However, as Shettleworth and Sutton show (this volume), most pigeons seem to understand that their performances decline at longer delays (they escape more to an easier task) even if they cannot 'report on' their metacognitions in the way monkeys can.

[37] Clayton *et al.*, this volume. In both William James's and Endel Tulving's definitions of episodic memory, the experience of remembering a past personal event involves the metacognitive, token-reflexive awareness of recollecting that event. Episodic memory is, in Tulving's (1985) words, the capacity 'to remember personally experienced events as such'.

[38] Heinrich 2000.

[39] On monkeys, see Cheney and Seyfarth 1990, Anderson and Gallup 1997. On dolphins, the evidence is unclear; see Tschudin this volume; see also Herman this volume.

We can conclude, firstly, that whereas by definition an animal endowed with a mind-reading capacity is able to monitor and predict the behaviour of others in a mentalistic way—that is, by attributing mental states to others—metacognition does not seem to require any mentalistic attribution. Nor does metacognition depend on metapresentation *per se*. Secondly, we can observe that while it is not the case that all species with metacognitive capacities are able to read minds, there is no species that can either mind-read or metarepresent but that lacks metacognition. What can such an asymmetry, if it is indeed confirmed by further evidence, teach us about the evolution of rationality? We will come back to this question in Section 12.4.3.

12.4.2 Metacognition as procedural reflexivity

We saw in Section 12.3.6 that metacognitive capacities belong to control systems, and might use internal dynamic models to select commands. Launching a specific search in memory, for example, might require an inverse model, while appreciating the correctness of the retrieved memory might depend on a forward model. In the particular case of signal detection tasks used by researchers in animal metacognition, reported above, a subject might use forward modelling to predict whether it can complete a task on a given perceptual material; a simulated command would be activated (top–down); (bottom–up) internal feedback might then be evaluated, and, say, be found insufficient for an actual command to be activated.

The fact that metacognitive systems have the architecture of a control system explains why metarepresentations are not needed to evaluate correctly the system's cognitive-informational status. Sending a command and using feedback to monitor it form a single functional cycle activated in a particular task. Command and feedback reflexively refer to each other: this command orients the system to look for a matching feedback; match or mismatch in feedback immediately retroacts on the command level. There is a causal (bidirectional) relationship between command and feedback; but there also is an internal informational relationship between the command (for example: discriminate P or Q) and a kind of feedback (for example: P or Q not discriminable). Each refers to the other in an implicitly reflexive way (it is part of the natural meaning of this feedback that it was triggered by this command, and, reciprocally, it is part of the natural meaning of this command that it produces this feedback). We will refer to the form of reflexivity inherent to control structures in general, and metacognitive control structures in particular, as 'procedural'.

A distinction between two levels of procedural reflexivity, however, is needed to account for the fact that simulation can be effected in a 'simply' representational or in a cognitive way (see Sections 12.2.2 and 12.3.6). In a 'simple', modular, internal model, the limited and specialized scope of the commands, as we saw in Section 12.2.2, prevents them for qualifying as semantic entities; a 'simple' internal model taken in isolation does not have access, however implicitly, to the reflexive character of its own processes. Implicit access to the reflexive structure of control loops only appears when 'simple' internal models (used in motor control, for example), are harnessed to higher-order forms of cognitive internal models. Procedural reflexivity in that case is reflected in the semantics of the higher-order internal models. It is procedural rather than explicit, however, because it does not need to be articulated in a specific semantic constituent to be actually part of what the system thinks.[40] In such cases, the informational content does not need to be represented as such by the animal to be effective.[41] It is, rather, an architectural outcome of the control structure.[42] A metacognitive mechanism is just that; it correlates the feasibility (probability of success) of a task with preselected types of dynamic cues (like the quantity or intensity of activation in the feedback neural population).

An example of how procedural reflexivity works can be borrowed from studies of human metamemory. How does a subject come to realize that she can retrieve a particular memory? Asher Koriat (1993) has offered a model—called the 'accessibility model'—of how this kind of metamemory works in human subjects. The basic idea can be reinterpreted in the following way. The subject launches the relevant search during a short interval. The dynamics of the control loop so created, still in simulatory mode, tell the subject whether success is possible. The ease of access to a specific memory content is not inferred from the contents to be retrieved: for this would not be a prediction, but the execution of the task itself. Rather, it is 'read' in the dynamic properties of the vehicle, that is, the dynamic properties of the neural networks involved in the retrieval process; higher activation levels predict success in retrieval in a reasonably short time. Thus the functional articulation between vehicle reading, prediction, and command, supply implicitly a sensitivity to the fact that the information acquired 'concerns' the decision to launch the search in memory.

To summarize: metacognition involves an implicit, procedural form of reflexivity, which can, but need not be, articulated as an explicit mental content. Thus the

---

[40] Perry 2000 discusses the role of 'unarticulated constituents' in thought.

[41] Editors' note: for a related point about whether or not simulation is monitored, see Hurley this volume, Ch. 6.4, 6.7.

[42] This view on reflexivity is close to John Searle's view (Searle 1983) on the kind of reflexivity present in perception, action, and memory. When an agent acts, she experiences what it is to be an efficient agent in the world. This experience is not *about* causality. It is rather about the property that her action brings about (switching the light, cooking a steak, etc.). She can, but need not, form a second-order thought about the causal aspect of her experience (reflect, for example, on her strength, her ability or lack thereof); this kind of thought is not part of her ordinary experience as an agent. Such awareness is a metacognitive feature that does not need to be explicitly represented in order to work effectively as a constraint on what counts as an experience of acting. This explains why small children or animals have access to that kind of causal self-referentiality. As Harman (1976) also observes, the agent does not need to form a 'metaintention' to the effect that the first intention should lead one to do A in a certain way. The first-order intention already contains a reference to itself. See also Perry (2000, Afterword), for an interesting discussion.

connection between a discriminatory process and a response of uncertainty, or any other metacognitive command–feedback coupling, does not require that a semantic link between feedback and command be *explicitly* recognized by the agent.

12.4.3 From metacognition as procedural reflexivity to metapresentation and explicit reason-giving

If these speculations are roughly on the right track, we have to understand how procedural reflexivity can be transformed into intentional reflexivity, that is how the gap between animal and human rationality can be bridged. How can an implicitly rational hierarchy of control structures be turned into an explicitly rational (reason-giving) agent?

Reporting one's reasons to act consists in providing explicit structure to a process that depends on implicit heuristics. Some of these heuristics may have been selected through evolution; for example, an agent innately recognizes whether she is acting or is just made to move. Others may be the result of implicit learning; for example the metamemory process that allows a subject to know 'non-inferentially' whether she can retrieve an item from her memory might depend on the practice of memory retrieval. How might linguistic (and additional conceptual) structure be superimposed on these implicit metacognitive states?

A plausible supposition is that, in human beings, the output or monitoring loop of the lower-level metacognitive control structure is used as the input to another higher-level control structure. This new structure controls the linguistic communication of cognitive agents. Indeed a recent hypothesis about the development of working memory (the structure that secures command execution on the basis of prior feedback) is that humans have two classes of such mechanisms available (Gruber 2003). A phylogenetically older system, also present in non-human primates, regulates behaviour on the basis of sensory information (visual and auditory feedback properties in relation to space and to objects). A later, more powerful system has developed exclusively in humans to recode linguistically the output of the first system. This recoding is what allows us to rehearse mentally our intentions to act, and to plan our delayed actions. Such a process of recoding—called 'redescription' by several authors—seems to be how primate mental architecture develops in phylogeny, from specialized modules to more flexible and robust control loops.[43] Now if metacognition relies on primitive forms of internal feedback, linguistic recoding might provide the subject with a rehearsal mechanism that helps her convert unarticulated, non-representational metacognitive knowledge into reflexive concepts. Metarepresentation would then result from an explicit, theoretically laden, recoding of metacognitive processing.

---

[43] This process of redescription is studied in more detail in Karmiloff-Smith 1992, Povinelli 2000, Proust 2003b, and in Pacherie and Proust in press.

Obviously, speaking of 'one' linguistic control structure is an idealization. For the linguistic capacity consists of a hierarchy of such control loops, and social communication has its own logic, pragmatics, and rules of relevance, independent of their linguistic expression. In addition, offering reasons taps on the 'personal' control loop that allows an agent to recognize herself as the same over time in terms of her own past and future plans along with their corresponding epistemic actions.[44]

Our idealization, however, seems justified insofar as what might work as the decisive factor for human rationality is the ability to couple two control levels, one having to do with actual execution, the other with social (linguistic) justification. This new control loop does not directly cause the action (physical or epistemic) at the lower level, but it provides an additional, stable and external representational format for storing the commands and analysing and storing the feedback. It also works, obviously, at its own level: it interprets incoming information concerning actions performed by self or others in folk-logical and folk-psychological terms, and generates new actions (in particular speech acts, but also other acts related to informational competition).[45]

Although I cannot develop this point here, the higher-level, speech-based control loop serves social needs. The essential features of human rationality are derived from the requirements of a social life where linguistic communication plays a major role. It is plausible that covert linguistic rehearsal of intentions has to do with internalizing verbal commands uttered by others; metarepresenting one's reasons to act in an explicit way may similarly be connected with social requirements in a Machiavellian world, such as proving one's good faith, or establishing one's reputation as a reliable ally and information carrier. Verbally expressed 'social knowledge', constitutive of folk psychology, might have been the most recent form of control, through which an agent is able to represent herself (to herself and to others) as rational, which as we have seen is a different matter from simply being rational.

12.4.4 Rationality without normativity

The present proposal raises a number of other issues, connected tightly to the social dimension of human rationality, some of which have been explored in Kim Sterelny's chapter (Chapter 14, this volume; cf. in particular his analysis of the status of folk logic). We will concentrate here on the concept of a norm; norms are often taken to be part and parcel of the concept of rationality. How does the definition of rationality as a disposition that tends to be realized by a cognitively-operated control system deal with this view?

One might claim that what makes rationality a matter of norms is that any cognitively operated control operation can be compared with the corresponding ideal

---

[44] Proust (2003c) discusses in the framework of control theory issues related to the constitution of a person that remains identical over time.

[45] See Dunbar 1997.

strategy, as determined by some *a priori*, mathematical reasoning, on the basis of the agent's utilities and costs (expressed in probabilities). In Signal Detection Theory, one speaks of a 'norm' of decision, that can be computed *a priori* in any noise and signal + noise probability distribution. Let us call this kind of normativity 'prescriptive normativity' (or P-normativity) as it suggests that, each time a cognitively operated control system is active, an optimal solution exists that the system *should* select. It seems, at least *prima facie*, that P-normativity so understood does not need to be restricted to agents who can indeed understand that their control operations can be compared to norms. As was shown above, strategies engaged in signal detection actually can be applied by agents who are: (i) not reflexive 'in an explicit way' (as we saw in Section 12.4.2); and (ii) neither equipped for, nor interested, in communicating their reasons to others. It therefore seems plausible that the P-normative character of rationality is a term that can apply to animal behaviour in general.

However attractive, this line of reasoning is contentious. It is indisputable that a behaviour may approximate more or less to a particular norm, and be regarded as more or less rational on that ground. But should we take the P-normative character of such an evaluation to be a natural inherent property of the rational behaviour under consideration? A natural property is one that derives from causal connections of a physical system with its environment, taken independently of an observer's attributional capacities. Could not P-normativity be rather a late and inessential feature of rationality, relying on specifically human metarepresentational capacities?

A well-known argument in favour of the view that rationality is naturally P-normative consists in invoking the additional fitness conferred on entities that are able to approximate ideal players. Ruth Millikan[46] has forcefully pressed the point that the notion of function brings with it the notion of normativity:[47] if having a property F helped the members of a reproductive family to proliferate (in relation to those that did not have it), and explains why it proliferated, then the bodily structure that implements this property is normatively taken to have F, that is its normal function; a heart *is supposed to* circulate blood because it is this effect that explains how and why a heart is present in a newborn. This point applies to cognitive functions as well as to other adaptations: normativity of content consists in the fact that the representational icons are *supposed to* map their referents; normativity of beliefs consists in the fact that they are *supposed to* indicate a state of affairs. Rationality, from this perspective, is naturally P-normative by virtue of the functional mechanisms of decision at work (either as explicit rules or innate decision mechanisms): observed strategies are 'supposed to' match (at least approximately, as heuristics normally do) ideal decision strategies.[48]

---

[46] Cf. Millikan 1993.

[47] Editors' note: also endorsed by Hurley, this volume.

[48] See Gigerenzer and Selton 2001.

One might still resist the view that this form of P-normativity is a natural one, that is inherent in the very nature of a biological function. What does it mean to say that the heart is supposed to circulate the blood, or that an animal is supposed to organize its foraging in a rational way? It just means that the heart, or the exploration/exploitation behaviour, have been selected for their consequences. But the selective value of an adaptation is relative to the biologist's explanatory concerns; it does not play any role in the causal structure in which the function is implemented. Rather, the causal effects of the physical structure drive its reproduction and selection. Functions are generally not properties that interact with other entities to cause events; they are historical supervenient properties (defeasible over time) that need a physical basis in order to become causal-explanatory.[49] A functional element taken in itself, in other words, does not have causal effects over and above the physical effects that it produces (history is generally not reflected in physics). The fact that an element has a function depends on its selective history, but its history carries no present causal weight.[50]

An objector might still insist that it is crucial for a rational system to be able to recognize that mistakes can occur.[51] Natural P-norms, in this view, are inherent in cognitive functions because an organism can only form mental representations if it has the capacity to misrepresent, for it is essential to representations that they be true or false.[52] The capacity to misrepresent presupposes that the animal has some practical way of recognizing its mistakes in perceptual categorization, and to correct them.

Now, as we saw above, control views of the mind can easily account for the practical capacity to correct mistakes. It is a condition of the dynamic coupling of a changing organism in a changing environment. Representational control views further hypothesize that what is corrected over time is an internal model of a given context. Such a model is falsified if an action based on it fails to produce the expected feedback. In that case, the model will be selectively modified as a function of the observed feedback. This correction has the minimal form of procedural reflexivity (see Section 12.4.2) that constitutes a precursor for full-blown reflexive belief revision.

The objector who claims that mistake correction is a source of inherent P-normativity, however, does not need only to show that speechless animals have revisable and implicitly reflexive internal models of action. She also needs to explain whether, and in what sense, such a capacity involves natural P-normativity. Again there is some suspicion that the

---

[49] It might be objected here (Susan Hurley, personal communication) that functional level properties can sometimes be causal explanatory while the physical properties on which they supervene are not, as for example, in the case of money. But money precisely belongs to a function - or rather to a control system - whose normativity is not natural, but rather established by convention.

[50] See Proust 1997, Ch. 7.

[51] See Hurley (2003, 2005, and this volume), who also discusses how the capacity to recognize mistakes could arise from a control perspective on the mind.

[52] See Dretske 1988.

kind of normativity involved in correction is not necessarily P-normative. It can be redescribed as an architectural fact about epistemic control systems; the system must adopt the means that are modelled through its prior interactions with the goal to obtain the expected feedback and reach the goal as intended.[53] The core of 'normativity' in control systems boils down to 'bending to constraints'. A control system, if viable at all, cannot choose which constraints have to be taken into account when acting. No type of action can be consistently successful if it does not appreciate the universal constraints that apply to each and every case, such as the signal-to-noise ratio, the intermodal coherence in perceived spatial properties, and the various ways of combining available information according to types and properties. The fact that any cognitive system is bound to reflect these constraints constitutes the closest *descriptive* equivalent for what is termed 'normativity' in the human case. Bending to constraints indeed is not necessarily P-normative. As Dretske (2000) puts it : 'Beliefs and judgements must be either true or false, yes, but there is nothing normative about truth and falsity'.[54] They would be essentially normative if false beliefs and judgements were necessarily and inherently bad; but they only are *described* as bad by an external observer. In sum: from a control theory viewpoint, representing correctly the environment, using an updated form of feedback, does generally serve the purposes of a cognitive control system, but purpose *per se* does not involve P-normativity.

How then does P-normativity appear in the evolution of rationality? The response is suggested by our former distinction between metacognition and metarepresentation. The explicit form of normativity required by P-normativity is only possible when the recognition that mistakes can occur becomes explicitly reflexive and publicly assessable. In this view, normativity is a feature of the interpretive framework through which an agent explicitly makes sense of her actions, and through which she justifies them to others. Normativity is engaged through the fact that agents explicitly interpret their conversational exchanges in terms of commitments—to truth, relevance, etc.[55] Let us briefly spell out these two fundamental dimensions of P-normative rationality.

First, norms do not exist at an implicit level. As long as a control model is constructed and used in an implicitly reflexive way, there is nothing more going on than using the architectural constraints of the informational variety rehearsed above. As we saw in Section 12.4.3, an explicit reflexive access to one's own practices cannot be secured in systems with no metarepresentational capacity. The existence of norms depends on having explicit ways of assenting to (or dissenting from) a particular control token.

---

[53] This argument is originally spelled out in Dretske (2000) and here adapted to control systems.

[54] Dretske 2000, p.247.

[55] This view on normativity is compatible with the observation that only a social environment can confer a determinate meaning to rules, and with the view that there are universal constraints (in reference, logical connections between propositions, semantic rules, or instrumental action) that allow a control system to work, if it is to work at all.

Norm-following behaviour presupposes a capacity for redescribing an instrumental practice as a system of rules. The evolutionary importance of having available a verbal redescription of implicit forms of control might precisely lie in the social co-ordination that is gained through explicit P-normativity.

Second, a rich interpretive conceptual framework is required for an agent to justify her actions, evaluate her deeds, and possibly correct her life plans. It relies in large part on metarepresentation and mind reading, (as well as on the larger executive capacities associated with redescription), and constitutes the core of folk psychology.[56] The mental–conceptual framework, with the set of justificatory associated practices, thus forms in turn a control structure that can be used to predict how good the justifications of others are (whether they are likely to co-operate) and to detect lies.

Evidence in favour of such a superposition of control systems in humans is provided by neurophysiology. As explained by Rolls (1999), there are two routes to control action; one is the explicit, conscious, and verbally reportable route followed by evaluative signals from the amygdala and the orbitofrontal cortex to linguistic structures; this route allows planning and conscious decision making. Folk psychology may find a proper domain for its explanations and causal impact in this route that reflects the requirements of social control. The other route, however, projects directly from the limbic structures to the basal ganglia and triggers implicit behavioural responses. Here folk psychology is at a loss. This route has a different time frame; it allows for rapid reactions based on the reward/punishment value of the stimuli, and it tends to ignore contextual and relational properties.

Both forms of control are aspects of human rationality. The direct route is shared with non-human animals, and does not involve any form of normative evaluation, but only efficient control procedures. Folk psychology on the other hand is normative by design; it is part of an explicit social control loop that regulates interpersonal behaviour and co-operation. This social control structure requires mental concept use and metarepresentation, which seems to exist only in verbal animals, that is human beings. Given that, as we saw, individual control and not justification is what matters for animal rationality, P-normativity is not inherent to it, although human beings have a hard time resisting the suggestion that it is.

12.5 **Summary**

This chapter tried to avoid some of the difficulties that a top–down approach to the evolution of rationality raises, by trying to identify the selective pressures that have shaped rationality throughout phylogeny. The environmental complexity hypothesis,

---

[56] We cannot deal here with the question whether mind reading is best understood in terms of theory or of simulation. A plausible view is that theory and simulation both play a role in metarepresentations used in mind reading. On this and related questions, see Proust 2002, 2003b, in press.

offered by Peter Godfrey-Smith (1996) and Kim Sterelny (2003b), was used as our starting point: behavioural flexibility has been crucial in the variable environments in which organisms have had to survive. As a result, a major selective pressure was exerted that favoured extracting information and creating alternative control procedures. The present chapter explored its implications for the evolution of rationality.

Two classes of adaptive control systems were discussed, according to the scope and nature of their representational means. Rationality was seen to emerge as a disposition realized by control systems that can adapt to changing circumstances and that rely on cognition to do so. Examination of the structure of informational control identified two major sources of evolutionary pressures on information users: one is how to decide where to locate the decision criterion (for detection and categorization) in new circumstances; the other is how to improve the signal to noise ratio, that is signal discriminability (for oneself and for friends), or how to impair it (for one's foes). Epistemic actions are an evolutionary response to these two problems. Epistemic actions are a new form of control through which an organism can bring about changes in the informational quality of its external or internal environments. We concentrated here on a subset of epistemic actions, metacognitive ones, that is actions that aim at restoring a reference level of informational quality in current information processes against the perturbations generated by external and internal noise. The relationship of metacognition and metarepresentation was discussed, and the different forms of reflexivity that they respectively exemplify were distinguished. A possible evolutionary scenario was suggested concerning how metacognition could have evolved into metapresentation through a redescription process. Finally, it was argued that normativity is not an inherent feature of rationality, but belongs rather to its latest metarepresentational development.

Acknowledgements

The research presented in this chapter is supported by the European Science Foundation EUROCORES programme 'The Origin of Man, Language, and Languages'. All my thanks to Susan Hurley for her very helpful comments on a previous version, to Sliman Bensmaia, Dick Carter, and Matthew Nudds who corrected my English and suggested welcome amendments on content. I am also very grateful to Jean-Pierre Aubin, Anne-Marie Bonnel, Ingar Brinck, and Gloria Origgi for useful comments. I thank Wolfgang Prinz and Asher Koriat for stimulating discussions on metacognition

References

Anderson, J. R. and Gallup, G. G. Jr. (1997). Self recognition in Sanguinus? A critical essay. *Animal Behaviour*, **54**: 1563–1567.

Ashby, F. G., ed. (1992). *Multidimensional Models of Perception and Cognition*. Hillsdale, NJ: Erlbaum.

Baron-Cohen, S. (1995). *Mindblindness, An Essay on Autism and Theory of Mind*. Cambridge: MIT Press.

Bermúdez, J. L. (2003). *Thinking Without Words*. New York: Oxford University Press.

Boyd, R., and Richerson, P. J., (1995). Why does culture increase human adaptability? *Ethology and Sociobiology*, **16**: 125–143.

Call, J., and Tomasello, M. (1999). A non-verbal theory of mind test: the performance of children and apes. *Child Development*, **70**: 381–395.

Cheney, D. L., and Seyfarth, R. M. (1990). *How Monkeys See the World*. Chicago: University of Chicago Press.

Conant, R. C., and Ashby, W. R. (1970). Every good regulator of a system must be a model of that system. *International Journal of Systems Science*, **1**: 89–97.

Decety, J., Perani, D., Jeannerod, M., Bettinardi, V., Tadary, B., Woods, R., Mazziotta, J. C., and Fazio, F. (1994). Mapping motor representations with PET. *Nature*, **371**: 600–602.

Dretske, F. (1988). *Explaining Behavior: Reasons in a World of Causes*. Cambridge: MIT Press.

Dretske, F. (2000). Norms, history and the constitution of the mental. In: *Perception, Knowledge and Belief: Selected Essays*, pp. 242–258. Cambridge: Cambridge University Press.

Dunbar, R. (1997). *Grooming, Gossip, and the Evolution of Language*. Cambridge, MA: Harvard University Press.

Evans, G. (1982). *The Varieties of Reference*. Oxford: Clarendon Press.

Frith, C. (1992). *The Cognitive Neuropsychology of Schizophrenia*. Hillsdale: Lawrence Erlbaum Associates.

Gigerenzer, G., and Selton, R. (2001). *Bounded Rationality: The Adaptive Toolbox*. Cambridge, MA: MIT Press.

Godfrey-Smith, P. (1996). *Complexity and the Function of Mind in Nature*. Cambridge: Cambridge University Press.

Godfrey-Smith, P. (2003). Folk psychology under stress: Comments on Susan Hurley's 'Animal action in the space of reasons'. *Mind and Language*, **18**: 266–272.

Hampton, R. R. (2001). Rhesus monkeys know when they remember. *Proceedings of the National Academy of Sciences (USA)*, **98**: 5359–5362.

Harman, G. (1976). Practical reasoning. *Review of Metaphysics*, **29**: 431–463.

Hauser, M. D. (1997). *The Evolution of Communication*. Cambridge, MA: MIT Press.

Heinrich, B. (2000). Testing insight in ravens. In: C. M. Heyes and L. Huber, eds. *The Evolution of Cognition*, pp. 289–309. Cambridge, MA: MIT Press.

Hurley, S. L. (2003). Animal action in the space of reasons. *Mind and Language*, **18**: 231–256.

Hurley, S. L. (2005). The shared circuits model: How control, mirroring, and simulation can enable initation and mindreading. Archived interdisciplines web forum on mirror neurons, at: www.interdisciplines/mirror/papers/5.

Inman, A., and Shettleworth, S. J. (1999). Detecting metamemory in nonverbal subjects: A test with pigeons. *Journal of Experimental Psychology: Animal Behavior Processes*, **25**: 389–395.

Jeannerod, M. (1999). To act or not to act, perspectives on the representation of actions. *Quarterly Journal of Experimental Psychology*, **52A**: 1–29.

Karmiloff-Smith, A. (1992). *Beyond Modularity: A Developmental Perspective on Cognitive Science*. Cambridge, MA: MIT Press.

Koriat, A. (1993). How do we know that we know? The accessibility model of the feeling of knowing. *Psychological Review*, **100**: 609–639.

Miller, G. F. (1997). Protean primates: the evolution of adaptive unpredictability in competition and courtship. In: A. Whiten and R. Byrne, eds. *Machiavellian Intelligence II, Extensions and Evaluations*, pp. 312–340. Cambridge: Cambridge University Press.

Millikan, R. (1993). *White Queen Psychology and Other Essays for Alice*. Cambridge: Bradford Books.

Moran, N. (1992). The evolutionary maintenance of alternative phenotypes. *American Naturalist*, **139**: 971–89.

Nelson, T. O., and Narens, L. (1992). Metamemory: a theoretical framework and new findings. In: T. O. Nelson, ed. *Metacognition, Core Readings* Boston: Allyn & Bacon. pp. 117–130.

Odling-Smee, F. J., Laland, K. N., and Feldman, M. W. (2003). *Niche Construction: The Neglected Process in Evolution*. Princeton: Princeton University Press.

Owings, D. (2002). The cognitive defender: how ground squirrels assess their predators. In: M. Bekoff, C. Allen, and G. Burghardt, eds. *The Cognitive Animal: Empirical and Theoretical Perspectives on Animal Cognition*, pp. 19–26. Cambridge, MA: MIT Press.

Pacherie, E., and Proust, J. (forthcoming). Neurosciences et compréhension d'autrui. In: L. Faucher, ed. *La Philosophie et les Neurosciences*. Montreal: Bellarmin.

Perry, J. (2000). Thought without representation. In J. Perry (ed.): *The Problem of the Essential Indexical and other Essays*, expanded edition, pp. 171–188. Stanford: CSLI.

Povinelli, D. J. (2000). *Folk Physics for Apes: The Chimpanzee's Theory of How the World Works*. Oxford: Oxford University Press.

Povinelli, D. J., and Vonk, J. (2003). Chimpanzee minds: suspiciously human? *Trends in Cognitive Science*, 7: 157–160.

Proust, J. (1997). *Comment L'esprit Vient aux Bêtes*. Paris: Gallimard.

Proust, J. (1999). Mind, space and objectivity in non-human animals. *Erkenntnis*, **51**: 41–58.

Proust, J. (2001). A plea for mental acts. *Synthese*, **129**: 105–128.

Proust, J. (2002). Are empirical arguments acceptable in philosophical analyses of the mind? In: U. Moulines and K. G. Niebergall, eds. *Argument and Analyse*, pp. 163–186. Paderborn: Mentis.

Proust, J. (2003a). Does metacognition necessarily involve metarepresentation? *Behavior and Brain Sciences*, **26**: 352.

Proust, J. (2003b). *Les Animaux Pensent-Ils?* Paris: Bayard.

Proust, J. (2003c). Thinking of oneself as the same. *Consciousness and cognition,* **12**: 495–509.

Proust, J. (in press). Agency in schizophrenia from a control theory viewpoint. In: W. Prinz and N. Sebanz, eds. *Disorders of Volition*. Cambridge: MIT Press.

Rolls, E. T. (1999). *The Brain and Emotion*. Oxford: Oxford University Press.

Searle, J. (1983). *Intentionality*. Cambridge: Cambridge University Press.

Shettleworth, S. J. (1998). *Cognition, Evoluton, and Behavior.* Oxford: Oxford University Press.

Simon, H. (1982) *Models of Bounded Rationality : Behavioral Economics and Business Organization*, vol. 2. Cambridge, MA: MIT Press.

Smith, J. D., Shields, W. E., Allendoerfer, K. R., and Washburn, D. A. (1998). Memory monitoring by animals and humans. *Journal of Experimental Psychology: General*, **127**: 227–250.

Smith, J. D., Shields, W. E., Schull, J., and Washburn, D. A. (1997). The uncertain response in humans and animals. *Cognition*, **62**: 75–97.

Smith, J. D., Shields, W. E., and Washburn, D. A. (2003). The comparative psychology of uncertainty monitoring and metacognition. *Behavioral and Brain Sciences*, **26**: 317–373.

Sober, E. (1994). The adaptive advantage of learning versus a priori prejudice. In: *From a Biological Point of View*, pp. 50–70. Cambridge: Cambridge University Press.

Sousa, R. de (2004). *Evolution et Rationalité*. Paris: Presses Universitaires de France.

Sterelny, K. (2003a). Charting control-space: A commentary on Susan Hurley's 'Animal action in the space of reasons'. *Mind and Language*, **18**: 257–266.

Sterelny, K. (2003b). *Thought in a Hostile World, The Evolution of Human Cognition.* Oxford: Blackwell.

Tomasello, Call, J., and Hare, B. (2003). Chimpanzees understand psychological states—the question is which ones and to what extent. *Trends in Cognitive Science*, 7: 153–156.

Tulving, E. (1985). Memory and consciousness. *Canadian Psychology*, 26: 1–12.

Whiten, A. (1997). The Machiavellian mind-reader. In: A. Whiten and R. Byrne, eds. *Machiavellian Intelligence II, Extensions and Evaluations*, pp. 144–173. Cambridge: Cambridge University Press.

Wolpert, D. M., Ghahramani, Z., and Flanagan, J. R. (2001). Perspectives and problems in motor learning. *Trends in Cognitive Sciences*, 5: 487–494.

Wolpert, D. M., Ghahramani, Z., and Jordan, M. I. (1995). An internal model for sensorimotor integration. *Science*, 269: 1880–1882.

Wolpert, D. M., Miall, R. C. and Kawato, M. (1998). Internal models in the cerebellum. *Trends in Cognitive Sciences*, 2: 338–347.

Chapter 13

Rationality, decentring, and the evidence for pretence in non-human animals

Gregory Currie

Abstract

I argue that pretence involves a capacity for what I call decentring. Because of its dependence on decentring, pretence constitutes an indication of rationality. I give a brief account of decentring, distinguish it from metarepresentation, and say something about the relations between pretence, deception and imitation. I make a suggestion about the kinds of evidence of pretence we should look for if we are considering whether a creature is capable of pretence. I consider whether Morgan's canon, or something like it, might help us in weighing the evidence. Finally, I suggest that the phenomenon of seeing-in may underlie pretence, and offer a speculation on the evolutionary history of the capacity for seeing-in.

13.1 **Two kinds of cognitive variation**

A creature may be capable of appropriate behaviour in a wide range of circumstances without our having much inclination to ascribe rationality to it. For that we need something more. Part of the more we need is evidence of a capacity to stand back, cognitively speaking, from the immediate environment—a point made by Ruth Millikan in her discussion of simulative modes of planning (Chapter 4, this volume). Pretence provides us with interesting examples of this capacity; in pretence, a creature may respond to the environment, but as it is transformed by imagination. Young children participating in a game of pretence will respond to the pouring of water over Teddy by taking a cloth and drying him; an appropriate response if ever there was one. But in fact it is merely a game; no water was poured, the container was empty, and

Teddy is not really wet, as the children well know. Their response is appropriate, but to the world as imaginatively transformed.[1]

In thinking about the cognitive underpinnings of this imaginatively mediated engagement with the world, some authors have reached for the idea of metarepresentation. George Botterill, Peter Carruthers, Alan Leslie and others have suggested that fundamental to understanding pretence is the capacity to represent thoughts and other mental states, our own and those of others. Only in this way, they suggest, are we able to explain the fact that people with autism, who suffer spectacular deficits of mind reading, also show absent or very impoverished pretend play.[2]

The capacity for metarepresentation signals, in any creature that possesses it, a significant degree of cognitive sophistication. Humans show spectacular facility with it, while the evidence for even glimmerings of it in other primates is thin. But metarepresentation is one dimension of cognitive variation. There are others, and one of them seems to me particularly important for both pretence and rationality. Think of metarepresentational capacity as something measured on a vertical axis; we often say that the transition from thinking 'P' to thinking 'She thinks that P' is an ascent. Our other axis is horizontal—at the origin you respond to the here and now as given from your perspective; shifts along this axis denote an increasing ability to represent the world in some other way: as it was for you yesterday, as it is for someone else now, as it might be for some character of fiction. Let us call this the capacity for decentring.[3]

Progress along this axis is, I suggest, central to pretence; the pretending creature represents the world, not as it is, but as it might be. And, conceptually at least, our two dimensions of variation are independent. In representing the world as it is given from your point of view I need not be representing your point of view.[4] So decentring need not involve metarepresentation. And if I represent the thoughts of a creature I'm currently confronted with, we have (or may have) metarepresentation without decentring.

----

[1] Vygotsky recognized pretence as a form of decentring: the pretending creature is guided 'not only by immediate perception...but by...meaning' (Vygotsky 1978, p.97). But Vygotsky's emphasis on the immediacy of perception does not seem quite right in view of the fact that honey bee dances and associated behaviours are elicited by conditions reflected in their past as well as their present experience: they normally dance just when they have recently travelled between the hive and a nectar source. Eliciting conditions can be richer than those describable in terms of immediate perception without the behaviour being decentred.

[2] See Leslie 1987, Botterill and Carruthers 1999. See also Whiten and Byrne 1991. For criticism see Currie and Ravenscroft 2002, Ch. 7.

[3] Susan Hurley has suggested, in her introductory remarks to the conference Rational Animals, that decentring helps to distinguish the rational creature from the merely intelligent one (see also the Introduction to this volume).

[4] Joëlle Proust (this volume) argues for a distinction between metacognition and metarepresentation. Decentring may be a form of metacognition in her sense; my claim is that it is not to be assimilated to metarepresentation.

Things conceptually distinct are sometimes closely connected in fact; perhaps creatures that do well on one of these dimensions tend to do well on the other. For example simulation theorists have argued that the mind-reading problem is much easier to crack if you can use your own mind to model those of your fellows than it is if you have to rely on developing a complex theory of behaviour and its mental causes. And modelling involves decentring: thinking about the world, not as it presents itself to you but as it presents itself to those others. Still, we do well to bear in mind the distinctiveness of these two forms of variation, and to look to special arguments and (where we can get them) empirical results to establish connections between them. This is not always recognized. Here is Robin Dunbar commenting, in an admittedly informal account of hominid cognitive evolution, on the capacity of humans to disengage from the immediate environment:

> The ability to step back from the immediacy of the world may be crucial in allowing us to assess the consequences of alternative courses of action. This, after all, is in effect what theory of mind is all about—the ability to step back from one's personal experience and imagine that the world could be other than it is, to imagine that someone else could have a false belief about the world. (Dunbar 2004, p. 68–69)

Twice Dunbar makes an unacknowledged shift from decentring to metarepresentation—from the idea of retreat from 'the immediacy of the world' to a point about 'theory of mind'; from imagining the world as other than it is to imagining someone with a false view of the world. A properly nuanced account of human cognitive evolution needs to keep these things distinct, if only to show how they interact.

I don't propose to sort out these interactions here. Metarepresentation has many advocates; here I urge the case for decentring as a significant and independent factor in the debate over animal cognition. Throughout I focus on the idea of pretence in primates. Recall the connection: A creature that pretends is one that responds to the world as that creature imaginatively transforms it; evidence for pretence is therefore evidence for decentring, and hence for something that helps build a case for seeing that creature as rational, if only to a degree.

13.2 **Behaviour indicative of pretending**

Creatures at play are not necessarily creatures that pretend. Puppies chase and grapple with each other, responding to and even using the topography of the landscape as they do so. They can be seen to be making appropriate use of a piece of terrain, using it as a place from which to spring, without our having to understand the landscape as having been transformed by them in any imaginative way: we need not think of the puppy as engaged in decentring, and his play is not, by my lights, pretence. What sorts of behaviour would give us strong grounds for thinking that some sort of imaginative transformation is going on? The point at which we are best placed to make a judgement about whether pretence is being engaged in is not the point at which, allegedly, the agent begins pretence.

For such behaviours can be very difficult to distinguish from mere imitation or from behaviour provoked because of confusion about whether the normal stimulus for such behaviour is present; a gorilla said to be pretending to 'nurse a baby' while holding a piece of bark may simply be responding to a suitably baby-like stimulus (Byrne 1995, p. 139). We do better to track the behaviour over time, looking for some later point where the pretence is elaborated.

Elaboration may take the form of a voluntary enrichment of the game, or it might consist in simply recognizing that something previously unnoticed is part of the game. Recognition is illustrated by an example of Kendall Walton's (Walton 1973). In a game of make-believe, where it is pretence that globs of mud are pies, it is facts about how many globs of mud there really are that determine how many pies there are according to the pretence, and hence what we are to do when it is time to 'share the pies'. There being n globs of mud makes it pretence that there are n pies, and its suddenly being discovered that there are two more globs of mud around than we thought might make it pretence that we should get an extra pie each. The resulting behaviour—sharing out quantities of wet soil—is seen as appropriate only when we acknowledge that this behaviour is a response to the (make-believe) discovery of pies.[5]

Enrichment can vary greatly in the extent to which it displays creative pretending. Here are two cases. In the first, minimally creative case, an experimenter simulates pouring tea into a cup that contains a lolly stick, saying 'Show me how you stir Teddy's tea with the spoon'. Young 2-year-old humans generally used the stick in a way appropriate to its role as a spoon, making a stirring motion, and even some older 1-year-olds did the same (Harris and Kavanaugh 1993, Experiment 3). Here the children produced a novel action—one not previously displayed in the game—and by doing so made it pretence that the tea is stirred. But they did so in response to a request that clearly indicated what action was required. More creative would be behaviour that took the game to a new stage, not implied in anything so far specified or enacted. Suppose children see a simulated act of squirting toothpaste on a toy pig's tail. They might enrich the game at this point by 'wiping' Pig's tail without any explicit instruction. Here recognition that Pig is 'dirty' goes with an intervention that gives the narrative a novel twist: Pig is now clean.[6] None of these behaviours would be easily explained as non-pretended acts of imitation (about which I shall say more in the next section). In humans, recognition and enrichment are

---

[5] How the pretend behaviour is brought about is, I think, an interesting question to which we have no clear answer. Elsewhere I have elaborated two models, one based on the idea that states of imagining drive the behaviour directly, and another based on the idea that the behaviour is the outcome of familiar belief-desire based practical reasoning, where the contents of the beliefs and desires themselves refer to pretence (Currie and Ravenscroft 2002, Ch. 6). The first model is attractive for explaining pretence in non-human animals, if any exists. See the discussion of Morgan's canon below.

[6] This is based on, but does not quite correspond to, Experiment 5 in Harris and Kavanaugh 1993.

evident in young 2-year-olds and sometimes even in younger children. It is probably the first clear indication of pretence proper.[7]

Are recognition and enrichment constitutive features of pretence? There are two ways to take the question. We can ask what is to count as pretend-behaviour, or we can ask what it is to be a creature capable of pretence, something which might, for accidental reasons, go with never engaging in any. Taking the question in the first way, I see three options. The strongest is to say that each episode of pretence must display one or both of recognition and enrichment; a weaker option is to say that, while no particular episode need display either in order to be pretence, some such episode must; the weakest option allows for the possibility that the creature exhibits pretence without ever showing either. The first option is surely too strong, since an agent may begin pretence that is then broken off before there is opportunity for recognition or enrichment. The second is certainly an uncomfortable stopping place; if any particular episode can be broken off, why cannot the same happen, improbably, to all such episodes? So I accept the weakest option, allowing that, while recognition and enrichment are occasions on which we are likely to get good evidence for pretending, neither is strictly required for pretending to take place.

Suppose we take the question in the second way. Should we say that a creature capable of pretence must be capable of either or both recognition and enrichment? Even this might be too strong a requirement. There is a sort of behaviour we find in humans that is only marginally voluntary or conscious and usually of very brief duration. Some examples of it have been highlighted in Velleman (2000) in his discussion of belief and pretence: enacting snatches of dialogue from an imagined conversation, or nodding in response to an imagined conversation. Perhaps a case can be made for saying that this is a minimal unit of pretence, on account of its being action (rather than mere reflex) in response to an imagined situation. But such behaviour need not go with a capacity to display recognition or enrichment. I don't seek to settle the constitutive question, in its capacities version. I'm content to make a purely evidential point; recognition and enrichment are the best kinds of evidence I know of that are indicative of genuine pretending.[8]

---

[7] Recognition and enrichment are capable of being observed in co-operative pretend play, but also in solitary and parallel play. Focusing on these aspects of play makes no assumptions about the degree of sociality there is in primate pretence, if it exists.

[8] Another indicator is a generally playful demeanour; Rakoczy and colleagues (2004) found that 3-year-old children distinguished between (otherwise behaviourally similar) instances of trying to pour water and pretending to pour it when the actor adopted a playful expression in the pretend case and a puzzled expression in the case of a failed attempt. (The actor also sometimes engaged in what I would call enrichment, as when he added sound effects to the pretend pouring.) But I am interested in pretence as it also might occur in deception, where we would not expect to find overt indications of playfulness. Certainly a playful demeanour can be at most evidence for pretence, even of a non-deceptive kind; it is not a constitutive feature.

13.3 **Evidence for pretence in non-human primates**

'Pretence' is often used in descriptions of animal behaviour. It will be as well to distinguish between three overlapping categories: imitation, deception, and pretence. I start with a general point about deception, and one about imitation.

While there is no consensus on this question, I take the view that deceptive behaviours in non-human primates can be explained without appeal to mind-reading skills.[9] A creature that hides food may be doing so on the basis of an association between dominants being around when he finds food, and that food being regularly taken away from him, and not because he thinks that he had better prevent the dominant knowing that he has food. You might conclude from this that non-human animals don't engage in deception, on the grounds that deception requires the intention to alter belief. I don't think that such an intention is strictly necessary for deception, though no doubt it is very useful. Behaviour may fairly be called deceptive when it is intentional, in the sense that it is directed towards some end desired by the creature concerned, and likely to achieve this end only by providing grounds for a false belief in another creature—a false belief not shared by the agent.[10] A mother rhesus macaque defends her son from attack by giving warning calls in a direction whence, apparently, no threat comes, with the result that the assailant breaks off his attack, peering in the indicated direction (Byrne and Whiten 1990, p. 28). On my account this is deception even though the behaviour is explicable in terms of the mother wanting to break off the chase, and believing that this behaviour will have the effect of breaking it off, without our needing to invoke beliefs about the mental states of conspecifics. But if the behaviour did not depend for its success on its bringing about changes to belief, or was not intentional, or was based on a mistaken belief that an attack was immanent, it would not count as deceptive. Deception requires first-order, not second-order, intentionality on the part of both deceiver and deceived.

It is important to distinguish between the imitation of behaviour and the imitation of goal-directed acts. It has been argued that the latter is almost exclusively a human capacity, and indeed the basis of almost everything that distinguishes us from other

---

[9] See Tomasello and Call 1997, Section 8.1.

[10] Without the final clause (that the false belief be not shared by the agent) we could not distinguish deception from behaviour based on a mistake: telling you that P when P is false is not deception unless I don't believe P myself. Developmentalists make a distinction between deception and sabotage (Sodian and Frith 1992). But the kind of behaviour in animals I am here labelling deception is not merely sabotage. In Sodian and Frith's sabotage case you lock the box so that your opponent can't get into it, while in their deception case you lie about whether it is locked. Frith and Sodian argue that children with autism are capable of sabotage but not of deception, which involves understanding the causes and consequences of false belief. Deception in my sense does not necessarily involve the intentional manipulation of belief, but it does involve behaving in a way that affects another's belief; sabotage does not depend on what the opponent believes. Thanks here to Hannes Rakoczy. See also, again, Joëlle Proust's discussion (this volume) of metacognitive versus metarepresentational capacities.

species (Tomasello 1999). The imitation of goal-directed acts requires a capacity for imaginative mental projection; one places one's self in the position of the target and comes to understand the motive for the behaviour. Here I shall be concerned simply with the imitation of behaviour, which has been defined as 'copying by an observer of a feature of the body movement of a model' (Heyes 2001, p. 253). Imitation of behaviour does not require even first-order intentionality, though it may require specialized learning capacities and experiences; there is good evidence of primate imitation only among chimps with experience of interaction with humans.[11]

Understood in these ways, imitation, deception and pretence can be pairwise co-occurrent: there is deceptive imitation, deceptive pretence, and imitative pretence. We can have all three (deceptive pretending that involves imitation), and we can have each on its own; which, if any, are exemplified in the behaviour of non-human animals is, of course, a contingent matter. I am particularly interested (for reasons that will be clear in the next section) in cases where activities that might be described as pretence are best seen as belonging to one of the other categories. Gorillas sometimes wipe or rub a surface, having seen humans doing this as part of a cleaning routine. This seems to be imitation rather than genuine pretence, for we do not see recognition or enrichment in the behaviour.[12] The behaviour is undifferentiated, and insensitive to details of the environment. But if we found the gorilla paying particular attention to places where dirt is likely to be, or, when the object is dropped during the operation, rubbing or wiping the part that had been in contact with the floor, we would have better reason to call this 'pretending to clean'.

Deceptive behaviours are particularly prone to be called pretence by investigators, but the tests I have proposed suggest that in many cases this is not an appropriate description. Go back to the case of the mother macaque. Is she pretending to detect a threat? It seems more likely that she simply reproduces a certain bit of behaviour, knowing how disruptive of conspecifics' activities this is likely to be. There is no evidence here of recognition or enrichment, and the behaviour is best explained as an appropriate response to the environment as it is. Suppose, however, that the behaviour extended into a narrative that included such things as apparent eye tracking across territory from whence the 'threat' came, as if watching a movement, or paying attention to a bit of suspiciously waving foliage that she could conveniently recruit to the pretence. Once again, we would then have better evidence for the attribution of pretence.

A puzzling case is that of eye-covering behaviour, observed in a range of primates. Anne Russon and colleagues have studied this behaviour, which includes 'blind' walking, running, spinning, hanging from a swing, exploring surfaces, wrestling, and merely sitting alone (Russon *et al.* 2002). Should we count this as pretence? Only, I think, if making sense

----

[11] See Heyes and Ray 2000.

[12] See Gomez and Martin-Andrade 2002, p. 264.

of the behaviour requires us to assume a make-believe transformation of the environment. This does not seem to be the case. On the contrary, eye-covering primates attempt to negotiate their way through the environment as it actually is. Against this it might be argued that they are behaving in a way that would be appropriate if they were seeing, which they are not. In that case we should count them as pretending to see. It is important to distinguish here between the idea that the creatures concerned imagine seeing their environment, and the idea that they negotiate their way through it with some assistance from mental imagery. A creature capable of using mental imagery—as primates may well be—need not be a creature capable of imaginative transformations in the sense relevant here. While the case is a genuinely puzzling one, I see no compelling reason to explain the behaviour by reference to pretending to see; it is at least as plausible that the creatures are simply negotiating their way through the environment without looking at it, perhaps because the resulting experience is challenging, interestingly novel, or both.[13] What does seem compelling is the idea that this is play. But as I have said, not all play is pretend play.

Do we find, among non-human primates, examples of the kind of recognition and enrichment that would be persuasive evidence of pretence? There are few examples, and those there are concern primates in human environments that have included language training and encouragement to imitate. In one noted case Vicki, a chimp raised with a human family, is said to have 'pulled' an invisible trolley through the house, seeking help from her adoptive mother in 'un-snagging it' from objects in its path—a remarkable exercise in enrichment (Hayes 1952, pp. 80–84).

Language greatly enhances a capacity to display recognition, through acts of reference that are appropriate to their objects only when imaginatively transformed. Koko, the signing gorilla, engages in some behaviour of this kind: she held a plastic toy alligator to her nipple and signed 'drink'; she put her finger into a toy lizard's mouth, pulled it out and shook her hand, signing 'bite'. (Matevia et al. 2002, pp. 289–291). Kanzi, a human-raised bonobo, and the most linguistically advanced non-human to date, appeared to be eating an (invisible) fruit, spitting out the pips, and signing 'bad' (Savage-Rumbaugh and McDonald 1988). This might count as enrichment, going from the pretence that there is fruit here to the pretence that the fruit is bad. Kanzi also shows what looks like recognition. He put a piece of (imaginary) food on the floor, grabbing it when someone reaches out to the spot; the grabbing seems to indicate awareness that something new is part of the game—the food is now about to be taken (Savage-Rumbaugh et al. 2001, p. 59). The best hypothesis currently seems to be that

---

[13] This is in line with the conclusion of Russon and colleagues; see Russon et al. (2002, p.254). Pretence is also a tempting description in the case of the play fighting of squirrel monkeys. It is noticed that dominance relations, while not absent in play, tend to be reduced (Biben 1998). Are dominant players pretending to be less dominant than they are? It seems preferable to describe this in terms of behaviour appropriate to the actual situation: self-handicapping makes for a more interesting bout.

pretence is like imitation and language, but unlike deception; beyond the resources of non-human primates except in artificially enriched environments.

13.4 **Pretence and Morgan's canon**

Am I being unreasonably cautious about attributions of pretence? There is some reason, I shall suggest, for caution, though it is easy to be confused about its nature. A natural idea is that pretence is one of those 'higher mental processes' that Morgan warned us not to attribute where it is possible to attribute a lower one.[14] Morgan's own understanding of 'higher' and 'lower', together with his evolutionary argument for the validity of his canon, are rather suspect. However, there is a sense in which pretence *is* a 'higher' mental process, and one we should be cautious in attributing to non-human primates, given what we see elsewhere of their behaviour. Elliott Sober suggests that we understand 'X is higher than Y' to mean:

> The behavioural capacities entailed by X properly include the behavioural capacities entailed by Y (see Sober, 1998, p. 236).

Higher capacities, in this sense, enable a creature to do more. To take an example of Sober's, we might explain the broken wing display of piping plovers by saying that they want to protect their young and believe that the display will have that effect by drawing the predator away. Or we might explain it metarepresentationally, by saying that they want to protect their young and believe that the display will induce a false belief in the predator. Morgan's canon tells us to prefer the first explanation. Why? Not, according to Sober, because we should assume that behaviour capable of being caused by first-order (i.e. non-metarepresentational) states really is caused in that way.[15] The reason is rather that the plover does not (as best we know) have at its disposal other behaviours that *only* metarepresentational states would generate. Thus construed, the canon amounts merely to the observation that we should not prefer an explanation in terms of a higher capacity as long as: (i) an explanation in terms of a lower one is available; and (ii) we have no reason to think that the creature has in its repertoire other behaviours explicable only in terms of the higher one.

How is it with pretence? I am not here appealing to metarepresentation in my account of pretence, so I am not going to treat this as strictly parallel with the plover case. But a similar argument is available. Recall that, in my terms, genuine pretending depends

---

[14] Morgan stated the canon in Morgan (1894). Against Groos (Groos 1898), Morgan urged that putative examples of pretence be explained without recourse to 'the conscious self-illusion of make-believe' (Morgan 1900, p. 281).

[15] Morgan (1894) held that 'In no case may we interpret an action as the outcome of the exercise of a higher psychical faculty, if it can be interpreted as the outcome of one which stands lower in the psychological scale'. Sober shows how certain facts about phylogenetic groupings can make it more probable that a species possesses the higher and the lower faculty than that it possesses the lower only.

on imaginative transformation of the world; the pretending agent responds to the world, not as it is, but as it is imagined. The capacity to respond to the world as it is imagined to be (A) is higher, in the Morgan–Sober sense, than is the capacity to respond to the world as it is (B). A creature able to respond to the world as it imagines it to be can respond to the world as it is, if only because the way the world is is surely one of the ways you can imagine it being. But the converse implication does not hold; many creatures, it is fair to assume, respond appropriately to the way the world is, and have no capacity to transform it in imagination.

Now we observe surface-rubbing in gorillas, or closed-eye play in orang-utans and we have the choice, so I have argued, of explaining this by reference to A, or by reference to B. Only by explaining it in terms of A do we count the behaviour concerned as pretence. But we ought not to explain it in this way *if* we can just as well explain it in terms of B, and assiduous investigation does not reveal other behaviours explicable in terms of A but not in terms of B. This highly conditional principle, and the facts that support its antecedent clauses, are all that I am invoking in support of the presumption against attributions of pretence.

13.5 **Pretence and seeing-in**

Walton, in describing recognitional behaviour, speaks of rules at least tacitly understood by game-players. Rules, tacit or not, have their place in many examples of pretence; the children playing mud pies acknowledge a rule according to which the number of pies is determined by the number of locally available globs of mud. But I doubt that they are a necessary feature of pretence. Saying that game-players understand rules suggests that they grasp propositions; indeed, one might take recognition of what is pretence itself to be a matter of grasping a proposition, though one imagined rather than believed. However, I think that some behaviours suggestive of a capacity for decentring may be driven by cognitively less sophisticated mechanisms. One consequence of this is that there may be a place for the idea of 'rationality in perception'.[16]

Aestheticians have drawn attention to the phenomenon of 'seeing-in'—seeing a woman in a picture or a face in the clouds.[17] Such seeing-in does not involve seeing a woman, nor does it involve the perceptual illusion of seeing one; neither is it a case merely of judging that the picture represents a woman; it is a genuinely perceptual phenomenon. Seeing-in

---

[16] This goes against some influential views about rationality. According to Donald Davidson, for example, to be rational just is to be the possessor of propositional attitudes (Davidson 1982). But Ruth Millikan also suggests that 'a certain kind of rationality may occur on the level of perception prior to cognition' (Ch. 4, this volume). Millikan emphasizes the role of looking for and seeing affordances in the environment. Though I have no space to develop the thought here, it seems to me that this might be a kind of seeing-in. (On play as the exploitation of affordances see Tomasello, 1999, Ch. 3.)

[17] See the account in Wollheim 1986, Lecture 2.

seems to be fast, mandatory, encapsulated, very little dependent on learning, and subject to characteristic impairment. Try *not* seeing a person in the picture next time you look at a painted portrait—or, for that matter, a photograph, or a televisual or cinematic image. The selective impairments we find in object recognition tend to reproduce themselves in perception of depictions of objects, so a good test of whether someone can no longer recognize faces is whether he can no longer recognize *pictures* of faces.

What is the relevance of this to pretence in non-human primates? Though I have been discouraging on the extent of pretence in these creatures, we have seen indications that, in the right sorts of environments, some of them are capable of it; and we have reason to believe that some of them are capable of seeing things in pictures. It is, of course, important to distinguish between the so-called 'confusion' and 'equivalence' models of this capacity. According to the former, creatures respond selectively, and in some ways appropriately, to the picture because they confuse it with its subject; according to the latter they respond selectively because, without confusing the two, they see the object *in* the picture.[18] Comparisons between reactions to cut-outs and whole pictures of objects (with the background included in the picture) suggest that baboons treat the pictures as equivalents, but confuse the cut outs with real objects.[19] The best evidence of a seeing-in reaction rather than simply confusion comes, again, from those home-reared great apes like Sarah, Vicky, and Koko, who seem to be able to sort pictures, including line drawings, by what they depict.[20] Chimps occasionally show a limited capacity to use pictures to find objects in the corresponding real environment.[21]

The next step is to suggest that, just as we can see things in pictures, we can see things in simple mimetic acts. When someone moves in a certain way we can see in their movements such acts as driving a car, hitting a cricket ball or nursing a baby; yet the person concerned need not really be doing any of these things. Moreover, it can be evident that he or she is not; there is no illusion of the reality of these events, just as there is no illusion of the presence of a woman when we see a woman in the picture. The movements might be exaggerated or stylized, but we can still see the action in the performance, just as we see a well-known face in its caricature. And a further generalization is in order, this time across sensory modalities. I can see Baroness Thatcher in a cartoon caricature, and I can hear her in any number of performances that lampoon her voice. (I choose caricature here because hearing a recording of Baroness Thatcher seems to be a case of

---

[18] Fagot, Martin-Malivel *et al.* 1999. Commenting on the capacity of very minimal resemblance in a substitute object to trigger behaviour appropriate to the resembled thing (parent, prey), Gombrich seems to suggest that confusion alone draws animals into the realm of 'images', though he is careful to use scare-quotes around the label (Gombrich 1963, p.6).

[19] Bovet and Vauclair 1998.

[20] Hayes and Nissen 1971, Premack and Woodruff 1978, Savage-Rumbaugh *et al.* 1978, Matsuzawa 1990, Itakura 1994, Patterson 1978.

[21] Boysen and Kuhlmeier 2002.

simply hearing her, and not of hearing her *in* anything.) It also generalizes to our awareness of action. In addition to being able to see an act of dancing-with-a-partner in a mime artist's solo performance, the artist herself can experience this act in her own performance. She does not (typically) do this by seeing herself in a mirror, but through her first-person sense of her own bodily action.

My suggestion is that seeing-in (or, more generally, experiencing-in) is an important aspect of pretence, of the recognition of pretence in others, and of the displays of recognition and enrichment that I have described above.[22] Seeing-in may constitute part of the primitive basis of pretence; it enables pretence to be enacted and communicated without the necessity for full-blown conceptual capacities. The child who sees water pouring in the act of the adult who tips up the cup will spontaneously and unreflectively pretend that the animal thus poured on is wet, and the same child will see the cup that has been tipped up as now empty. If non-human animals engage in pretence at all, it is may be that their pretence is similarly structured.

If pretence and depiction both depend on seeing-in we may suspect that there will be developmental relations between them, though it is hard, given the kinds of evidence available at present, to assess the extent to which this is so in animals. (While primates may see things in depictions they do not *produce* depictions.) Lillard and colleagues found that young children are willing to describe a troll character Moe as 'pretending to hop like a rabbit' even though Moe does not know anything about rabbits.[23] The result of Lillard's experiment was generally thought surprising. This made me wonder about an analogous experiment with depiction: We tell the children that Moe knows nothing about rabbits. However, Moe draws something that looks very rabbit-like. Would the children then say that Moe was drawing a rabbit?

As it turns out, Lillard and colleagues have done the experiment.[24] In their study 75 per cent of 4-year-olds said that Moe was drawing a rabbit, which is exactly the same as the proportion who, in the earlier experiment, said that Moe was pretending to be a rabbit. Both sets of results would be explained if we thought that the views of young children about what is pretended/depicted were dependent on what they found themselves seeing-in those acts of pretence and those depictions.

In an essay already overloaded with speculation I'll close with another. If we think of pretence as closely related to seeing-in, we may have the beginnings of a story about the biological evolution of make-believe.

Those behaviours that I called deceptive are ones which involve first-order intentionality. Deceptive behaviour emerges from a much longer evolutionary history of

---

[22] Hannes Rakoczy suggested to me that it also plays a role in co-operative endeavours such as the use of gestural communication.

[23] See Lillard 1993.

[24] Richert and Lillard 2002.

signalling, much of which has the proper function of manipulating the behaviour of other creatures, including conspecifics (Krebs and Dawkins 1984). For example a creature 'plays dead', or is disguised as some other, more dangerous creature. What I am calling deception is a certain stage in this process whereby the deceptive signal functions, not to manipulate behaviour directly, but to manipulate belief—and hence potentially to influence a whole range of behaviours.

We would expect that the evolution of so advantageous a trait as a capacity to produce manipulative signals would be countered by the evolution of progressively better ways of dealing with them. How might this happen? Consider manipulative signalling along the lines of eye-shaped patches on a butterfly's wing; if the signal works, the predator's perceptual content will be 'dangerous creature up front' (or something of the sort). If it does not, the idea of a dangerous creature fails to register at all. Now, suppose we consider intelligent and highly social creatures, such as many non-human primates are, and as our own ancestors millions of years ago presumably were. If you are one of these creatures, your conspecifics are likely to put on a variety of deceptive performances; performances that will attain their objects by getting you to believe something false. It will be a good thing, from your point of view, if you can avoid being taken in. In addition, it would be good from your point of view to be able to register the performance as deceptive. You may then be able to apply sanctions that will discourage future deception, or be generally more sceptical of that agent's behaviour in future. For that you need to be able to do something that the butterfly's predator presumably cannot do—to see a warning gesture *in* the behaviour, without treating this *as* a genuine warning.

13.6 **Conclusion**

Intelligent creatures vary in the degree to which they are constrained by their current environment. Behaviour which is best described as a response, not to how things are but to how they are imagined to be, is evidence for a high degree of independence in this regard; such decentring may be a sign that the creature concerned is to some degree rational. Evidence for such behaviour in languageless creatures is hard to come by, and there are good reasons for treating any claims to have found it with some scepticism. I have focussed on evidence of a particular kind of decentring, pretence, and of two sorts of actions within the scope of a pretence, recognition and enrichment. Recognition is the discovery of a make-believe state of affairs; enrichment is the exploitation of one piece of make-believe to create another. Certain kinds of behaviours, in a context where pretence is suspected, would be strongly indicative that we were seeing recognition or enrichment. In fact, almost no evidence of either kind is available for non-human primates, and the little there is, is confined to a few individuals much exposed to human language and encouraged to imitate. Given this failure to find convincing evidence for pretence, it makes sense to look for precursor states in our closest relatives. One is likely to be a capacity for seeing-in, a notion which I have generalized somewhat from the account given of it by aestheticians. I have suggested that this capacity for seeing-in was driven partly by an arms-race between deception and deception-detection.

Acknowledgements

Thanks to Paul Noordhof and Shaun Nichols for comments on a much earlier version, to Angeline Lillard for discussion of pretence, and to Penny Patterson for an exchange about Koko. Susan Hurley and Matthew Nudds—exemplary editors—forced me to think and, I hope, to write, more clearly. Special thanks to Hannes Rakoczy for many insightful comments on a close to final draft. An earlier version of some of this material appeared in a chapter entitled 'Pretence and rationality' in my *Arts and Minds* (Oxford: Oxford University Press, 2004). It has been revised for the present volume.

References

Biben, M. (1998). Squirrel monkey play fighting: making the case for a cognitive training function for play. In: M. Bekoff and J. Byers, eds. *Animal Play*. Cambridge: Cambridge University Press.

Botterill, G., and Carruthers, P. (1999). *The Philosophy of Psychology*. Cambridge: Cambridge University Press.

Bovet, D., and Vauclair, J. (1998). Functional categorization of objects and their pictures in baboons. *Learning and Motivation*, **29**: 309–322.

Boysen, S., and Kuhlmeier, V. (2002). Representational capacities for pretense with scale models and photographs in chimpanzees (*Pan troglodytes*). In: R. Mitchell, ed. *Pretending and Imagination in Animals and Children*, pp. 210–228. Cambridge: Cambridge University Press.

Byrne, R. (1995). *The Thinking Ape*. Oxford: Oxford University Press.

Byrne, R. and Whiten, A. (1990). "Taclical Deception in Primates: the 1990 database", *Primate Report*, **27**, 1–101.

Currie, G., and Ravenscroft, I. (2002). *Recreative Minds: Image and Imagination in Philosophy and Psychology*. Oxford: Oxford University Press.

Currie, G. (2004). *Arts and Minds*. Oxford: Oxford University Press.

Davidson, D. (1982). Rational Animals. *Dialectica*, **36**: 317–327.

Dunbar, R. (2004). *The Human Story*. London: Faber and Faber.

Fagot, J., Martin-Malivel, J., and Dépy, D. (1999). What is the evidence for an equivalence between objects and pictures in birds and nonhuman primates? *Cahiers de Psychologie Cognitive*, **18**: 923–949.

Gombrich, E. (1963). *Meditations on a Hobby Horse*. London: Phaidon.

Gomez, G., and Martin-Andrade, B. (2002). Possible precursors of pretend play in captive gorillas. In: R. Mitchell, ed. *Pretending and Imagination in Animals and Children*, pp. 255–268. Cambridge: Cambridge University Press.

Groos, K. (1898). *The Play of Animals*. New York: D. Appleton and Co.

Harris, P. L., and Kavanaugh, R. D. (1993). Young children's understanding of pretense. *Monographs of the Society for Research in Child Development*, **58** (1, Serial No. 231).

Hayes, C. (1952). *The Ape in our House*. London: Victor Gollancz.

Hayes, C., and Nissen, C. (1971). Higher mental functions of a home-raised chimpanzee. In: A. Schrier and F. Stollnitz, eds. *Behavior of Nonhuman Primates*, p. 4. New York: Academic Press.

Heyes, C. M. (2001). Causes and consequences of imitation. *Trends in Cognitive Science*, 5: 253–261.

Heyes, C., and Ray, E. (2000). What is the significance of imitation in animals? *Advances in the Study of Behaviour*, 29: 215–245.

Itakura, S. (1994). Recognition of line-drawing representations by a chimpanzee. *Journal of General Psychology*, 121: 189–197.

Krebs, J., and Dawkins, R. (1984). Animal signals: mind-reading and manipulation. In: J. Krebs and N. Davies, eds. *Behavioural Ecology: an Integrated Approach*. Oxford: Blackwell Scientific.

Leslie, A. (1987). Pretense and representation: the origins of 'theory of mind'. *Psychological Review*, 94: 412–426.

Lillard, A. (1993). Young children's conceptualization of pretense: Action or mental representational state? *Child Development*, 64: 372–386.

Matevia, M., Patterson, P., and Hillix, W. A. (2002). Pretend play in a signing gorilla. In: R. Mitchell, ed. *Pretending and Imagination in Animals and Children*, pp. 285–306. Cambridge: Cambridge University Press.

Matsuzawa, T. (1990). Form perception and visual acuity in a chimpanzee. *Folia Primatologica*, 55: 24–32.

Morgan, C. L. (1894). *An Introduction to Comparative Psychology*. London: Walter Scott.

Morgan, C. L. (1900). *Animal Behaviour*. London: Edward Arnold.

Patterson, F. (1978). Conversation with a gorilla. *National Geographic*, 154: 438–465.

Premack, D., and Woodruff, G. (1978). Does the chimpanzee have a theory of mind? *Behavioral and Brain Sciences*, 4: 515–526.

Rakoczy, H., Tomasello, M., and Striano, T. (2004). Young children know that trying is not pretending—a test of the 'behaving-as-if' construal of children's early concept of pretence. *Development Psychology*, 40: 338–399.

Richert, R., and Lillard, A. (2002). Children's understanding of the knowledge prerequisites of drawing and pretending. *Developmental Psychology*, 38: 1004–1015.

Russon, A., Vasey, P., and Gauthier, C. (2002). Seeing with the mind's eye: eye-covering play in orangutangs and Japanese macaques. In: R. Mitchell, ed. *Pretending and Imagination in Animals and Children*, pp. 241–254. Cambridge: Cambridge University Press.

Savage-Rumbaugh, E., and McDonald, K. (1988). Deception and social manipulation in symbol-using apes. In: R. Byrne and A. Whiten, eds. *Machiavellian Intelligence*. Oxford: Clarendon Press.

Savage-Rumbaugh, E., Rumbaugh, D., *et al.* (1978). Sarah's problem in comprehension. *Behavioral and Brain Sciences*, 1: 555–557.

Savage-Rumbaugh, S., Shanker, S., and Taylor, T. (2001). *Apes, Language, and the Human Mind*. New York: Oxford University Press.

Sober, E. (1998). Morgan's canon. In: D. Cummins and C. Allen, eds. *The Evolution of Mind*. Oxford: Oxford University Press.

Sodian, B., and Frith, U. (1992). Deception and sabotage in autistic, retarded, and normal children. *Journal of Child Psychology and Psychiatry*, 33: 591–606.

Sterelny, K. (2003). Charting control-space: a commentary on Susan Hurley's 'Animal action in the space of reasons'. *Mind and Language*, 18: 257–266.

Tomasello, M. (1999). *The Cultural Origins of Human Cognition*. Cambridge, MA: Harvard University Press.

Tomasello, M., and Call, J. (1997). *Primate Cognition*. New York: Oxford University Press.

Velleman, D. (2000). *The Possibility of Practical Reason*. Oxford: Oxford University Press.

Vygotsky, L. (1978). *The Mind in Society*. Cambridge, MA: Harvard University Press.

Walton, K. (1973). Pictures and make-believe. *Philosophical Review*, **82**: 283–319.

Whiten, A., and Byrne, R. (1991). The emergence of metarepresentation in human ontogeny and primate phylogeny. In: A. Whiten, ed. *Natural Theories of Mind*. Oxford: Blackwell.

Wollheim, R. (1986). *Painting as an Art*. London: Thames and Hudson.

Part IV

Social behaviour and cognition

Chapter 14

Folk logic and animal rationality

Kim Sterelny

Abstract

In this paper I argue that there are two different strategies for thinking
about rationality, both of human and of non-human agents. We could
treat rationality as a measure of overall cognitive efficiency: of the
capacity of an agent to respond adaptively to the challenges its
environment poses, even when adaptive response to those challenges
must take into account quite subtle or recondite features of that agent's
circumstances. Alternatively, we could develop a narrow conception of
rationality: rational thinking is an evolutionary response to the dangers of
misinformation and deception in social environments in which agents rely
heavily on information provided by others. In developing some ideas of
Dan Sperber, I argue that folk logic is a response to this specific feature of
human environments, and that theorizing about this narrow conception
of rationality is more likely to be fruitful than theorizing about rationality
as overall cognitive efficiency. However, the positive thesis about the
evolutionary origins of folk logic is independent of the sceptical, negative
thesis about rationality as cognitive efficiency. The speculation about folk
logic could be wrong while the sceptical thesis is right, and *vice versa*.

14.1 **Rational animal agents?**

Animals often succeed in acting adaptively in their environments, choosing means that
are admirably suited to their biological ends. Surprisingly often, such agents must be
sensitive to subtle features of their environment in solving the problems with which
their environments confront them. To meet their biological challenges, these agents
must have the capacity to use the information their environments make available to
them. An important recent collection (Bekoff *et al.* 2002) makes clear that such
informational sensitivity is widespread in many animal clades. Donald Owings, for
example, reviews his work on the problems predation poses to ground squirrels and
discusses the ways such squirrels trade the immediacy of danger against their need to
assess and respond to threat. Raptors pose a very urgent threat, and hence their
predominant response is simply to take cover. Snakes pose a more enduring but less

urgent danger. Adult squirrels need to assess the level of threat, and rather than simply flee, they respond by harassing the snake, warning their pups, probing the snake's capacity to strike, and so on. Owings shows that these decision problems are not simple, for the threat depends on the size, temperature, and species of the snake as well as the terrain. No simple response—like that of taking cover against raptors—will do, because the threat is both variable and enduring (Owings 2002). The last 30 years of study of animal agents in both field and laboratory has made it clear that there is no simple dichotomy between rational, informationally sensitive human agency and the mere blind, mechanical blunderings of the brutes. So it is natural to consider the extent to which animals are rational intentional agents (Dennett 1996; Hurley 2003 and Chapter 6, this volume).

It is indeed important to identify the rich variety of systems for the adaptive control of behaviour, rather than squeezing this richness into a few boxes. We need to recognize both the variety of systems for the cognitive control of adaptive behaviour and to chart the relationships between such systems. But I shall argue that these projects are not best pursued by asking about the extent of animal rationality. The argument develops in three stages. The first outlines a picture of the selective regimes that drive the evolution of the sophisticated use of information by animal agents. The second argues that hominid cognition has evolved in response to a somewhat different set of challenges and that (as a consequence) the transmission of social information and skill has come to be both a critical and an unusual feature of hominid selective and developmental environments. The third draws upon the ideas of Dan Sperber and others in arguing that the social transmission of information introduces (or makes much more important) a vetting problem. I shall suggest that we should see rationality as an evolved response to this vetting problem.

14.2 **Information, detection, action**

Peter Godfrey-Smith has developed a general framework for the analysis of the use of information to control behaviour (Godfrey-Smith 1996). Consider, for example, one of the standard illustrations in behavioural ecology: the foraging behaviour of starlings (Kacelnik 1984). When starlings bring food back to their young, the default expectation is that they maximize their fitness by maximizing their rate of food delivery. For better-fed chicks fledge earlier and heavier, and have better survival prospects. But starlings cannot maximize their rate of food delivery by following a simple behavioural rule of (say) hunting until their beak is full and then returning to their nest. For God neglected to provide them with handy pouches in which to carry their prey. Since starlings must keep what they have caught in their beak, their efficiency declines as they capture more. Hence they must trade the cost of declining efficiency against transit costs. A starling that varies its load in response to variation in its environment—in particular, in response to the distance between its foraging site and its nest—will do better than a starling that always takes the same load home.

When travel costs are low because the bird is foraging close to its nest, a lighter load is more efficient. When travel costs are high, it is worth paying the cost of foraging longer and carrying more per load. This is the key theme in Godfrey-Smith's analysis: information is important when behaviour that is not sensitive to variation in the environment is punished.

Actual starling behaviour seems to be flexible in this way; in experimental contexts, they succeed in making the right trade-offs between hunting efficiency and travel costs. Such nuanced behaviour depends on the ability of the agent to register and use information about the relevant feature of its environment. If starlings find their way back to their nest just by flying high and recognizing a landmark near their nest, they will not know how far they are from home. To take advantage of the economies of fitting its load to the distance it is to travel, the starling must remember how far it is to the nest. But no detection and response system is perfect. Hence, as Godfrey-Smith emphasizes, *there is a cost of error*. Optimizing starlings rely on the accuracy of their spatial representation of their environment. If their memory for distance often fails them, they will find themselves carrying an absurdly light load a long way. They will do worse than a starling that always carries an average load—one that would be optimal for the mean distance it travels. The economic advantage of optimizing the load to specific circumstances must outweigh the costs of error. Thus as Godfrey-Smith frames these issues, there will be selection for mechanisms that vary behaviour in response to variation in the world only if:

1. The organism's environment varies in ways that matter to that organism.
2. The organism has relevant variation in its repertoire; different actions have different payoffs in different environments.
3. The organism has access to information about its environment.
4. The benefit of optimizing behaviour to the specific state of the environment outweighs its costs.

However, not all information is created equal. Some decision problems require information that is readily available and easily used, given the inherited perceptual and cognitive equipment of the clade. Other informational environments are more challenging. The analysis of animal behaviour will often be misleading if the costs of error, and other aspects of the cognitive burden of the behavioural model, are ignored. Some decision situations impose only a light cognitive load on the actions of an agent. Starling foraging, at least in controlled experimental circumstances, is probably such a case. The starling decision model has a low cognitive load: the information required for correct strategy choice is minimal and it is potentially available to starlings as a by-product of their routine activity. These starlings gathered only one kind of food; they were sensitive only to one cost (time); and their fitness was maximized by maximizing a fairly simple and immediate physical quantity: the rate of food delivery.

If all this is right, there are unlikely to be serious cognitive constraints on starling strategy choice. But this is by no means always the case. Academic legend has it that in circumstances where co-operation is advantageous; where there is a risk that one of the co-operators will gain more than their fair share; and when the agents in question repeatedly interact with one another, Tit-for-Tat is a robustly successful strategy. And so it is, if the costs and reliability of information are ignored. Tit-for-Tat depends on the ability of each player to recognize whether others have co-operated. Yet in many social circumstances, signals that another agent has defected are unreliable, expensive, or both. Moreover, it takes a surprisingly low error rate to undermine co-operation between those playing Tit-for-Tat (Sigmund 1993).

In *Thought in a Hostile World*, I have argued that the informational characters of environments vary in ways that bear on this explanatory schema (Sterelny 2003). Some environments, though heterogeneous, are informationally transparent.[1] They offer the agent a reliable and discriminable cue to the relevant world state. In such environments the costs of information are low, and agents will be able to find their optimum strategy. Informationally transparent environments vary in ways that are relevant to our agent, but there is a reliable (enough) cue of that variation available to that agent. If the environment is informationally benign, the cognitive problems posed by environmental heterogeneity are tractable, and agents will often succeed in choosing an optimal strategy. Many shorebirds are migratory, breeding in the high arctic summers, taking advantage of the flush of invertebrate life of those summers, then flying south to winter. Such birds face an optimization problem. They must choose both the right time to arrive at their breeding grounds, and also the right time to leave. Linger too long, and they risk being caught in adverse weather. Leave too early, and they waste some of the resource boom of the arctic summer. No signal in these circumstances will be perfectly reliable. But there is a good correlation between weather and day length. The environment makes this information available for free, and we would expect over time (evolutionary time, in this case) each wader species to leave at the right time for it; the time that optimizes the trade-off between resources and risks.

However, many agents live in environments that are less tractable. Sometimes the information an agent would like is simply not available to it at all. And often environments are translucent. Information is available, but it is of lower quality, hence making error more likely, or it is more expensive, or both. In translucent environments, information and its costs cannot be ignored. These constraints on access to information are often the result of hostile biological interactions, for these degrade an agent's

---

[1] Strictly speaking, various aspects or domains of an environment are transparent or translucent, rather than an agent's environment as a whole.

informational environment, making the task of selecting the right action more difficult. And, of course, in hostile environments, the cost of error can be very high.[2] Competitive interactions, interactions between predator and prey, and those between parasite and host, degrade agents' informational environments. They do so in three ways: (i) by pollution; (ii) by agent-sensitive responses; and (iii) by increasing the cost of epistemic action.

14.2.1 Pollution

Often agents can determine the appropriate response to its current circumstances using a single cue. Much classic ethology of Tinbergen and Lorenz reported such cue/response systems. Herring-gull chicks' food begging is cued by a single and simple visual cue. Hostility creates informationally darker environments, environments in which reliance on a single cue is risky. Prey and predators hide, thus creating false negatives. And they also disguise themselves. Thus hostility accentuates the discrimination problem. There are differences between a mere tract of grass and scrub, and a tract of grass and scrub with a lion, but the lion's tawny skin makes those differences less easy to detect. A potential lion-meal must look longer, or from closer, to pick up the subtle differences in shape and shading between the two scenes. The lion will not 'pop out' in the way a lion with warning colours would—say one with bright red and black bands. Moreover, well-resolved local cues no longer covary so reliably with features of the world. A particular colour band pattern on a snake ceases to be an invariant sign of danger—all you need to notice—once harmless mimics evolve the same pattern. Thus hostility changes the informational character of local environments, degrading the covariation between easily discriminated cues and the functional properties they signal. Mimics are rarely perceptually identical to their models. Camouflage is rarely perfect. But agents relying on cryptic or mimicking colour schemes are not easily identified, especially in less than optimal circumstances.

14.2.2 Agent-sensitive response

When the target of one agent's actions is another agent, that target will respond subversively. In hostile interactions, the target will anticipate or observe the agent's actions and respond to subvert, block or thwart those actions. Gerd Gigerenzer and his colleagues have argued that intelligent action is not dependent on inferentially elaborate use of rich models of the world. Instead, 'fast and frugal' heuristics often suffice. One of their stock examples of such a heuristic involves catching a baseball by

---

[2] Compare, for example, the strategy of playing discriminating-hawk (playing hawk when you judge you can defeat in prospective competitor in a fight, and otherwise playing dove) with the strategy of playing all-dove. Discriminating-hawk has obvious temptations as a strategy. But clearly the error costs—entering a fight you cannot win—are high, and other agents are under selection to degrade the accuracy of your judgements.

attempting to keep the angle of gaze (the angle between eye and ball) constant rather than by predicting the place the ball will land, and attempting to reach that spot before the ball (Gigerenzer and Selton 2001). No doubt this is an effective heuristic for catching balls on the fly. But creatures attempting to avoid capture pose much less tractable problems, for their behavioural trajectories are not ballistic and that massively increases the difficulty of intercepting them. In non-cooperative interactions, each agent is trying to subvert the other whilst not themselves cueing their own response to their antagonist. Think, for example, of taking a penalty shot in soccer. The shooter avoids cueing his direction of shot, and the goalie tries to delay his commitment to blocking on one side as long as possible. In fighting, in predatory strikes against vigilant or dangerous prey, it must often be the case that each agent has an interest in choking off the flow of information to the other. It will also be the case that epistemic action—such as trying to avoid moving first so that you can see and respond to the move of the opponent—becomes very risky. So information that you would like to have is less available and more expensive.

14.2.3 Epistemic action and its costs

Agents engineer their epistemic environments as well as their physical environments. In doing so, they often simplify their informational environment, eliminating noise and ambiguity in a signal. But hostility increases the costs of epistemic action to make the agent's world more transparent. That is most obvious with predator inspection, which is certainly not risk free. But such costs are probably imposed even more often in competitive interactions within a social group, interactions in which each competitor is monitoring the other. Much energetically expensive male–male mate competition—particularly ritualized displays that can escalate into actual fights—is probably mutual probing. Each agent is trying to extract a response from the other which carries information about fighting capacities and stamina. This is expensive. But the costs of error in committing to fight are often even more extreme.

In *Thought in a Hostile World*, I argued that response to the epistemic consequences of hostility was the main selective force driving cognitive evolution beyond systems of mere discrimination and response. Over time, agents' cognitive systems evolve in response to these challenges. The declining reliability of single, easily discriminated cues selects for the evolution of robust tracking—for the ability to track functionally salient features of the environment by more than one cue. For similar reasons, there is selection for decoupled representation; that is, for a less automatic connection between environmental registration and action. If an agent with whom you are in competition does not react stereotypically to a particular stimulus, you cannot afford to either. However, despite evolved changes in cognitive equipment, the cognitive burden on decisions often remains significant, imposing important epistemic constraints on the set of available strategies. Thus reed warblers have only partially effective methods of discriminating between their own offspring and those of cuckoos. They do not use the fact that cuckoo eggs are a little larger than their own, nor do they use the differences between cuckoo chicks and their own chicks.

14.3 **Human cognition**

The epistemic consequences of hostility played a significant role in hominid cognitive evolution too. There is good reason to think that group selection has played a very significant role in human evolution, and hence competitive interactions between human groups (quite often violent ones) have played a critical role in our evolutionary history. Moreover, our lineage depends on the knowledge-intensive extraction of resources from our habitat. For example hominid diets depend, and have long-depended, on high value but difficult to process resources (Hill and Kaplan 1999; Kaplan *et al.* 2000). These resources are difficult to process because they are defended. Animals defend themselves morphologically and behaviourally; and some plant foods—especially tubers and other underground storage organs—defend themselves chemically.

Nonetheless, while hostile biological interactions formed an important part of the matrix that drove our cognitive evolution, the temporal, geographical, and ecological heterogeneity of hominid habitats has also been important. It is likely that hominid cognitive capacities evolved in response to environmental fluctuation as well as to the challenges of competition and predator/prey relationships. Richard Potts, for one, has argued that the African genesis of the hominid clade took place in conditions of increasing climatic instability, on both seasonal and generational time scales (Potts 1996, 1998). In addition, human environments were destabilized by features internal to the hominid lineage itself. One such mechanism is geographical and ecological expansion. As early hominids expanded from their areas of origin in east African grasslands and woodlands, their physical and ecological circumstances changed. Any behavioural response that decoupled their lifeways from the physical parameters of their origin would have permitted further expansion, and selection for further decoupling. As a species' range expands so that its habitat comes to be physically and biologically heterogeneous, it can respond in two ways. The species can fragment into a mosaic of differentiating populations, each evolving specific adaptations to suit local circumstances. Or it can evolve adaptive plasticity. Both responses characterize hominid evolution. The morphological differences between the Nuer and the Inuit show the existence of local adaptation to heat and to cold. But our lineage is also characterized by an impressive degree of adaptive plasticity, and that plasticity is largely cognitive and behavioural rather than morphological.[3]

Moreover, humans modify their habitats through their own activities. They partially construct their own niches (Odling-Smee 1994; Laland *et al.* 2000; Odling-Smee *et al.*

---

[3] Though there is certainly adaptive morphological plasticity too: for example, physiological response to life at high altitudes.

2003). This is true of many species. But human niche construction is important and pervasive because it is often cumulative. One generation inherits the modifications of its predecessor, and makes further changes. Even humans living in times of little externally forced variation can experience significant change in their habitat in a few generations. Human agents modulate the effects of physical parameters through the construction of shelters, fires, clothes, water containers, and food stores. They also influence their bioscapes. Most obviously, their foraging activities have a tendency to deplete the resources available to them, especially as populations expand and/or become more sedentary. But they also induce other changes, sometimes on a continental scale. Firestick farming is now believed to have caused fundamental changes in Australian flora over the last forty thousand years or so (Flannery 1994). Humans induce fundamental changes in their own microbial bioscape too, especially as they change from a nomadic to a sedentary existence. Different pathogens have different transmission modes, and these are very sensitive to lifestyle. Sleeping inside and by smoky fires can make people less vulnerable to some mosquito-born infections while making them more vulnerable to contamination of their water supply. Moreover, many pathogens will go locally extinct if their host population falls below a minimum size. So both the size of local populations and the frequency with which they are in contact with others bears importantly on whether diseases can sustain themselves in a particular area (Ewald 1994). In short, for millions of years, hominids have profoundly reshaped their interactions with their biological world, and their children inherit the results of these modifications. The shift from foraging to agricultural life styles, with their dependence on domestic plants and animals is an especially pervasive and recent example of environmental change through niche construction (though perhaps not only niche construction; Richerson *et al.* 2001). But it is by no means the only such example: the invention of cooking may be another. Cooking is crucial and pervasive, for it changed human resource ecology profoundly, making available important and abundant plant food that was otherwise toxic, increasing the food value of meat, and changing the physiological demands on teeth, tongue, and jaws. And it may have been much more ancient (Wrangham *et al.* 1999).

Cumulative niche construction also profoundly changes hominid social environments. Most of the crucial transitions in hominid and human social organization were the result of multigenerational, cumulative niche construction. I have in mind such transitions as: the transition from individualized, feed-as-you-go foraging to central place foraging and food sharing; transitions from hierarchical, bully-dominated social orders characteristic of our chimp relatives and, very probably, early hominids to the more egalitarian social orders of foragers; the transition from nomadic and foraging economies to sedentary and agricultural economies (Mithen 1990, 1996; McBrearty and Brooks 2000). Transitions from one such lifeway to another transforms the cognitive demands on individual agents. For example central place foraging reduces daily variance through sharing, and hence reduces the pressure on individual risk reduction. A lone forager must often trade maximizing daily catch against reducing variance in his/her daily catch. If a hunter encounters a very valuable but hard to catch animal he must trade its high value if caught against the risk of getting nothing. But as one cognitive demand eases, others will increase. Central place foraging increases the risk that you will not get your fair share of a shared resource. Likewise, mate choice strategies for women change as we shift from egalitarian social environments to those in which men differ markedly in the resources they control.

Particular social and ecological problems were cognitively challenging in hominid social worlds. But it is still more important that those problems, and the information needed to solve them, change on time scales faster than those of genetic adaptation. For instance, there is evidence, at least in Europe, that transitions from foraging life styles to agricultural life styles took tens to hundreds of years, not thousands of years. Likewise transitions from egalitarian social orders to chiefdomships and other more hierarchical social orders seem to have been relatively rapid (Shennan 2002). Environmental change on such time scales selects for the cultural transmission of information. Highly stable environments favour mechanisms that buffer behavioural capacities from the vagaries of individual experience, and which reduce the error costs of trial and error learning. If environments change very rapidly, the skills and information of one generation are out of date for the next, and each generation must learn for itself the specific features of its world. Intermediate rates of change select for the vertical (i.e. intergenerational) transmission of information, and hominid social worlds often changed at such intermediate rates (Laland 2001).

So climatic instability, geographic expansion, and cumulative niche construction, all jointly ensure that the information needed to control action varied very significantly in space and time. This feature of the hominid lineage selects for the vertical transmission of information. That selection seems to have been effective. There are many social mechanisms that make the intergenerational transfer of skills and lore more reliable. There has been a shift in the hominid lineage from the primate norm to intensive, accurate vertical transmission of information and of skill.[4] This flow of information is based on the combination of: (i) imitation learning; (ii) adult adaptations for teaching—these adaptations include those psychological mechanisms that facilitate joint attention, and perhaps even mind reading itself (Brockway 2003); (iii) the use of elements of existing material culture as templates; (iv) changes in human life history that extend childhood and adolescence and hence extend the period available for the flow of information from the elder to the younger; (v) language and perhaps other public representational media: mime, gesture and depictive representations; and perhaps (vi) the division of cognitive labour. Not everyone needs to know everything. That is important if memory constraints would otherwise restrict the total information

---

[4] Though I also think Avital and Jablonka are right to point out that the extent of vertical transmission of information in non-human animals is much understated (Avital and Jablonka 2000).

available to a group. There are, of course, many uncertainties about the evolutionary histories of these mechanisms, but collectively they make an impressive array of devices that facilitate the vertical flow of information (Tomasello 1999).

Thus humans have genetic adaptations that facilitate social learning. Some of our genes are selected for their roles in language, imitation, and other devices of social learning. But cumulative niche construction is self-amplifying. The mechanisms of cumulative niche construction are themselves made much more powerful by cumulative niche construction. The social institutions of teaching; the coining of an appropriate rich vocabulary to describe the resources and dangers of local environment; and the widespread availability of templates for the production of the local material culture are all products of cumulative niche construction. In short, while there are clearly genetic adaptations that facilitate the intergenerational transmission of information, the transition from a social species to a cultural species is itself partly mediated by cultural evolution.

Time to summarize the state of play. I began by sketching a framework for thinking about cognitive systems and their evolution. Strategy choice involves cognitive burdens of differing weights. Relevant information is relatively cheap in transparent environments, and in those environments access to information is not a severe constraint on an agent's set of available strategies; in less tractable environments that situation changes. In general, the cost of information is increased by hostile biological interactions, for in such interactions agents degrade one another's epistemic environment. In many lineages, the evolution of sophisticated cognitive capacities has been driven by response to such subversion. In part, human cognitive evolution has, likewise, been a response to hostility. But it has also been a response to both the externally forced and to the self-induced instability of human habitats.

So how does all of this connect to the existence of rational animal agents? It seems to me that there are two ways we might pursue the problem of animal rationality. One way would be to abstract away from the details of cognitive mechanism. Animal agents would be rational to the extent that their capacity to choose the optimal action in their situation was not subverted by constraints on their capacity to access and use information. So understood, rationality would be an aspect of optimal design. There is at least one alternative approach, one focusing on the evolution of a specific cognitive mechanism (or set of mechanisms). We do not just talk and think. We explicitly assess, not just represent, the thought and talk of others. We have a *folk logic*. For reasons I discuss in the next section, my bet is that we will do better stalking the evolution of rationality by stalking the evolution of folk logic than by stalking the evolution of optimal cognition. The key idea is that instability of hominid habitats has resulted in runaway selection for the social transmission of information, and while that transmission is an amazingly powerful means of enabling fast adaptation to a rapidly changing environment, it also risks extreme error costs. Folk logic is both a mechanism that amplifies the effects of social transmission and is a partial insurance against the error costs of extensive social learning.

14.4 **The hazards of information transfer**

Human cognition is very inference hungry. Not much human action is a direct response to a specific, perceptual feature of the immediate environment. We often rely on integrating information from a range of sensory modalities.[5] For example, in co-ordinating actions with others, we rely on what others are saying as well as what we can hear, see and sometimes feel them do, and on what we can see, hear and feel about the effects of their actions. Think of two or three people manoeuvring a large and clumsy desk from one office to another. Such a manoeuvre will be managed by a mix of mental rehearsal; actual rehearsal; instructions; grinding sounds from door frames; the feel of the desk as its point of balance changes. All this will happen and the movers will respond as the desk is gradually shuffled from one location to another. And, of course, as with other animals, much of our action rests partially on information stored in memory. In managing our actions, we rely heavily on inference, broadly conceived.

We do not just practice inference, though. We represent inference and its results. A distinctive feature of human cognition is the existence of a 'folk logic',[6] that is, a set of more or less explicit norms about reasoning and the representations that result from reasoning.[7] Folk logic presupposes metarepresentational capacities, but it involves more than metarepresentation.[8] Metarepresentational capacities might evolve for reasons quite unconnected to cognitive efficiency. The fact that another agent has signalled the presence of trees in fruit in the next valley is in itself a datum of importance, and hence is a feature of the situation worth representing. For, independently of its truth, this signal is a source of information about what that agent is likely to do (and what other parties that receive that signal are likely to do). To take advantage of this window into other agents' motivations, an utterance must be represented as information. It must be represented as a signal of trees in fruit, rather than just having a tendency, more or less strong, more or less contextually invariant, to cause the belief that there are trees in fruit in the next valley. If we were only interested in others' signals as extensions of our own sensory apparatus, we might have no interest in representing their signals as signals. They might be unrepresented elements in a causal chain from their perceptual registration of an environmental state to our belief about that state. Others' utterances would be like my retinal images. Like that image, their existence and character would be crucial to belief formation, but not themselves the

---

[5] Editors' note: see also Millikan, this volume.

[6] Or folk logics: there may well be significant variation in folk logic across time and space.

[7] Editors' note: see also Bermúdez (this volume) for a sketch of how folk logic might be assembled incrementally.

[8] Editors' note: see also chapters by Bermúdez and by Proust, this volume, for discussions relevant to the relationship between logic and metarepresentation.

targets of my beliefs.[9] So metarepresentation without representational evaluation can be important as a tool for predicting the actions of others and might evolve for that reason. But this would not explain the evolution of folk logic. For folk logic involves the *assessment* of representations. Most obviously, we have the notions of truth, falsity, and their cognates. But we also have a variety of more or less explicit ways of assessing inferential transitions as well.

The existence of such norms poses obvious questions: what is the connection between good reasoning and explicit systems for the representation and assessment of reasoning and its results? Do norms of reasoning promote inferential efficiency? Does having the concept of truth make it more likely that an agent's representations will be veridical? Such norms have made important human reasoning projects possible that would otherwise be beyond us, for the whole of modern science depends both on the invention of a variety of public representational media (diagrams, models, and the like) and on methodological explicitness. Every branch of science has specialists whose concern is with calibrating the tools—physical, experimental, and mathematical—of their specialities. Such calibration makes contemporary science possible. However, these enormous advances in cognitive power are the result of long-continued cumulative niche construction. Presumably folk logic long predated these late triumphs. So what role did it play when it was first assembled?

Folk logic is certainly not a precondition of inference. We do not need an explicit representation of representation and information processing in order to use inference.[10] As Fodor pointed out in *The Language of Thought*, some computational, inferential processes must be hard-wired. They must be the result of the physical structure of our reasoning devices, not the result of what those devices know about reasoning. Moreover, children and animals reason without folk logic. Tony Dickinson has argued that experimental rats have beliefs about the causal structure of their environment and those beliefs are sensitive to evidence. If they are provided with evidence that the delivery of food to their cages ceases to depend on their performance of specific actions, they cease to perform those actions (Balleine and Dickinson 1998; Dickinson and Balleine 2000). To the extent that rats and other animals represent and infer, they do so without the benefit of a folk logic. We do not need to have the notion of truth to have true beliefs, nor the notion of a warranted inference to make warranted inferences. Instead, we need well-designed and hence reliable computational procedures.

In the light of this, to take up the thread at the end of Section III, we face two ways of thinking about rationality. We could do so by thinking about the extent to which an

---

[9] This is brutally over-simplified, of course. My perceptual states might become targets of representation as part of a system involving feedback or self-correction.

[10] Editors' note: See and compare Proust, this volume, on metacognition without metarepresentation, and on metarepresentation and normativity.

agent's computational mechanisms are well designed: whether (for instance) they reliably deliver veridical and relevant representations in a wide range of the circumstances in which the animal finds himself/herself. And we can probe the selective environments that lead to good design. Rationality would then be a measure of overall cognitive efficiency; of the effectiveness with which agents manage their information gathering and information-processing problem.

I doubt that this approach is likely to be fruitful. For I think the notion of good cognitive design is like the notions of fitness or adaptedness. The idea that we can measure the overall fit between organism and its environment, its overall level of adaptedness, has been enormously intuitively appealing. Nonetheless, no one has succeeded in formulating and applying such a notion of adaptedness. The reasons for this failure seem to be principled rather than accidental. Organisms are adapted to their salient local circumstances. Local circumstances are very varied, and different aspects of those circumstances are salient to different organisms in virtue of their specific properties. Bees and bee-eaters may live in the same physical environment, but they will experience that environment very differently. The array of resources and threats for the two species will hardly overlap. This local, highly specific nature of selection and adaptation undercuts the project of defining a general notion of goodness of fit, applicable to organisms of different kinds in different environments (Beatty 1992; Godfrey-Smith 1996, pp. 188–192). It undercuts it technically, by making such a metric hard to define. And more important still, it undercuts it theoretically. For it is far from clear what explanatory role within evolutionary biology such a notion of overall adaptedness could play. The success or failure of organisms in their local environment is not explained by their level of adaptedness. Rather, it is explained by the *specific features* of their phenotypes in relation to their local habitat. For example a particular bee-eater population might be expanding because a change in the local environment makes safe breeding sites abundant. The population expansion is not explained by the bee-eaters' adaptedness. Rather, the expansion is a consequence of facts about the bee-eater phenotype (they breed in burrows they dig in steep banks) and their environment. These first-order facts about their phenotype and their environment explain their local biological success. The bee-eater's fitness, adaptedness, supervenes on its phenotype and its environment.[11] This, a least, is more or less the accepted wisdom on fitness and adaptedness, though there are still dissenters from this line of thought (Dennett 1995), and I have often had doubts about it myself.

If these reasons for scepticism about overall adaptedness are compelling, a similar scepticism would be appropriate about overall cognitive efficiency. What counts as fitness enhancing is specific to organism and environment. Likewise, what counts as efficient information use is specific to organism and environment, and for similar reasons. Cognitive power has costs, including error costs. Sometimes apparently good

---

[11] This is why fitness is not defined as a relationship between organism and environment but as a reproductive propensity: the expected reproductive success of an organism.

design would be over-design. Fitness is multidimensional. It cannot be optimized in every dimension simultaneously. No organism can simultaneously maximize investment in somatic growth and in reproductive effort. There must be trade-offs between potential excellences, and the right trade-offs will depend on the specifics of lineage and environment. For this reason, Richard Lewontin, for one, is very sceptical about the very idea of an optimal phenotype. Is an elephant's phenotype suboptimal because its skin is not bullet-proof and its flesh is not poisonous? No; constraints on basic biology make some conceivable organic designs not really possible. Organisms are optimal only given a realistic specification of the space of possibility. However, on Lewontin's view, there is no objectively optimal phenotype, because there is no objective way of specifying that space (Lewontin 1987).

Similar considerations apply to cognitive efficiency. The life of the mind is a life of trade-offs.[12] No agent can exhibit all the epistemic virtues simultaneously. Information is not free, so there is no global answer to questions about the acceptable level of risk in acting on incomplete information. It is rarely possible to minimize both false negatives and false positives. Learning involves similar dilemmas. If in trial and error learning, an agent's search space—the array of hypotheses he or she considers—is too large, experience may never converge on an answer. If too small, the true solution may lie outside the agent's search space. Social learning raises analogous problems. How should social creatures choose between finding out for themselves by individual learning (finding the fruiting trees yourself) and social learning (accepting advice)? Whether to trust another will depend on the costs of individual learning; the price of advice; the error profile of individual learning versus that of social learning; and the costs of different kinds of error. The right trade-off between these various power/ reliability dilemmas will depend on specific features of agents and circumstances.

Putting all these considerations together: if Lewontin is right, there is no optimal cognitive phenotype because there is no way of specifying objective limits to biological possibility. Even if Lewontin is wrong, good cognitive design will be circumstance-specific. An agent cannot be rational to the extent that it has a high overall level of cognitive efficiency, because efficiency is multidimensional, and strength in one dimension often precludes strength in others. There is no global metric of cognitive efficiency from low to high. Yet if rationality is just good cognitive design in the specific circumstances of the agent, it will quite often be rational to be stupid and dumb, for sometimes the costs of finding and using information exceed its benefits. That is, if we think of rationality as cognitive optimality given the circumstances and biological constraints on the lineage, there is no reason to expect that rational animal agents will be cognitively sophisticated cognitive agents.

----

[12] A fact reflected in folk wisdom through its plethora of inconsistent maxims: 'look before you leap' versus 'he who hesitates is lost'. Both maxims offer good advice, but they cannot simultaneously be followed.

These considerations do not rule out general theories of cognitive evolution, of general environmental conditions that select for global properties of cognitive systems. Thus I have myself argued for a connection between hostility-induced informational translucence, robust tracking, and decoupled representation. Nonetheless, I think overall cognitive efficiency is probably too multidimensional and context sensitive to be a good target of general theory, and hence in my view, a good alternative is to probe folk logic, investigating its evolutionary history. I think it is more productive to think of rationality as a feature of the hominids: it is an evolved response to the dilemmas in the construction and use of socially transmitted information.

I emphasized in Section 14.3 the crucial role the social transmission of information has played in cognitive evolution. Social learning is enormously powerful. It is the foundation of our ecological, geographical, and then technological expansion. But it is also very risky. These costs have recently been emphasized by those sympathetic to a meme-based conception of cultural evolution. The distinctive feature of memes is that they can grow in popularity—in the proportion of a population that house them—despite damaging the biological fitness of their hosts. Thus Dawkins has described religion as a virus-like meme. It propagates in human minds and spreads from one to another, despite the damage religious ideas do to those who have religious beliefs—most spectacularly, by causing celibacy or martyrdom. Putting the problem in the language of memes: an agent who is adapted to absorb information transmitted socially thereby runs the risk of infection by outlaw memes. Willingness to accept advice about dangers of a kind you have never experienced automatically makes you vulnerable to accepting advice about dangers which are not real, and following that advice can be very expensive indeed. But though the language of memes makes these costs very vivid, we do not need that language to make the point. Any feature that makes an agent's behaviour sensitive to the signals of another puts that agent at risk of exploitation. Learning from others is such a trait, and it does seem to have a potentially high error cost. For example it makes sense to study and imitate the actions of very successful members of your group, for their success will often be a result of the special skills and expertise. But a 'copy the best' learning rule can go horribly wrong, too. The preference for male children found in many cultures may be a maladaptive consequence of such a rule. A preference for sons is adaptive for the successful. Male reproductive variance is greater than that of females, and the successful have the resources to invest heavily in sons. So their sons are likely to be amongst the successes of their generation. But for just this reason the less successful should, if anything, prefer daughters.

Language poses an especial risk of infection. Consider the contrast between learning through language and learning through watching a model demonstrate a skill. Learning from a demonstration by others poses a significant cognitive challenge. It may well be very hard to decompose the skills of an expert hunter or an expert potter into their functional components. Moreover, what the experts say about what they do may be not much help. Expert practitioners are by no means invariably good at analysing their own performances. But deception and the assessment of reliability are rarely an issue when capacities are transmitted from models by imitation. There are exceptions. Shamans, priests, and the like claim expertise that they do not have.[13] However, in general, it is very hard to fake genuine expertise, especially in small communities where individual success and failure are usually widely known.

Moreover, the social transmission of skill is rarely purely social. Rather, skills are often acquired by guided trial and error—by a mixture of observation, trial, and instruction. The apprentice sees an expert make a fire, or a pot, and tries to do it himself, sometimes under active guidance from the model. This makes the social acquisition of skill very cognitively demanding. For the apprentice has to integrate information across social and physical domains. However, the apprentice will often get immediate feedback from the world that serves as a check on what he sees and believes. Likewise, in situations of cultural contact the products of material culture can sometimes be reverse-engineered partially or completely from their exemplifications. It may well be possible to work out how to fix a spearhead to a spear by examining a functional spear. The history of construction may not be recoverable from pots, fires, or the use of foodstuffs that must be detoxified before being eaten. But those without these technologies can at least find out that pots can be made that hold water when heated; controlled fires can be made; palm hearts can be eaten if suitably prepared. Once again, though the cognitive challenges posed by the diffusion of technology are significant, deception is a minor problem.

In short, information that travels via cycles of models and mimics or through physical templates comes with an intrinsic warrant of reliability. None of this is true of language. Linguistic signals carry no intrinsic marks of reliability, for they are both arbitrary and independent of the immediate environment of speaker and audience. Thus Dan Sperber[14] has argued that folk logic plays a crucial filtering role. If memes are viruses, folk logic is our immune system (Origgi and Sperber 2000; Sperber 2001).[15]

We may not be very good at detecting bullshit—especially in contemporary societies where language is often our only source of information about a person or a topic—but we would be even worse off without folk logic. At least we have the concept of being conned to help make us wary of con-jobs. Sperber's suggestion gains some plausibility

---

[13] It is striking that many (perhaps all?) such cases both involve ritual knowledge—information that is not openly available—and linguistically encoded information.

[14] And in a somewhat different vein, Cosmides and Tooby (2000) emphasize the importance of metarepresentation (rather than folk logic) as a tool for keeping track of where information came from. This point, too, is well taken. Others vary widely in their reliability. In listening to another, an agent always has to consider: would she know that? And the answer will vary with speaker, time, and subject.

[15] It is perhaps no accident then that Sperber is a strong critic of the received evolutionary-psychology view of the Wason selection task experiments. Cosmides and Tooby famously interpret these as showing both our grave limitations as general inference engines, and the powerlessness of explicit norms of logic to improve inferential performance. Sperber argues that the experimental basis of these judgements is deeply flawed (Fiddick et al. 2000; Sperber and Girotto 2003).

from an otherwise curious fact. Folk logic is mostly a system for the assessment of linguistic representation, not representation in general. Some terms of folk logic can be used to assess models, displays, demonstrations, and exemplars. 'Inaccurate', 'misleading', and the like can be applied to all species of public representation. But folk logic has much richer resources for linguistic assessment. Thus Locke was able to joke that God had not made man, and then left it to Aristotle to make him rational. For a couple of thousand years, logic as an academic discipline was not much more than a systematization of folk judgements about particular inferences. The same is still true in much of semantics and philosophy of language. Our apparatus for describing the semantic properties of sentences is still to a very considerable extent a systematization of folk metarepresentational categories. That is not true of the study of other representational media. For example the development of the theory of perspective in visual representation owed nothing to folk metarepresentational thought. Perhaps this difference between the rich folk metarepresentational apparatus for discussing language and the relatively sparse folk framework for discussing pictorial representation is a consequence of the urgent need for filtering information that comes via language.

Obviously, we are here in the realm of speculation not demonstration. Sperber's speculation *is* plausible. Nevertheless, in accounting for the existence of folk logic I think he somewhat overweighs the issue of manipulation and defence against it. For that logic is, I think, also an evolved response to hominid cognitive plasticity. One very striking feature of human cognition is our capacity to develop—often through intensive learning, long practice, and a highly structured developmental environment—automatized, module-like skills. Driving a car, playing chess, and identifying birds all initially depend on explicit, self-conscious, and attention-demanding processes. A novice birder, despite being slow and inaccurate, can do nothing else at the same time. They cannot talk, listen to cricket on the radio, or drive a car while identifying a wedge-tailed eagle perched on a roadkill. Long practice both improves skill levels and makes them more automatic.

Amongst the capacities we can acquire this way are those with new inferential techniques and unfamiliar representational media. With appropriate environmental structuring and practice (which may need to be very intensive), these skills, too, can become automated and routine. Good engineers can come to be able to 'read' circuit diagrams at a glance, just as a good birder can recognize a distant Pomeranian skua at a glance. Likewise, in contemporary societies, many of us have made routine mathematical reasoning capacities that once would have been rare. Even basic arithmetical operations are difficult without the scaffolding provided by Arabic notation. The development of reasoning and representation skills, if their development is like other automated skills, goes via an initial stage of conscious, reflective awareness of representations and correct procedures for manipulating them.

If Tony Dickinson is right, rats are quite good at probabilistic inference. Suppose a particular operation (pressing a leaver) is causally effective in producing food. Rats will rapidly learn that fact. And they will unlearn it, too, if conditions change, and leaver pressing ceases to be effective. They will unlearn it more slowly if the leaver press/food delivery connection was probabilistic rather than invariant. And in this they show their inferential acumen. In these circumstances, it takes more evidence to show a change in causal regime. But for better or worse, while rats are good probabilistic reasoners, it is my hunch that they cannot adaptively change their reasoning dispositions. To learn about reasoning, you need to be able to represent and evaluate your own reasoning capacities. My guess is that, in part, folk logic had its origin in selection for cognitive plasticity in learning and in the use of arbitrary representational media.

In this final section, I have developed both a positive and a negative case; a positive thesis about the origins and function of folk logic; a negative thesis about the idea of identifying rationality with overall cognitive efficiency. It is worth noting that these two ideas are not a package deal. They are independent. The speculation about folk logic could be wrong while the sceptical thesis is right, and *vice versa*.

References

Avital, E., and Jablonka, E. (2000). *Animal Traditions: Behavioural Inheritance in Evolution*. Cambridge: Cambridge University Press.

Balleine, B., and Dickinson, A. (1998). Consciousness: the interface between affect and cognition. In: J. Cornwell, ed. *Consciousness and Human Identity*, pp. 57–85. Oxford: Oxford University Press.

Beatty, J. (1992). Fitness: theoretical contexts. In: E. Fox Keller and E. Lloyd, eds. *Keywords in Evolutionary Biology*. Cambridge: Harvard University Press.

Bekoff, M., Allen, C., and Burghardt, G. M. (2002). *The Cognitive Animal: Empirical and Theoretical Perspectives on Animal Cognition*. Cambridge: MIT Press.

Brockway, R. (2003). Evolving to be mentalists: the 'mind-reading mums' hypothesis. In: J. Fitness and K. Sterelny, eds. *From Mating to Mentality: Evaluating Evolutionary Psychology*, pp. 95–123. Hove: Psychology Press.

Cosmides, L., and Tooby, J. (2000). Consider the sources: The evolution of adaptation for decoupling and metarepresentation. In: D. Sperber, ed. *Metarepresentation*, pp. 53–116. New York: Oxford University Press.

Dennett, D. C. (1995). *Darwin's Dangerous Idea*. New York: Simon and Shuster.

Dennett, D. C. (1996). *Kinds of Minds*. New York: Basic Books.

Dickinson, A., and Balleine, B. W. (2000). Causal cognition and goal-directed action. In: C. M. Heyes and L. Huber, eds. *The Evolution of Cognition*, pp. 185–204. Cambridge, MA: MIT Press.

Ewald, P. W. (1994). *Evolution of Infectious Disease*. Oxford: Oxford University Press.

Fiddick, L., Cosmides, L., and Tooby, J. (2000). No interpretation without representation: the role of domain-specific representation and inference in the Wason selection task. *Cognition*, 75: 1–79.

Flannery, T. (1994). *The Future Eaters*. Sydney: Reed.

Gigerenzer, G., and Selton, R. (2001). *Bounded Rationality: The Adaptive Toolbox*. Cambridge, MA: MIT Press.

Godfrey-Smith, P. (1996). *Complexity and the Function of Mind in Nature*. Cambridge: Cambridge University Press.

Hill, K., and Kaplan, H. (1999). Life history traits in humans: theory and empirical studies. *Annual Review of Anthropology*, 28: 397–430.

Hurley, S. L. (2003). Animal action in the space of reasons. *Mind and Language*, **18**: 231–256.

Kacelnik, A. (1984). Central place foraging in starlings (*Sturnus vulgaris*). I Patch residence time. *Journal of Animal Ecology*, **53**: 283–299.

Kaplan, H., Hill, K., Lancaster J., and Hurtado, A. M. (2000). A theory of human life history evolution: diet, intelligence and longevity . *Evolutionary Anthropology*, **9**: 156–185.

Laland, K. (2001). Imitation, social learning, and preparedness as mechanisms of bounded rationality. In: G. Gigerenzer and R. Selton, eds. *Bounded Rationality: The Adaptive Toolbox*, pp. 233–248. Cambridge, MA: MIT Press.

Laland, K. N., Odling-Smee, F. J., and Feldman M. W. (2000). Niche construction, biological evolution and cultural change. *Behavioral and Brain Sciences*, **23**: 131–175.

Lewontin, R. C. (1987). The shape of optimality. In: J. Dupre, ed. *The Latest on the Best*, pp. 151–159. Cambridge, MA: MIT Press.

McBrearty, S., and Brooks, A. (2000). The revolution that wasn't: a new interpretation of the origin of modern human behavior. *Journal of Human Evolution*, **39**: 453–563.

Mithen, S. (1990). *Thoughtful Foragers: A Study in Prehistoric Decision Making*. Cambridge: Cambridge University Press.

Mithen, S. (1996). *The Prehistory of the Mind*. London: Phoenix Books.

Odling-Smee, F. J. (1994). Niche construction, evolution and culture. In: T. Ingold, ed. *Companion Encyclopedia of Anthropology*, pp. 162–196. London: Routledge.

Odling-Smee, F. J., Laland, K. N., and Feldman, M. W. (2003). *Niche Construction: The Neglected Process in Evolution*. Princeton: Princeton University Press.

Origgi, G., and Sperber, D. (2000). Evolution, communication and the proper function of language. In: P. Carruthers and A. Chamberlain, eds. *Evolution and the Human Mind*, pp. 140–169. Cambridge: Cambridge University Press.

Owings, D. (2002). The cognitive defender: how ground squirrels assess their predators. In: M. Bekoff, C. Allen, and G. Burghardt, eds. *The Cognitive Animal: Empirical and Theoretical Perspectives on Animal Cognition*, pp. 19–26. Cambridge, MA: MIT Press.

Potts, R. (1996). *Humanity's Descent: The Consequences of Ecological Instability*. New York: Avon.

Potts, R. (1998). Variability selection in hominid evolution . *Evolutionary Anthropology*, **7**: 81–96.

Richerson, P., Boyd, R., and Bettinger, R. L. (2001). Was agriculture impossible during the pleistocene but mandatory during the holocene? A climate change hypothesis. *American Antiquity*, **66**: 387–411.

Shennan, S. (2002). *Genes, Memes and Human History: Darwinian Archaeology and Cultural Evolution*. London: Thames and Hudson.

Sigmund, K. (1993). *Games of Life: Explorations in Ecology, Evolution and Behaviour*. London: Penguin.

Sperber, D. (2001). An evolutionary perspective on testimony and argumentation. *Philosophical Topics*, **29**: 401–413.

Sperber, D. and Girotto, V. (2003). Does the selection task detect cheater-detection? In: J. Fitness and K. Sterelny, eds. *From Mating to Mentality: Evaluating Evolutionary Psychology*, pp. 197–226. Macquarie University Series in Cognitive Psychology.

Sterelny, K. (2003). *Thought in a Hostile World*. New York: Blackwell.

Tomasello, M. (1999). *The Cultural Origins of Human Cognition*. Cambridge, MA: Harvard University Press.

Wrangham, R. W., Jones, J. H., Laden, G., *et al.* (1999). The raw and the stolen: cooking and the ecology of human origins. *Current Anthropology*, **40**: 567–594.

Chapter 15

Rationality in capuchin monkey's feeding behaviour?

Elsa Addessi and Elisabetta Visalberghi

Abstract

Capuchin monkeys forage in social groups. To what extent is foraging by monkeys affected by social influences and, in particular, which are the proximate processes involved? We carried out a series of experiments aimed to investigate whether capuchin monkeys (*Cebus apella*) learn what to eat and what to avoid from the behaviour of other group members. Since capuchins are omnivorous and very socially tolerant, they are particularly well suited for this kind of study. Our studies show that capuchins' feeding behaviour is socially biased. In particular, although in the short run capuchins do not seem to take into account the full set of information provided by others' behaviour, in the long run their behaviour ends up being adaptive. In short, there is evidence that they learn *with* others rather than *from* others.

We first show both that novel foods are eaten less than familiar foods and that they are eaten more when individuals are with group members than when they are alone. The role of social influences is then evaluated in more detail. If an individual learns about food palatability from others, we should expect the novel food, consumption of which is socially facilitated, should be of the same kind (in our experiment, of the same colour) as the food eaten by group members. We found, on the contrary, that: (1) eating was socially facilitated regardless of colour difference between the food eaten by the individual and by group members; and (2) even when given a choice between two foods, only one of which matches the colour of the food eaten by group members, the individual does not consume more of the matching than of the non-matching food. Although an individual is more likely to eat novel foods when group members are around, and thus to learn from its own feeding experience the palatability of a novel food, it does not eat more of a novel food when it matches the food that group members are eating. Therefore, the assumption that capuchins learn about food palatability from others is unwarranted.

In another experiment, we found that the nutrient content of foods plays a major role in determining individual preferences about novel foods, and that social influences do not affect these nutrient-based preferences. These and similar findings about primates suggest that individuals are equipped independently of social context with the behavioural and physiological tools necessary to select energy-rich foods and to avoid deleterious foods. Preferences for sweet foods and aversion to bitter substances (bitterness is often associated with toxic compounds), coupled with a neophobic response and food aversion learning, are effective tools that enable individuals to choose successfully among foods. To learn what to ingest and what to avoid, an individual does not need to observe other group members, who after all may often be absent or not behaving informatively in a given situation. Nevertheless, other social influences that do not take into account the whole set of information present in the behaviour of others are still at work. Social facilitation and local and stimulus enhancement increase the chances that a naïve individual will feed at the same time and place as its group members, with the result that its food choices are similar to theirs. This scenario and that expected if social learning implies the whole set of information provided by group members (i.e. that they are eating a specific food) look similar. Moreover, they are impossible to distinguish unless *ad hoc* experiments are carried out. The scenario described above, in which individual learning is socially biased by a subset of the information present in the behaviour of group members, serves biological fitness quite well.

15.1 **Introduction**

In this chapter we report the results of a series of experiments carried out on capuchin monkeys (*Cebus apella*) in order to assess the assumption, frequently made by both laymen and scientists, that monkeys learn from others which foods to eat and which to avoid. This assumption reflects an attribution of rationality based on rational psychological processes (i.e. reasoning) that would be natural to attribute to human beings.

However, our experiments show that this assumption is unjustified when applied to capuchins. We offer an account in which capuchins' feeding behaviour is socially biased by some (but not all) of the information present in the behaviour of others. The resulting behavioural output is nevertheless adaptive.

According to Rozin (1996, p. 244) 'food is a social instrument for humans by virtue of the fact that more than one person is almost always involved with any food, from harvesting to ingestion' and in all human cultures the social context strongly affects what people eat. It has seemed obvious to assume that human and non-human primates' feeding habits in social contexts are influenced by similar social learning processes, so primatologists have attributed to social learning a major role in shaping diet and food preferences in non-human primates. In particular, it has been assumed that non-human primates learn which foods to eat from others and that dietary convergence or diffusion of new feeding habits in wild groups result from social learning (e.g. Kummer 1971; Nishida 1987; for a critical review see Visalberghi 1994). Observation of experienced group members has been considered an especially important factor in improving juveniles' foraging skills (Janson and van Schaik 1993); the occurrence of the infants' first experiences with food in close proximity with their mothers and other group members was considered to have a formative role on infants' later feeding behaviour (Box 1984; Fedigan 1982; Goodall 1986; Kummer 1971; Watts 1985). King (1999) argues that primate infants seem to have been selected to be information extractors. In popular science the claim is even more straightforward. According to de Waal (2001), animals learn from one another what to eat and what to avoid. Monkeys and apes look so much like us that it is difficult to suppose that simple behaviours like feeding in social contexts result from different processes. In particular, it is very tempting (though we shall argue, unjustified) to attribute to them what would be for human beings a rational process of learning from others about food choices.

For a naïve individual, learning from knowledgeable group members seems a safe way of selecting foods adequate for its diet while avoiding foods that are poisonous or lack a good balance of nutrients. The acquisition of information about food from group members seems particularly beneficial in omnivorous species whose diets include foods that might contain toxic substances, such as rats (Galef 1993) and capuchin monkeys (Visalberghi and Fragaszy 1995). Moreover, since 'social dynamics influence the likelihood of social learning' (Coussi-Korbel and Fragaszy 1995, p. 1446) social learning should occur more easily in species in which individuals feed close to one another than in species in which individuals do not do so. However, these tempting attributions to animals of rational psychological processes, or reasoning, in learning about food from others need careful scrutiny. Social foraging in the wild that looks superficially like the result of such a social learning process may, with experimental probing, be revealed to result from individual's preferences and aversions, associative and trial and error learning, and simple forms of social bias on the individual's learning. The combination of these processes ends up serving the animals' biological fitness very well.

As cognitive primatologists, we consider it appropriate to label animals' behaviour 'rational' only if it is based on psychological processes involving reasoning (more or less what Kacelnik, this volume, describes as PP-rationality). Nevertheless, in this chapter we sometime refer to B-rationality, as this term is defined by Kacelnik ('B-rationality emphasizes outcome rather than process...' and '...a B-rational individual can be defined as one whose actions maximize its inclusive fitness', Kacelnik, Chapter 2, this volume), in order to facilitate the readers' comparison of our thinking with that of other contributors of this volume who refer to Kacelnik's framework.

Our studies were designed to probe the nature of the social influences on foraging in the tufted capuchin, a species that is both omnivorous and socially tolerant. Capuchins feed on a wide variety of food items and manage seasonal changes in food availability by exploiting novel foods, some of which (e.g. insects, leaves) contain toxic substances

(Brown and Zunino 1990; Kinzey 1997; Sussman 2000; Terborgh 1983). Capuchins exhibit a high degree of inter-individual tolerance, especially towards infants and juveniles, in the wild (Izawa 1980; Janson 1996; Perry and Rose 1994) as well as in captivity, where it has been observed that food is sometimes transferred from one individual to another (de Waal *et al.* 1993; Fragaszy *et al.* 1997; Thierry *et al.* 1989). Since learning about food from group members is particularly expected in omnivorous and socially tolerant species, its investigation in tufted capuchins is pertinent and potentially promising.

15.2 Experimental evidence of social influences on feeding behaviour in capuchin monkeys

It is not possible to determine the contribution of social context to learning from field data; laboratory studies are required. To assess social bias on individual learning, some of our experiments were designed to include an individual condition (in which subjects were tested alone and social influences were removed) and a social condition (in which subjects were tested together with their group members). In other experiments, an individual faced one food or a choice between two foods while observing group member(s) eating a food.

15.2.1 Social influences on the response to novel foods

In the wild capuchins are neophobic (i.e. novel foods are eaten less than familiar foods) and only a few individuals taste a novel food when they encounter it for the first time (Visalberghi *et al.* 2003). Captive tufted capuchins are also neophobic and acceptance of novel foods is socially facilitated (i.e. a capuchin eats novel foods more when it observes its group members eating nearby than when it is alone) (Visalberghi and Fragaszy 1995; Visalberghi *et al.* 1998). Do these results mean that the naïve observer becomes less cautious toward the novel food because it observes its group members eating this same food and learns thereby that the food should be palatable? Or can this pattern of social foraging be explained in other terms, without invoking learning from the behaviour of others whether a novel food is palatable?

To learn socially about a safe diet, a naïve capuchin should pay attention to *what* other group members are eating and should eat more of a novel food *only if* its own food matches the food that group members are eating. In a series of four experiments we investigated whether the increase in acceptance of a novel food by a naïve observer occurs only when its novel food matches in colour the food eaten by group members (demonstrators), as would be expected if the observer is able to learn about food palatability from others.

Primates rely extensively on vision to understand the world around them and to forage (Fleagle 1999; Dominy *et al.* 2001). Since individuals, with exception of mother–infant pairs, usually feed at some distance from one another, vision is likely to play a primary role in acquiring information about food from others' behaviour (Santos *et al.* 2001). In all species, an individual can observe others' feeding activities whereas only in some species an individual tolerates a group member approaching, sniffing, and tasting its own food. Therefore in our experiments, based on an observer–demonstrator paradigm, the input provided to the observer by the demonstrator(s)' eating activities was primarily visual. This setting aimed to mimic foraging activities in which capuchins have visual access to one another's food but not direct access to the food itself, or to its odour. The experimental setting included a transparent panel separating the naïve observer (who was in one cage) from its demonstrator(s) (who was/were in the adjacent cage). The transparent panel allowed the observer to see its demonstrator(s) (and *vice versa*) but prevented the observer from smelling the demonstrator(s)' muzzles and/or food, as well as preventing food transfers.

In Experiments 1 and 2, the observer received a novel food in a box attached to the panel, and the demonstrators received a food in a larger box on their side of the panel. Both the familiar and the novel food were dyed with artificial colourings to obtain foods of strikingly different colours, or of the same colour, according to the experimental condition (see below).[1] By contrast, when the familiar and the novel foods were coloured the same, a human observer (and very likely the capuchin observer) could not discriminate them. Both foods were mashed to make their texture alike, so that colour was the only visible different feature between them. Since capuchins are cautious towards novel foods and the experimental design required demonstrator(s) to eat their food, the food given to the demonstrator(s) was familiar and desirable (mashed potatoes). This procedure was successful and the demonstrator(s) provided a continuous eating input to the observer throughout the 5-minute trial.[2]

Experiment 1 consisted of three experimental conditions: (1) Alone: a subject was in one cage and its box was filled with the novel food, while the group members were not in the adjacent cage and their food box was empty; (2) Group present: the observer was in one cage and its box was filled with the novel food, while the group members were in the adjacent cage and their food box was empty; (3) Group plus food: the observer was in one cage and its box was filled with the novel food, while the group members were in the adjacent cage and their food box was filled with the

----

[1] The different colours were easily discriminated by trichromatic human observers. Capuchins' colour vision is polymorphic: whereas males are always dichromatic, females can be either dichromatic or trichromatic (Jacobs 1996). Therefore their colour discrimination is not completely like our own. Consequently, before carrying out the experiments, we determined that both sexes could discriminate the colours of plastic chips with the same colours as the dyed foods (Addessi 2003).

[2] Although our capuchins sometime receive mashed foods, this previous experience did not make their response towards the mashed novel foods and the mashed potatoes similar. When facing the novel foods (for example, mashed lentils, beans, green peas, and chick peas) capuchins were cautious and slowly ate small amounts; in contrast, when facing mashed potatoes, they enthusiastically approached and scooped as much of this food from the box as they could.

Fig. 15.1 Average amount of food eaten (±SE) in the Alone and the Group plus food conditions ($t_{14} = -3.19$, $p < 0.01$) (modified from Visalberghi and Addessi 2000).

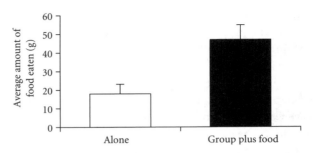

Fig. 15.2 Average amount of food eaten (±SE) in the Different colour and the Same colour conditions ($t_{15} = 0.04$, NS) (for further details see Visalberghi and Addessi 2001).

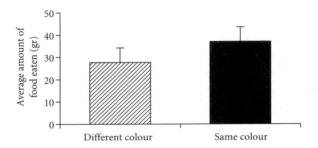

familiar food. The findings of Experiment 1 demonstrate that the observer's consumption of novel foods is significantly lower when the observer is alone (Alone Condition) than when the observer can see its demonstrators eating their food on the other side of the panel (Group plus Food Condition) (Fig. 15.1; Visalberghi and Addessi 2000). In addition, there is a positive significant correlation between how many demonstrators are eating and the frequency of the observer's eating behaviour (for details see Addessi and Visalberghi 2001). However, the demonstrators' eating behaviour increases the observer's acceptance of novel foods even if demonstrators eat a food that is strikingly different in colour (for example, whitish versus red) from that given to the observer.

Experiment 2 consisted of two conditions: (1) Different colour: the demonstrators ate food of a colour different from the observer's novel food; (2) Same colour: the demonstrators ate food of the same colour as the observer's novel food. If capuchins were PP-rational in taking into account the demonstrators' behaviour in order to learn about a safe diet, the observer would eat more of the novel food when its own food matched in colour that eaten by the demonstrators. In contrast to this prediction, in Experiment 2, the observers did not ingest significantly different amounts of the novel foods that matched and did not match the colour of the food eaten by the demonstrators (Fig. 15.2; Visalberghi and Addessi 2001). Since Experiments 1 and 2 show that eating novel foods was socially facilitated regardless of what was eaten, they do not support the hypothesis that capuchins learn about a safe diet from observing what others eat.

However, it could be argued in reply that wild capuchins often encounter several foods at the same time from which to choose, not just a single food, as in the preceding experiments. It is possible that, given a choice between two novel foods only one of which matches in colour (or some other property) the food that demonstrator(s) are eating, a capuchin would prefer the matching food. Experiments 3 and 4 were designed to address this question. In particular, we addressed the question of whether a capuchin, given a choice between two novel foods, only one of which matches in colour the food its group member(s) are eating, would direct its preference towards the food matching in colour the food eaten by them.

In Experiment 3, the naïve observer had a choice between novel food in two different colours, each presented in a separate box attached to the transparent panel. The food in one box was the same colour as the food eaten by demonstrators, while the food in the other box was a different colour. If the observer's acceptance of the novel food is affected by the information provided by its demonstrators, namely, that the food they eat is palatable, within a PP-rational framework one would expect the observer to choose the colour-matching food and to eat more matching than non-matching food. Our results do not fulfil this expectation: neither the observer's latencies to eat, the first food chosen, nor the amount of food ingested by the observer differed between the matching and the non-matching food (Visalberghi and Addessi 2001).

In Experiment 4, the input provided by a demonstrator to the naïve observer became more salient. The demonstrator was presented with food of the same two colours as the novel foods presented to the observer; before the experiment started, the demonstrator was trained to eat food of one colour and to avoid food of the other colour. The input provided to the observer was therefore two-fold: the demonstrator systematically both avoided one food and ate the other food (Fig. 15.3). Again, within a PP-rational framework one would expect the observer to choose the colour-matching food and to eat more matching than non-matching food. However, regardless of this two-fold input from the demonstrator, there was no evidence that social influences were directed to a specific food target: again, neither observer's latencies to eat, the first food chosen, nor the amount of food ingested by the observer differed between the matching and the non-matching food (Addessi and Visalberghi, unpublished results).

These last findings strengthen those of Experiment 3 by showing that the additional input provided by the demonstrator in the present study (i.e. the demonstrator itself was making a choice by eating only one of two available foods) did not tune the observer's choice to that of the demonstrator. Moreover, Hikami *et al.* (1990) and Hikami (1991) demonstrated that observing a conspecific avoiding a food does not decrease an observer's consumption of the same food, even when the aversively conditioned models are particularly salient individuals, such as the dominant male for group members,

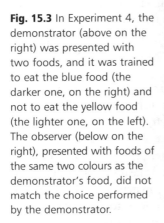

Fig. 15.3 In Experiment 4, the demonstrator (above on the right) was presented with two foods, and it was trained to eat the blue food (the darker one, on the right) and not to eat the yellow food (the lighter one, on the left). The observer (below on the right), presented with foods of the same two colours as the demonstrator's food, did not match the choice performed by the demonstrator.

or a mother for her offspring. Not even in rats (the species in which food avoidance behaviour has been most thoroughly investigated) or in chickens is food avoidance learned observationally (Galef 1985; Galef, 2003; Galef *et al.* 1990; Johnston *et al.* 1998).

15.2.2 The influence of nutrients on food choices

By contrast, feedback from ingested nutrients does affect primates' food preferences (Conklin-Brittain *et al.* 1998; Laska *et al.* 2000; Laska 2001; Visalberghi *et al.* 2003). Recently, we investigated the acquisition of preferences in relation to seven novel foods presented to 26 individual capuchins (Visalberghi *et al.* 2003). We found that, after eating each food only a few times, capuchin's preferences correlated positively with glucose and fructose and negatively with total fibre content. Later, after several further encounters, food preferences changed and became positively correlated with total energy. Therefore nutrients begin to play a major role in determining food preferences towards unknown foods very early after ingestion of very small portions of each novel food. Furthermore, this nutritional feedback is not affected by social influences.

15.3 Is capuchin monkeys' feeding behaviour PP-rational or B-rational?

Although capuchins are a promising species in which to look for the influence of social learning on feeding behaviour, the experimental findings obtained in our laboratory[3] do not support the common assumption that monkeys learn from one another which foods to eat and which to avoid. However natural it may be, this assumption is not justified.

Are capuchins neglecting social information about food palatability 'irrational'? Perhaps we would be justified in regarding human beings as irrational if they displayed a similar lack of social learning. Since capuchins are a very successful species it means that individuals do indeed learn what to eat. By means of which processes do capuchins acquire a safe diet? In fact, mechanisms other than learning from the behaviour of group members are available to capuchins. At least three important tools work together to influence food choice: taste perception, neophobia, and food aversion learning.

There is increasing evidence that taste perception is an adaptive response to the animals' need to assess foods' nutritional contents (Hladik *et al.* 2002; Simmen and Hladik 1998; Laska *et al.* 2000; Visalberghi *et al.* 2003). Primates tend to prefer and to ingest foods that taste sweet and to avoid bitter tasting substances. Taste perception leads primates to prefer fruits containing sugars and to avoid leaves or other items containing toxic compounds (that often taste bitter, as in the case of alkaloids). Moreover, taste acuity serves species' energy requirements by improving foraging efficiency. Primates with large bodies that need to ingest large amounts of food have higher taste acuity (i.e. low thresholds for sugars); in this way, these species perceive a wide range of sugar concentrations (i.e. a wide range of foods) as palatable, and consequently can also exploit food items with low sugar concentration. In contrast, small-bodied species with a high metabolism have high thresholds for sugars that allow the detection of the more energy-rich foods. In contrast, taste perception is not a reliable cue to avoid ingesting food that contain toxic substances which may cause illness and/or digestive problems. Although taste allows primates to avoid the bitter tasting foods (that often contain toxic compounds, such as alkaloids), toxicity is not always advertised by a bitter taste. For example a lethal alkaloid such as the dioscine of *Dioscorea dumetorum* is almost tasteless (Hladik and Simmen 1996).

Primates are extremely good at perceiving novelty, and novelty elicits an ambivalent response, characterized by avoidance and interest (Greenberg and Mettke-Hoffman 2001). The success of an omnivorous species depends both on its propensity to explore and sample novel foods and to include them in its diet, and on its caution toward them, so as to detect and avoid the risk of ingesting poisonous foods (Glander 1982; Milton 1993). Many primate species exhibit neophobia toward novel (potential) foods, either avoiding them completely or eating only small quantities at first (Rozin 1976). By eating a small amount of a novel food the individual succeeds both in avoiding fatal risks (since in general a small amount of a food does not contain enough toxic

---

[3] For other experiments not mentioned here see Visalberghi and Addessi, 2003.

substances to be lethal) and in assessing whether incorporating the novel food into its diet is possible. In fact inclusion does indeed occur if the very first few tasting experiences with novel food are not paired with negative post-ingestive consequences (see below). The predominance of neophobia over exploration, or *vice versa*, is affected by the degree of novelty (Hughes 1997) and by the ecology of the species (Greenberg 1990a, b). Moreover, a powerful factor influencing consumption of a novel food is how often it has been encountered. For captive capuchins, a food remains unfamiliar (i.e. they respond to it neophobically) only for the first few eating experiences. Visalberghi *et al.* (1998) showed that the temporal course of capuchins' neophobia towards novel foods was such that after several short presentations, the novel foods were eaten to a similar extent as familiar ones.

If ingestion of a novel food has noxious consequences, an individual will associate them with its consumption and that food will not be eaten anymore. Food aversion learning (Garcia *et al.* 1955; Garcia and Koelling 1966), first documented in rats, is widespread among animal species (e.g. hamsters, Zahorik and Johnston 1976; herbivorous mammals, Zahorik and Houpt 1981; Japanese macaques, squirrel monkeys, vervet monkeys, for a review see Visalberghi 1994). This is a robust learning mechanism that operates at the individual level and primarily concerns novel foods, and to a lesser extent familiar foods. In food aversion learning, the ingestion of a food that is associated with strong negative experience(s), such as gastrointestinal illness, leads to complete avoidance of the noxious food. The avoidance persists when the food is no longer noxious, and avoidance learning is quicker when the food is novel than when the food is familiar. If the ingestion of a novel food does not have negative postingestive consequences, it gradually becomes familiar and becomes part of the animal's diet.

The strong preferences for sweet foods and the dislike for bitter substances which are often associated with toxic compounds, and the neophobic response coupled with food aversion learning efficiently reduce the risk of making a fatal mistake. The individual does not need, therefore, to observe others to learn what to ingest and what to avoid, but is already equipped with the behavioural and physiological tools necessary to select energy-rich foods and to avoid deleterious ones. These tools are available to the individual all the time and do not require group members to be present exactly when the information is needed and to behave in informative ways. In short, the tools capuchins do rely on to choose foods are extremely adaptive.

15.4 **Socially biased individual learning**

Are we denying social influences an important role on individual capuchin's feeding? No, not at all. However, the experimental evidence shows that social influences on capuchins' behaviour are not as goal-oriented and specific as has been assumed and calls for a parsimonious interpretation in which behavioural synchrony (social facilitation *sensu* Clayton 1978) and local and stimulus enhancement[4] (see Galef 1988) operate. Let us briefly examine the inherent power of these learning mechanisms when they act together.

Field primatologists describe feeding as an activity that group members perform together. Laboratory evidence shows that in several primate species eating is socially facilitated (i.e. observing others eating increases eating behaviour) (humans, de Castro 1990; de Castro and Brewer 1992; capuchins, Visalberghi and Fragaszy, 1995; common marmosets, Vitale and Queyras 1997). Capuchins eat more as the number of individuals eating with them increases (Addessi and Visalberghi 2001; de Castro and Brewer 1992); satiated capuchins resume eating if nearby individuals start eating themselves (Galloway *et al.*, 2005). It has been suggested that mirror neurones responding to oro-facial actions provide a neurophysiological basis for feeding synchronization (Ferrari *et al.* 2003; Ferrari *et al.*, 2005). Mirror neurones are sensorimotor neurones located in the premotor area (F5) of macaque cortex; their distinctive feature is firing both when the monkey performs an action of a specific type and when it observes another individual performing an action of the same type. Mirror neurones may allow an individual to recognize that its own action and that of another individual match (Rizzolatti *et al.* 1999).

Individuals moving together are likely to encounter the same food sources. If this is the case, feeding synchronization increases the probability that knowledgeable individuals, as well as naïve ones, eat the same foods. However, social facilitation of eating occurs regardless of the correspondence between the observer's novel food and the food its group members are eating (Visalberghi and Addessi 2000). Social facilitation works non-specifically, by supporting the synchronization of feeding activities rather than the learning of what to eat. Nevertheless, the combination of feeding with group members, local and stimulus enhancement, and individual propensities (taste perception, neophobia, and food aversion learning) is undoubtedly effective in increasing the chances that a naïve individual's feeding activities are canalized in the same way as those of its group members. While the underlying processes do not depend on social learning of a kind that would be natural to attribute to human beings under relevantly similar conditions, the result is biologically rational and may perhaps be advantageous for capuchins, since (as already indicated) it does not depend critically on the presence and informative behaviour of others.

A final comment concerns whether monkeys can learn from what others *do not* do. Do capuchins treat avoidance of a food as a cue that this food is toxic? Years ago Hearst argued that 'recognizing and learning from absence, deletion, and non-occurrence are

---

4 Editors' note: Stimulus enhancement occurs when observing another's act on an object draws the observer's attention to that object, which then evokes an independently determined response in the observer, which may be innate or have been independently learned.

surprisingly difficult. Animals and people, it seems, accentuate the positive' (1991, p. 432). He reviewed the relevant literature and showed that animals detect presence much more easily than absence and that they learn more readily from the arrival than from the removal of a stimulus. In a social context, an observed individual can enhance the observer's attention to or interest in what it is eating, but not to what it is not eating (that is, what it is avoiding). Consequently, it seems obvious that animals do not learn what to avoid from watching others avoid particular foods. As Galef (1985) pointed out when explaining his many experiments in which rats avoid the same food as that avoided by conspecifics, socially induced avoidance of one food may be the indirect result of socially induced preference for an available alternative.

It cannot be ruled out *a priori* that an animal can learn to avoid a food from watching another spit out the food, vomit, or show distress and a facial expression of disgust. However, an observer would be extremely lucky to have an opportunity to learn by observing such behaviour. Not only are such behaviours rare, because individuals quickly learn not to repeat their wrong food choice again (as indicated in the discussion of the efficiency of food aversion learning above), but they typically have brief duration and so in order to learn from them an observing individual would need to be present and paying attention at precisely the right time period. Again, relying on individual learning has an advantage over social learning, since opportunities for the former are easier to come by.

15.5 **Concluding remarks**

Our interpretation of social foraging by capuchins, based on the experimental evidence obtained with these monkeys, is less demanding than the common assumption of them as actively learning what to eat from others. At present this common assumption of learning from others lacks supporting data (see Visalberghi 1994), and further *ad hoc* studies should carefully evaluate how other primate species deal with novel foods and incorporate them into their diet. Similarly, studies are needed to assess the possibility that monkeys learn what to avoid from others, by comparing the effects of providing as input to the observer mere passive avoidance by the observed animal with the effects of providing stronger observed behaviours, such as spitting out the food, vomiting, or distress and facial expressions of disgust.

We have contrasted two interpretations of social influences on capuchin feeding preferences: a more demanding interpretation, in terms of social learning from others' behaviour, and an interpretation in terms of socially-biased individual learning (Fragaszy and Visalberghi 2004). Although the former interpretation has been widely assumed since it corresponds to what would be PP-rational for humans in analogous circumstances, our data support the latter. Nevertheless, capuchin feeding behaviour ends up being adaptive (i.e. Kacelnik, Chapter 2, this volume, would label it as 'B-rational').

Acknowledgements

We acknowledge financial support by a grant from the FIRB/MIUR (RBNE01SZB4) to E. Visalberghi and a fellowship from CNR to E. Addessi. We are also grateful to the Bioparco SPA for hosting the laboratory where the experiments were carried out and to our keepers S. Catarinacci and M. Bianchi for their help with the monkeys.

References

Addessi, E. (2003). Ruolo delle influenze sociali sulle neofobia alimentare nel cebo dai cornetti (*Cebus apella*) e nello scimpanzé (*Pan troglodytes*), Ph.D. thesis, Roma.

Addessi, E., and Visalberghi, E. (2001). Social facilitation of eating novel foods in tufted capuchin monkeys (*Cebus apella*): Input provided, responses affected and cognitive implications. *Animal Cognition*, **4**: 297–303.

Box, H. O. (1984). *Primate Behaviour and Social Ecology*. London: Chapman and Hall.

Brown, A. D., and Zunino, G. E. (1990). Dietary variability in *Cebus apella* in extreme habitats: Evidence for adaptability. *Folia Primatologica*, **54**: 187–195.

Clayton, D. A. (1978). Socially facilitated behavior. *Quarterly Review of Biology*, **53**: 373–392.

Conklin-Brittain, M. L., Wrangham, R., and Hunt, K. D. (1998). Dietary responses of chimpanzees and cerchopithecines to seasonal variation in fruit abundance: II. Macronutrients. *International Journal of Primatology*, **19**: 971–998.

Coussi-Korbel, S., and Fragaszy, D. (1995). On the relation between social dynamics and social learning. *Animal Behaviour*, **50**: 1441–1453.

de Castro, J. M. (1990). Social facilitation of duration and size but not rate of the spontaneous meal intake in humans. *Physiology and Behavior*, **47**: 1129–1135.

de Castro, J., and Brewer, E. M. (1992). The amount eaten in meals by humans is a power function of the number of people present. *Physiology and Behavior*, **51**: 121–125.

de Waal, F. (2001). *The ape and the sushi master. Cultural reflections by a primatologist*. New York: Basic Books.

de Waal, F. B. M., Luttrell, L. M., and Canfield, M. E. (1993). Preliminary data on voluntary food sharing in brown capuchin monkeys. *American Journal of Primatology*, **29**: 73–78.

Dominy, N. J., Lucas, P. W., Osorio, D., and Yamashita, N. (2001). The sensory ecology of primate food perception. *Evolutionary Anthropology*, **10**: 171–186.

Fedigan, L. M. (1982). *Primates Paradigm*. Montreal: Eden Press.

Ferrari, P. F., Gallese, V., Rizzolatti, G., and Fogassi, L. (2003). Mirror neurons responding to the observation of ingestive and communicative mouth actions in the monkey ventral premotor cortex. *European Journal of Neuroscience*, **17**: 1703–1714.

Fleagle, J. G. (1999). *Primate Adaptation and Evolution*, 2nd edn. San Diego: Academic Press.

Fragaszy, D, and Visalberghi, E. (2004). Socially biased learning in monkeys. *Learning and Behavior*, **32**: 24–35.

Fragaszy, D. M., Feuerstein, J. M., and Mitra, D. (1997). Transfers of food from adults to infants in tufted capuchins (*Cebus apella*). *Journal of Comparative Psychology*, **111**: 194–200.

Galef, B. G., Jr. (1985). Direct and indirect behavioral pathways to the social transmission of food avoidance. In: P. Bronstein and N. S. Braveman, eds. *Experimental Assessments and Clinical Applications of Conditioned Food Aversions*, pp. 203–215. New York: New York Academy of Sciences.

Galef, B. G., Jr. (1988). Imitation in animals: history, definition, and interpretation of data from the psychological laboratory. In: T. R. Zentall and B. G. Galef, Jr., eds. *Social Learning. Psychological and Biological Perspectives*, pp. 3–28. Hillsdale, NJ: Lawrence Erlbaum Associates.

Galef, B. G., Jr. (1993). Function of social learning about food: a causal analysis of diet novelty on preference transmission. *Animal Behaviour*, **46**: 257–265.

Galef, B. G., Jr. (2003). 'Traditional' foraging behaviors of brown and black rats (*Rattus norvegicus* and *Rattus rattus*). In: D. Fragaszy and S. Perry, eds. *Traditions in Non-human Animals: Models and Evidence*, pp. 159–186. Cambridge: Cambridge University Press.

Galef, B. G., Jr., McQuoid, L. M., and Whiskin, E. E. (1990). Further evidence that Norway rats do not socially transmit learned aversions to toxic baits. *Animal Learning and Behavior*, **18**: 199–205.

Galloway, A. T., Addessi, E., Fragaszy, D., and Visalberghi, E. (2005). Social facilitation of eating familiar food in tufted capuchin monkeys (*Cebus apella*): Does it involve behavioral coordination?. *International Journal of Primatology*, **26**: 181–189.

Garcia, J., and Koelling, R. A. (1966). Relation of cue to consequence in avoidance learning. *Psychonomic Science*, **4**: 123–124.

Garcia, J., Kimeldorf, D. J., and Koelling, R. A. (1955). A conditioned aversion towards saccharin resulting from gamma radiation. *Science*, **122**: 157–158.

Glander, K. E. (1982). The impact of plant secondary compounds on primate feeding behavior. *Yearbook of Physical Anthropology*, **25**: 1–18.

Goodall, J. (1986). *The Chimpanzees of Gombe*. Cambridge, MA: Harvard University Press.

Greenberg, R. (1990a). Ecological plasticity, neophobia and resource use in birds. *Studies in Avian Biology*, **13**: 431–437.

Greenberg, R. (1990b). Feeding neophobia and ecological plasticity: a test of the hypothesis with captive sparrows. *Animal Behaviour*, **39**: 375–379.

Greenberg, R., and Mettke-Hoffmann, C. (2001). Ecological aspects of neophobia and exploration in birds. *Ornithology*, **16**: 119–178.

Hearst, E. (1991). Psychology of nothing. *American Scientist*, **79**: 432–443.

Hikami, K. (1991). Social transmission of learning in Japanese monkeys (*Macaca fuscata*). In: A. Ehara, T. Kimura, O. Takemaka, and M. Iwamoto, eds. *Primatology Today*, pp. 343–344. Amsterdam: Elsevier.

Hikami, K., Hasegawa, Y., and Matsuzawa, T. (1990). Social transmission of food preferences in Japanese monkeys (*Macaca fuscata*) after mere exposure or aversion training. *Journal of Comparative Psychology*, **104**: 233–237.

Hladik, C. M., and Simmen, B. (1996). Taste perception and feeding behavior in non-human primates and human populations. *Evolutionary Anthropology*, **5**: 58–72.

Hladik, C. M., Pasquet, P., and Simmen, B. (2002). New perspectives on taste and primate evolution: the dichotomy in gustatory coding for perception of beneficent versus noxious substances as supported by correlations among human thresholds. *American Journal of Physical Anthropology*, **117**: 342–348.

Hughes, R. (1997). Intrinsic exploration in animals: motives and measurement. *Behavioral Processes*, **41**: 213–226.

Izawa, K. (1980). Social behavior of the wild black-capped capuchin (*Cebus apella*). *Primates*, **21**: 443–467.

Jacobs, G. H. (1996). Primate photopigments and primate color vision. *Proceedings of the National Academy of Science*, **93**: 577–581.

Janson, C. H. (1996). Toward an experimental socioecology of primates: Examples from Argentine brown capuchin monkeys (*Cebus apella nigrivittatus*). In: M. Norconk, P. Garber, and A. Rosenberger, eds. *Adaptive Radiations of Neotropical Primates*, pp. 309–325. New York: Plenum Press.

Janson, C. H., and van Schaik, C. P. (1993). Ecological risk aversion in juvenile primates: Slow and steady wins the race. In: M. E. Pereira and L. A. Fairbanks, eds. *Juvenile Primates: Life History, Development, and Behavior*, pp. 57–74. Oxford: Oxford University Press.

Johnston, A. N. B., Burne, T. H. J., and Rose, S. P. R. (1998). Observation learning in day-old chicks using a one-trial passive avoidance learning paradigm. *Animal Behaviour*, **56**: 1347–1353.

King, B. J. (1999). New directions in the study of primate learning. In: H. O. Box and K. R. Gibson, eds. *Mammalian Social Learning: Comparative and Ecological Perspectives*, pp. 17–32. Cambridge: Cambridge University Press.

Kinzey, W. G. (1997). *New World Primates: Ecology, Evolution and Behavior*. Hawthorne, NY: Aldine.

Kummer, H. (1971). *Primate Societies*. Arlington Heights, IL: Harlan Davidson.

Laska, M. (2001). A comparison of food preferences and nutrient composition in captive squirrel monkeys, *Saimiri sciureus*, and pigtail macaques, *Macaca nemestrina*. *Physiology and Behavior*, **73**: 111–120.

Laska, M., Hernandez Salazar, L. T., and Rodriguez Luna, E. (2000). Food preferences and nutrient composition in captive spider monkeys, *Ateles geoffroy*. *International Journal of Primatology*, **21**: 671–683.

Milton, K. (1993). Diet and primate evolution. *Scientific American*, **269**: 70–77.

Nishida, T. (1987). Local traditions and cultural transmission. In: B. Smuts, D. Cheney, R. Seyfarth, R. Wrangham, and T. Struhsaker, eds. *Primate Societies*, pp. 462–474. Chicago: University of Chicago Press.

Perry, S., and Rose, L. (1994). Begging and transfer for coati meat by white-faced capuchin monkeys, *Cebus capucinus*. *Primates*, **35**: 409–415.

Rizzolatti, G., Fadiga, L., Fogassi, L., and Gallese, V. (1999). Resonance behaviors and mirror neurons. *Archives Italiennes de Biologie*, **137**: 85–100.

Rozin, P. (1976). The selection of foods by rats, humans, and other animals. In: D. Lehrman, R. A. Hinde, and E. Shaw, eds. *Advances in the Study of Behavior*, vol. 6, pp. 21–76. New York: New York Academic Press.

Rozin, P. (1996). Sociocultural influences on human food selection. In: E. D. Capaldi, ed. *Why We Eat What We Eat: The Psychology of Eating*, pp. 233–263. Washington, DC: American Psychological Association.

Santos, L. R., Hauser, M. D., and Spelke, E. S. (2001). Recognition and categorization of biologically significant objects by rhesus monkeys (*Macaca mulatta*): The domain of food. *Cognition*, **82**: 127–155.

Simmen, B., and Hladik, C. M. (1998). Sweet and bitter discrimination in primates: Scaling effects across species. *Folia Primatologica*, **69**: 129–138.

Sussman, R. W. (2000). *Primate Ecology and Social Structure*, vol 2: New World Monkeys. Boston, MA: Pearson Custom Publishing.

Terborgh, J. (1983). *Five New World Primates*. Princeton, NJ: Princeton University Press.

Thierry, B., Wunderlich, D., and Gueth, C. (1989). Possession and transfer of objects in a group of brown capuchins (*Cebus apella*). *Behaviour*, **110**: 294–305.

Visalberghi, E. (1994). Learning processes and feeding behavior in monkeys. In: B. G. Galef, M. Mainardi, and P. Valsecchi, eds. *Behavioural Aspects of Feeding: Basic and Applied Research on Mammals*, pp. 257–270. Chur: Harwood Academic.

Visalberghi, E., and Addessi, E. (2000). Seeing group members eating a familiar food enhances the acceptance of novel foods in capuchin monkeys. *Animal Behaviour*, **60**: 69–76.

Visalberghi, E., and Addessi, E. (2001). Acceptance of novel foods in *Cebus apella*: Do specific social facilitation and visual stimulus enhancement play a role? *Animal Behaviour*, **62**: 567–576.

Visalberghi, E., and Addessi, E. (2003). Food for thoughts: social learning and the feeding behavior in capuchin monkeys. Insights from the laboratory. In: D. Fragaszy and S. Perry, eds. *Traditions in Non-human Animals: Models and Evidence*, pp. 187–212. Cambridge: Cambridge University Press.

Visalberghi, E., and Fragaszy, D. (1995). The behaviour of capuchin monkeys, Cebus apella, with novel foods: The role of social context. *Animal Behaviour*, **49**: 1089–1095.

Visalberghi, E., Valente, M., and Fragaszy, D. (1998). Social context and consumption of unfamiliar foods by capuchin monkeys (*Cebus apella*) over repeated encounters. *American Journal of Primatology*, **45**: 367–380.

Visalberghi, E., Janson, C. H., and Agostini, I. (2003). Response towards novel foods and novel objects in wild *Cebus apella*. *International Journal of Primatology*, **24**: 653–675.

Visalberghi, E., Sabbatici, G., Stammati, M, and Addessi, E. (2003). Preferences towards novel foods in *Cebus apella*: The role of nutrients and social influences. *Physiology and Behavior*, **80**: 341–349.

Vitale, A., and Queyras, A. (1997). The response to novel foods in common marmoset (*Callithrix jacchus*): The effects of different social contexts. *Ethology*, **103**: 395–403.

Watts, D. P. (1985). Observation on the ontogeny of feeding behavior in mountain gorillas (*Gorilla gorilla beringei*). *American Journal of Primatology*, **8**: 1–10.

Zahorik, D. M. and Houpt, K. A. (1981). Species differences in feeding strategies, food hazards, and the ability to learn food aversions. In: A. C. Kamil and T. D. Sargent, eds. *Foraging Behavior: Ecological, Ethological and Psychological Approaches*, pp. 289–310. New York: Garland STPM Press.

Zahorik, D. M. and Johnston, R. E. (1976). Taste aversions to food flavors and vaginal secretions in golden hamsters. *Journal of Comparative and Physiological Psychology*, **90**: 57–66.

Chapter 16

Social cognition in the wild: Machiavellian dolphins?

Richard Connor and Janet Mann

Abstract

Bottlenose dolphins have large brains and exhibit impressive cognitive abilities in captive studies. Observations of wild dolphins in Shark Bay, Western Australia suggest that these abilities are important for solving the problems dolphins face in the social and foraging domains. Dolphins must keep track of a large number of social relationships while associating in groups that often vary in composition, be able to navigate nested within-group alliances, and learn how, where, and when to forage on a wide range of prey. The social problems dolphins encounter appear especially daunting, suggesting that the Machiavellian Intelligence hypothesis might apply to dolphins as well as to large-brained terrestrial mammals. With cetaceans, we are presented with a group of large, long-lived mammals that live in a habitat strikingly different from the terrestrial sphere and that exhibit striking diversity in brain size among species of comparable body size. We here review field studies of wild dolphin behaviour that is potentially relevant to the Machiavellian intelligence hypothesis, describing general features of dolphin society, multilevel male alliances, female relationships, and affiliative interactions. We then explain why it is more plausible that brain evolution in dolphins was driven social demands than by foraging demands.

16.1 Introduction: general features of dolphin society

For 3 days in 1976, 30 false killer whales,[1] including 17 females and 13 males, remained in the shallows along the shore of the Dry Tortugas off the Florida coast. Oceanographer

---

[1] *Pseudorca crassidens*, a large member of the dolphin family *delphinidae*, which also includes the bottlenose dolphin.

James Porter described this unusual but fascinating mass-stranding where, because there was little tidal movement, the whales were not actually stranded but floating in the shallows (Porter 1977). The whales flanked a large male that lay on his side, blood seeping from his right ear. When well-meaning people attempted to push them seaward, separating them from the group, the whales became agitated. Otherwise, they did not react when people rubbed sunscreen on their backs, but a female did bare her formidable teeth when a person ventured too close to her calf. Strangely, when Porter entered the water with a mask and snorkel, a flanking whale broke rank, approached him and pushed him shoreward. He tried this three to four times on each side with the same result. Nothing happened when he swam without the snorkel. After the male died (he had a severe nematode infestation in his ear) people were able to push the whales offshore; indeed, a few had already left the night before.

On the face of it, such behaviour seems anything but rational—especially in an adaptationist sense (as in Kacelnik's 'biological rationality', this volume). Certainly an extreme degree of mutual dependence is indicated—not surprising perhaps for mammals that live in a predator-rich but refuge-poor habitat. But why would these animals take such a risk to remain in the company of a dying male? He might be a relative or an alliance partner to the other males, but why would the females remain? Indeed, it would be difficult to generate a sensible adaptationist hypothesis if all we had to go on were studies of terrestrial mammals. One hypothesis is suggested by results from studies of killer whales (*Orcinus orca*), another large delphinid. In the 'resident' population of killer whales off south-western Canada, neither males nor females leave their mothers (reviewed in Baird 2000). This degree of not only geographic but also social philopatry[2] has no match on land. It suggests the reasonable and testable hypothesis that both the male and female false killer whales may have been taking risks to help a relative. Of course, even if he were a relative, we would still like to know why his relatives valued him to that degree. Do adult males help defend offspring from predators or infanticide? Are old individuals reservoirs of social and foraging knowledge? Unfortunately, we cannot begin to address these questions with what little we know of the social lives of false-killer whales.

While the social systems of killer whales and false killer whales may be highly unusual, for present purposes we are more interested in whether the individuals navigating those social systems face unusually severe cognitive challenges. The social system has a number of components, including social structure (pattern of social interaction, dispersal, nepotism, tolerance), social organization (size, sex ratio, spatiotemporal organization), and mating system (social and genetic components

---

[2] Philopatry, 'love of home', describes the tendency of animals to remain in the place they were born (or more exactly, natal philopatry). In almost all mammals, members of one sex, usually males, disperse further than individuals of the other sex. Social philopatry describes the tendency to remain with the group you are born in, irrespective of place.

involved in reproduction; Kappeler and van Schaik 2002). The Machiavellian Intelligence hypothesis about the evolution of large brains posits that analysis of social systems is critical for understanding selective pressures favouring greater cognitive capacity. Such an analysis requires examination of dyadic interactions and how these relationships are influenced by further relationships with others—that is how dyadic interactions fit into the broader social system.

Only four species of cetaceans have been studied well enough that we understand the basics of their social system (Mann *et al.* 2000b). Even among these few species, our knowledge of individual social relationships is reasonably advanced in only one, the bottlenose dolphin. And of the numerous locales where bottlenose dolphins have been studied, we have learned the most about their social relationships in a population in Shark Bay in Western Australia.

As we observe wild bottlenose dolphins in Shark Bay we can see what they do and, with the logic of natural selection, sometimes infer why. That inference can then be used to generate hypotheses and predictions suitable for testing. But at present we have no knowledge of what sort of mental or cognitive processes guide their decisions or whether they are conscious of their decisions, goals or beliefs, or the beliefs of others[3]. Our studies in Shark Bay are thus relevant to what Kacelnik (Chapter 2, this volume) calls 'biological rationality', but not yet to psychological rationality in a sense that requires actions to result from reasoning or from other specific cognitive processes. Whether the behaviours we observe can be interpreted in terms of an intermediate 'economic' sense of rationality is an interesting further question that we do not try to answer here. However, our observations *can* help us understand how the cognitive abilities demonstrated in captivity (see Herman and Tschudin, Chapters 19 and 20, this volume) might be put to use in the wild, not only in the pursuit of social advantage but in resource acquisition as well. As was the case with primates, our studies of the complex behaviour of wild dolphins will, in due course, lead to better understanding of the functions of dolphins' cognitive capacities and to further development and assessment of the Machiavellian Intelligence hypothesis (see Byrne and Whiten 1988).

After providing some general background, we will review evidence of the social bonds among dolphins in Shark Bay, focussing on male alliances, both first-order and second-order, on female relationships, and on affiliative interactions. We have no data on the role that social rank and reconciliation play in the dolphins' lives, but we infer that they may be important on the basis of captive studies (Samuels and Gifford 1997; Weaver 2003; Samuels and Flaherty 2000). Other Machiavellian favourites, including tactical deception and knowledge of third-party relationships, have not been explored systematically in captive or wild dolphins.

....

[3] Metarepresentation—the ability to 'know what one knows' and to attribute belief or 'know what another individual knows' (for discussions, see Call, Shettleworth and Sutton, Proust, Currie,

Some features of Shark Bay dolphin behaviour, such as a fission–fusion grouping pattern, within-group alliances, and parental care lasting several years, invite comparisons with primates (e.g. Connor *et al.* 1992; Connor *et al.* 2000; Whitehead and Mann 2000). By contrast with the case of non-human primates, the social, cognitive, and developmental features of bottlenose dolphins cannot be attributed to a common phylogenetic history with human beings. Dolphins thus provide a window into the underlying social and ecological conditions that might favour convergent evolution of complex forms of behaviour and learning. Like the resident killer whales, the Shark Bay dolphins exhibit features, such individual foraging specializations, that are not common in terrestrial mammals (Connor 2001; Mann and Sargeant 2003). These areas of apparent behavioural convergence and divergence among species should be reflected in the differential cognitive demands on dolphins in the wild.

In the final section, we will return to the Machiavellian hypothesis in light of our review of the social cognition and behaviour of wild dolphins. We will argue that cetaceans present an extraordinary opportunity to examine the relationship between advanced cognitive skills, complex behaviour, and the evolution of large brains. Compared to any group of terrestrial mammals, cetaceans have a greater number of large brained species and, more importantly, more size-matched species that exhibit extreme differences in brain size.

Before we provide details of dolphin relationships in Shark Bay, it will be helpful to have some general background on dolphin life history and reproduction and on Shark Bay dolphin society.

16.1.1 **Life history and reproduction**

Like primates, delphinids such as the bottlenose dolphin, have 'slow' life histories and delayed maturity in comparison with most other mammals. Females in Shark Bay have their first calf no earlier than 11 or 12 years of age (Mann *et al.* 2000a) and may live for

---

Tschudin, his volume)—is often thought to result from selection for social or 'Machiavellian' intelligence. Deceit and its use and detection are also thought to be important signs of social intelligence (Whiten and Byrne 1988). Some authors, notably Trivers (1985) and Alexander (1979), have taken the social intelligence hypothesis to its logical conclusion and suggested that self deception is the result of an arms race between skills at deception and at detecting deception. To put it simply, if you are not conscious of your deceit then others will not be able to detect subtle signs that you are lying. Alexander (1979, 1987) has argued that we may not have conscious access to considerable social information, including (some of) our intentions and motives. If self-deception is a hallmark of human social intelligence, then other large-brained animals (or at least some hominid in our past) might have greater conscious access to their social knowledge than humans; they would 'know more about what they know' than do modern humans. Moreover, it would be interesting to ask if human beings self-deceive not only about their intentions and beliefs, but also about what they know or believe about others. In many social interactions it would pay to conceal subtle clues about what you are able to read in the minds of others.

several decades (Connor *et al.* 2000). Males may be capable of sexual reproduction at a younger age but do not appear to solidify alliances until their mid-to-late teens (Connor *et al.* 2000). A mature male that has not achieved whatever size, rank, or experience is required for alliance membership is likely to employ a different, perhaps more opportunistic, tactic (Connor *et al.* 2000). Indeed, recent paternity analyses suggest that subadult or 'juvenile' males may sometimes sire calves (Krützen *et al.* 2004).

After a 12-month gestation period, a female gives birth to a single calf she will nurse for an average of 4 years (range 3–8 years; Mann *et al.* 2000a). Variation in weaning age may be related to sex-specific strategies. Daughters are weaned, later than sons, with most males weaned at age 3, and most females weaned after age 4 (Mann *et al.* 2003). The respective advantages of early weaning for sons and late weaning for daughters may reflect differences in social advantages to males (developing strong male–male bonds and potential alliance partners) and to females (integration into her mother's network and the development of foraging strategies like the mother) (Mann and Sargeant 2003).

During the third year of her calf's life (but ranging from 1.5 to older) a female will be followed and consorted by males that co-operate in alliances to consort females. We have often used the term 'herding' to describe consort associations because many, and possibly all, are initiated and maintained by threats and aggression. Reproduction is 'diffusely seasonal' in Shark Bay, with a peak in the Austral spring/early summer (September to January), but births can occur during any month (Connor *et al.* 1996; Mann *et al.* 2000). Given the 12-month gestation period, this means that there is a peak mating season as well, but that conceptions, like births, may occur during any month. One of the more interesting puzzles in Shark Bay is the fact that females that conceive in the primary spring/summer breeding season will often become attractive to males during the preceding Austral winter (July to August). Consortships during the winter months are typically shorter than during the breeding season and may play a role in allowing females to mate with many males, confusing paternity and reducing the risk of infanticide (Connor *et al.* 1996). We suggest two possible explanations of why males initiate consortships with non-attractive females. First, as we will describe below, some consorting may have more to do with male–male bonds than female attractiveness. Second, some consortships might be anticipatory. Consortships sometimes last longer than the estimated cycle duration of bottlenose dolphins (30 days; Yoshioka *et al.* 1986) and especially the 5–7-day period of rising oestrogen levels reported by Schroeder (1990; see Connor *et al.* 1996). Goodall (1986) suggested that chimpanzees may consort 'flat' females (with non-tumescent sex skins) to avoid being pre-empted by other males; likewise, dolphins may consort a female that they anticipate will become attractive (Connor *et al.* 1996). Such pre-season consorting might also allow males to assess the female's response to capture, aggression, and other behaviours associated with consorting or to impress females with their vigour.

16.1.2 **General features of Shark Bay dolphin society**

The Shark Bay dolphins live in an open 'fission–fusion society', that is individual dolphins associate in small groups that change in composition, often many times a day (Connor *et al.* 2000). The fission–fusion characteristic is one found in a number of other social mammals with complex social relationships. Among primates, the chimpanzees and spider monkeys have fission–fusion systems similar to that found in bottlenose dolphins (Wursig 1978; Goodall 1986; Symington 1990; Smolker *et al.* 1992). Smolker *et al.* (1992) argued that the constantly changing social milieu of a fission–fusion society can select for social intelligence:

> In some species an individual is almost always in association with the same set of conspecifics, while in others, an individual may rarely associate with the same conspecifics from day to day. Bottlenose dolphins exhibit an intermediate pattern, associating very consistently with a few others, but within the context of a wide range of different party types, containing individuals of varying degrees of familiarity drawn from an extremely large social network. Thus, each dolphin's social relationships are maintained within a constantly changing social environment, perhaps placing a premium on the evolution of cognitive abilities. (pp. 65–66).

This hypothesis has been further developed by Barrett *et al.* (2003) to explain differences in the cognitive abilities of monkeys and apes.

The 'open society' Shark Bay dolphins inhabit is dissimilar to primate societies in that primates almost invariably live in groups with strong social boundaries. In our current 200 km^2 study area off the east side of Peron Peninsula in Shark Bay, there are no boundaries demarcating closed or semiclosed groups such as one would find between two baboon troops or chimpanzee communities. Rather, we find a pattern of variably overlapping home ranges extending from one end of the study area to the other and offshore. Of course, we do not know what occurs beyond our study area.

A large number of individuals live in our study area; we have over 600 individuals in our current identification catalogue. It is unlikely that all of these individuals know each other, as the home ranges of many do not overlap. What must happen often in such a society is that dolphin A knows dolphin B who knows dolphin C who does not know A because A and C's ranges do not overlap or overlap only slightly. This again is very different from the case of primates, who are likely to know all of the members of their semiclosed groups. Primates might be able to develop a reasonably complete knowledge of relationships between group members and use this information in strategically advantageous ways (e.g. Seyfarth and Cheney 2001). From a given dolphin's point of view, living in an open society can exacerbate a problem facing individuals living in fission–fusion societies: uncertainty about changes in third-party relationships may occur out of sight and sound or 'off-camera'. Imagine, for example, that during occasional forays to the edge of their range a male dolphin alliance occasionally encounters a potential rival, but has no knowledge as to whether the rival is supported by two or twelve alliance partners.

The number of individuals an individual knows has been used as an indirect measure of social complexity in order to test the Machiavellian Intelligence hypothesis (Dunbar

1992, 1998; but see van Schaik and Deaner 2003). Counting the number of relationships in non-primates is as easy as counting the number of group members; but how do you count associates in an open society? Without following individuals constantly, you can't. However, Connor et al. (in prep) took a 4-year 'snapshot' of the number of associates an individual has in resting, travelling, and socializing groups combined. This number is likely to underestimate even an 'instantaneous' measure of associations, since we observe such a tiny fraction of the dolphins' dynamic social lives. Two interesting results emerged; the number of associations (typically 60 to 70) were similar to those found in the largest large primate societies (Dunbar 1992; Barton 1996), and the largest (and smallest) numbers of associations were found in females (a few were over 100). The variation in group size for females may be related to foraging strategies, thus raising the possibility of trade-offs between costs and benefits of foraging and social strategies in females.

The Shark Bay dolphin society also differs dramatically from primate societies in exhibiting natal philopatry by both sexes, a phenomenon that may be ubiquitous in cetaceans, or nearly so (Connor 2000). Males may have larger ranges than females, and while we cannot exclude emigration by some individuals, it is clear that normally members of both sexes continue to maintain their natal range in their adult range (Connor et al. 2000; Krützen et al. 2004). From a cognitive perspective, this means that individuals of both sexes can potentially begin negotiating relationships from infancy that may be important when they are reproductive adults.

Shark Bay dolphins maintain their strongest social affiliations with members of the same sex (Smolker et al. 1992). We begin our discussion of social relationships by exploring the cognitive challenges facing a male dolphin trying to negotiate the Shark Bay network of alliances.

16.2 **Male alliances in Shark Bay: structure and complexity**

Male dolphins in Shark Bay exhibit multilevel alliances within their social network. Groups of two to three males (first-order alliances) co-operate to sequester and consort females for periods of minutes to over a month (Connor et al. 1992a, b, 1996). Teams of first-order alliances make up second-order alliances, which co-operate to take females from other alliances and to defend against such attacks. Males bonds within some first-order alliances show extraordinary short and long-term stability. In a given year, certain males will be observed together in most or all sightings, yielding half-weight coefficients of association (COAs)[4] of 80–100 (Smolker et al. 1992;

---

[4] COA = coefficient of association, a measure of how often two individuals are found together. We employ the 'half-weight' method (Cairns and Schwager 1987), defined as $100 \times 2Nt/(N_A + N_B)$ where Nt is the number of groups containing both individuals A and B, and N_A and N_B are the total number of groups for each individual, respectively. COA's range from 100 for two individuals that are always in the same group, to 0 for two individuals that are never together.

Connor *et al.* 1992a). Bonds between males in stable alliances have endured for as long as 20 years. Stable first-order alliances typically form second-order alliances with one or two other stable first-order alliances. Occasional transient relationships among male dolphins cut across alliance levels (Connor *et al.* 1992).

Males in a fourteen-member second-order alliance called the 'super-alliance' (Connor *et al.* 1999) formed first-order alliances that were labile, but variably so. Males in this group participated in 17 to 57 per cent of their consortships of females with their most common first-order alliance (all trios). This result suggests a general correlation between first-order alliance stability and second-order alliance size. In 2001, we began a study of over 100 adult males in the community, including several other large groups, and current trends suggest that, in general, the predicted correlation will be found. Why might such a relationship exist? A simple explanation is that in larger groups more males are available (not already in a consortship) at any given time to form a trio and consort a female. A more complex and interesting explanation would be that forming first-order alliances with more males helps to maintain second-order alliance bonds (Connor *et al.* 2001). However, while there may be elements of truth to both of these explanations, neither is sufficient to explain the complexities revealed by more detailed analysis of first-order alliance formation in the super-alliance (Connor *et al.* 2001, below). The males exhibited striking preferences (and avoidances!) in whom they formed first-order alliances with, and males in more stable first-order alliances spent more time in consortships with females (Connor *et al.* 2001). Consistent choice of alliance partner, however, did not yield the same advantage (i.e. a male forms first-order alliances with the same partner but the third member of the trio varies; Connor *et al.* 2001). These results suggest a complex social structure, perhaps including important dominance relationships, within the super-alliance.

16.2.1 **Alliance levels**

Alliances within social groups, common in primates and some large social carnivores, but infrequent elsewhere, are a canvas for complex social relations (Harcourt 1992; Cords 1997; de Villiers *et al.* 2003; Engh *et al.* 2005). Why is this so? Firstly, within-group relations are triadic (Kummer 1967) so individual A might form an alliance with B against C. Second, relationships are mediated by affiliative interactions, so individual B might recruit A as an alliance partner against C by grooming her. But imagine that C is also trying to recruit A (perhaps A is high ranking) and that there is competition for allies (e.g. Seyfarth 1976). This combination of characteristics, triadic interactions mediated by affiliative behaviour, so prevalent in within-group interactions, is not found in interactions between groups in non-human primates. While two groups of monkeys might simultaneously threaten or chase a third (for example members of two bands of hamadryas baboons on a sleeping cliff might threaten a third band), there is no evidence that such interactions reflect a higher-level allegiance between groups that is maintained by grooming between members of different bands.

While threats from other groups might temper aggression within groups (e.g. if you kill a member of your group you may be more vulnerable to attacks from other groups), imagine how much more complex decisions would become if relations between groups were triadic and mediated by affiliative interactions. Suppose that members of one social group use affiliative interactions to recruit allies in contests against other groups. Now when individual B considers recruiting A against C, B might also have to consider potential consequences at the next level of alliance, when her second-order alliance ABC tries to recruit alliance DEF against alliance GHI. For example it is possible that conflict within their group—if C is injured—might make their group a less attractive ally for alliance DEF who would then be more likely to ally with GHI against ABC. If this has a ring of familiarity, it is because the one primate to which such nested or 'hierarchical' alliances are extremely important is *Homo sapiens* (Boehm 1992). Human beings routinely navigate a landscape of nested alliances that range from kin factions to nation states (see discussions in Connor 1992a; Connor and Krützen 2003).

The potential of nested alliances to exacerbate demands on cognition in the social realm has long been under-appreciated. Among non-human primates, we find the best example of nested within-group alliances in female-bonded old-world monkeys. Matrilineal relatives form first-order alliances and members of matrilines share adjacent ranks. Most of the monkeys' alliance behaviour can be explained by two simple rules—ally with members of your matriline against non-members and with high-ranking non-relatives against low-ranking non-relatives. Additional complexity is suggested by the occasional reversal of the second rule, in which lower-ranking matrilines unite to overthrow a higher-ranking matriline (Samuels *et al.* 1987; Chapais 1992; Gygax *et al.* 1997) and the existence of more transient social relationships among females from different matrilines (Seyfarth and Cheney 2001). Nonetheless, if an individual's relationships at the first alliance level (e.g. matrilines) are entirely predictable (or nearly so), then a second alliance level might not introduce a significant burden on social calculations—compared, say, with the burden for a species that negotiates only one level but of relatively unpredictable alliances (Connor and Krützen 2003).

An alliance landscape can be considered complex to the extent that it entails strategic options and risk. This will be true if individuals can benefit by developing or changing alliances strategically and if the bonds they form are always at risk because of the strategic options available to others. If a female monkey has no real options other than to ally with her matrilineal kin, then her alliance landscape is simplified. On the other hand, increasing the number of options (e.g. choosing allies from among a large number of non-relatives) may not, by itself, increase complexity unless there is also significant risk involved in decision making. In general, a high-risk alliance is one where alternative choices can result in a significant increase or decrease in reproductive success. The latter might result from injury, a loss of rank, or expulsion from the social group. High-risk alliances, such as those that might occur among non-relatives and/or where the consequences of losing are severe, would place a premium on social intelligence. Choosing the wrong ally or the wrong time to form an alliance could be very costly. Our female monkey who has limited options as to her choice of first-order allies—the matriline she is born into—may engage in high risk alliance behaviour with other matrilines when they join forces to overthrow the top-ranking matriline. In addition to the second order alliances between female old world monkeys, other examples of high-risk alliances in primates are first order alliances between individual male chimpanzees and New World cebus monkeys (de Waal 1982; Nishida 1983; Perry 2003).

Sometimes hierarchical association patterns are interpreted too generously as hierarchical alliances. Kitchen and Packer (1999) claim that 'discrete social groups' of African savannah elephants and Hamadryas baboons 'show second-, third-, and fourth-level alliances'. There is no evidence, however, of repeated agonistic interactions between groups at each level of association for either of these species (see Connor *et al.* 1992a for a discussion of male alliances in Hamadryas baboon). While further study might reveal hierarchical alliances in some cases, hierarchical social structure can clearly exist for other reasons. For example Mitani and Amsler (2003) report male subgroups, but not alliances, among male chimpanzees in an unusually large community of 140 individuals in Uganda. They suggest that the subgroups, distinguished by the age and rank of their members, are an outcome of peer relations in a large community where integration into the adult social network is difficult. By contrast, both first- and second-order dolphin alliances operate within the social network and are mediated by affiliative interactions (Connor *et al.* 1992a, b, 2000, submitted, and see below).

To assess how complex the dolphin alliances are, we need to consider each level of alliance. Pairs and trios that are highly stable and include relatives (Krützen *et al.* 2003) might be the most predictable and risk free. However, some stable alliances are not composed of relatives; it is not clear that even a majority are. Moreover, over a period of years we have seen shifts that suggest a significant risk even in stable alliance relations.

16.2.2 Shifting alliance relationships

The following sort of interaction has been reported in chimpanzees: a beta male attempts to overthrow the alpha male by forming an alliance with the third ranking male, only to have the alpha and third ranking male join forces against him, leading to a drop in rank. Here the beta male took a significant risk and lost. We might expect relationships to be less risky if they are at either extreme of stability. A perfectly stable alliance relationship likely entails less risk for its members, and highly fluid alliance relations suggest that members are interchangeable, so interactions should pose little risk or cognitive challenge for that matter—if you fail to acquire one alliance partner you simply switch to another. Some male dolphins' alliance relationships are highly stable and some are quite fluid. In the rest of Section 16.2 we examine patterns of shifts in alliance membership closely for evidence of strategic options and risk.

16.2.3 Shifting relations in 'stable alliances': a 17-year history

The potential for strategic behaviour among males in stable alliances is shown by alliance shifts over time. In Box 1 we illustrate this with a 17-year history (1985–2001) of two males whose alliance affiliations intersected during this period but who have very different histories; Real Notch (Rea) and Lucky (Luc, who apparently isn't very lucky!). Three salient points emerge from the review of Luc and Rea's social histories. First, stable alliances aren't really very stable; rather, they are at the stable end of a continuum. Although infrequent, the changes that occur (such as Luc's apparent expulsion from the alliance), suggest that stable alliance bonds carry a significant risk. Second, when observing these interactions between dolphin *alliances*, it is difficult not to see parallels with interactions between *individual* primate males. We can speak of social relation-ships between alliances, not simply between individuals. Third, the history suggests how changes at one level of alliance might impact another. When the trio of Rea and Hii and Poi quit associating with Luc in 2000, they were left without a second-order alliance partner. While they increased their association with some members of two large second-order alliances (achieving COAs of 20–30 with several of these males in 2001), as of 2003 they still did not associate consistently with any other alliance.

Box 1 **Real Notch and Lucky: a 17-year history**

During 1985–89 Lucky (Luc), who appeared to be a small adult, maintained a mutually strong association with Poi, a juvenile. Each also associated with another juvenile, Lod (Table 16.1). During 1985–87, Luc and Poi maintained modest associ-ations with members of a second-order alliance composed of the trio Snu and Sic and Bib and the pair Wav and Sha. Luc and Poi and Lod did not consort females in the 1980s.

Meanwhile, by at least 1986, Rea and his associates, all adults, were consorting females on a regular basis. In 1985, Rea was in a pair with Hac and Hii was in a pair with Pat; these two pairs associated closely (Smolker *et al.* 1992). These four males associated only occasionally with two stable trios, Tri and Bit and Cet, and Cho and Bot and Lam (Table 16.1).

An important shift occurred in 1986 with the disappearance of Hac and Pat. Before disappearing, Pat received aggression from his own alliance partners, including a co-ordinated attack with Cho and Bot and Lam. In the ensuing attack, Pat was bitten and chased away. With Hac and Pat gone, Rea and Hii formed a very strong bond (COA = 100) and from early 1987 to April 1989, they split their time between the trio Tri and Bit and Cet, and the trio Cho and Bot and Lam. This shift was associated with a sharp decline in the association between the latter two trios (whose between-alliance COAs dropped from 55–71 in 1985–1986 to 7–16 in 1987–88).

Box 1 **Real Notch and Lucky: a 17-year history** (*Continued*)

On rare occasions when all three alliances were together, aggression and avoidance revealed conflict. To illustrate, in one sequence Rea and Hii dropped behind the group, surfaced side-by-side synchronously, then charged into the group and, with Cho and Bot and Lam, proceeded to chase off Tri and Bit and Cet. About half an hour later, Cho and Bot and Lam separated from Rea and Hii and a few minutes later captured the female Yog. In another 2 minutes, Rea and Hii bolted into the trio controlling Yog and there was aggression between Rea and Cho. This was followed by an intense petting session among the males, including a bout between males from the different alliances. The two alliances later separated, and shortly afterwards Cho and Bot and Lam released Yog. Then the other trio, Tri and Bit and Cet, while petting each other, approached and joined Cho and Bot and Lam for about 15 minutes, before departing again. Five minutes after their departure, Rea and Hii approached and joined Cho and Bot and Lam. Again, there was an interalliance petting bout.

There was another transition year in 1989, as Cho and Lam vanished and Bot joined Rea and Hii to form a new trio. Rea and Hii and Bot continued to associate with Tri and Bit and Cet through 1989, after which the latter trio disappeared.

It is at this point, in early 1990, the alliance histories of Rea and Luc intersect. From 1990–1994, the trio Rea and Hii and Bot enjoyed a second-order alliance relationship with Luc and Poi, who also began consorting females during this period.

The first indication of Poi switching allegiance from Luc to Bot occurred during a month-long period in 1994. Rea and Hii consorted a female for 35 days from 21 Jul–24 Aug. Bot, excluded from this consortship, paired with Poi for the period at the apparent expense of Luc. On each of the 8 days they were observed during this span, Bot and Poi consorted a different female. We suspect that some of the consorting by Bot and Poi had more to do with their new, and probably fragile, first-order alliance bond than with the reproductive state of the consorted female (one of the females they consorted had a 1.5-year-old calf and was unlikely to be receptive). During this period Lucky was observed 'shadowing' the other males at a distance and leaving female groups before they joined. After the 35-day consortship, Bot reformed the trio with Rea and Hii, who associated often with the pair, Luc and Poi.

In 1995, Luc associated infrequently with Poi and the other males. Luc spent more time with Poi *et al.* in 1996; three times Luc formed a trio with Bot and Poi to consort a female. An interesting and unusual sequence later that year suggests how some pair switches may simply be expedient. One day the female Squ was being consorted by Rea and Hii and Bot, who were also in the company of Luc and Poi. The next day on they were observed, Luc was gone and the pair Rea and Bot were consorting Squ, while the novel pair Hii and Poi consorted the female Try. A few days later Hii and Poi were consorting the female Puc, who they consorted for nearly a month. When Rea

Box 1 **Real Notch and Lucky: a 17-year history** (*Continued*)

and Bot ceased consorting Squ after 3 weeks, Hii and Poi still had Puc. Given that pair changes rarely, if ever, occur during a consortship (Connor *et al.* 1996; Connor and Smolker 1995), it not surprising that Rea and Bot proceeded to consort more females. This sort of observation suggests that availability can play an important role in explaining alliance shifts. On the other hand, availability does not explain why Hii left the trio to ally with Poi instead of Poi's usual partner Bot. Luc rejoined Poi for the last 2 months of observation (October to December), occasionally consorting females and associating with the trio Rea and Hii and Bot.

From 1997 to 1999, Luc associated infrequently with the others, who associated as two pairs, Rea and Hii and Bot and Poi. Luc's association coefficient was slightly higher in 1997 (Tables 16.1) because of an increase in his association for 2 months late in the breeding season (November and December). However, Luc failed to pull off a similar 'late season rally' in 1998. As Luc's association with Poi and the others declined during 1997 to 1999, he was increasingly found with older (9 to 11-year-old) juvenile males (Coo and Smo and Jse and Urc).

Bot was gone by 2000, which coincided with a return to the 1990–1994 pattern: Luc and Poi, associating with Rea and Hii. Concomitantly, Luc's association with the older juveniles fell sharply (Table 16.1). Unfortunately for Luc, in 2001 Poi abandoned him to join Rea and Hii and form a new trio. Luc did not return to associating with the older juveniles, who were now 11 to 13-years-old. Instead, Luc has since been found almost exclusively in female groups.

16.2.4 Shifting relations in the super-alliance

Whereas partner shifts in stable alliances were usually associated with disappearances (or the lingering presence of Luc in the mid-1990s), shifts were the norm in the 14-member super-alliance. Based on 100 consortships of females observed during the 3-year study (1995–1997), we documented 39 different first-order alliances in the group. Male trios were involved in 95 of the consortships and only five consortships were by pairs of males (we have never conclusively documented a consortship by more than three males). No consortships involved males from outside the 14-member group. Each male was observed in 10–30 consortships with 5–11 different alliances and 5–11 alliance partners.

As with stable alliances, partner shifts in the super-alliance occurred between, not during, consortships. The frequency of the shifts in this group might suggest that for the purposes of consorting females, males in the super-alliance have no preferences about male partners. Cognitively, the males could be following a simple 'equivalence' rule (Shusterman *et al.* 2000), consorting with any available male in the group. Connor *et al.* (2001) tested and rejected this hypothesis; super-alliance males exhibited strong

Table 16.1 Alliance affiliations of Luc and Rea from 1985–2001

	Year														
	1985	1986	1987	1988	1989	1990	1991–1993	1994	1995	1996	1997	1998	1999	2000	2001
Real Notch															
Hii	81	100	100	100	100	100	95	100	100	92	98	96	98	95	95
Hac	79	64	64												
Pat	100	85	85												
Cho	10	16	35	46	33										
Bot	10	19	38	43	79	81	94	96	90	94	91	76	68		
Lam	10	15	34	46	33										
Tri	14	26	42	39	56										
Bit	14	27	39	37	51										
Cet	15	16	31	31	51										
Luc	0	0	6	0	0	52	50	50	0	36	19	10	4	52	0
Poi	0	6	5	0	0	48	59	76	79	90	89	74	68	69	95

Lucky

Poi	**79**	**64**	**73**	**76**	**82**	**86**	**77**	**76**	0	**48**	22	10	15	**73**	0
Lod	**39**	**34**	**66**	**50**	**76**										
Rea	0	0	6	0	0	**52**	**50**	**50**	0	**36**	19	10	4	**52**	0
Hii	0	0	6	0	0	**52**	**47**	**50**	0	**41**	19	10	4	**52**	0
Bot	8	3	12	13	5	**38**	**41**	**44**	0	**43**	19	9	4		
Snu	**37**	**38**	**27**	0											
Sic	24	**35**	**32**	0											
Bib	**29**	**25**	**27**	0											
Wav	**36**	**40**	**31**	0	10	0	0	0							
Sha	**29**	**39**	**37**	0	10										
Jse								0	**40**	5	20	**32**	**50**	10	11
Urc								0	13	0	**39**	**29**	**47**	6	17
Coo								10	**38**	0	18	**28**	**54**	9	16
Smo								**38**	**40**	0	18	19	**46**	13	11

Numbers in cells are halfweight association coefficients (COAs). A blank cell indicates that a male has died or disappeared. Following Smolker et al. 1992, all coefficients over 24 are in bold.

and significant partner preferences and avoidances within the group. Alliance relationships within the super-alliance are thus individually differentiated (a more complex hierarchical equivalence model might do it justice, Schusterman and Kastak 2002, but see Seyfarth and Cheney 2003). The sheer size of the super-alliance is also of interest in light of the correlation in primates between neocortex size and the number of primary social associates an individual maintains (Kudo and Dunbar 2001). Whether one uses the total group size (14) or the number of first-order alliance associates in the super-alliance (5 to 11), the number is larger than that found in most primates.

16.2.5 Shifting relations among the provisioned males and their allies

Frequent alliance partner shifts also characterized relations among three provisioned males that made daily visits to a beach where they accepted dead fish from humans (Connor and Smolker 1985; Connor *et al.* 1992a). This trio of provisioned males were second-order allies of a pair in 1987, which itself became a trio in 1988. Two of the three provisioned males were paired for any given consortship, and the excluded male sometimes paired with a male from their 'buddy' alliance. We suspect that the free food played a relatively indirect role in the extraordinary rate that the provisioned males consorted females (over 250 cases in 1987–1988!). A substantial reduction in fish fed had no apparent affect on the frequency of consorting. Feeding may have exacerbated conflict between the males, who had to pay attention to fish buckets arriving from the shore as well as to the female, who was often trying to escape in the opposite direction.

We suspect that much of the consorting by the provisioned males had more to do with maintaining fragile male–male bonds than the proximity to female ovulation.[5] Observations suggest that some of the females they consorted may have not been receptive. For example, only the provisioned males were observed to consort females with newborn calves and 53 per cent of the females they consorted in 1987 were back with them in 1988 (having not produced a surviving calf), compared to 11 per cent for the non-provisioned alliances during the same period (Connor *et al.* 1996).

For two consecutive years before the provisioned males died in 1989, the frequency of partner changes among them was high from May to August (3 to 11 changes per month), before dropping significantly during the breeding season months of September to December (1 to −2 changes per month; see Connor *et al.* 1992a; Connor and Smolker 1995). Again, although consortships lasted longer during the breeding season, reduced availability of male partners during the breeding season could not

---

[5] As in the month in 1994 when the pair Bot and Poi were consorting a different female during each sighting; see Box 1.

explain this result (Connor and Smolker 1995). Rather, we suggest that frequent shifts prior to the breeding season may function in testing or competing for potential alliance partners and forming alliances (Connor and Smolker 1995).

16.2.6 Shifting relations among second-order alliances

Interactions between males that belong to different second-order alliances may be hostile or amicable and such relationships can change over time. This is illustrated by relationships between three large second-order alliances that have extensively overlapping home ranges, the original super-alliance (the WC group), and the PD and KS groups. During the original super-alliance study, the WC group had 14 members that shared COA's of 34 to 97 (81 per cent were in the 40 to 70 range). PD group members were juveniles in the mid-late 1980s and formed their second order alliance during the mid-1990s; COA's in the group ranged from 26 to 73 during 1994–95, 56–96 during 1996, and 75 to 91 during 1997; that is the lowest COA during 1997 was greater than the highest in 1994–1995. The large (at least 11 members) KS group was more of a mixed bag during this time with mostly older juvenile/maturing males, but at least one was clearly an adult that had been apparently excommunicated from his original group (but unlike Luc, he found a new home).

Interactions between the WC group and the KS and PD groups during the 3 years of the WC study were infrequent and hostile. By 2001 the WC group was down to 10 males; three of the original 14 had disappeared and one joined a different second-order alliance. The KS group (now a coherent second-order alliance with 14 members) and the PD group were intact and, importantly, several years older. In 2001, the PD group was observed associating (sometimes with affiliative interactions) several times each with the KS group, with the WC group and even with the trio Rea and Hii and Poi; however, the WC and KS groups did not associate with each other.

Two observed interactions between the second-order alliances WC, KS, and PD suggest a third level of alliance formation. The more interesting of these interactions involved all seven members of the PD group, five members of the KS group, and all ten members of the WC group. We first encountered the PD and KS groups together. Two trios in the PD group were consorting females and four KS members had a female. It was not clear which KS males were consorting the female, which may have been a source of conflict in the group as two KS males chased and fought. Then a trio of WC males blasted into the group and the aggression level escalated, with much chasing and aggressive vocalizing. One of the KS members involved in the fight was surrounded and attacked by at least six to eight males. Later the WC trio left briefly to meet seven other WC males that were leaping in from hundreds of meters away, so all ten WC males surged past our boat into what was now a very tight group of KS and PD males. This entrance resulted in a group of 22 males engaging in an especially intense and chaotic period of chasing, fighting, and splashing, with a cacophony of aggressive vocalizing that at one point, incredibly, stopped abruptly. Finally, the five members of the KS group left, minus their female, which was now with the first WC trio that joined. After the altercation, the PD group still had their two females. Neither of the two later-arriving WC trios had a female when they entered and they evidently did not try to take the females from the PD group. Critically, in neither this nor the other observed interaction were we able to determine if the PD group 'took sides' or remained neutral.

Second-order alliances occur when two first-order alliances co-operate against another first-order alliance. Similarly, co-operation between two second-order alliances against another second-order alliance would constitute a remarkable third level of within-group alliance in the Shark Bay society. If, in all the chaos of the fight, we had been able to determine that the PD group sided with the WC group against the KS group, this criterion would be met. Our observations fell short of that, but the affiliative associations we see between second-order alliances and the presence of three second-order alliances in a fight is certainly suggestive. Thus, at this stage we cannot claim that males form third-order alliances, but we can say that there are generally affiliative relationships between some second-order alliances, as indicated by associations between members and even complete groups that sometimes include petting between males from the different groups.

16.3 Female relationships in Shark Bay

Female relationships display greater flexibility and lability than male relationships. While members of a male alliance are frequent associates, females are likely to spend less that 30 per cent of their time with their closest female associate (Smolker *et al.* 1992). The average group size for adult females is four to five animals, but they range from being highly solitary to highly social (Smolker *et al.* 1992; Mann *et al.* 2000a; Connor *et al.* 2000). While the competitive nature of male relationships is obvious from the striking agonistic interactions among males, observations of female–female agonistic interactions are rare. Of the few cases that were observed, nearly all occurred in the provisioning area in Monkey Mia, Shark Bay where three to four adult females have been hand-fed fish by tourists standing in knee-deep water over the last 30 years (Mann and Smuts 1999; Mann and Kemps 2003). Thus, given our present state of knowledge, we classify females as tolerant in their social relationships (Scott *et al.* 2005).[6]

Access to food is likely to be the limiting factor in female reproductive success (Mann *et al.* 2000a; Mann and Watson–Capps 2005). Thus, one might expect females to compete over food. However the nature of dolphins' food source—single, mobile, difficult-to-capture prey—may limit competition. Each dolphin catches and typically quickly swallows her prey whole. Shark Bay dolphins, whether female or male, almost never

---

[6] Having been observed only a few times in 15 years and 1960 hours of observation on focal adult females (Scott *et al.* submitted).

steal each other's fish. This 'ownership' rule extends even to fish tossed several meters (Connor *et al.* 2000). Such respect for ownership is notable. Moreover, unlike many carnivores, females do not share prey, even very large fish catches, with their offspring (Mann and Sargeant 2003). Thus dolphins are both 'polite' and 'selfish'; they neither steal nor share food. This characteristic pattern may further support tolerance amongst female dolphins. They stay close when resting, socializing, or travelling, but can disperse easily for foraging, which is more of a scramble than contest competition. Nevertheless, it remains possible that dolphins may be excluded by others from particular foraging areas, or from active-feeding groups (see Wilson *et al.* 1997). We have seen dolphins leap toward a group actively feeding on a fish school, only to stop short and simply watch without joining.

Each female has a distinctive foraging profile with a limited range of foraging techniques (Mann and Sargeant 2003). These may well dictate the patterning of her social relationships, since some of those techniques (such as sponge-carrying, a form of tool-use) seem to require a high proportion of her activity budget (Smolker *et al.* 1997; Mann and Sargeant 2003). Some females use techniques rarely used by other members of the population and/or they may become specialists, using predominantly one foraging tactic. For example four females regularly beach themselves to catch prey in a specific area of Shark Bay, despite the apparent risk of stranding (Sargeant and Mann 2003; Sargeant *et al.* in press). Although dozens of dolphins regularly associate with the beaching dolphins and can clearly view this behaviour, they have never been observed beaching (Sargeant *et al.* in press). Thus, these females engage in a highly specialized foraging tactic that may require years of practice. The five calves born to these beaching females have been observed engaging in intermediate stages of the behaviour (Sargeant *et al.* in press). At least two of the beaching females have been using this foraging method for over 10 years. Females who engage in similar foraging tactics may be more likely to associate, if only because of extensively overlapping home ranges; the beaching females commonly associated even when away from the beach (Sargeant *et al.* in press).

Female associations may be related to reproductive state (Connor *et al.* 2000), but evidence on this is currently lacking. Females in the same reproductive condition are likely to experience similar energetic constraints and needs. During lactation, association with other lactating females may reduce predation risk and provide social opportunities for the calf. Pregnant females are typically still nursing their previous calf until mid-way through the 12-month gestation period (Mann *et al.* 2000a), and the presence of a large dependent calf may have a greater influence on her activity budget than pregnancy *per se*. Cycling females may co-operate to reduce the costs of harassment by juvenile males (Connor *et al.* 1992; Connor *et al.* 2000). However, since cycling females are relatively rare (dolphins operate with a skewed operational sex ratio, given 4 to 5-year inter-birth intervals), most of the available female associates for cycling females would be lactating.

Females have been observed jointly mobbing sharks (Mann and Watson-Capps 2005; Mann and Barnett 1999), but not attacking males. However, there is some anecdotal evidence illustrating the potential of female tactics even against adult males. For example while Puck was being consorted by three males (Rea and Hii and Bot), Puck suddenly sped up to join an all female group. As soon as the males joined the group of females, the females began petting and rubbing with all three adult males. Puck was flanked by two females and slowly escorted away from the group. When Puck was about 50 meters outside of the group, she bolted and the two escorts slowly returned to the group. Moments later, the males apparently noticed Puck's absence and suddenly broke from petting and rubbing with the females. They bolted in three different directions, but failed to find Puck until the next day.

It is tempting to interpret these events as evidence for tactical deception rather than as a confluence of coincidences, since all the behaviours exhibited are infrequent. It is not common for three males to be involved in petting females at the same time or for a single female to be flanked by two other females. By flanking Puck, the females essentially concealed her, both visually and acoustically, from the males. The slow movement away from the group prior to the rapid bolt appeared 'intentionally' deceptive; but in any case, the males were clearly fooled.

Most of a female's life is consumed with calf care and balancing the trade-offs between maternal care and foraging. During foraging, females accelerate to chase fish and cannot easily maintain contact with their calves. Calves begin to learn to hunt as early as 4 months of age; their hunting increases steadily throughout dependency (Mann and Sargeant 2003). Mothers may reduce the costs of lactation by facilitating foraging skill in their calves. Although some foraging tactics are probably learned individually without social exposure, most of the techniques probably involve social learning, primarily from the mother (Mann and Sargeant 2003). Social learning (broadly defined as learning that is influenced by conspecifics through mechanisms such as imitation, social facilitation, local or stimulus enhancement) is clearly implicated in dolphin foraging, although the precise mechanisms are not well understood. There is ample field evidence for social learning in the acoustic domain for cetaceans (e.g. see Janik and Slater 1997; Noad et al. 2000; Deecke et al. 2000), but evidence for social learning of gestures and motor movements has been limited to captive studies of bottlenose dolphins (Janik 1999) and one field study (Mann and Sargeant 2003).

Foraging presents an appropriate avenue for investigating social learning in bottlenose dolphins because they exhibit diverse foraging techniques both within and between populations (Shane 1990; Connor et al. 2000; Mann and Sargeant 2003). In Shark Bay we are examining the matrilineal patterns of foraging, the ontogeny of foraging among calves, and foraging patterns of the larger population. Foraging might represent the best examples of socially learned motor (as opposed to vocal) activity and the development of calf foraging might be a critical factor in determining the length of nursing or nutritional dependency on the mother (ranging from 3 to 8 years in Shark Bay;

Mann *et al.* 2000a). An understanding of skill development during infancy informs theoretical models about life history schedules. Our current work demonstrates that at least four foraging strategies qualify as 'traditions,' socially-mediated learning that is transmitted across generations. For example, sponge-carrying, the only form of tool use in any wild dolphin or whale (Smolker *et al.* 1997), is a behavior infants learn from their mothers (Krutzen *et al.* 2005) Sponge-carrying emerges between 2 and 4 years of age, and only among calves born to sponge-carrying mothers (Mann and Sargeant 2003). This specialization is exclusive to approximately 33 (mostly female) dolphins, less than 10 per cent of the female study population. It involves finding and tearing off conical marine sponges from the substrate, placing the sponge on the rostrum (beak) and using the sponge to ferret fish from the sea floor (Smolker *et al.* 1997). Sponge-carrying is used in the search process and ceases prior to the final chase and prey capture. Sponge-carrying and other foraging techniques probably involve both social and individual learning, and are relevant to ecological models of learning which predict, for example, that intergenerational transmission of learned techniques is favoured when environmental variability is moderate (see Laland and Kendal 2003).

16.4 **Affiliative interactions**

One of the key features of intragroup bonds in primates is that they are established, negotiated, maintained, and repaired by affiliative interactions. Affiliative interactions, such as grooming in primates, may have beneficial effects, such as removal of parasites or stress reduction; but close physical contact may also act as a potential stressor. In an intriguing but under-appreciated paper, Zahavi (1977) argued that the function of affiliative contact is to 'test the bond' between the individuals. Since physical contact of some kinds, such as kissing or embracing in human beings, would be stressful or otherwise unacceptable between two individuals that do not enjoy a particular bond, it can be used by one party to test the strength of their bond with the other.

16.4.1 **Petting**

Dolphins don't have hands with which to groom each other, but they do 'pet' each other with their flippers and can rub other parts of their body against each other (Connor *et al.* 2000). Petting may reduce stress but may also allow for testing of bonds. A category of petting that occurs often enough to rate special mention in our ethogram is 'mutual face–genital' petting where one dolphin is being petted on the genitals while the other receives petting around the face. Dolphins will also pet each other on the flukes and fins and it is not uncommon to see one dolphin stroke another directly on the blowhole. That touching in these areas is a potential stressor is obvious from watching humans interact with the provisioned dolphins. The dolphins are quite particular about where they will allow people to touch them—basically limiting contact to a stroke down the side. Petting the dolphin on top of the head (where the blowhole is), anywhere around the face or on the dorsal fin or flukes usually results in a warning (head jerk) followed by a bite or hit if the offending party persists.

A key distinction between inter- and intragroup relationships in primates is that only in the latter are bonds formed and mediated by affiliative interactions, allowing triadic interactions (Kummer 1967). Here and elsewhere (Connor *et al.* 1992a; Connor and Krützen 2003, we have argued that having both alliance levels within a social network distinguishes male dolphins from non-human primate males that form two levels of alliance, but only one within a social group. We should expect to find affiliative interactions not just between males in the same first-order alliance but also between males in the same second-order alliance but different first-order alliances. And this is indeed the case; petting occurs both within and between alliances that are members of the same second-order alliance, and even between different but affiliating second-order alliances. However, since a male's primary affiliation is with his first-order alliance we should expect to see males pet preferentially with members of their first-order alliance when two alliances are together.

The most intensively studied males were the three stable alliances that formed the social triangle in the 1980s: the pair Rea and Hii and their relationships with the trio Cho and Bot and Lam and with the trio Tri and Bit and Cet in 1987–1988, and then the relationships of the trio Rea and Hii and Bot with the trio Tri and Bit and Cet in 1989. We examined affiliative interactions when all members of two alliances were together. Although petting was frequently detected, telling who is stroking whom is difficult when the principles remain underwater. Thus sample sizes are unfortunately small and the results equivocal. Petting interactions between the pair Rea and Hii and the trio Tri and Bit and Cet ($n = 49$) were strongly correlated with alliance membership, but interactions between Rea and Hii and the different trio Cho and Bot and Lam were not ($n = 40$) (Connor *et al.* submitted). While petting between alliances is observed, especially during excited interactions with females, and is probably very important for maintaining co-operation between alliances, we suspect that with a larger sample size it will be shown that males pet more with their first-order alliance partners.

16.4.2 Synchrony

In addition to the elaborate synchronous displays males perform around females (Connor *et al.* 2000), male dolphins often 'synch' (surface side-by-side synchronously; see also Herman's discussion, this volume, of synchrony in captive dolphins). Since this behaviour occurs at the surface, individual participants are easier to identify than in petting. We found a strong relationship between first-order alliance membership and synchrony (Connor *et al.* submitted). Further, synchrony and petting were highly correlated in the relationship between the two alliances for which we had the largest sample size of both petting and synchs (Connor *et al.* submitted).

Synchs recorded between males of different alliances were more common when the males were socializing than when they were travelling or resting. This interesting result also holds when we remove petting from the analyses and focus specifically on 'excited socializing' that includes splashing, displays, chasing, and sexual behaviour. Tension between alliances should be greatest during these conditions and approximately 75 per cent of interalliance synchs occurred during excited socializing with the female consort or, occasionally, another female in the group. Thus, the relationship between interalliance synchrony and social behaviour may indicate that synchrony serves to reduce tension and/or to signal co-operation, as in some forms of primate affiliative behaviour (Aureli *et al.* 1999). Synchronous surfacings begin at birth and newborn calves have very high rates of synchronous surfacing with their mothers (Mann and Smuts 1999). This may be where calves learn the relationship between movement and bond formation that is so important also in adulthood.

16.4.3 Contact swimming: a female-specific affiliative behaviour

Because females do not form the strong alliances found in males, are rarely observed behaving aggressively, and have a much 'looser' network of same-sex associates, one might be tempted to conclude that social bonds are not that important to females. However, not only do females pet each other, but a striking category of pectoral fin contact behaviour is almost exclusively conducted between females (Richards 1996, Connor *et al.* submitted). In *contact swimming*,

> ...one dolphin (actor) rests its pectoral fin against the flank of another dolphin, behind the other dolphin's pectoral fin and below or just posterior to the dorsal fin. Two individuals swimming in this close staggered fashion are highly visible. (Connor *et al.* submitted)

A common context for contact swimming is mixed sex groups with males consorting or harassing females. These observations indicate that contact swimming may function to signal co-operation between females and/or reduce stress (Connor *et al.* submitted). Additionally, the staggered position of contact swimming suggests that the trailing female may enjoy a brief 'free ride', but we suspect the real importance of this slight altruism is its value as an honest signal from the ride-giver.

16.4.4 Sociosexual behaviour

Sexual behaviour in bottlenose dolphins is observed in a wide range of contexts, from aggression to affiliation (Connor *et al.* 2000). In addition to male–female sexual behaviour, observations among males range from one alliance member mounting another in a very relaxed manner, to one alliance herding another alliance. In the only clearly observed case of its kind, the pair Rea and Hii herded the second-order alliance partners of their rivals, the provisioned males, for 65 minutes (Connor and Smolker 1996). The sexual and aggressive behaviours directed at the male pair included charging, biting, chasing, mounting, and other contact with erections and the 'pop' vocalization that is associated with female consortships (Connor and Smolker 1996).

Studies of dependent infants reveal that male–male relationships are preferential from an early age. This is particularly evident in sociosexual behaviour: male calves prefer to interact with other male calves, when the availability of partners for each age-sex class is controlled for (Mann in press). Male calves were typically the actors, rather than recipients of sociosexual behaviour and were commonly involved in triadic interactions involving three males (Mann in press). Synchronous sociosexual behaviours (simultaneous mounts or beak-to-genital pokes or pushes) were conducted by males almost exclusively, although the recipients could be male or female. Homosexual behaviour is likely to be important in mediating the development of these male bonds, possibly by establishing reciprocity, 'trust,'(cf. Zahavi 1977) and assessing the manoeuvrability and social skills of potential alliance partners. For example male partnerships in sociosexual activities could mediate the development of long-term bonds through taking turns as actor and recipient (symmetrical relationships) and practicing synchronous movement in chasing, mounting, displaying, and goosing (where one dolphin pokes its rostrum into the genital area of another) other males or females. The recipient of sociosexual behaviour is vulnerable by exposing the belly and genital area to one or more males in the advantaged rear position. Role exchanges may be important for establishing trusted allies (Mann in press).

16.4.5 **Rational or emotional dolphins?**

The affiliative and aggressive interactions we observe make it seem obvious to us that emotions play an important role in dolphin social relationships (whether they are conscious of them or not). What role *should* emotions play in a rational dolphin? If a dolphin is interacting with another and assessing their relationship in an economically or adaptively rational way, it should recall all of its interactions with the other individual, whether they were positive or negative, and weight each interaction by its value (for example whether, as a result of the interaction, the dolphin gained a fish or lost an oestrus female). This history should be integrated in some fashion and the output of that integration, in combination with the value of the present interaction, should be used to make a decision about how to behave toward the other dolphin (e.g. pet it, smack it, or ignore it). Aureli and Schaffer (2002) suggest that emotions provide just such a bookkeeping and integration system. Emotions, in their view, function to provide a timely assessment that can guide social decisions. This is very similar to Damasio's view (1994) about the role of emotions in human decision making in social and other arenas. Ironically, as he describes, if affect is removed (say, by a stroke) human beings become incapable of making rational choices (assuming that these must take efficiency and the value of outcomes into account).

16.5 **Discussion: brains, cognition, and behaviour**

Most readers of this volume will be aware that bottlenose dolphins and other delphinids have large brains—larger than great apes of similar body size (Connor *et al.* 1992a).

Herman (Chapter 20, this volume) describes an impressive range of cognitive abilities that probably relate to the complex social lives and foraging tasks described in this chapter. We would like to understand why these attributes evolved in a group of aquatic mammals and if the same selective pressures were at work in the evolution of large brains on land. While these are not easy questions to answer, we nevertheless think that cetaceans have a great deal to bring to the table of comparative studies. This view is based on something that readers of this volume are probably *not* familiar with: the huge variation in brain size among cetaceans.

Delphinids have brains that are two to three times the size of some of the other small toothed whales of similar body size—a ratio similar to that distinguishing humans from the great apes (Connor *et al.* 1992a). Table 16.2 displays some interesting comparisons. For example the first listed member of the Delphinidae (*Sotalia*), the Phocoenidae (*Neophoenaena*), and the Pontiporidae (*Pontoporia*) are quite similar in body size but vary markedly in brain size. Or one can compare species of similar brain size but vastly different body size, such as the 5.5 m killer whale (*Orcinus*) and the 12 m humpback whale (*Megaptera*).[7]

We cannot do justice to the myriad hypotheses that have been forwarded to explain large brain evolution in dolphins, so we will limit ourselves to a few issues pertinent to perception, cognition, and brain evolution in animals.

16.5.1 Food and brains: energetics, resource distribution, and echolocation

Several hypotheses emerge from a consideration of dolphin prey acquisition. First, it is important to note that the energy-rich foods consumed by dolphins support an overall energy budget that renders a large brain much more affordable. A useful contrast is the relatively small-brained herbivorous manatee (from the only other mammalian order, Sirenia, to evolve a fully marine existence), which has a low metabolic rate and whose low-quality forage spends 6 days passing through its digestive tract (Lomolino and Ewel 1984). Five male and eight female manatees from Florida that ranged in length from 281 to 376 cm and mass from 449 to 1620 kg, had an average brain size of only 364 g (range 309–455; Pirlot and Kamiya 1985; O'Shea and Reep 1990). By comparison, at a length of only 2.5 m, the 'small' brained Chinese river dolphin (*Lipotes vexellifer*) has a brain 50 per cent larger than the manatee. A bottlenose dolphin (*Tursiops truncatus*), at a length of 245 cm, has a brain over 1550 g. Larger delphinids similar in length (320–340 cm; but smaller in weight) to manatees have brains in the 2000–2500 g range

Table 16.2 Cetacean brain size comparisons

Taxon	N	Body length (cm)	Body weight (kg)	Brain weight (g) or volume (cm^3)	Coefficient of variation for brain weight
Suborder odontoceti (toothed whales)					
Superfamily Delphinoidea					
F. *Delphinidae*					
Sotalia (c)	1	158	42.2	688.0	
Stenella l. (e)	9	178	—	643.6	
Delphinus (g)	10	193	67.6	835.6	0.10
Lagenorhynchus (g)	2	208	99.5	1256.5	
Steno (j)	1	215	—	1369.0	
Stenella c. (k)	18	226	137.8	937.2	
Tursiops (g)	19	246	167.4	1587.5	
Grampus (g)	1	320	400.0	2551.0	
Globicephala (m)	2	545	—	2711.0	
Globicephala (j)	1	—	1200.0	3050.0	
Pseudorca (m)	1	550	—	3650.0	
Orcinus (n)	3	564	2262.0	6143.3	0.01
F. *Phocoenidae*					
Neophocaena (a)	4	151	37.3	471.3	0.08
Phocaeana (d)	3	162	59.7	500.7	0.02
Phocaenoides (f)	10	187	86.4	871.1	0.12
F. *Monodontidae*					
Dephinapterus (g)	1	340	636.0	2083.0	
Superfamily Platanistoidea					
F. *Pontoporiidae*					
Pontoporia (b)	9	153	39.0	227.0	
F. *Platanistidae*					
Platanista (h)	4	197	59.6	295.3	
F. *Iniidae*					
Inia (h)	2	212	62.4	617.5	

	n				
F. *Lipotidae*					
Lipotes (l)	2	252	230.5	570.0	
Superfamily Physeteroidea					
F. *Kogiidae*					
Kogia (g)	1	320	248.0	999.0	
F. *Physeteridae*					
Physeter (o)	15	1530	—	7913.0	0.09
Superfamily Ziphioidea					
F. *Ziphiidae*					
Ziphius (g)	1	549	2273.0	2044.0	
Suborder Mysticeti (baleen whales)					
F. *Balaenidae*					
Balaena (p)	5	953	—	2774.2	
Eubalaena (q)	4	1568	—	2850.0	
F. *Balaenopteridae*					
Megaptera (r)	6	1268	—	6100.3	
Balaenoptera b. (c)	1	1585	—	4900.0	
Balaenoptera p. (s)	11	1869	—	6746.4	
Balaenoptera m. (r)	1	2552	—	6500.0	

Sources: (a) Pilleri and Chen 1982, Pilleri and Gihr 1972; (b) Pilleri and Gihr 1971, Kamiya and Yamasaki 1974; (c) Morgane and Jacobs 1972; (d) Weber 1897, Warncke 1908, Pilleri and Gihr 1970; (e) Ridgway and Brownson 1979; (f) Ridgway and Johnston 1966, Pilleri and Gihr 1970; (g) Ridgway and Brownson 1984; (h) Pilleri and Gihr 1970a and 1970b; (i) Greunberger 1970; (j) Pettit 1905; (k) Miyazaki *et al.* 1981; (l) Gihr *et al.* 1979; (m) Elias and Schwartz 1969; (n) Morgane and Jacobs 1972, Ridgway unpublished data; (o) Kojima 1951; (p) Ridgway 1981; (q) Quiring 1945, Morgane and Jacobs 1972, Omura *et al.* 1969; (r) Pilleri and Gihr1972; (s) Jansen 1952, Quiring 1945, Morgane and Jacobs 1972.

(Table 16.2). At 7.5 m, the recently extinct Stellar sea cow was as long as a large killer whale, but had a brain 1/5 as large.[8]

The significant costs of large brains must be exceeded by the benefits they provide. Large brains might be useful for animals facing the potentially difficult problem of predicting where food will be found in space and time and extracting it (e.g. Milton 1988). The spatial distribution of dolphin prey species that live on or near the seafloor, the temporal and spatial distribution of schooling prey and, especially, we are finding, the particular methods used to procure prey, may require a considerable amount of learning.

Dolphins employ an extremely sophisticated echolocation system to find food and navigate in their habitat (Au 1993, Thomas *et al.* 2004). Perhaps selection for sophisticated echolocation abilities led to the enlargement of the delphinid cortex. Herman (Chapter 20, this volume) rejects this common but 'specious' argument, based partly on the echolocation abilities of much smaller-brained bats. Comparative data on neural structures associated with acoustic processing in similar-sized dolphins also fail to support the echolocation hypothesis (Table 16.3). Differences in the size of cranial nerves and the colliculus that readily distinguish dolphins with different visual capabilities do not appear in acoustic comparisons. Thus, Morgane and Jacobs (1972) obtained an optic nerve fibre count of 15 500 for *Inia* (n = 1) compared to 147 118 for *Tursiops* (n = 8), but only 19 500 for the delphinid *Sotalia* (n = 1) which inhabits the same murky river waters as *Inia*. In contrast, Morgane and Jacobs (1972) report fibre counts in the VIIIth nerve for *Tursiops* of 116 414 ($\pm$4014) and for *Inia* of 120 000. Even with a much smaller brain, *Pontoporia* has an acoustic nerve and inferior colliculus as large as the common dolphin, *Delphinus delphis* (Table 16.3). Behavioural studies of the perceptual abilities of smaller-brained odontocetes are required to test the hypothesis that greater acoustic discrimination accounts for the larger cortex in delphinids (Worthy and Hickie 1986). Worthy and Hickie's hypothesis also predicts that there is no correlation between the volume of acoustic tracts and primary auditory cortex in dolphins, unlike, for example, the correlation between the optic tract and primate primary visual cortex (Dunbar 2003).

16.5.2 Machiavellian intelligence in Dolphins

Herman (1980) was the first person to suggest that the key to understanding large brain evolution in dolphins might be found in their social lives. The complexity of the social lives of dolphins in Shark Bay does nothing to undermine this perspective. Perhaps what makes the Machiavellian intelligence hypothesis heuristically more attractive than the hypothesis that foraging demands have driven brain evolution is a

---

[8] Cranial capacity of 1100–1225 ml, compared to a brain weight of over 6000 g for the killer whale; O'Shea and Reep 1990.

Table 16.3 Subcortical visual and acoustic structures in the odontocete brain

Genus	Number	Brain weight (g)	Superior colliculus		Inferior colliculus		Cranial nerve II optic nerve (mm)	Cranial nerve VIII 'auditory' nerve (mm)
			length (mm)	width (mm)	length (mm)	width (mm)		
Pontoporia (a)	3	229.3	3.0	6.0 (1)	8.0	12.0	2.7	6.0
Platanista (a)	4	295.3	3.5	2.2	15.8	13.8	0.5–0.8	6.8
Lipotes (b)	1	550.0	5.5	5.0	10.0	13.0	2.0	7.0
Inia (c)	2	617.5	5.0	7.0	17.5	15.5	2.0	6.0
Delphinus (d)	3		9.6	10.0	14.6	14.3	5.0 (16)	6.3 (16)
Tursiops (d)	8						4.6	6.0
Stenella (d)	2						7.0	8.0
Grampus (d)	1						5.0	6.0

Sources: (a) Pilleri 1972; (b) Chen 1979, Gruenberger 1970; (c) [please add]; (d) Pilleri and Gihr 1970. Numbers in parentheses indicate where sample sizes vary from those given under 'number'. Pilleri and Gihr (1970) did not state the size of the delphinid from which the data were taken. However, the range of brain sizes in Pilleri's data were: 635–875 g (Delphinus); 785–980 g (Stenella); 1930–2240 g (Tursiops).

fundamental difference between food and foes: individuals in your social group try to outwit you, but your food does not.

Why did dolphins evolve complex social lives? In his deservedly famous essay on the social function of intellect, Humphrey (1976) linked social complexity to technical knowledge; 'the open sea is an environment where technical knowledge can bring little benefit and thus complex societies—and high intelligence—are contraindicated (dolphins and whales provide, maybe, a remarkable and unexplained exception).' But if not technical knowledge, then what? Given that the social complexity hypothesis assumes a strong dependence on group living, we can ask what there is about living in the ocean that may have fostered a strong mutual dependency in some cetaceans. There are two obvious candidates: predators and each other.

Prior to the recent focus on intragroup alliance formation in primates, a leading theory of the evolution of human brain size and intelligence focused on intergroup alliances, that is warfare (Alexander 1979, 1989; Alexander and Tinkle 1968; Bigelow 1969). Intergroup conflict places a premium on social cognition because individuals in a group are in reproductive competition with the same individuals with whom they must co-operate against a most formidable adversary, other humans (Alexander 1979). In other words, as the danger of intergroup conflict escalates, so does the mutual dependence of group members (Alexander 1979; Connor and Norris 1982), and consequently, 'individual reproductive success would depend increasingly on making the right decisions in complex social interactions involving self, relatives, friends and enemies' (Alexander 1979, p. 214).

The theories of intragroup alliances in non-human primates and intergroup alliances in humans have rarely crossed paths; most papers on the evolution of social complexity in non-human primates neither discuss nor cite the theory that intergroup conflict promoted social complexity in human evolution (e.g. Harcourt 1988, 1992; but see Rodseth *et al.* 1991; Manson and Wrangham 1991). The multiple levels of alliance formation in dolphins provide a conceptual link between theories of intragroup conflict in non-human primates and theories of intergroup conflict in humans. The key issue is whether individuals must base decisions at one level of interaction at least partly on the impact that those decisions will have at other levels.

Even for species with only one level of alliance formation within the group, the interaction between within-alliance conflicts and between-alliance conflicts can be very important. Levels of aggression within alliances will probably be influenced by the magnitude of threat from neighbouring alliances (see Vehrencamp 1983). In humans and chimpanzees, for example, this threat can be substantial (Goodall 1986; Manson and Wrangham 1991; Alexander 1979) and it is exacerbated by a fission–fusion social system that can produce encounters with an imbalance of numbers on opposing sides. This can reduce the cost of escalated aggression for those 'with the numbers' thereby increasing the risk and dependency for all involved (see Manson and Wrangham 1991). Dolphins often separate widely from their alliance partners during foraging bouts where they might find themselves at risk, as might a lone pair or trio in the range of a large second-order alliance. While no lethal aggression has been observed among males in Shark Bay, cases where several males line up head-to-head against one in an almost ritualistic fashion (in one case the fifth dolphin came from behind the target to do so), suggests the possibility.

Predators may have (at least) started dolphins on the road to more complex social lives. Sharks prey on dolphins and can threaten the calves of larger odontocetes and baleen whales (Wood *et al* 1970, Chapter 3). Cetaceans at sea inhabit a three-dimensional environment in which they cannot climb up a tree, crawl down a burrow, or hide behind a rock; they have nothing to hide behind except each other. Connor and Norris (1982) pinpointed birth and early nurture of offspring at sea as the critical factor increasing mutual dependence in dolphin societies. Vulnerable offspring that require significant investment would have favoured selection for those life history characteristics (long periods of dependency, late maturation, and longevity) associated with large brain size in mammals (see van Schaik and Deaner 2003). A focus on vulnerable offspring explains why mutual dependence may be high in mammals that are relatively invulnerable to predators as adults, such as sperm whales (Best 1979; Whitehead 2003) and elephants (Douglas-Hamilton 1975; Lee and Moss 1986); and it offers a partial explanation for their convergent social systems and large brains (Weilgart *et al.* 1996). The typical ungulate strategies of 'hiding' and 'following' are not available to elephant calves, which are slow and hard to conceal. Likewise, sperm whale calves have nowhere to hide and may not be able to follow their mothers to great depths. Sperm whale calves approach other older sperm whales at the surface while their mothers dive; thus they are not left unprotected at the surface when their mothers are feeding (Whitehead 1996).

16.5.3 Concluding remarks

Our observations in Shark Bay have revealed a society of great size with complex social relationships including nested alliances, affiliative behaviours that range from the expected (gentle touching) to the surprising (synchrony and the sex-specific 'contact swimming'), and individual foraging specializations and tool use. After over 20 years of observation at this 'Dolphin Gombe', we are still learning how the cognitive skills discussed by Herman (Chapter 20, this volume) are employed in the wild. The studies of synchrony reported in Herman's chapter and ours illustrate the potential exchange that may occur between captive and wild studies. It is unfortunate that there are not more institutions with captive dolphins supporting long-term research on cognition and behaviour (apart from the heavily funded but narrowly focused studies on echolocation).

Combined with results from captive studies and the remarkable variation in brain size among species of similar body size, our Shark Bay discoveries suggest strongly that, apart from primates, no animal group offers more potential for productive exploration of the relationship between brain size, cognitive skills, and behaviour than cetaceans.

The difficulties of studying many wild cetaceans that once appeared insurmountable have been mitigated to a significant degree by advances in technology (such as acoustic localization techniques and dive 'tags' that record everything from vocalizations to swimming speed; see Whitehead *et al.* 2000a).

Here we have compared the Shark Bay dolphins to terrestrial species, especially primates. However, it is also essential that we should be able to compare the Shark Bay society to other species of cetaceans such as the Baiji, the Susu, and Pontopora that have much smaller brains—if those poorly known species can be saved from their current race to extinction (Whitehead *et al.* 2000a). Only then will we be able to understand the selective forces that produced such remarkable brains, societies, and cognitive skills in a habitat so strikingly different from the terrestrial habitats in which primate brains evolved.

Acknowledgements

The National Geographic Society has supported the male alliance project since its inception in the mid-1980s. Observations in the 1980s were also supported by the NSF and in the 1990s by an NIH post-doctoral fellowship. Generous support for the recent study on alliance size and stability was provided by the Eppley foundation. The Monkey Mia Dolphin Resort has generously provided accommodation to members of the Monkey Mia Dolphin Research Project for over 12 years.

References

Alexander, R.D. (1979). *Darwinism and Human Affairs*. Seattle: University of Washington Press.

Alexander, R.D. (1987). *The Biology of Moral Systems*. New York: Aldine de Gruyter.

Alexander, R.D., and Tinkle, D.W. (1968). Review of On Aggression by Konrad Lorenz and The Territorial Imperative by Robert Audrey. *Bioscience*, **8**: 245–248.

Au, W. (1993). *The Sonar of Dolphins*. New York: Springer-Verlag.

Aureli F., Preston S.D., and de Waal F.B.M. (1999). Heart rate responses to social interactions in free-moving rhesus macaques: A pilot study. *Journal of Comparative Psychology*, **113**: 59–65.

Aureli F., and Schaffner C.M. (2002). Relationship assessment through emotional mediation. *Behaviour* **139**: 309–420.

Baird, R. W. (2000). The killer whale—foraging specializations and group hunting. In: J. Mann, R. C. Connor, P. L. Tyack and H. Whitehead, eds. *Cetacean Societies: Field Studies of Dolphins and Whales*, pp. 127–153. Chicago: University of Chicago Press.

Barrett, L., Henzi, P. and Dunbar, R. (2003). Primate cognition: from 'what now?' to 'what if?'. *Trends in Cognitive Sciences*, **7**: 494–497.

Barton, R. A. (1996). Neocortex size and behavioural ecology in primates. *Proceedings of the Royal Society, London, B,* **263**:173–177.

Berggren, P. (1995). Foraging behavior by bottlenose dolphins (*Tursiops* sp.) in Shark Bay, Western Australia. Abstract, *Eleventh Biennial Conference on the Biology of Marine Mammals*, Orlando, FL.

Best, P. B. (1979). Social organization in sperm whales, *Physeter macrocephalus*. In: H. E. Winn and B. L. Olla, eds. *Behavior of Marine Mammals: Current Perspectives in Research*, vol.3: *Cetaceans*, pp. 227–289. New York: Plenum Press.

Bigelow, R. (1969). *The Dawn Warriors*. Boston: Little Brown and Co.

Boehm, C. (1992). Segmentary warfare; and the management of conflict: Comparison of East African chimpanzees and patrilineal-patrilocal humans. In: A. H. Harcourt and F. B. M. de Waal, eds. *Us Against Them: Coalitions and Alliances in Humans and Other Animals*, pp. 137–173. Oxford: Oxford University Press.

Byrne, R., and Whiten, A., eds. (1988). *Machiavellian Intelligence*. New York: Oxford University Press.

Cairns, S. J., and Schwager, S. (1987). A comparison of association indices. *Animal Behaviour*, 3: 1454–1469.

Chapais, B. (1992). The role of alliances in social inheritance of rank among female primates. In: A.H. Harcourt and F.B.M. de Waal. eds. *Us Against Them: Coalitions and Alliances in Animals and Humans*, pp. 29–59. Oxford: Oxford University Press.

Chen, Y. (1979). On the cerebral anatomy of the Chinese river dolphin, *Lipotes vexillifer* Miller. *Acta Hydrobiologica Sinica*, 6: 365–372.

Connor, R. C. (2000). Group living in whales and dolphins. In: J. Mann, R. C. Connor, P. L. Tyack, and H. Whitehead, eds. *Cetacean Societies: Field Studies of Dolphins and Whales*, pp. 199–218. Chicago: University of Chicago Press.

Connor, R. C. (2001). Social relationships in a big-brained aquatic mammal. In: L. A. Dugatkin, ed. *Model Systems in Behavioral Ecology*, pp. 408–432. Princeton: Princeton University Press.

Connor, R. C., and Krützen, M. (2003). Levels and patterns in dolphin alliance formation. In: F. B. M. de Waal and P. L. Tyack, eds. *Animal Social Complexity Intelligence, Culture, and Individualized Societies*, pp. 115–120. Cambridge, MA: Harvard University Press.

Connor, R. C., and Norris, K. S. (1982). Are dolphins reciprocal altruists? *American Naturalist*, 119: 358–374.

Connor, R.C., and Smolker, R.A. (1985). Habituated dolphins (*Tursiops* sp.) in Western Australia. *Journal of Mammology*, 36: 304–305.

Connor, R.C., and Smolker, R.A. (1995). Seasonal changes in the stability of male-male bonds in Indian Ocean Bottlenose dolphins (*Tursiops* sp.). *Aquatic mammals,* 21: 213–216.

Connor, R. C., and Smolker, R. A. (1996). 'Pop' goes the dolphin: A vocalization male bottlenose dolphins produce during consortships. *Behaviour*, 133: 643–662.

Connor, R.C., Heithaus, R.M., and Barre, L.M. (1999). Super-alliance of bottlenose dolphins. *Natre*, 371: 571–572.

Connor, R.C., Heithaus, M.R., and Barre, L.M. (2001). Complex social structure, alliance stability and mating access in a bottlenose dolphin 'super-alliance'. *Proceedings of the Royal Society of London*, Series B, 268: 263–267.

Connor, R. C., Mann, J., Tyack, P. L., *et al.* (1998). Social evolution in toothed whales. *Trends in Ecology and Evolution*, 13: 228–232.

Connor, R.C., Mann, J., and Watson, J. (Submitted). A sex-specific affiliative contact behavior in Indian Ocean bottlenose dolphins (*Tursiops aduncus*).

Connor, R. C., Read, A. J., and Wrangham, R. W. (2000). Male reproductive strategies and social bonds. In: J. Mann, R. C. Connor, P. L. Tyack, and H. Whitehead, eds. *Cetacean Societies: Field Studies of Dolphins and Whales*, pp. 247–269. Chicago: University of Chicago Press.

Connor, R. C., Richards, A. F., Smolker, R. A., *et al.* (1996). Patterns of female attractiveness in Indian Ocean bottlenose dolphins. *Behaviour*, 133: 37–69.

Connor, R.C., Smolker, R.A., and Bejder, L. (Submitted). Synchrony, social behaviour and alliance affiliations in male Indian Ocean bottlenose dolphins (*Tursiops aduncus*).

Connor, R. C., Smolker, R. A., and Richards, A. F. (1992a). Dolphin alliances and coalitions. In: A. H. Harcourt and F. B. M. de Waal. eds. *Coalitions and Alliances in Humans and Other Animals*, pp. 415–443. Oxford: Oxford University Press.

Connor, R. C., Smolker, R. A., and Richards, A. F. (1992b). Two levels of alliance formation among bottlenose dolphins (*Tursiops* sp.). *Proceedings of the National Academy of Sciences*, **89**: 987–990.

Connor, R. C., Wells, R. S., Mann, J., *et al* (2000). The bottlenose dolphin: social relationships in a fission-fusion society. In: J. Mann, R.C. Connor, P.L. Tyack, and H. Whitehead. eds. *Cetacean Societies: Field Studies of Dolphins and Whales*, pp. 91–126. Chicago: University of Chicago Press.

Cords, M. (1997). Friendships, alliances, reciprocity and repair. In: A. Whiten and R. W. Byrne. eds. *Machiavellian Intelligence II: Extensions and Evaluations*, pp. 24–49. Cambridge: Cambridge University Press.

Damasio, A. (1994). *Descartes' Error*. New York: G.P. Putnam's Sons.

Deecke, V. B., Ford, J. K. B., and Spong, P. (2000). Dialect change in resident killer whales: implications for vocal learning and cultural transmission. *Animal Behaviour,* **60**: 629–638.

Douglas-Hamilton, I., and Douglas-Hamilton, O. (1975). *Among the Elephants*. New York: Viking Press.

Dunbar, R. I. M. (1992). Neocortex size as a constraint on group size in primates. *Journal of Human Evolution*, **20**: 469–493.

Dunbar, R. I. M. (1998). The social brain hypothesis. *Evolutionary Anthropology*, **6**: 178–190.

Dunbar, R.I. M. (2003). Why are apes so smart?. In: P.M. Kappeler and M.E. Periera, eds. *Primate Life Histories and Socioecology*, pp. 285–298. Chicago: University of Chicago Press.

Elias, H. and Schwartz, D. (1969). Surface areas of the cerebral cortex of mammals determined by stereological methods. *Science,* **166**: 111–113.

Engh, A.L., Siebert, E., Greenberg, D., and Holekamp, K. (2005). Patterns of alliance formation and post-conflict aggression indicate spotted hyaenas recognize third party relationships. *Animal Behaviour*, **69**: 209–217.

Gaskin, D. E. (1982). *The Ecology of Whales and Dolphins*. London and Exeter, New Hampshire: Heineman.

Gihr, M., Pilleri, G., and Zhou, K. (1979). Cephalization in the Chinese river dolphin, *Lipotes vexillifer* (Platanistoidea, Lipotidae), *Investigations on Cetacea*, **10**: 257–274.

Goodall, J. (1986). *The Chimpanzees of Gombe*. Cambridge, MA: Harvard University Press.

Gross, M. R. (1996). Alternative reproductive strategies and tactics: Diversity within sexes. *Trends in Ecology and Evolution*, **11**: 92–98.

Gruenberger, H.B. (1970). On the cerebral anatomy of the Amazon dolphin, *Inia geoffrensis*. *Investigations on Cetacea*, **2**: 129–144.

Gygax, L., Harley, N., and Kummer, H. (1997). A matrilineal overthrow with destructive aggression in *Macacca fascicularis*. *Primates*, **34**: 149–158.

Harcourt, A.H. (1988). Alliances in contests and social intelligence. In: R. Byrne and A. Whiten. eds. *Machiavellian Intelligence*, pp. 132–152. Oxford: Clarendon Press.

Harcourt, A. H. (1992). Coalitions and alliances: are primates more complex than non-primates. In: A. H. Harcourt and F. B. M. de Waal, eds. *Coalitions and Alliances: Are Primates More Complex than Non-primates?*, pp. 445–471. New York: Oxford University Press.

Harcourt, A. H., and de Waal, F. B. M., eds. (1992). *Coalitions and Alliances in Humans and Other Animals*. Oxford: Oxford University Press.

Herman, L. M. (1980). Cognitive characteristics of dolphins. In: L. M. Herman, ed. *Cetacean Behavior: Mechanisms and Functions*, pp. 363–429. New York: Wiley Interscience.

Hinde, R. A. (1976). Interactions, relationships and social structure. *Man*, 11:1–17.

Humphrey, N. K. (1976). The social function of intellect. In: P. P. G. Bateson and R. A. Hinde, eds. *Growing Points in Ethology*, pp. 303–317. Cambridge: Cambridge University Press.

Janik, V. (1999). Origins and implications of vocal learning in bottlenose dolphins. In: H. O. Box and K. R. Gibson, eds. *Mammalian Social Learning: Comparative and Ecological Perspectives*, pp. 308–326. Cambridge: Cambridge University Press.

Janik, V., and Slater, P. J. B. (1997). Vocal learning in mammals. *Advances in the Study of Behavior*, 26: 59–99.

Janik, V., and Slater, P. J. B. (2000). The different roles of social learning in vocal communication. *Animal Behaviour*, 60: 1–11.

Jansen, J. (1952). On the whales brain with special reference to the weight of the brain of the fin whale (*Balaenoptera physalus*). Norsk Hvalfongst-tidende, 9: 480–486.

Kamiya, T., and Yamasaki, F. (1974). Organ weights of *Pontoporia blainvillei* and *Platanista gangetica* (Platanistidae). *Scientific Reports of the Whale Research Institute*, 26: 265–270.

Kappeler, P. M., and van Schaik, C. (2002). Evolution of primate social systems. *International Journal of Primatology*, 88: 707–740.

Kitchen, D. M., and Packer, C. (1999). Complexity in vertebrate societies. In: L. Keller, ed. *Levels of Selection in Evolution*, pp. 176–196. Princeton: Princeton University Press.

Kogima, T. (1951). On the brain of the sperm whale (*Physeter catadon* L.). *Scientific Reports of the Whales Research Institute, Tokyo*, 6: 49–72.

Krützen, M., Barre, L. M., Connor, R. C., *et al.* (2004). O father: where art thou? Paternity assessment in an open fission-fusion society of wild bottlenose dolphins (*Tursiops* sp.) in Shark Bay, Western Australia. *Molecular Ecology*.

Krützen, M., Mann, J., Heithaus, M.R., Connor, R. C., Bejder, L., and Sherwin, W.B. (2005). Cultural transmission of tool use in bottlenose dolphins. Proceedings of the National Academy of Sciences. 102(25): 8939–8943.

Kudo, H., and Dunbar, R. I. M. (2001). Neocortex size and social network size in primates. *Animal Behaviour*, 62: 711–722.

Kummer, H. (1967). Tripartite relations in hamadryas baboons. In: S.A. Altman, ed. *Social Communication Among Primates*. Chicago: University of Chicago Press.

Laland, K. N., and Kendal, J. R. (2003). What the models say about social learning. In: D. Fragaszy and S. Perry, eds. *The Biology of Traditions: Models and Evidence*. Cambridge University Press.

Lee, P. C., and Moss, C. J. (1986). Early maternal investment in male and female African elephant calves. *Behavioral Ecology and Sociobiology*, 18: 353–361.

Lomolino, M. W., and Ewel, K. C. (1984). Digestive efficiencies of the West Indian manatee (*Trichechus manatus*). *Florida Scientist*, 47: 176–179.

Mann, J. (in press). Homosexual and sociosexual behaviour among Indian Ocean bottlenose dolphins. In: P. Vasey and V. Sommer, eds. *Homosexual Behaviour in Animals: An Evolutionary Perspective*. Cambridge: Cambridge University Press.

Mann, J., and Barnett, H. (1999). Lethal tiger shark (*Galeocerdo cuvieri*) on a bottlenose dolphin (*Tursiops* sp.) calf: Defense and reactions by the mother. *Marine Mammal Science*, 15: 568–575.

Mann, J., and Kemps, C. (2003). The effects of provisioning on maternal care in wild bottlenose dolphins, Shark Bay, Australia. In: N. Gales, M. Hindell, R. Kirkwood, eds. *Marine Mammals: Fisheries, Tourism and Management Issues*, pp. 304–320. CSIRO Publishing

Mann, J., and Sargeant, B. (2003). Like mother, like calf: The ontogeny of foraging traditions in wild Indian Ocean bottlenose dolphins (*Tursiops* sp.). In: D. Fragaszy and S. Perry. eds. *The Biology of Traditions: Models and Evidence,* pp. 236–266. Cambridge University Press.

Mann, J., and Smuts, B.B. (1999). Behavioral development in wild bottlenose dolphin newborns (*Tursiops* sp.). *Behaviour,* **136**: 529–566.

Mann, J., and Watson–Capps, J. (2005). Surviving at sea: Ecological and behavioural predictors of calf mortality in Indian Ocean bottlenose dolphins (*Tursiops* sp.). *Animal Behaviour,* **69**: 899–909.

Mann, J., Barre, L., Connor, R. C., *et al.* (2003). Female-biased investment in wild bottlenose dolphins. *Fifteenth Biennial Conference on the Biology of Marine Mammals,* Greensboro, NC.

Mann, J., Connor, R. C., Barre, L. M., *et al.* (2000a). Female reproductive success in bottlenose dolphins (*Tursiops* sp.): Life history, habitat, provisioning, and group size effects. *Behavioral Ecology,* **11**: 210–219.

Mann, J., Connor, R. C., Tyack, P. L., and Whitehead, H., eds. (2000b). *Cetacean Societies: Field Studies of Dolphins and Whales.* Chicago: University of Chicago Press.

Manson, J. H., and Wrangham, R. W. (1991). Intergroup aggression in chimpanzees and humans. *Current Anthropology,* **32**: 369–390.

Milton, K. (1988). Foraging behavior and the evolution of primate intelligence. In: R. Byrne and A. Whiten, eds. *Machiavellian Intelligence: Social Expertise and the Evolution of Intellect in Monkeys, Apes And Humans,* pp. 285–305. Oxford: Oxford Science Publications.

Mitani, J. C., and Amsler, S. (2003). Social and spatial aspects of male subgrouping in a community of wild chimpanzees. *Behaviour,* **140**: 869–884.

Miyazaki, N., Fujise, Y., and Fujiyama, T. (1981). Body and organ weight of striped and spotted dolphins off the Pacific coast of Japan. *Scientific Reports of the Whales Research Institute,* **33**: 27–67.

Morgane, P.J., and Jacobs, M.S. (1972). Comparative anatomy of the cetacean nervous system. in R.J. Harrison, ed. *The Functional Anatomy of Marine Mammals,* vol. 1, pp. 117–244. London: Academic Press.

Nishida, T. (1983). Alpha status and agonistic alliance in wild chimpanzees (*Pan troglodytes schweinfurthii*). *Primates,* **24**: 318–336.

Nishida, T., ed. (1990). *The Chimpanzees of the Mahale Mountains: Sexual and Life History Strategies.* Tokyo: University of Tokyo Press.

Noad, M. J., Cato, D. H., Bryden, M. M., *et al.* (2000). Cultural revolution in whale songs. *Nature,* **408**: 537.

Noë, R. (1990). A veto game played by baboons; a challenge to the use of the Prisoners Dilemma as a paradigm for reciprocity and cooperation. *Animal Behaviour,* **39**: 78–90.

Omura, H., Ohsumi, S., Nemoto, T., Nasu, K., and Kasuya, T. (1969). Black right whales in the North Pacific. *Scientific Reports of the Whale Research Institute,* **21**: 1–78.

O'Shea, T.J., and Reep, R.L. (1990). Encephalization quotients and life-history traits in Sirenia, *Journal of Mammology,* **71**: 534–543.

Perry, S. (2003). Coalitionary aggression in white-faced capuchins. In: F. B. M de Waal and P. L. Tyack, eds. *Animal Social Complexity: Intelligence, Culture, and Individualized Societies,* pp. 111–114. Cambridge, MA: Harvard University Press.

Pettit, A. (1905). Description des encephales de Grampus griseus Cuv., de *Steno frontatus* Cuv., et de *Globicephalus melas* Traill, provenant des campagnes du yacht Princesse-Alice. Resultats des campagnes scientifiques accomplies sur son yacht par Albert 1er Monaco.

Pilleri, G., and Chen, P. (1982). The brain of the Chinese finless porpoise *Neophocaena asiaeurientalis* (Pilleri and Gihr, 1972): 1) Macroscopic anatomy. *Investigations on Cetacea,* **13**: 27–78.

Pilleri, G., and Gihr, M. (1970a). Brain-body weight ratio of *Platanista gangetica*. *Investigations on Cetacea*, **2**: 79–82.

Pilleri, G., and Gihr, M. (1970b). The central nervous system of the mysticete and odontocete whales. *Investigations on Cetacea*, **2**: 89–127.

Pilleri, G., and Gihr, M. (1971). Brain-body weight ratio in *Pontoporia blainvillei*. *Investigations on Cetacea*, **3**: 69–73.

Pilleri, G., and Gihr, M. (1972). Contribution to the knowledge of the cetaceans of Pakistan with particular reference to the genera *Neomeris, Sousa, Delphinus*, and *Tursiops* and description of a new Chinese porpoise (*Neomeris asiaeorientalis*). *Investigations on Cetacea*, **4**: 107–162.

Pirlot, P., and Kamiya, T. (1985). Qualitative and quantitative brain morphology in the Sirenian *Dugong dugong* Erxl. *Zeitschrift fuer Zoologische Systematik und Evolutionsforschung*, **23**: 147–155.

Porter, J. W. (1979). Pseudorca strandings. *Oceans*, **10**: 8–15.

Quiring, D. P. (1943). Weight data on five whales. *Journal of Mammology*, **24**: 39–45.

Richards, A. F. (1996). *Life history and behavior of female dolphins in Shark Bay, Western Australia*. PhD Dissertation, University of Michigan, Ann Arbor, MI.

Ridgway, S.H. (1981). Some brain morphometrics of the bowhead whale. *Tissue structural studies and other investigations on the biology of endangered whales in the Beaufort Sea*. Final Report for the Period April 1, 1980 through June 30, 1981, II, pp. 837–844. US Department of the Interior, Alaska OCS Office, Anchorage.

Ridgway, S. H., and Brownson, R. H. (1979). Brain size and symmetry in three dolphin genera. *Anatomical Record,* **193**: 664.

Ridgway, S. H., and Brownson, R. H. (1984). Relative brain sizes and cortical surface areas in odontocetes. *Acta Zoologica Fennica*, **172**: 149–152.

Ridgway, S. H., and Johnston. D. G. (1966). Blood oxygen and ecology of porpoises of three genera. *Science*, **151**: 456–458.

Rodseth, L., Wrangham, R. W., Harrigan, A. M., and Smuts, B. B. (1991). The human community as a primate society. *Current Anthropology,* **32**: 221–254.

Samuels, A., and Flaherty, C. (2000). Peaceful conflict resolution in the sea? In: F. Aureli and F. de Waal. eds. *Natural Conflict Resolution*, pp. 229–231. Berkeley: University of California Press.

Samuels, A., and Gifford, T. (1997). A quantitative assessment of dominance relations among bottlenose dolphins. *Marine Mammal Science,* **13**: 70–99.

Samuels, A., Silk, J., and Altman, J. (1987). Continuity and change in dominance relations among female baboons. *Animal Behavior*, **35**: 785–793.

Sargeant, B. L., and Mann, J. (2003). Beaching by wild bottlenose dolphins in Shark Bay, Western Australia. *Fifteenth Biennial Conference on the Biology of Marine Mammals*, Greensboro, NC.

Sargeant, B. L., Mann, J., Berggren, P. and Krützen, M. (in press). Specialization and development of beach hunting, a rare foraging behavior, by wild Indian Ocean bottlenose dolphins (*Tursiops* sp.) *Canadian Journal of Zoology*.

Schroeder, J. P. (1990). Breeding bottlenose dolphins in captivity. In: S. Leatherwood and R. R. Reeves, eds. *The Bottlenose Dolphin*. Orlando, FL: Academic Press.

Schusterman, R. J., and Kastak, D. (2002). Problem solving and memory. In: A. R. Hoelzel, ed. *Marine Mammal Biology*, pp 371–387. Oxford: Blackwell.

Schusterman, R. J., Reichmuth C. J., and Kastak, D. (2000). How animals classify friends and foes. *Current Directions in Psychological Science*, **9**: 1–6.

Scott, E., Mann, J., Watson-Capps, J. J., Sargeant, B. L., and Connor, R.C. (2005). Aggression in bottlenose dolphins: Evidence for sexual coercion, male-male competition, and female tolerance through analysis of tooth-rake marks and behaviour. *Behaviour*, **142**: 21–44.

Seyfarth, R. M. (1976). Social relationships among adult female baboons. *Animal Behavior*, **24**: 917–938.

Seyfarth, R. M., and Cheney, D. L. (2001). Cognitive strategies and the representation of social relations by monkeys. In: J. A. French, A. C. Kamil, and D. W. Leger, eds. *Evolutionary Psychology and Motivation,* vol. 47, Nebraska Symposium on Motivation, pp. 145–178. Lincoln: University of Nebraska Press.

Seyfarth, R. M., and Cheney, D. L. (2002). The structure of social knowledge in monkeys. In: M. Bekoff, C. Allen, and G. M. Burghardt, eds. *The Cognitive Animal: Empirical and Theoretical Perspectives on Animal Cognition*, pp. 379–384. Cambridge, MA: MIT Press.

Seyfarth, R. M., and Cheney, D. L. (2003). Hierarchical structure in the social knowledge of monkeys. In: F. B. M. de Waal and P. L. Tyack, eds. *Animal Social Complexity: Intelligence, Culture, and Individualized Societies*, pp. 207–229. Cambridge, MA: Harvard University Press.

Shane, S. H. (1990). Comparison of bottlenose dolphin behavior in Texas and Florida, with a critique of methods for studying dolphin behavior. In: S. Leatherwood and R. R. Reeves, eds. *The Bottlenose Dolphin*, pp. 541–558. San Diego: Academic Press.

Smolker, R. A., Richards, A. F., Connor, R. C., Mann, J., *et al.* (1997). Sponge-carrying by Indian Ocean bottlenose dolphins: Possible tool-use by a delphinid. *Ethology,* **103**: 454–465.

Smolker, R. A., Richards, A. F., Connor, R. C., and Pepper, J. W. (1992). Sex differences in patterns of association among Indian Ocean bottlenose dolphins. *Behaviour*, **123**: 38–69.

Symington, M. M. (1990). Fission-fusion social organization in Ateles and Pan. *International Journal of Primatology*, **11**: 47–61.

Thomas, J. A., Moss, C. F., and M. Vater, eds. (2004). *Echolocation in Bats and Dolphins*. Chicago: University of Chicago Press.

Trivers, R. L. (1985). *Social Evolution*. Menlo Park, CA: Benjamin/Cummings.

van Schaik, C. P., and Deaner, R. O. (2003). Life history and cognitive evolution in primates. In: F. B. M. de Waal and P. L. Tyack, eds. *Animal Social Complexity: Intelligence, Culture, and Individualized Societies*, pp. 5–25. Cambridge: Harvard University Press.

Vehrencamp, S. (1983). Optimal degree of skew in cooperative societies. *American Zoologist*, **23**: 327–335.

de Villiers, M. S., Richardson, P. R. K., and van Jaarsveld, A. S. (2003). Patterns of coalition formation and spatial association in a social carnivore, the African wild dog (*Lycaon pictus*). *Journal of Zoology*, **260**: 377–389.

de Waal, F. B. M. (1982). *Chimpanzee Politics*. New York: Harper and Row.

Warncke, P. (1908). Mitteilung neuer Gehirn und Korpergewichtsbestimmungen bei Saugern, nebst Zusammenstellung Der gesamten bisher beobachtelen absoluten und relativen Gehirngewichte bei den verschiedenen spezies. *Zeitschrift Fur Psychologie und Neurologie*, **13**: 335–403.

Weaver, A. (2003). Conflict and reconciliation in captive bottlenose dolphins, *Tursiops truncatus*. *Marine Mammal Science*, **19**: 836–846.

Weber, M. (1897). Vorstudien uber das hirngewicht der Saugethiere. In: *Festschrift zum Siebenzigsten Geburtstage von Carl Gegenbaur, AM 21 August 1986*, pp. 105–123. Leipzig: Verlag von Wilhelm Engelman.

Weilgart, L., Whitehead, H., and Payne, K. (1996). A colossal convergence. *American Scientist*, **84**: 278–287.

Whitehead, H. (1996). Babysitting, dive synchrony, and indications of alloparental care in sperm whales. *Behavioral Ecology and Sociobiology*, **38**: 237–244.

Whitehead, H. (2003). *Sperm Whales: Social Evolution in the Ocean*. Chicago: University of Chicago Press.

Whitehead, H., and Mann, J. (2000). Female reproductive strategies of cetaceans: life histories and calf care. In: J. Mann, R. C. Connor, P. L. Tyack, and H. Whitehead, eds. *Cetacean Societies: Field Studies of Dolphins and Whales* , pp. 219–246. Chicago: University of Chicago Press.

Whitehead, H., Christal, J., and Tyack, P. L. (2000). Studying cetacean social structure in space and time: innovative techniques. In: J. Mann, R.C. Connor, P. L. Tyack, and H. Whitehead, eds. *Cetacean Societies: Field Studies of Dolphins and Whales*, pp. 65–87. Chicago: University of Chicago Press.

Whitehead, H., Reeves, R. R., and Tyack, P. L. (2000). Science and the conservation, protection and management of wild cetaceans. In: J. Mann, R. C. Connor, P. L. Tyack, and H. Whitehead, eds. *Cetacean Societies: Field Studies of Dolphins and Whales*, pp. 308–332. Chicago: University of Chicago Press.

Whiten, A., and Byrne, R. W. (1988). Tactical deception in primates. *Behavioral and Brain Sciences,* **11**: 233–273.

Wilson, B., Thompson, P. M., and Hammond, P. S. (1997). Habitat use by bottlenose dolphins: Seasonal distribution and stratified movement patterns in the Moray Firth, Scotland. *Journal of Applied Ecology,* **34**: 1365–1374.

Wood, F. G. Jr., Caldwell, D. K., and Caldwell, M. C. (1970). Behavioral interactions between porpoises and sharks. In: G. Pilleri, ed. *Investigations on Cetacea*, vol. 2, pp. 264–279. Berne: Institute of Brain Anatomy.

Worthy, G. A. J., and Hickie, J. P. (1986). Relative brain size in marine mammals. *American Naturalist,* **128**: 445–459.

Würsig, B. (1978). Occurrence and group organization of Atlantic bottlenose porpoises (*Tursiops truncatus*) in an Argentine bay. *Biological Bulletin,* **154**: 348–359.

Yoshioka, M., Mohri, E., Tobayama, T., Aida, K., and Hanyu, I. (1986). Annual changes in serum reproductive hormone levels in the captive female bottle-nosed dolphins. *Bulletin of the Japanese Society of Scientific Fisheries*, **52**: 1939–1946.

Zahavi, A. (1977). The testing of a bond. *Animal behaviour*, **25**: 246–247.

Part V

Mind reading and behaviour reading

Chapter 17

Do chimpanzees know what others see—or only what they are looking at?

Michael Tomasello and Josep Call

Abstract

A variety of recent experiments suggest that apes know what other individuals do and do not see. The results of each experiment may be explained by postulating some behavioural rule that individuals have learned that does not involve an understanding of seeing. But the postulated rule must be different in each case, and most of these do not explain more than one experiment. This patchiness of coverage gives this kind of explanation a very *ad hoc* feeling, especially since there is rarely any concrete evidence that animals actually have had the requisite experiences to learn the behavioural rule—there is just a theoretical possibility. It is thus more plausible to hypothesize that apes really do know what others do and do not see in many circumstances. Moreover, and more generally, there is no reason to assume—other than some kind of blind allegiance to behaviourism—that just because an animal has learned something, no cognitive processes are involved.

17.1 **Introduction**

Observations of naturally occurring behaviour are always open to multiple interpretations, especially when the issue is the underlying cognitive processes involved. Some theorists prefer richer interpretations (so-called boosters) while others prefer leaner interpretations (so-called scoffers). The ultimate boosters are those who think there are no significant differences between non-human animal and human cognition, and the ultimate scoffers are radical Behaviourists who do not believe it is useful to talk of cognitive processes at all. Most of the contributors to this volume fall somewhere in between these two extremes. With particular reference to primate social cognition ('theory of mind'), there are boosters who think that non-human primates understand much about the psychological states of other organisms, and scoffers who think they understand little or nothing. In effect, the scoffers think that non-human animals are behaviourists who see and respond to the behaviour of others but do not conceive of any cognitive processes underlying this behaviour.

Not so many years ago, the data on non-human primate social cognition was so thin that it was possible for sensible people to espouse fairly extreme booster or scoffer positions. But in the last few years a number of new studies have produced data that constrain, in significant ways, the range of interpretive options. In this chapter, we look at these new data and compare richer and leaner interpretations of them. In particular, we focus on data concerning what chimpanzees know about what others can and cannot see—do they have such knowledge, or are they only capable of determining where others are looking, and what they are looking at? There are a number of ways in which the difference between these two interpretations—understanding seeing versus responding to looking behaviour—could be characterized, but for the moment let us proceed as if it were clear enough. We will explicate the intended distinction further by reviewing data from four different experimental paradigms: gaze following, competing for food, begging and gesturing, and self-knowledge. For each paradigm we start with a basic finding and then contrast the booster and scoffer alternatives with the data available in the literature.

17.2 **Gaze following**

Tomasello *et al.* (1998) conducted a very simple experiment. A human experimenter (E) waited until a pair of chimpanzees was spatially arranged so that one was facing toward him in the observation tower (the looker) and another was oriented to that looker (the subject). E then held up a piece of desirable food, inducing the looker to look up at him, and observed how the subject responded to the looker's looking behaviour. In control trials E displayed the food in an identical manner but when the subject was alone. The finding was that subjects reliably followed the gaze direction of the looker to the food a greater number of times in experimental than in control trials (Fig. 17.1). Other studies have found that chimpanzees also follow the gaze direction of human beings, and they can even do this on the basis of eye direction alone and independently of the experimenter's head direction (e.g. Povinelli and Eddy 1996a).

There are of course many ways to interpret straightforward gaze-following behaviour, but other studies have helped to reveal how chimpanzees understand the behaviour they are observing. First, chimpanzees follow the gaze direction of human beings to a specific location even if they have to look past and ignore other novel objects along the way in order to fixate the target location. This would seem to indicate that they are not just turning in the same general direction as the looker and then searching randomly for something interesting; they are seeking the target of the looker's perceptual activity (Tomasello *et al.* 1999). Second, if adult chimpanzees track the gaze of another

Fig. 17.1 Chimpanzees gaze following.

individual to a location and find nothing interesting there, they quite often look back to the individual's face and track her gaze direction a second time (Call *et al.* 1998; Bräuer *et al.* submitted). This 'checking back'—which only adult chimpanzees do—is a key criterion used to assess human infants' understanding of the visual experience of others, since it would seem to indicate that the subject expects to find the looker's perceptual orientation directed at a target (Bates 1979). Interestingly, if a looker looks repeatedly to a location with no salient target adult (but again not juvenile) chimpanzees stop responding to that individual's looking activities (Tomasello *et al.* 2001) indicating acquired expectations about when it is likely that following the gaze direction of another is likely to lead to an interesting target.

But perhaps the most telling situation occurs when a human being looks behind a barrier. Following a suggestive finding of Povinelli and Eddy (1996a), Tomasello *et al.* (1999) had a human E look around various types of barriers (or look straight ahead in a control condition). In this case, a simple gaze following response (turning head to look in the direction E is looking) would not be enough—that would simply lead to the subject fixating the barrier itself. To track E's gaze to its target, subjects needed to move a few meters so as to attain the appropriate viewing angle enabling them to look behind the barrier. And this is just what they did (much more often than in the control condition) for all four of the barrier types investigated.

The booster interpretation of these findings on gaze following is simply that chimpanzees follow the gaze direction of others—checking back when they find nothing there (and eventually stopping if they repeatedly find nothing there)—because they want to see what the other is seeing. When there is a distractor they ignore it; when there is a barrier, they move themselves in order to see what the other is seeing. The scoffer interpretation avoids positing that the subject knows that the other is seeing something and instead relies on some combination of biological predispositions and individual learning. The first proposal is simply that chimpanzees are biologically predisposed to orient in the direction in which another individual orients, and to account for the barrier findings this predisposition must be geometrically quite specific. The second proposal is that individuals learn, through personal experience, that when they look in the direction toward which another individual is oriented, they often find something interesting or important (i.e. that is rewarding in some way). It is not clear in these lean interpretations exactly how to account for the facts that chimpanzees check back with the looker when they find nothing and that they eventually stop looking if they repeatedly find nothing, but presumably some account in terms of learning could be constructed.

17.3 **Competing for food**

In a recent series of studies, Hare *et al.* (2000) placed a subordinate and a dominant chimpanzee into rooms on opposite sides of a third room. Each had a guillotine door leading into this middle room which, when opened at the bottom, allowed them to observe two pieces of food at various locations—and to see the other individual looking under her door. After the food had been placed, the doors for both individuals were opened and they were allowed to enter the third room. The basic problem for the subordinate in this situation is that the dominant will take all of the food she can see. However, in some cases things were arranged so that the subordinate could see one piece of food that the dominant could not see, for example by placing it on the subordinate's side of a small barrier (with the other piece of food in the open) (Fig. 17.2). The question was thus whether the subordinate knew that the dominant could not see a particular piece of food, and so it was safe to go for it. The basic finding was that the subordinates did indeed go for the food that only they could see much more often than they went for the food that both they and the dominant could see.

One scoffer possibility is that subordinates in these studies may have been monitoring the behaviour of dominants, rather than dominants' perceptual access to the food, and reacting to that. But this possibility was ruled out in some of the studies in this series by giving subordinates a small headstart and forcing them to make their choice between the two pieces of food before the dominant was released into the middle cage.

Fig. 17.2 A subordinate and a dominant chimpanzee compete over food. (a) The subordinate sees two pieces of food while the dominant sees only one (and an opaque barrier). (b) The subordinate chimpanzee takes the hidden piece while the dominant takes the visible one.

Moreover, in one additional control condition the dominant's door was lowered before the two competitors were let into the room (and again the subordinate got a small headstart), so that the subordinate could not see the dominant at all at the moment of choice—and so could not react to her behaviour. Subordinates still targeted the piece of food the dominant could not see. Another possibility is that subordinates viewed the barrier as a physical impediment to the dominant's access to the food, in that it might slow her down by requiring her to reach around the barrier for the object of her desire. So in another control condition food was placed on the subordinate's side of a *transparent* barrier. In this case, subordinates chose equally between the two pieces of food, seeming to know that the transparent barrier was not blocking the dominant's visual access to the food—and seeming not to care about any possible blocking of physical access.

Another scoffer interpretation of this line of research was suggested by Povinelli and Giambrone (2001). They suggested that subordinates were attracted to the food behind the barrier because in general they prefer to forage in the vicinity of barriers rather than out in the open where dominants might easily see them (the peripheral feeding hypothesis). But this hypothesis is ruled out by a second line of similar experiments. Hare *et al.* (2001) used the same basic procedure but with only one piece of food, which was hidden behind one of two barriers; they varied whether or not the dominant witnessed the hiding process. Thus, in some trials the dominant did not witness the food being hidden because her door was down, whereas in other trials she witnessed the hiding process under her partially open door (subordinates always saw the entire baiting procedure and could monitor the visual access of the dominant competitor as well). Subordinates in this case preferentially went for the food that dominants had not seen hidden. This indicates that they know not only what others can and cannot see at the moment, but also what others have just seen in the immediate past. Moreover, since the food was always hidden behind one of the two barriers and the only difference between conditions was whether the dominant had or had not witnessed the baiting the peripheral feeding hypothesis does not apply. It is also interesting that if the dominant individual was switched with another dominant individual (who had seen nothing) just before the moment of choice, subordinates now felt free to go for the food no matter what had transpired earlier. Chimpanzees seemingly know which particular individuals have and have not seen important events. This is important because it rules out the possibility that subordinates have used the mere presence/absence of a dominant during the baiting.

One final scoffer interpretation of these food competition experiments is the so-called evil eye hypothesis. Perhaps subordinates believe that any piece of food observed by a dominant is 'contaminated'—it is forbidden once the dominant has put the evil eye on it—and so the only safe food is food that she cannot see and indeed has never seen. In one final study in Hare *et al.* (2001), both the dominant and the subordinate watched the food being hidden behind one of the two barriers; the dominant's evil eye was thus placed on it, and so on this interpretation it should be avoided at all costs. But then in one experimental condition only the subordinate watched it being moved to a new location (dominant's door down), whereas in another condition they both watched it being moved. Subordinates went for the food when they alone had watched the moving process, but *not* when both competitors had watched the moving process. It is thus clear that they did not believe in any dominant evil eye, since they went for the food whose movement to a new location the dominant had not witnessed (even though she had seen it earlier).

These food competition studies are especially important in determining what chimpanzees know about seeing because in these studies subordinates did not just follow the gaze direction of another to a location, rather they demonstrated by their behaviour that they knew something about the actual *content* of what the dominant could see and had seen in the immediate past. This is clearly indicated by the fact that subordinates made inferences about the dominant's impending action; if the dominant could see the food or had seen it just before, subordinates could infer that she would go for it. They would not make this inference if what the dominant could see or had seen was a rock. This means that they knew not just where the other was looking, but they knew the content of what the others saw—the food—because only with this knowledge could they predict her action. Of course they may have had learning experiences in the past in which they noted a dominant looking at food and then going for it, but conditions involving such things as the transparent barrier and the food whose location was moved after the dominant had seen it hidden are not likely to be common experiences in chimpanzees' lives, and so not things about which they could have learnt.

17.4 **Begging and gesturing**

Povinelli and Eddy (1996b) trained young chimpanzees to approach a Plexiglas barrier and extend their hand toward one of two human experimenters (each with a hole in the Plexiglas in front of them) to request the food that was on a table between them. The attentional state of the two humans was varied, and it was found that, for example, chimpanzees consistently gestured toward the human who was facing toward them, in preference to one facing away from them (thus confirming the naturalistic findings of Tomasello *et al.* 1994, 1997). However, they did not gesture differentially for a human who wore a blindfold over his eyes (as opposed to one who wore a blindfold over his mouth), or for one who wore a bucket over his head (as opposed to one who held a bucket on his shoulder), or for one who held his hands over his eyes (as opposed to one who held his hands over his ears), or for one who had his eyes closed (as opposed to one who had his eyes open), or for one who was looking away (as opposed to looking at the subject), or for one whose back was turned but who looked over his shoulder to the subject (as opposed to one whose back was turned and he was looking away) (Fig. 17.3). To explain these results, Povinelli and Eddy proposed the valence hypothesis: chimpanzees have a general notion of perceptual access (communicative access) based

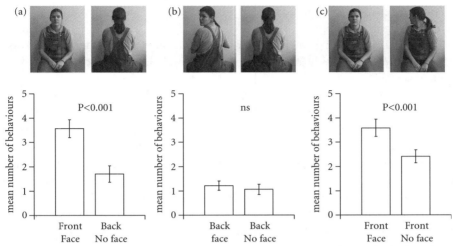

Fig. 17.3 Mean number of behaviours used for food begging depending on the body and face orientation of the human experimenter. Three comparisons are depicted: (a) body and face orientation; (b) face orientation when the experimenter has her back turned; and (c) face orientation when the experimenter has her body oriented to the subject.

mainly on body orientation, with the face and eyes playing basically no role. Thus in this scoffer hypothesis they know little, if anything, about seeing as a psychological process.

Kaminski *et al.* (2004) used a related but different methodology to test this hypothesis in a modified way (see also Call and Tomasello 1994). The difference was that chimpanzees did not have to choose between human communicators, but were always faced with only one communicator. Moreover, unlike Povinelli and Eddy (1996a, b) there was no training required at the beginning of the test. We simply took advantage of the begging behaviour that apes display in these situations. In different experimental conditions, experienced by the subject at different times, the communicator was oriented in different ways. For example in one condition the human communicator faced the subject, and in another she turned her body away and looked over her shoulder back at the subject. In this new one-communicator design, chimpanzees clearly demonstrated that they were sensitive to whether the communicator was looking at them (perhaps because the situation involving only one potential recipient of a communicative act is more natural). Specifically, when the communicator was bodily facing them, chimpanzees gestured differently depending on whether her face was also oriented toward them. This sensitivity to the face contradicts the valence hypothesis. Kaminski *et al.* (2004) argued, therefore, that body orientation and face orientation indicate two different things for an ape when it begs food from a human being. Whereas the human being's body orientation indicates his disposition to give the subject food (i.e. when it is oriented so as to transfer food effectively), face orientation encodes information about whether the human being is able to see the subject's begging gesture.

Similarly, Hare *et al.* (in press) found that when chimpanzees attempt to steal food from a human being whose body is facing one piece of food but whose face is directed at a second piece of food (he is looking over his shoulder), they take the food in front of the body and avoid the one in front of the face. The finding of sensitivity to the face orientation in this case also contradicts the valence hypothesis, and strongly suggests that chimpanzees know which piece of food the human being can see. In a related study, Hare *et al.* (in press) used the same competitive set up but put one piece of food behind a transparent barrier and one behind an opaque barrier while the human competitor stared straight ahead. Chimpanzees preferentially selected the piece that was behind the opaque barrier, the one that the competitor could not see, thus providing further confirmatory evidence for the findings of the food competition studies of Hare *et al.* (2000, 2001). Moreover, and importantly, chimpanzees in this competition-with-a-human-being paradigm sometimes attempted to hide the beginning of their approach to the food by using circuitous routes that hid them from the human being's view (behind the barrier) from as early in their approach as possible—which would seem to indicate a fairly sophisticated understanding of the viewing angle of the human experimenter.

The scoffer hypothesis for these competition-with-a-human-being results is that chimpanzees do not compute what the human competitor can and cannot see, but rather they simply avoid approaching food if they themselves can see a competitor's face. In a third study, Hare *et al.* (in press) found something different. In this study, subjects had to choose to approach food either behind: (1) an opaque barrier that obstructed the human competitor's view of the subject and the food; or (2) a split barrier (each part half the size of the barrier on the other side, so that overall area covered was the same) that occluded the subject's view of everything at eye level (including the human competitor's face) and also at lower body level—but with a gap allowing the competitor to see the subject's body approaching. This condition thus compared the hypothesis that subjects were simply going to the side where they themselves could not see the human competitor—in which case they would show no preference—with the hypothesis that they were going to the side where the competitor could not see them—in which case they should approach behind the opaque barrier. The finding was that chimpanzees preferred approaching the food that was behind the opaque barrier, indicating their knowledge that the competitor could see them approaching behind the split barrier. It is possible that in this condition chimpanzees could see the competitor's torso or feet (through the split) and so they simply tried to avoid any sight of their competitor at all. But note that this explanation does not apply to the Hare *et al.* (2000, 2001) studies in which chimpanzees' competed with one another while looking directly at one another the whole time; seeing the competitor in these studies did not prevent their approaching the food when they thought the other could not see it. Moreover, seeing the dominant looking during the baiting period did not prevent subordinates from approaching and taking the food when the original dominant was switched by a naïve dominant.

17.5 **Self-knowledge**

There is one last paradigm indicating something about chimpanzee's understanding of seeing—as opposed to just looking behaviour. Call and Carpenter (2001) presented chimpanzees and orang-utans with two hollow tubes perpendicularly oriented toward the subjects and then they placed a reward inside one of the tubes. To get the reward, subjects had to touch the baited tube (at which point the experimenter gave them the contents, if any). There were two baiting conditions. In the visible condition, the experimenter placed the food inside the tube in full view of the subject. In the hidden condition, the experimenter baited one of the tubes but prevented the subjects from witnessing the baiting process—so that subjects knew there was food in one of them but they did not know which one. Subjects looked into the tubes before choosing more often when they had not witnessed the baiting (as was necessary for them to be successful) than when they had witnessed the baiting (Fig. 17.4).

From a scoffer point of view, one could argue that somewhere in their past or in experiments subjects learned a conditional discrimination of the kind: if the location

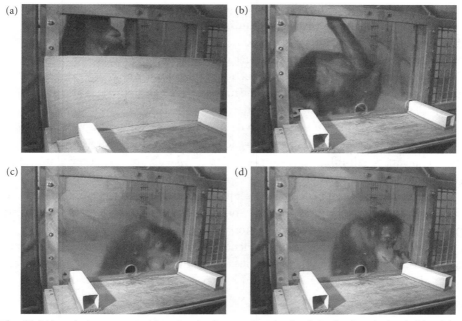

Fig. 17.4 Upon being prevented by a barrier from seeing the food destination during baiting (a), an orang-utan inspects two hollow tubes in succession (b, c) before selecting the baited one (d).

of the food is unknown, then bend down and look inside the tubes until you find the reward. However, in another condition it was found that when the tubes are blocked from the front (and the subjects sees this barrier), apes do not attempt to look (Call, unpublished data). One could still argue that although the sight of the tubes paired with not witnessing the baiting makes subjects bend down, this response is cancelled if something is blocking the front of the tubes; from a behavioural point of view, the blockage is a cue that modulates the conditional discrimination. This is possible, but of course the alternative to this complex conditional discrimination (a skill not easily mastered by chimpanzees), based on hypothesized but not observed learning experiences is that subjects know what they have and have not seen and they look to seek information when they need it. They know something about their own seeing.[1]

17.6 So what do chimpanzees really understand about seeing?

The current review has shown that chimpanzees follow the gaze of conspecifics and humans, follow it past distractors and behind barriers, 'check back' with humans when gaze following does not yield interesting sights, stop following the gaze of a looker who keeps looking at nothing, use gestures appropriately depending on the visual access of their recipient, know when they themselves have seen something, and select different pieces of food depending on whether their competitor has visual access to them—understanding transparent barriers and split barriers in the process. Taken together, these findings make a strong case for the booster hypothesis that chimpanzees have some understanding of what other individuals can and cannot see. The scoffer position, on the other hand, requires about a dozen different hypotheses to account for all of these phenomena—no one of which accounts for more than a few phenomena. Table 17.1 summarizes the main experimental findings, the scoffer explanations needed to account for them, and the one booster explanation that, if correct, explains them all. Note that the booster explanation is analogous to the concept of intervening variable used by Whiten (1994, 1996) to discuss the issue of 'mind reading' in animals. Whiten argued that one viable alternative to postulating a one-to-one correspondence between sets of stimuli and responses, is to invoke a common intervening state (produced various different stimuli) that regulates the various responses.

In the debate between boosters and scoffers, parsimony is often raised as an issue. But parsimony is a notoriously protean concept. Inspecting Table 17.1, it would seem that the boosters clearly have parsimony on their side. The *number* of different explanations required to explain the evidence available is sensibly smaller (12 versus 1). Of course, looking at it another way one could say that the scoffers need only invoke behaviouristic principles of learning, whereas the boosters need to posit an understanding of various psychological constructs—perhaps in addition to principles

---

[1] Editors' note: see also Chapters 10, 11, and 12, this volume.

Table 17.1 Experimental findings related to chimpanzees' understanding of seeing, along with the associated booster and scoffer theoretical explanations (see text for detailed discussion of evidence for and against particular scoffer hypotheses)

Phenomenon	Scoffer explanation	Booster explanation
1 (i) Gaze follow; (ii) Ignore distractors	Natural or learned co-orientation	
2 (i) Check back; (ii) Habituate when no target	Learned contingencies w/r/t gaze direction and rewarding targets	
3 Follow gaze around barriers	Geometric gaze follow (perceptual stimulus enhancement)	Chimpanzees
4 Go for hidden food in competition	Assume barrier blocks access	
5 Avoid transparent barrier	Peripheral feeding hypothesis	
6 Know when competitor saw hiding	Evil eye hypothesis	understand
7 Know when competitor saw moving	Competitor gives cues while watching moving	
8 Beg from E who can see you	Valence hypothesis	
9 Beg from E who is both oriented and can see	Beg when both face and body oriented	seeing
10 Steal from E not looking	Avoid seeing E's face	
11 Steal from behind split barrier	Avoid seeing E's body	
12 Know when to move so as to see something	Learned conditional discrimination	

of learning. But what seems to be really going on when a scoffer claims parsimony here is that they are making an appeal to something akin to Morgan's Canon: learning is a simpler mechanism—that is, a lower level mechanism shared with many animal species—whereas understanding psychological states is a more complex or higher level mechanism. However, the distinction between lower and higher mechanisms is plagued with interpretative problems. For one thing, there is no objective basis for classifying learning as a lower or simpler mechanism and insight or representational as higher or complex. In general, we are not strong proponents either of parsimony (unless one clearly defines the criteria for parsimony) or of Morgan's Canon—certainly not as substitutes for grappling with data when there is plenty of it.

It is also important to note that when learning explanations are applied to the range of phenomena in Table 17.1, the scoffer must propose many and extremely complicated learning scenarios for which there is no direct evidence. For example learning that transparent barriers do not block the visual access of others would require a series of experiences in which *each* subject approached food behind a barrier, took it while the dominant was looking, and was punished by the dominant when the barrier was transparent but not when the barrier was opaque. Similarly, to learn to stay away from food that has been moved from behind one barrier to another while a dominant watches, but to go for the food if the dominant was not watching, requires another set of possible but extremely unlikely learning experiences *for each subject*. And it is important that there is no evidence for any of these hypothesized learning experiences. They are just raised by scoffers as theoretical possibilities to account for certain experimental results. Indeed, if the true situation was that chimpanzees individually learn relations between the looking behaviour of others in certain situations and their subsequent behaviour, one would expect to see some individuals learning important things during the experiments themselves. But in none of the studies reported above are there any signs of individuals improving over trials within the experiment.

We believe that imagining possible learning scenarios in a subject's past, or even during a particular experiment, is an important exercise for making sure that experimental designs aimed at cognitive processes measure what they aim to measure and also take account of other potential hypotheses. But when we have many different experiments using many different methods—as we have in the current case—scoffer hypotheses positing all kinds of different learning accounts for different studies begins to become implausible. It is much more plausible to simply credit chimpanzees with understanding seeing.

One final point—learning explanations and cognitive explanations are often presented as mutually exclusive alternatives to explain some range of behavioural phenomena. We believe this is a theoretical mistake. For example it is often supposed that if an organism could have learned something on the basis of personal experience, then it is *ipso facto* not cognitive. If chimpanzees could have learned about transparent barriers through rewards and punishments, for instance, then we do not need to invoke an understanding of seeing to account for their behaviour with respect to them. But many of the most complex, sophisticated, and abstract human cognitive skills are acquired through learning. For example children take many years to learn a language, algebra, reading, and so on. When we view learning more broadly in interaction with other cognitive skills, then we can see that the real issue is: when an organism learns something, does it do so on the basis of blind associations or on the basis of some causal or intentional understanding of the situation? In the current case, it is very likely that chimpanzees individually learn some things about what others can and cannot see, but this does not mean that they do not understand seeing. This might be tested by conducting experiments to see whether an individual learns something (for example to avoid food behind a transparent barrier) more quickly or more slowly than it would learn some arbitrary contingency (for example to avoid food behind a red barrier).

We are willing to bet that a naive chimpanzee in the presence of a dominant would learn to avoid food behind a transparent barrier much more quickly (if indeed learning was required at all) than it would learn to avoid food behind a red barrier if it was rigged so that the dominant could actually see the food (say, if there were a red one-way mirror) and so punish the subordinate accordingly when she took it. In fact, there is some evidence suggesting that this preference for causal as opposed to arbitrary connections is also found in physical cognition (see Call 2003, Chapter 10, this volume; Premack and Premack 1994).

Science is open-ended, and the case is certainly not closed on the issue of whether chimpanzees understand seeing. But now that the evidence is overwhelmingly in favour of the booster position, we believe that that they do. Of course negative evidence— for example failure to infer what another could see in some situation—would provide support for the scoffer position. Leaving aside the thorny issue of interpreting negative evidence, currently there are more positive than negative results, only a motley collection of alternative hypotheses—no small set of which explain all of the data—and a generic appeal to possible learning experiences that individuals may have had previous to their participation in an experiment. There are also the negative findings of Povinelli and Edy (1996b), but these have been to some degree superseded by the positive findings of Kaminski *et al.* (2003). We are therefore boosters on this specific question. At the same time, however, we must reiterate that understanding seeing does not mean that chimpanzees understand other psychological states; this might be the only one. The scientific situation is that we must look at each psychological state—perceiving, intending, desiring, believing, and so on—separately, and with the same kind of variety of experimental methods reported here. Whether chimpanzees understand any particular psychological state is, in each case, an empirical question.

References

Bates, E. (1979). *The emergence of symbols: Cognition and communication in infancy*. New York: Academic Press.

Bräuer, J., Call, J., and Tomasello, M. (submitted). All four great apes follow gaze, including around barriers.

Call, J. (2003). Beyond learning fixed rules and social cues: Abstraction in the social arena. *Transactions of the Royal Society,* **358**: 1189–1196.

Call, J., and Carpenter, M. (2001). Do apes and children know what they have seen? *Animal Cognition,* 4: 207–220.

Call, J., Hare, B., and Tomasello, M. (1998). Chimpanzee gaze following in an object choice task. *Animal Cognition,* 1: 89–100.

Call, J., and Tomasello, M. (1994). The production and comprehension of referential pointing by orangutans. *Journal of Comparative Psychology,* **108**: 307–317.

Hare, B., Call, J., Agnetta, B., and Tomasello, M. (2000). Chimpanzees know what conspecifics do and do not see. *Animal Behaviour,* **59**: 771–785.

Hare, B., Call, J., and Tomasello, M. (2001). Do chimpanzees know what conspecifics know? *Animal Behavior,* **61**: 139–151.

Hare, B., Call, J., and Tomasello, M. (in press). Chimpanzees deceive a human by hiding. *Cognition*.

Kaminski, J., Call, J., and Tomasello, M. (2004). Body orientation and face orientation: Two factors controlling apes' begging behavior from humans. *Animal Cognition*, 7, 216–23.

Povinelli, D. J., and Eddy, T. J. (1996a). Chimpanzees: Joint visual attention. *Psychological Science*, 7: 129–135.

Povinelli, D. J., and Eddy, T. J. (1996b). What young chimpanzees know about seeing. *Monographs of the Society for Research in Child Development*, 61: 1–152.

Povinelli, D. J. and Giambrone, S. (2001). Reasoning about Beliefs: A Human Specialisation? *Child Development*, 72: 691–695.

Premack, D., and Premack, A. J. (1994). Levels of causal understanding in chimpanzees and children. *Cognition*, 50: 347–362.

Tomasello, M., Call, J., and Hare, B. (1998). Five primate species follow the visual gaze of conspecifics. *Animal Behaviour*, 55: 1063–1069.

Tomasello, M., Call, J., Nagell, K., Olguin, R., and Carpenter, M. (1994). The learning and use of gestural signals by young chimpanzees: A trans-generational study. *Primates,* 37: 137–154.

Tomasello, M., Call, J., Warren, J., Frost, T., Carpenter, M., and Nagell, K. (1997). The ontogeny of chimpanzee gestural signals: A comparison across groups and generations. *Evolution of Communication*, 1: 223–253.

Tomasello, M., Hare, B., and Agnetta, B. (1999). Chimpanzees follow gaze direction geometrically. *Animal Behaviour*, 58: 769–77.

Tomasello, M., Hare, B., and Fogleman, T. (2001). The ontogeny of gaze following in chimpanzees and rhesus macaques. *Animal Behaviour*, 61: 335–343.

Whiten, A. (1994). Grades of mind reading. In: C. Lewis and P. Mitchell, eds. *Children's Early Understanding of Mind*, pp. 47–70. Hillsdale, NJ: Lawrence Erlbaum Associates.

Whiten, A. (1996). When does smart behaviour-reading become mind-reading? In: P. Carruthers and P. K. Smith, eds. *Theories of Theories of Mind*, pp. 277–292. Cambridge: Cambridge University Press.

Chapter 18

We don't need a microscope to explore the chimpanzee's mind

Daniel Povinelli and Jennifer Vonk

Abstract

The question of whether chimpanzees, like humans, reason about unobservable mental states remains highly controversial. On one account, chimpanzees are seen as possessing a psychological system for social cognition that represents and reasons about behaviours alone. A competing account allows that the chimpanzee's social cognition system *additionally* construes the behaviours it represents in terms of mental states. Because the range of behaviours that each of the two systems can generate is not currently known, and because the latter system depends upon the former, determining the presence of this latter system in chimpanzees is a far more difficult task than has been assumed. We call for recognition of this problem, and a shift from experimental paradigms that cannot resolve this question, to ones that might allow researchers to determine when it is necessary to postulate the presence of a system which reasons about both behaviour *and* mental states.

18.1 **Emergence of a gentle controversy**

Are humans alone in their ability to interpret behaviour in terms of unobservable mental states—things like feelings, beliefs, desires, emotions, and intentions? Or do we share the ability to reason about mental states (at least to some degree) with other species? Premack and Woodruff (1978) coined the phrase 'theory of mind' to isolate and draw attention to the class of psychological systems that have the property of reasoning about such states: 'A system of inferences of this kind,' they noted, 'may properly be regarded as a theory because such [mental] states are not directly observable, and the system can be used to make predictions about the behaviour of others' (p. 515). It is important to note that we use the term 'theory of mind' in the broadest sense, as shorthand for the ability to represent and reason about mental states.

Whether this ability is implemented through an intentional system that is truly 'theory-like' in the sense implied by the 'theory–theory' view, or whether one of the 'simulationist' accounts is correct, does not affect the arguments we make in this chapter. In this chapter, we do not endorse a particular theoretical position on this topic, but we avoid the term 'mind reading' as shorthand for the capacity to represent and reason about mental states because of its historical association with phenomena related to parapsychology (e.g. extrasensory perception).[1]

For several years in the late 1990s, there appeared to be an emerging convergence of evidence which supported the idea that humans might indeed be alone in possessing a 'theory of mind' system; in other words, that the capacity to conceive of mental states might have evolved hand-in-hand with the human lineage (e.g. Povinelli and Prince 1998; Tomasello and Call 1997). To be sure, there were (and continue to be) plenty of opinions to the contrary (those who believe that chimpanzees, at least, also possess a system for reasoning about mental states). Nonetheless, the vast majority of experimental tests had suggested that chimpanzees (and other non-human primates) might not possess this ability (reviews by Heyes 1998; Tomasello and Call 1997). Alas, as is so often the case in the cognitive sciences, even this limited consensus proved transitory. The question of whether chimpanzees possess a 'theory of mind' system excites as much controversy now as it did when Premack and Woodruff (1978) first proposed the possibility a quarter century ago.

In particular, the two research groups that have published most widely in this area—our own group, based in the United States in Lafayette, Louisiana, and a group led by Michael Tomasello in Leipzig, Germany—have now parted ways in their assessment of the evidence related to 'theory of mind' capacities in primates (see Povinelli and Vonk 2003; Tomasello *et al.* 2003a, b). Of importance, however, is that the conclusions reached by these laboratories are not of the same kind. The Leipzig group has asserted a factual claim, proposing that by 'turning up the microscope' recent data have more or less definitively established that chimpanzees possess at least 'parts' of a 'theory of mind' system: '... although chimpanzees almost certainly do not understand other minds in the same way that humans do (e.g. they apparently do not understand beliefs) they do understand some psychological processes (e.g. seeing)' (Tomasello *et al.* 2003b, p. 239). In contrast, our group has reached a different interim conclusion: although we do not deny the possibility that chimpanzees possess an ability to represent and reason about mental states, we contend that *the research paradigms that have been heralded as providing evidence that*

---

[1] Editors' note: In the rest of this volume, 'mind reading' is used in the generic sense for which Povinelli and Vonk prefer in this chapter to use 'theory of mind'. 'Mind reading' is used in the rest of the volume in order clearly to distinguish the generic capacity to understand the minds of others and attribute mental states to others from a particular account of how that is done, the theory-theory. Obviously, from use of 'theory of mind' in the generic sense that simply refers to mind-reading capacity, nothing can be inferred about the correctness of theory-theory as opposed to simulation theory.

they do reason about such mental states, do not, in principle, have the ability to provide evidence that uniquely supports that hypothesis. Thus, we show that so long as we rely upon the current paradigms, 'turning up the microscope' will not help. And, in a further move, we have also argued that key aspects of the data point toward the possibility that if chimpanzees do have a 'theory of mind' system, it must be radically different from our own. In what follows, we dissect this 'gentle controversy' and offer productive suggestions for how to make progress toward resolving it.

18.2 **Is 'theory of mind' anthropocentric?**

First, let us address a question that always evokes some worry when discussing the question of 'theory of mind' capacities in other species: 'Why devote so much energy to trying to determine if chimpanzees have a human-like 'theory of mind' system? Why not try to figure out what makes them chimpanzees, instead?'

At first glance, this would seem to be a legitimate concern. In trying to reconstruct the evolution of certain forms of cognition, we should never lose sight of the fact that the human mind is not the only psychological system in town (a point we have repeatedly stressed in recent years: e.g. Povinelli 2000; Povinelli and Bering 2002; Povinelli and Prince 1998). Surely, then, in comparing the psychologies of humans and chimpanzees, we should not overlook fascinating questions concerning the unique abilities of chimpanzees.

A moment's reflection, however, will show that this is only a single side of one of the many coins in the purse of comparative psychology. Furthermore, the flip side of this particular coin is that just as understanding the unique nature of chimpanzees can and should interest us as a project with its own intrinsic merits, so too should the task of discovering the unique abilities of humans. Indeed, a comparative psychology would surely be comparative: it would embrace the evolutionary notion of diversity, and should therefore be a science that asked equally about similarities *and* differences among species. Applied to the present problem—the question of the evolutionary history of a 'theory of mind' system—a mature comparative psychology should give equal scrutiny to each of the logical possibilities: (a) that the capacity to reason about mental states may be shared by many species; (b) that it may be unique to primates; (c) that it may be unique to some primates; (d) that it may be unique to humans alone; (e) that different aspects of the system may be present in different species (see Povinelli and Eddy 1996a, Chapter 2). In any event, it would never commit so egregious a sin as to exclude questions about what makes the human mind uniquely human in the first place. After all, this, too, is a fascinating question.

So, to our way of thinking, the criticism that it is anthropocentric to ask whether chimpanzees possess the capacity to reason about mental states is almost irrelevant. There is even a sense in which the validity of the charge of anthropocentrism will depend on the outcome of the science. If it turns out that 'theory of mind' capacities are shared by at least some other species, then the endeavour would not have been anthropocentric at all. On the other hand, if the ability to reason about mental states turns out to be a unique specialization of the human mind, then, yes, of course, these efforts would have been profoundly anthropocentric. But the overarching point is that the motivation to ask whether other species possess these abilities is not an anthropocentric one, and, in any event, we will have learned more about both chimpanzees and humans for having done so.

18.3 Concepts about behaviour versus concepts about behaviour and mind

For the moment, let us place in abeyance the empirical question of whether chimpanzees actually possess an ability to reason about mental states, and instead address another frequently asked question, 'How can you ever determine if a language-less organism reasons about the hidden, subjective mental states of others?'

The standard approach to answering this question goes something like this: 'If the organism is reasoning about a particular mental state (for example, an <intention>) then we ought to be able to devise a behavioural test that can tap into it. So, (a) conceive of a situation in which a subject would need to respond in some fashion to the behaviour of another organism, then, (b) devise an experiment that will sort out whether the subject is reasoning about just the behaviour of another, or about both their behaviour *and* the relevant underlying mental state. Finally, make the design clear enough so that a distinct response 'r' can be predicted if the subject is reasoning strictly about behaviour, whereas distinct response 'q' can be predicted if the subject is also reasoning about the unobservable mental state.'

At first glance, this approach to constructing non-verbal 'theory of mind' tasks for an organism like the chimpanzee seems remarkably easy. But let us examine the logic of this general scheme and expose its fundamental flaws. First, explicit in this approach is the assumption that reasoning about the underlying mental state in such paradigms would *inevitably* lead to a fundamentally different response on the part of the subject than reasoning about surface behaviour alone. Second, there is an implicit assumption that the humans who design the experiments can use their folk psychology to successfully intuit which responses can be produced only by reasoning about the underlying mental state. In other words, it relies upon the experimenters' use of their own folk psychology to posit a causal relationship between reasoning about mental states and subsequent behaviour. Anyone who doubts that this is in fact the current practice need only examine published scientific reports; typically, no formal demonstrations (logical or empirical) are provided to show that response 'q' is possible *only* if the subject is reasoning about the particular mental state under consideration. Rather, intuitions about how the human 'theory of mind' system works are used as a basis for designing experimental situations for other species.

A simple example from our own laboratory may help to clarify the role that our folk psychology plays in designing a non-verbal 'theory of mind' task, and certain logical flaws therein. Imagine that you wished to know if chimpanzees represent the psychological experience of <*seeing*> in others. You might reason as follows: 'If I confronted them with two people—one of whom could see them, the other of whom could not—and created a situation where they had to use their natural begging gesture (a visually-based signal), and then if they gestured to the person who could see them, this would indicate an understanding of <*seeing*> on their part.' We conducted precisely such a set of studies by confronting our chimpanzees with the simple situation depicted in Fig. 18.1a–c. First, the chimpanzees encountered only one experimenter, either on the right or left, approached her, and requested a piece of fruit by using their begging gesture. In testing, the situation was different: two experimenters were present, one facing the subject, the other facing away (Fig. 18.2a). The chimpanzees paused, but then proceeded directly to the experimenter who was facing them (the one who could see them) and gestured. Here, the purported response 'q' was consistently approaching the experimenter who could see them across trials, whereas purported response 'r' was approaching the experimenters equally often. The empirical results show that the chimpanzees, by the way, consistently produce response 'q' (see Povinelli and Eddy 1996a). As shall become clear, however, one of our central conclusions is that for the class of experiments in question, neatly carving up the response space into 'q-' and 'r-types' may be impossible. For instance, it will

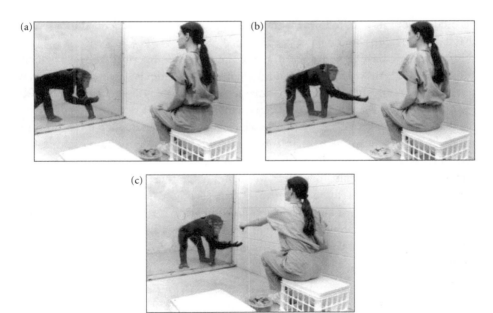

Fig. 18.1 A chimpanzee (a) approaches the experimenter, (b) requests a food reward with a species typical begging gesture, and (c) receives the food reward.

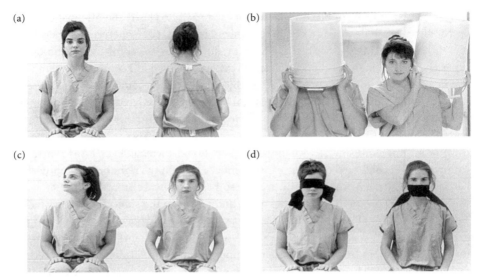

Fig. 18.2 Simple choice situations used to test chimpanzees' understanding of seeing: (a) front versus back condition, (b) buckets condition, (c) attending versus distracted condition, (d) blindfolds condition. Chimpanzees were successful in condition (a) from the first trial onward, but were at chance in conditions (b)-(d) until they received many trials with differential feedback.

become obvious in what follows that response 'q' could easily be generated by a system that does not reason about mental states.

Elsewhere, we have provided a critique of the approach described above (Povinelli and Vonk 2003). However, because our previous verbal critique was so general, it may not have been fully understood (see Tomasello *et al.* 2003b). Here we shall be slightly more formal (and hence more explicit). Let S_b stand for a psychological system dedicated to social cognition, but one which forms and uses concepts about only 'behaviours' which can, in principle, be observed. Further, let us suppose that this system is every bit as sophisticated as other cognitive systems already known to exist in humans and other animals. To be clear, we are not suggesting that a system that reasons about behaviour alone is not a 'cognitive' system. Clearly such a system would depend on the ability to represent and reason about complex intervening variables. The point is merely that the variables about which the organism is reasoning do not include representations of mental states. In particular, we conceive of S_b as having three main components:

1. a database of representations of both specific behaviours and statistical invariants which are abstracted across multiple instances of specific behaviours (representations that may be formed either by direct experience with the world, or may be epigenetically canalized);

2. a network of statistical relationships that adhere between and among the specific behaviours and invariants in the database;

3. an ability to use the statistical regularities to compute the likelihood of the specific future actions of others.

Important for this characterization of S_b is that it also interacts with the organism's representations of the physical layout of the world. (Baird and Baldwin (2001) have proposed that a system similar to this is fully operational in human infants, and Povinelli (2001) has described its operation in other species.)

This system (S_b) can now be properly contrasted with a psychological system that, in addition, reasons about the mental states of other organisms. Our use of the qualifier 'in addition' is crucial to understanding our argument, because in its traditional characterization, the human 'theory of mind' system cannot be thought of as operating in isolation from an organism's representations of behaviour; the system does not generate inferences about mental states in others at random. Rather, it uses information about ongoing, recent, or even quite temporally distant behaviours, to generate inferences about the likely mental states of others. Thus, we believe that the kinds of representations we posit to exist in the database of S_b exist in humans and interact with the 'theory of mind' system. *Indeed, it seems likely that much human social interaction is supported solely by the features of S_b that we have just described.* Indeed, Baird and Baldwin (2001) propose that human infants may initially rely precisely upon such a system.

So, to return to the case of the chimpanzee using its begging gesture to request food from the experimenter facing them as opposed to the one facing away, we can clearly see that our common-sense intuitions about what should qualify as response 'q' and what should qualify as response 'r' are highly problematic. We can see how response 'q' could have easily been generated by S_b and, furthermore, that response 'r' may not have been a reasonable prediction of the behaviour that would be generated by this system. Although the chimpanzee may or may not attribute the mental experience of <*seeing*> to the person facing them, they almost certainly know from previous experience that organisms that are facing them have a non-zero probability of responding to their visually based gestures, whereas those facing away do not. Indeed, in this situation we can clearly see that the notion of <*seeing*> is clearly secondary to the detection of the observable invariants associated with 'facing forward.'

What all of this means is that on the standard interpretation, the 'theory of mind' system can be considered to have a mutualistic relationship with S_b. Thus, because it is a system which must perform joint computations about both the behaviour and the mental states of other organisms in order to successfully predict future behaviour (and hence assist the organism in determining what actions it should take), we describe the 'theory of mind' system as S_{b+ms}. The inescapable implication of this is that making inferences about mental states does not allow an organism to skip the step of having to detect the abstract categories of behaviour and compute the regularities among them.

With these formalisms in mind, we can now ask a more difficult question: 'Can the research paradigms that are currently in use with chimpanzees (and other animals) effectively distinguish between the operation/presence of S_{b+ms} versus S_b?' We submit that they cannot.

To show why, let us begin with an example. Faced with the indeterminacy of the results of the 'front-versus-back' test described above, suppose one wanted to further pursue the question of whether chimpanzees reason about who can and cannot <see> them. In this case, we allow chimpanzees to approach two individuals, one who is wearing blindfolds over her eyes and one who is not (Fig. 18.2d). Now (contrary to the actual empirical results), let us imagine that the chimpanzees immediately and consistently deploy their (visually-based) begging gestures in front of the person whose eyes are not covered by blindfolds. One interpretation is that the chimpanzee behaves in this manner because he or she knows that this person can <see> her, whereas the other person cannot. The logic of this interpretation is as follows:

(a) chimp observes Suzy with eyes blindfolded (eyes not visible);

(b) chimp observes Mary with eyes not blindfolded (eyes visible);

(c) chimp concludes that because Mary's eyes are uncovered, she can <see>, whereas Suzy cannot;

(d) chimp gestures to Mary (because only people that can <see> the gestures respond appropriately to begging gestures).

But, given that (a) and (b) are both observable regularities, and (d) is a contingent outcome that stems from (c), we must critically ask which of the two aspects of Mary that the subject represents in (c) are causally related to generating the response of gesturing to her: the invariant associated with the observable feature of 'unobstructed eyes,' or the additional attribution of <seeing>? In short, if we substituted the perceptual invariant for the mental state in (d), would the same response occur? We submit that it would:

(a) chimp observes Suzy with eyes blindfolded (eyes not visible);

(b) chimp observes Mary with eyes not blindfolded (eyes visible);

(c) chimp gestures to Mary (because only people whose eyes are visible respond appropriately).

One might object that this possibility could be easily ruled out by determining if chimpanzees would respond in the same way if the eyes were not merely visible, but were oriented in an appropriate versus inappropriate direction (e.g. Fig. 18.2c; Povinelli and Eddy 1996b). So, in this case Suzy's eyes are open and uncovered, but are directed away from the subject, but Mary's eyes are open and oriented in the subject's direction. Imagine now that the chimpanzee still prefers to gesture to Mary over Suzy. Surely, then, this is not a simple case of assessing the presence or absence of the eyes, so the chimpanzee must be reasoning about who can <see> them—right?

Unfortunately, this logic does not hold. The reason is simple: the same kind of contingent dependencies between the observable features of others and the inference to a mental state still exist. The inference about *<seeing>* that would be generated by S_{b+ms} depends in the first place upon the orientation of the observable feature of eye direction. But the predicted outcome upon which the chimpanzee is basing its decision (the other person responding or not responding to their gesture) also depends on the orientation of the eyes. For instance, in the previous example, even assuming that the chimpanzee subject has an understanding of *<seeing>*, if the subject did not have a robust representation of the relevant perceptually invariant aspect of the other agent's pupil orientation, how could he or she ever compute what the other agent was *<seeing>*?

The general difficulty is that the design of these tests necessarily presupposes that the subjects notice, attend to, and/or represent, precisely those observable aspects of the other agent that are being experimentally manipulated. Once this is properly understood, however, it must be conceded that the subject's predictions about the other agent's future behaviour could be made either on the basis of a single step from knowledge about the contingent relationships between the relevant invariant features of the agent and the agent's subsequent behaviour, or on the basis of multiple steps from the invariant features, to the mental state, to the predicted behaviour. Without an analytical specification of what additional explanatory work the extra cognitive step is doing in the latter case, there is nothing to implicate the operation of S_{b+ms} over S_b alone.

Some researchers will object on the grounds of parsimony, claiming that positing the presence of S_b alone requires the existence of an intractably large number of specific rules by which the chimpanzee subject, for example, would need to behave (perhaps even drawing an analogy to the historical rejection of behaviourism on the grounds of parsimony; see Tomasello and Call, Chapter 17, this volume). As we have made clear, however, this is misleading. A hypothetical chimpanzee subject, endowed with a full-blown, human-like 'theory of mind' system, would still need the ability to detect every behavioural category that is relevant to a proper 'theory of mind' inference (regardless of whether one ascribes to theory–theory or simulation accounts of 'theory of mind' abilities). This is a key point: as originally conceived by Premack and Woodruff (1978), reasoning about mental states must entail observing and reasoning about behaviour (in all its subtleties) and, on the basis of such observed features, generating and reasoning about representations of unobserved mental states. Thus, the capacity to reason about mental states does not somehow relieve the burden of representing the massive nuances of behaviour or the statistical invariances that sort them into more and less related groups. In either event, these behavioural abstractions must be represented. With respect to parsimony, then, the question becomes a simple one: 'Is a system (S_b) that represents the invariant spatiotemporal aspects of behaviour which are the purported and observable manifestations of *<wanting>* a banana, for example, any more or less parsimonious than a system that represents the invariant aspects of the same class of behaviours, but, in addition, generates a mental state concept to go along with it?'

18.4 **Parsimonious illusions?**

In light of the preceding discussion, let us briefly examine how the concept of parsimony has been deployed in the current controversy. At several points, the Leipzig group has asserted that although it is possible that chimpanzees form concepts solely about behaviour, the case for this is unproven, and they seem to imply that parsimony should push us toward assuming that they do, in fact, represent mental states. For example Tomasello *et al.* (2003b) assert: 'Of course it is possible that human beings are the only species that understand any psychological processes in others, and we ourselves held this position not so very long ago. But evidence is mounting that it is simply not the case' (p. 239–240). Referring to us (i.e. Povinelli and Vonk 2003), they go on: 'We cannot dismiss this evidence by noting that simpler explanations are hypothetically possible with no supporting evidence' (p. 240).

The best interpretation of such statements is that the Leipzig group has not yet addressed the very heart of our analysis: namely, that because the current studies that are held up as evidence for 'theory of mind' capacities in chimpanzees presuppose that the subjects can form very (very!) subtle and abstract concepts of behaviour, dependent measures which then show that they know what to do when confronted with situations that are empirically linked to such behaviours cannot provide evidence in support of the hypothesis that they also form concepts about mental states purportedly linked to such behaviours. Why not? As we have shown above, the experiments do not specify the unique causal work that concepts about mental states do above and beyond the work that can be done by representations of the invariant aspects of behaviour. Indeed, if our analysis is correct, there is no sense in which a system that makes inferences about behavioural concepts alone provides a *less* parsimonious account of behaviour than a system that must make all of those same inferences *plus* generate inferences about mental states.[2] Although it is possible to

---

2 Perhaps the sense of parsimony that is being invoked in such discussions pertains more to linguistic or explanatory parsimony—the idea that one can or should *describe* behaviour in the simplest *linguistic* manner possible, regardless of the underlying complexity of the behaviour itself. It is in some sense 'easier' for us to describe the chimpanzees as understanding <*seeing*> then it is to *explain that they understand all of the unique and specific behavioural regularities that must be computed first in order to then represent a concept of* <*seeing*> (see Dennett's 1987, notion of the 'intentional stance'). This sense of the term parsimony is misleading in the current discussion, however. Making use of a single term or concept such as <*seeing*> *to subsume all of the necessary abstractions does not change the fact that the psychological system itself cannot skip the step of representing the behavioural abstractions necessary to extract the concept of* <*seeing*>.

imagine situations in which responding appropriately in relatively novel situations might be facilitated by a system that reasons about mental states, we contend that a system that reasons about behavioural abstractions alone suffices to explain the data that currently exists (see also Povinelli and Vonk 2003, p. 159). Again, we are not suggesting that S_b is a simpler or lower-level system than S_{b+ms} in the sense implied by Morgan's canon (see Tomasello and Call, Chapter 17, this volume), despite the undeniable fact that the ability to reason about mental states depends upon an ability to reason about behaviour and not *vice versa*. In this context, it is unfortunate that we have previously referred to 'high-level' and 'low-level' models when testing for the presence of S_b versus S_{b+ms}, respectively (e.g. Povinelli 2000).

Povinelli and Vonk (2003) applied this analysis to several recent experiments that are frequently championed by the Leipzig group as establishing that chimpanzees reason about mental states (e.g. Hare *et al.* 2000, 2001). In responding to our analysis, Tomasello *et al.* (2003b) cry foul, asserting that we 'ignore' (p. 239) certain 'control conditions' present in their studies and that we simply assert alternative explanations with no data to support them (p. 240).

Unfortunately, these objections highlight the Leipzig group's failure to address our analysis which suggests that in the context of the class of experiments we have indicted, no control conditions can ever help to establish the presence of S_{b+ms} over S_b (see Povinelli and Vonk 2003). By way of illustration, let us examine their specific objection (Tomasello *et al.* 2003b). In one set of their published studies (Hare *et al.* 2000), a subordinate and dominant chimpanzee are placed in separate enclosures, facing each other, with an arena between them. Inside the arena are two pieces of food, one on the right, one on the left, with one of the pieces visible only to the subordinate because it is behind a small barrier (Fig. 18.3). These studies are designed to elucidate whether the subordinate can reason about which piece of food the dominant animal is able to see by measuring where the subordinate heads first when released into the arena (the dominant is released immediately thereafter). In some studies, the subordinate and dominant are allowed to look into the enclosure at the same time, raising the worry that the subordinate's future behaviour may be the result of reasoning about the

Fig. 18.3 An experimental condition used in Experiment 6 of Karin-D'Arcy and Povinelli (2002), an unsuccessful attempt to replicate key findings of Hare *et al.* (2000). One piece of food is fully occluded (visible only to the subordinate animal). The other piece of food is fully visible to both the subordinate and the rival.

observable behaviour of the dominant, avoiding the food toward which the dominant was oriented.

Apparently thinking that reasoning about behaviour must occur on-line, Tomasello *et al.* (2003b) note that a control condition was implemented in which the dominant's door was down 'and so there was no behaviour to read' (p. 239). However, even for their own logic to go through, the subordinate subject must be able to store a representation of the previous observable orientation of the dominant's body/face in relation to the food items—otherwise how could a subject with an understanding of <*seeing*> ever make the inference that 'the dominant was able <*to see*> the food over there, in the open'?

But, of course, once this point is granted, then one must also grant that the subordinate's reaction (heading away from the food that is visible to the dominant animal) can be explained either by the subject's possession of a concept about the statistical invariants that exist in head/eye/body orientation toward food, on the one hand, and future behaviour, on the other, or all of that, *plus* a representation of an unobservable mental state. To put a finer point on this issue, would the Leipzig group really wish to deny that chimpanzees have concepts about the invariant aspects of the observable past and current behaviour of their conspecifics, or, for that matter, their future behaviour? Of course not. And, if the Leipzig group concedes this, then in the absence of the production of an analytical proof demonstrating the work of S_{b+ms} over S_b, they must also concede our broader point that the outcome responses they believe to be of type 'q' can be explained in terms of the operation of S_b or S_{b+ms}.

Thus, the problem we face is not primarily an empirical one. Instead, the most pressing problem is to come to grips with the fact that the experimental results from the kinds of techniques that are currently in vogue cannot add a single bit of evidence in unique support of the conclusion that chimpanzees reason about mental states—*any* mental states.

18.5 Retooling our research paradigms

Of course, there is more to the current controversy than a debate over parsimony. For example it seems possible (even likely) that an organism possessing an S_{b+ms} wields certain predictive and explanatory abilities over and above an organism possessing only an S_b, irrespective of the issue of parsimony. As presently conceived and implemented, however, research paradigms do not analytically cope with this. There is no formal or informal demonstration associated with current research techniques which shows that S_b cannot generate the exact same behavioural predictions (without having to generate mental state inferences) as S_{b+ms}. Instead, interested scholars are simply asked to accept the folk psychological intuitions of the researchers in question.

If our current methods cannot distinguish between the presence of S_b versus S_{b+ms} in other species, then where can we turn for help? The core issue to emerge from the above discussion is that the current generation of experiments do not offer an *a priori* way of demonstrating the additive causal impact of reasoning about mental states when it is combined with a system that already reasons about behaviour. Thus, the way out of this trap must be to either: (a) develop a general, formal, analytical equation specifying the theoretical limits on selected parameters for social complexity that are achievable by S_b, and further specifying the exact alteration of social complexity that the introduction of any particular S_{b+ms} could achieve; or (b) develop specific empirical tests which perform a logical end-run around S_b; tests which have as at least one of their possible outcomes, q-type responses that logically cannot be generated by a system which reasons about behaviour alone (tests which deprive the subjects of the historical linkages between the relevant observable features of others and particular behavioural invariances). Tomasello *et al.* (2003a) correctly point out that this is, conceptually, what they and others (ourselves included) have been trying to do; as our analysis shows, however, they are incorrect in assuming that they (or we) have been successful.

Because we believe that the kind of general purpose analytical solution to the problem referred to in (a) will have to begin with assumptions about the computational limits of S_b and/or the unique impact of S_{b+ms}, on the complexity of a given social system, and because we further believe that such assumptions will be extraordinarily difficult to empirically validate, we have focussed our attention on option (b) above. In particular, we have proposed pursuing a class of behavioural tasks which have, among their possible outcomes, behaviours that can be generated only by mapping self-experience onto the experience of others (Povinelli and Vonk 2003). Although not motivated by precisely the same theoretical concerns, Gallup (1988) proposed using such a class of tasks to assay whether other species might be capable of using their own mental states to model the experiences of others. For example he suggested allowing organisms to receive extensive first-person experience wearing sound-dampening ear muffs, and to then determine if they would alter the volume of their vocalizations when attempting to gain the attention of another individual wearing them. We believe that this class of self → other inference tasks might achieve precisely the circumvention of S_b that is desired—as long as certain key stipulations are met.

To illustrate this approach, we recently suggested that a task offered by Heyes (1998) could be modified to provide unique evidence in favour of the idea that chimpanzees are reasoning about the mental experience of *<seeing>* in others. Heyes proposed a task in which chimpanzees would be given extensive experience covering their own eyes with two pairs of goggles, one of which they would be able to see through, the other of which they would not. Critically, the goggles would need to appear visually identical with the exception of an arbitrary cue, such as their colour. Also of critical importance, the subjects would not be allowed to observe others interacting with the goggles. This requirement ensures that the only relevant experience with the two pairs of goggles is the subject's first person, subjective experience of *<seeing>* or *<not seeing>* through the goggles. Eliminating the subject's opportunity to observe the distinct behavioural invariances associated with others wearing the two goggle types, negates the possibility that correct discrimination between two such individuals at test is based upon such invariants. In other words, S_b has no information in its database on how others behave when such goggles are covering their eyes. In contrast, an organism that represents mental experiences would have the possibility of mapping its own first person experience of <*seeing*> or <*not seeing*> onto the other agents, and from that knowledge, make an inference about how to behave.

Heyes' (1998) proposed task involving these goggles was to be implemented as an extension of the guesser/knower procedure developed by Povinelli *et al.* (1990): after subjects obtained first-person experience with the sensory properties of the goggles, they would (a) observe two persons wearing the goggles who in turn are 'watching' a third person hide food in one of two locations, and then (b) receive contradictory advice from these persons (by pointing) about the location of the food. Although there is nothing in principal wrong with Heyes' proposal, this particular test requires subjects to discriminate between the *knowledge* states of others (that is, first the subjects would have to infer who can <*see*> the hiding process, then, from that, they have to infer who <*knows*> where the food is located). However, given the apparent consensus that chimpanzees do not appear to reason about epistemic states (see Tomasello *et al.* 2003a, p. 156; 2003b, p. 239), we proposed a variation of this test (again based on one of our earlier procedures) in which the subjects would be required to make only a putatively 'simpler' inference regarding the experimenters' ability to <*see*> (see Povinelli and Vonk 2003; for discussions of the development of 'level one' and 'level two' visual perspective taking in young children, see Flavell *et al.* 1981).

Our task is simply a variant of the one proposed by Heyes (1998) to focus it strictly on the question of <*seeing*>. Subjects would first be exposed to the subjective experience of wearing two buckets containing visors which look identical from the outside, but one of which is see-through, the other of which is opaque (Fig. 18.4a). The buckets would be of different colours and/or shapes in order to provide the arbitrary cue to their different experiential qualities. Then, at test, subjects are given the opportunity to use their begging gesture to request food from one of two experimenters, one wearing the <*seeing*> bucket and the other wearing the <*not seeing*> bucket (Fig. 18.4b). Here, response 'q' would be the subjects' gesturing to the experimenter wearing the bucket with the see-through visor from the first trial forward. By definition, S_b has no information that would lead the subjects to generate this response. In contrast, a system that first codes the first-person mental experience, and then attributes an analogue of this experience to the other agent (in other words, S_{b+ms}) could have relevant information upon which to base a response.

Tomasello *et al.* (2003b) have responded to our proposal in a curious fashion. They have not denied the analytical problem we have outlined, nor indicted the task as an invalid measure of the attribution of <*seeing*>. Instead, they suggested that the

(a)    (b)

Fig. 18.4 (a) A subject being familiarized with the buckets. (b) A subject gestures to the incorrect experimenter.

proposed test has 'very low ecological validity' (p. 239) and, in support of this, they cite a brief commentary by Kamawar and Olson (1998) who report that the results of their unpublished pilot study using Heyes' (1998) original test did not correlate well with other 'theory of mind' tasks in preschool children.

But the difficulties in their reasoning here are apparent. First, why should the standard, laboratory-based 'theory of mind' tasks used by Kamawar and Olson (and other developmental psychologists)—tests such as false belief, appearance-reality, etc.— be considered any more ecologically valid than Heyes' original test? Second, the Heyes/Kamawar/Olson task is not the one we suggest. Heyes' test was designed to assay *<knowledge>* attribution (something that Tomasello *et al.* 2003b, already assert that chimpanzees cannot do). Our modification is a test for their understanding of *<seeing>* (i.e. level-one visual perspective taking), precisely the ability that is currently under scrutiny and is the focal point of much of the current controversy with non-human primates. Finally, and most importantly, even if the results of our proposed test of understanding of *<seeing>* turned out not to correlate perfectly with other laboratory-based tests for this ability, what should we conclude? Should we follow the logic of Tomasello *et al.* and conclude that it is inferior to other non-verbal tests developed to assay this ability in chimpanzees? But wait: if the other tests could be 'solved' by either S_{b+ms} or by S_b, then why should the results of such indeterminate tasks be favoured? The analytical challenge we have offered would remain: we need tests that can, in principle, distinguish between S_b on the one hand, and S_{b+ms}, on the other. At the very least, this would be a test whose results *could* uniquely implicate the presence of S_{b+ms} even if it were not the 'easiest' test of its kind (as inferred by the results obtained with human children).

We are not trying to sell any task as the definitive 'acid test' for reasoning about mental states in general, or about *<seeing>* in particular; in fact, our task has its own potential pitfalls and limitations. Nor are we trying to suggest that there are no other approaches that will prove capable of distinguishing between the two systems in question. Rather, we have used this example to emphasize that if we want to address the question of whether chimpanzees (or any other non-verbal creatures) have the ability to reason about mental states, we need to use tests that have the resolving power to discriminate between the work of S_b versus S_{b+ms} (see Povinelli and Vonk 2003).

18.6 Do we really need a more powerful microscope?

If current comparative methods are, *by their nature*, insufficient to address the question of 'theory of mind' capacities in chimpanzees, then we must take a harder look at Tomasello *et al.*'s (2003a) assertion that the 'way forward in research on chimpanzee social cognition is to "turn up the microscope"' (p. 156), a phrase repeated word for word in Tomasello *et al.* (2003b, p. 240, see also Tomasello and Call 2004). The logic of their analogy would appear to be that we can continue using our old microscopes as long as we crank up the magnification factor. And, apparently, they believe that by doing so, we will reveal previously hidden (apparently microscopic) elements of the chimpanzees' 'theory of mind' system. Indeed, Tomasello *et al.* (2003a, b) allude to a group of unpublished studies from their laboratory which they argue further establish the presence of an ability to reason about certain mental states in chimpanzees.

Alas, if we are right, simply generating more studies within the current paradigm will not help. As we noted above (and see especially Povinelli and Vonk 2003), we do not need more experiments that have no ability to distinguish between the presence of S_b versus S_{b+ms}; we need experiments of a conceptually different nature. Turning up the power on our existing microscopes, to pursue their metaphor, will merely confirm what we already know: that our minds are very good at automatically construing certain behaviours in terms of mental states (Fig. 18.5). Scaling up current methods

Fig. 18.5 Current experimental techniques reveal more about the workings of the human mind, than the chimpanzee's mind.

can do nothing but scale up the indeterminacy of the results. (Incidentally, this will be true whether the testing frameworks are 'co-operative' or 'competitive' (e.g. Hare 2001) because that is not the axis along which the conceptual conflation of S_b and S_{b+ms} occurs; see also Item 6 in Appendix I where we address the criticisms of our studies about <*seeing*>.)

But then why does the 'microscope metaphor' appeal so much to the Leipzig group that they would use it three times? One possibility is that there may be a general proclivity to see 'the theory of mind system' as the 'holy grail' of comparative cognition, a view which inevitably creates a kind of narrow focus in which researchers come to believe that with just the right design, or just the right control condition, the long sought-after bejewelled cup will be found. Coupled with a conceptual framework that has difficulty acknowledging the evolution of novel cognitive innovations (see Povinelli and Vonk 2003), such an approach creates the illusion that we already know the answer, and that it's just a matter of coming up with the right test to prove it. Such dynamics appear throughout their recent opinion piece (Tomasello *et al.* 2003a). For example, experiments that would appear to confirm the presence of abilities to reason about certain mental states are hailed as 'breakthroughs' whereas our own, carefully conducted, programmatic set of over two dozen studies following a group of chimpanzees over their life-span, is dismissed as part of a general pattern of 'negative evidence' (p. 153). An objective scientific approach, however, would see progress toward providing evidence that uniquely supports either possibility as a breakthrough, and thus place the greatest emphasis on the resolving power of the methods used, not the results obtained.

In the end, there may be an even more fundamental danger in adopting the microscope metaphor of Tomasello *et al.* (2003a, b). The very idea of needing a microscope, let alone a more powerful one, to successfully characterize the nature of the chimpanzee's mind, resonates with the already widespread antievolutionary idea that the minds of other species are simply smaller, more watered-down versions of our own. We don't need a microscope to explore the chimpanzee's capacity to reason about mental states, we need experimental techniques that can distinguish between the operation of S_b and S_{b+ms}.

18.7 **'All or none'?**

There is an additional confusion that must be clarified if the current controversy is to be resolved. Tomasello *et al.* (2003b) assert that we believe that chimpanzees either have the entire human 'theory of mind' system or none of its components: 'Povinelli and Vonk argue that human beings have a theory of mind and chimpanzees do not. But this black and white picture is exceedingly misleading' (p. 239). Indeed, such a view *is* highly misleading; fortunately, it is not our own.

To begin, even a cursory glance at our laboratory's theoretical papers makes the falsity of this claim apparent. For example in our laboratory's early monograph,

What young chimpanzees know about seeing (Povinelli and Eddy 1996a), we devoted an entire chapter to issuing a plea that researchers consider the possibility of breaking down the 'theory of mind' system as it exists in adult humans into component parts, and then think about reconstructing the evolutionary timing of these separate parts:

> ... [U]sing the techniques outlined by Premack and Woodruff (1978), and applying them to the questions that have emerged from investigations of theory of mind in young children, it is now possible to determine whether theory of mind represents a psychological innovation unique to the human lineage or whether it is a more primitive innovation, perhaps one that evolved sometime after the divergence of the great ape-human lineage from other primates... *Additionally, it is quite possible that the psychological innovations responsible for theory of mind dispositions were not, in fact, a single innovation at all but rather evolved in a number of discrete steps... Thus... it is quite possible that transitions in theory of mind dispositions identified by developmental psychologists represent the retention of discrete ontogenetic innovations during the course of primate evolution.* (italics added, p. 14)

We concluded the chapter with the following suggestions:

> It is clear from the above considerations that reconstructing the evolution of theory of mind will proceed through three distinct phases. To begin, *researchers must use the methods of comparative psychology to identify which species possess which aspects of mental state attribution and at what point in development.* The second step will be for researchers to use the methods of phylogenetic reconstruction to infer what the likely features of theory of mind were in each common ancestor... Once this reconstruction has occurred, the exact timing and order of each of the features will be known.... (italics added, p. 15–16).

Indeed, over the past several years we have clearly noted that our strategic retreat from asking about chimpanzees' understanding of epistemic states (e.g. <*knowing*>) in the late 1980s and early 1990s, to perceptual states (e.g. <*seeing*>) in the mid to late 1990s, was motivated by our recognition that the human 'theory of mind' system might not be evolutionarily hegemonic (see Povinelli and Eddy 1996a; Povinelli and Prince 1998; Povinelli and Giambrone 2000). In doing so, we were merely following the lead of developmental researchers who were quite comfortable in thinking of various components of the human 'theory of mind' system emerging at different points in development.

Given that we most definitely do not believe that the 'theory of mind' system as found in adult, western cultures must be thought of as an indivisible psychological unit, why have Tomasello *et al.* (2003b) and others attributed this belief to us? The answer may be that it is easy to conflate our claim that the current evidence does not exclude the possibility that the capacity to conceive of mental states (possibly all hypothetical entities) is a unique feature of the human mind, with the very different claim that if any component of the human system is absent in chimpanzees, then the whole system must be absent.

But this cannot be the complete explanation, because Tomasello *et al.* (2003b) believe that it is unproductive to even entertain the hypothesis that the entire 'theory of mind' system is uniquely human: '... to repeat our earlier, more general point, we are certainly never going to make progress on questions concerning the evolution and ontogeny of social cognition if we think in terms of a monolithic 'theory of mind' that species either do or do not have' (p. 239). Such an assertion could only have force if one already knew (from independent evidence) that the ability to conceive of mental states was not unique to the human species.

In the hope of diffusing this issue, let us break down the reasoning behind the Leipzig group's proscription and examine its logic. First, the point regarding human ontogeny is a red herring: no one is disputing the claim that human adults have the ability to reason about mental states. Thus, given that the system exists in some form or another in humans from all cultures, one very real possibility is that the system 'develops' in component parts (a possibility that has been urged by various researchers for many years; e.g. Leslie 1987; Wellman 1990). But this has no bearing on whether other species possess limited aspects of the system. Members of other species are not, after all, immature adult humans.

Next, with respect to the assertion about the *evolution* of the 'theory of mind' system, why is it not possible (or even likely) that a system for reasoning about mental states is indeed a uniquely derived feature of the human lineage? Why will we make 'no progress' by seriously considering this possibility? By way of analogy, could we make no progress toward understanding the evolution of echolocation in bats unless we assumed that closely related species (e.g. primates) have at least some parts of this echolocation system? Of course not.

In this sense, then, our laboratory's strong experimental emphasis on chimpanzees' understanding of <*seeing*> was a 'test case'—a point we have repeatedly stressed (e.g. Povinelli and Prince 1998). In humans, because our reasoning about visual perception is a context in which two worlds commingle—the world of observable things (gaze direction, head movements, eye movements, orientation of the torso, position of the eyelids, direction of movement, etc.) and the world of the private, unobserved features we infer in others—it seemed like an excellent place to examine possible evolutionary associations—and *dissociations*—between S_b and S_{b+ms}.

Indeed, we believe the results of the studies from our laboratory and elsewhere concerning non-human primates' understanding of visual perception have produced general lessons for trying to assess the ability of an organism to understand mental states. For example no one currently disputes that chimpanzees reason about the observable aspects of others that are relevant to visual perception (their face, eye direction, etc.). Indeed, numerous studies have shown that chimpanzees understand that they should direct their gestures preferentially to others who are facing them (Fig. 18.2a; Povinelli and Eddy 1996a; see also Hostetter *et al.* 2001; Tomasello *et al.* 1998). Other studies have revealed that they will follow gaze of others, even in response to simple eye movements (Povinelli and Eddy 1996b), and will even prefer to approach someone who makes eye contact with them (Povinelli and Eddy 1996c). Clearly, then, their knowledge about observable features related to the folk psychological notion of <*seeing*> is impressive.

18.8 **Current tests can (and do) implicate the presence of S_b alone**

Finally we can now turn to the second part of our claim, that, at the same time, the very same chimpanzees who provided us with the evidence for the abilities just described responded as if they knew nothing about <seeing> in such simple situations as those outlined in Fig. 18.2b–d. Why should this be so? In other words, if S_b is powerful enough to extract the kinds of information that we have suggested, then why do chimpanzees not show immediate evidence of understanding the implications of all of the regularities that exist in our tests? Furthermore, if we are right that current tests are inadequate to demonstrate the presence of S_{b+ms}, then how can we claim that the very same tests could uniquely implicate the presence of S_b alone?

First, there is no logical problem with our argument. We are simply proposing that a pattern of results of type 'x' could be produced by either S_{b+ms} or S_b, but that a pattern of results of type 'y' would be expected for S_b, but not for S_{b+ms}. In the case of their understanding of <seeing>, for example, the generalizations that our chimpanzee made across the carefully planned experiments were highly sequential and specific (first, generalizations about the front and back, then the face, then the eyes, with the initial generalizations being more important than the ones that were learned later), not an overarching generalization encapsulating the concept of <seeing> (see Povinelli and Eddy 1996a; Reaux et al. 1999). We have interpreted this pattern of results as showing that if S_b alone is present, although much of the organism's behaviour will look strikingly similar to one that possesses S_{b+ms}, an experimental, microgenetic analysis will detect the tell-tale indicators of this fact (see especially Reaux et al. 1999, Exp. 4). Indeed, it is the complex interplay of what chimpanzees do and do not do in the same context (pattern 'y') that has lead us to suggest that one viable hypothesis is that they have a powerful S_b, but no S_{b+ms}.

One reasonable way of thinking about this is to suppose that the chimpanzee's S_b is wired up to spontaneously detect and exploit numerous relevant regularities that exist in the behavioural interactions of themselves and others, but that this system computes the fewest number of abstractions needed. Only if forced to make a distinction between two patterns that normally co-vary (e.g. eye orientation versus face orientation), will S_b bother to extract these distinct regularities, and then will do so only according to the contingencies to which it is exposed. We propose that the best interpretation of the combined experimental results from multiple laboratories is that S_{b+ms} and S_b are dissociable, and that chimpanzees may be living proof of how S_b will detect and store for future use only those relationships minimally necessary to uncover the predictive relationship between the current and past behaviour of others (broadly construed) and their future behaviour. To be clear, our results do not force this conclusion, they merely allow for it.

A distracting, but important, side-issue is the objection of numerous critics that our studies about <seeing> are flawed in one way or another. Because these criticisms are numerous, varied, and even contradictory, we present them along with our responses in Appendix I. Suffice it to say here, however, that these criticisms merely highlight the fact that the most fundamental point of all of these studies has been frequently misunderstood. Our central conclusion has *not* been that chimpanzees 'cannot do x', but to the contrary, that under the right set of contingent experiences, S_b is powerful enough to abstract out the spatiotemporal invariances relevant to the given situation.

Again, to be absolutely clear, we possess no privileged information as to whether chimpanzees have the ability to reason about mental states. Thus, we assert no factual claim. However, we do believe that no current evidence uniquely provides evidence in favour of the idea that they do, and considerable evidence suggests that they are not making such inferences in situations where humans (children and adults) would readily do so. This is not a truth-claim: it is our assessment of how the current evidence bears upon the hypotheses at stake.

18.9 **Will the real sceptic please stand up?**

We end with a general message to the current generation of students who are fascinated by the question of whether other species possess the capacity to reason about mental states: 'Do not lose your fascination with this problem, but, at the same time, do not be dissuaded from pursuing a more rational approach to investigating the question of whether other minds reason about mental states. Do not be blown into one camp or the other by jeers that you are a 'sceptic' or that you have produced 'negative' findings. Realize that the scholar who doubts that chimpanzees have a 'theory of mind' system, and the scholar who doubts that it is a uniquely human trait, are both sceptics, and that without scepticism there can be no such thing as science. Yes, *be* sceptical. Pursue multiple working hypotheses simultaneously, and be ruthless in your tests of the hypothesis that, in your heart, you know you truly favour (see Chamberlin 1897). In a single mind, embody *both* sceptical natures. Recognize that the hardest path is pursued by those who constantly challenge their beliefs, but also recognize that this is the most intellectually rewarding path. And most of all, do not be afraid of differences if that is where the evidence leads you. Differences among species are real. They're what evolution is all about.'

Appendix 18.1 Criticisms of Povinelli and colleagues' seeing-not seeing studies with rebuttals

1. The fact that the chimpanzees were responding to human experimenters (as opposed to conspecifics) invalidates the findings.

In order to be consistent, those who raise this objection to our experiments (the results of which consistently provided evidence inconsistent with the view that the chimpanzees were reasoning about <*seeing*>) would need to object just as strongly to the experiments from their own (or other) laboratories that also make use of human experimenters, even those which seem to provide evidence confirming the presence of 'theory of mind' skills. For instance, results from experiments that require chimpanzees to distinguish between the intentions (accidental versus intentional), 'line-of-sight', deceptive actions, and attentional status of human experimenters have all recently been marshalled as strong support for the idea that chimpanzees reason about psychological states. The fact that the objection to using human experimenters only appears when the results *disconfirm* the presence of 'theory of mind' abilities, reveals a powerful underlying confirmatory bias in which experiments that seem to produce one class of evidence are not held to the same scrutiny as experiments producing a different class. If chimpanzees are expected to reveal evidence for inferring the mental states of humans in some situations, why should we ignore data from different situations in which they appear not to do so?

Indeed, if we disregarded the data from all studies in which chimpanzees were asked to infer the mental states of human experimenters we would be left with only observational studies (inferences from which are severely problematic; see Povinelli and Vonk 2003), and a very small set of experiments, the data from which has not uniquely confirmed the presence of one psychological system over the other (e.g. Call *et al.* 2000; Hare *et al.* 2000, 2001; Karin-D'Arcy and Povinelli 2002).

At a more conceptual level, the existing data robustly supports the view that chimpanzees respond appropriately to virtually the entire range of social signals from humans that are the visible manifestations of <*seeing*>. Indeed, time and time again, in our own studies we have shown that our chimpanzees attend and respond to these social signals (for a recent example see Povinelli *et al.* 2003). In fact, some researchers have taken their propensity to follow human gaze, their attraction to humans making eye contact with them, their tendency to use different forms of communication with humans instantiating differing attentional states, etc., as indicative of an ability to read the mental states of humans, as well as their behaviours. Thus, the existing data shows that chimpanzees respond to human social signals of eye/head direction and movement in the same manner (at least at our current level of resolution) that they respond to those of their fellow chimpanzees.

Finally, why should a 'theory of mind' system be so narrowly functional in chimpanzees that they can infer only the mental states of members of their own species? Although, we are not closed-minded to this possibility, (indeed we were among the first to suggest it), if it were the case, this narrow ability would clearly be radically different from that which humans invoke (see discussion of this issue in Povinelli 1996). Humans, for instance, attribute mental states not only to other species, but even to inanimate objects. Given the evidence that chimpanzees are avid psychological consumers of at least the behaviour of others, even when the behaviour is being performed by humans, it seems highly suspect to propose that the system for inferring mental states, which depends upon the system for reading behaviours, would be activated only when the object of perception has the exact same physical features of a chimpanzee.

2. The pretraining procedures in the Povinelli studies, in effect, trained their chimpanzees not to attend to the faces and eyes of the experimenters.

The fact that this objection is so often made, and yet so obviously false, makes it hard to know how to respond to it. For the archival record, our chimpanzees robustly demonstrated an inherent predisposition to attend to the most subtle cues regarding the faces and eyes of the experimenters, such as a slight deflection of the pupils, and required no training from us to do so (see for example Povinelli and Eddy 1996a, b, c, 1997; Povinelli *et al.* 1997; Theall and Povinelli 1999; Povinelli *et al.* 1999; Povinelli *et al.* 2002; Povinelli *et al.* 2003). Importantly, our chimpanzees exhibited these sensitivities on the very same trials in which they made no discrimination in the choice of gesturing to an experimenter who was visually attending to them, and one who was not (see especially, Povinelli and Eddy 1996a, Exp. 12). Thus, the claim that something about our procedures trained our subjects not to attend to the relevant social cues exhibits a severe lack of familiarity with our studies and their results.

3. The Povinelli studies reveal only that chimpanzees do not understand the eyes in particular as the portals of visual attention.

Again, this claim is empirically false. First, the data suggest that our chimpanzees minimally did not appreciate the specific relevance of the entire face to visual attention. In other words, they initially did not discriminate between conditions in which the experimenter's face was or was not visible (e.g. looking-over-the-shoulder, buckets, screens; see Povinelli and Eddy 1996a). Furthermore, Povinelli and Eddy (1996a, pp. 137–138) went to great lengths to outline a possible system in which chimpanzees might have concepts about an 'amodal' psychological state of <*attention*> without understanding the unique relevance of the eyes, or the face, in determining such attentional experience. However, as they pointed out, it is difficult to imagine how the hypothesis that chimpanzees have some notion of <*seeing*> that is linked only to the general frontal features of another organism, could ever be rigorously separated from the idea that they understand the importance of directing visually-based gestures to the fronts of others (see Povinelli 2001, for an extended discussion of this problem).

Tomasello and Call (this volume, Chapter 17) cite a paper by Kaminski *et al.* (2004) whose results they claim 'supersede' the 'negative' findings of Povinelli and Eddy (1996a). However, while Kaminski *et al.* apparently show that some apes are sensitive to the orientation of an experimenter's face when the experimenter's body is facing forward, notably the apes did not respond differently when the experimenter's body was turned away from the subjects, even though the experimenter's facial orientation was equally relevant in both cases. Thus, it is not clear that their apes have a general understanding of the significance of facial orientation; they may simply be responding to the salience of the cues, body orientation first and presence of the face second, exactly as suggested by Povinelli and Eddy (1996a). Furthermore, Kaminski *et al.* measured the behavioural response in terms of frequency of communicative signals.

It is not clear that such measures are comparable to those which require an ape to make a discriminative choice between two experimenters—one of whom can see them and the other of whom can not.

4. The fact that the experimenters in the Povinelli studies did not look the chimpanzees directly in the eyes invalidates the conclusions.

The core misconception of this objection seems to be that the scrupulous choreography used in *one* set of our experiments in which we did not allow the experimenters to make direct eye contact with the chimpanzees, was an oversight or flaw when the experiments were designed. In fact, this was a deliberate and crucial aspect of the procedure (for an extensive discussion of this issue, see Povinelli and Eddy 1996a, pp. 34–36). In those studies, eye contact was deliberately neutralized so as to preclude a different interpretation of the chimpanzees' possible correct choices; namely, that they may simply be attracted to the salience of direct eye contact in the correct conditions without necessarily drawing inferences about the experimenters' underlying mental experience of <seeing>. Significantly, analogous experiments with 2.5-year-old children, which provided even less salient direct eye contact cues, resulted in the children performing correctly from trial one forward (Povinelli and Eddy 1996a, Exp. 15).

Furthermore, additional studies specifically explored the role of direct eye contact, among other cues, in the exact same setting (see Povinelli and Eddy 1996b). And, indeed, in those studies, our chimpanzees *did* preferentially choose to respond to the experimenter who made eye contact with them. Of course, they also preferred to gesture to someone whose eyes were closed, but who made subtle head bobbing movements resembling chimpanzee behaviour, as opposed to someone whose eyes were open! Regardless of how one wishes to interpret these findings, they certainly vindicate our conceptual concern that eye contact (among other signals) may simply be a 'hot' social cue that has a high valence, quite independent of any understanding of <seeing>.

5. The reason for the difficulty that Povinelli's chimpanzees encountered was that they may simply not have the mental capacity to track and compute the cues related to the psychological states of two experimenters simultaneously.

This criticism would seem to begin with the assumption that chimpanzees have a concept of <seeing>, deploy and use it in their everyday natural social lives, but that they cannot keep track of who can and cannot see them. On the face of it, this seems implausible. In the relevant studies, the chimpanzees in actuality had only to interpret the attentional status of a single experimenter in order to succeed in either task. For example, they could enter the test unit and determine that the first experimenter they looked at did not <see> them and then approach the other. Or they could ascertain that the first experimenter they looked towards did <see> them and thus proceed to approach him or her. At the very least, this critique seems to embrace the idea that chimpanzees have a 'weaker' or more 'watered-down' version of the human system for reasoning about mental states, an idea we find strongly suspect (see Povinelli 1996; Povinelli and Vonk 2003). Furthermore, we submit that data from analogous single-experimenter studies (see Theall and Povinelli 1999; Hostetter *et al.* 2001; Kaminski *et al.* 2004) simply confirm the findings of the two-experimenter studies.

6. The Povinelli studies were tests that occurred in a co-operative context, and co-operative settings are not appropriate for assaying the chimpanzees' ability to reason about mental states.

This criticism can be thought of as the 'ecological validity' complaint. Hare (2001) and colleagues have levelled this criticism, championing the use of competitive paradigms instead. They believe that competitive tests may be more ecologically valid and may thus provide a more conducive and natural context for the chimpanzees to engage in visual perspective taking, for example. Although this line of reasoning is possible, from a number of perspectives it seems questionable (see Povinelli 1996).

First and foremost, as we have already pointed out, the results from competitive tests can provide no better evidence for 'theory of mind' capacities than co-operative ones so long as the exact causal power of S_{b+ms} and of S_b continues to be unknown. Worse yet, because of the critical role of competition in the survival of an organism, competitive situations are precisely those for which evolution might have prepared an organism's cognitive system to respond in an intelligent, fast manner in the context of what may be highly costly situations. A system such as a 'theory of mind' system that is presumably 'designed' to allow for highly abstract interpretations of behaviour may not facilitate rapid responding, and may be least adaptive in such circumstances in which behavioural contingencies are relatively invariant. On the other hand, cognitive systems which exploit abstract representations of classes of behaviours, such as 'approach', 'facial expression type x', and 'aggression', may allow for less error, and may have been shaped in the evolutionary history of the species. The use of such representations may well appear to be evoked spontaneously (i.e. without evidence of 'learning'). For example socially isolated infant macaques that have never seen adults of their own species react appropriately to threat versus neutral facial expressions in photographs of adult monkeys (Sackett 1966). Thus, situations of 'high ecological relevance' may be among the worst contexts in which to seek responses generated by S_b that differ from those generated by S_{b+ms} (and hence, the worst circumstances to distinguish between the two systems).

Second, and perhaps more to the point, the argument that co-operative situations are unnatural for chimpanzees is unfounded. Every highly social species will have specifically evolved mechanisms that balance co-operative and competitive tendencies (see de Waal 1986). Hence, 'co-operation' is every bit as important to chimpanzee social ecology as is 'competition,' even though in the final analysis both have evolved in a manner that serve individual fitness. In fact, from the perspective of this criticism, it is somewhat ironic that observations of chimpanzees co-operatively hunting in the wild have routinely been cited as suggestive of their mental perspective taking abilities (Boesch and Boesch-Acherman 2000). More specifically, the chimpanzees' begging gesture that is the focal point of this criticism was not specifically trained—it emerges in the course of normal social development. Because chimpanzees use this gesture to obtain food—both from their conspecifics and their human counterparts (both in captivity and in the wild)—it can hardly be construed as an unnatural behaviour. As it pertains to our own group of captive chimpanzees, they have an extensive history of using this gesture to obtain foods and other desirables from humans. Thus, utilizing scenarios in which they beg for food from humans in an experimental setting hardly seems ecologically invalid.

7. Povinelli's apes suffer from an impoverished rearing history and living environment that cripple their ability to respond in species-specific ways to tests about their social cognitive skills.

Finally we hook a red herring, an argument that is often used to deflect attention away from the real issues at hand. First, our chimpanzees have: (a) undergone extensive behavioural enrichment in the form of highly diverse cognitive tests in which they have participated two to three times a day for the past decade and a half; (b) been housed in social groups, replete with toys and other enriching objects, for their entire lives; and (c) have had extensive contact with humans (an experience often speculated to facilitate the acquisition of the very abilities we probed in our tests; i.e. Tomasello and Call 1996).

Second, our chimpanzees have led the way in demonstrating the complexities of their species' behaviour including the ability to follow human gaze, engage in mirror-guided self exploratory behaviours, joint attention, tool use, use social cues to determine object choice, etc. (for a recent example see Povinelli *et al.* 2003). In no way has their performance on our tasks, or their spontaneous behaviour, indicated that they deviate from the patterns of behaviour observed in wild chimpanzees or in other captive settings. In addition, preliminary data from our lab indicate that home-reared chimpanzees are no more likely than our own lab-reared chimpanzees to gesture preferentially to experimenters who can see them versus those who can not.

Third, if the concern about our chimpanzees' environment was valid, it would apply to any laboratory raised chimps, including those that other researchers claim show evidence of an understanding of mental states. Curiously, however, this complaint seems reserved for those cases in which the overall pattern of results is 'unpopular'. Would the same criticism have been launched at our chimpanzees had we been satisfied to accept indeterminate data as evidence of the ability to reason about mental states? We doubt it.

Thus, the surface behaviour of our chimpanzees is identical to that of chimpanzees in other labs, and indeed in the wild. But, what our studies have achieved has been to probe beneath the surface of their behaviour to examine in detail what social cues the chimpanzees were and were not using in deciding which experimenter to gesture towards. In doing so, we have been led to conclude that there is no compelling evidence that they understand <*seeing*> in others.

References

Barid, J. A., and Baldwin, D. A. (2001). Making sense of human behavior: Action parsing and intentional inference. In *Intentions and intentionality: Foundations of social cognition*. Malle, Bertram F.; Moses, Louis J., Cambridge, MA, US: The MIT Press. pp. 193–206.

Boesch, C., and Boesch-Acherman, H. (2000). *The Chimpanzees of the Tai Forest: Behavioral Ecology and Evolution*. Oxford: Oxford University Press.

Call, J., Agnetta, B., and Tomasello, M. (2000). Social cues that chimpanzees do and do not use to find hidden objects. *Animal Cognition*, 3: 23–34.

Chamberlin, T. C. (1897). The method of multiple working hypotheses. *Journal of Geology*, 5: 837–848.

de Waal, F. B.(1986). The Integration of dominance and social bonding in primates. *Quarterly Review of Biology*, 61: 459–479.

Dennett, D. C. (1987). *The intentional stance*. Cambridge, MA, MIT Press.

Flavell, J. H., Everett, B.A., Croft, K., and Flavell, E. R. (1981). Young children's knowledge about visual perception: Further evidence for the level 1–level 2 distinction. *Development Psychology*, 16: 10–12.

Gallup, G. G., Jr (1988). Toward a taxonomy of mind in primates. *Behavioral and Brain Sciences*, 11: 255–256.

Hare, B. (2001). Can competitive paradigms increase the validity of experiments on primate social cognition? *Animal Cognition*, 4: 269–280.

Hare, B., Call, J., Agnetta, B., and Tomasello, M. (2000). Chimpanzees know what conspecifics do and do not see. *Animal Behaviour*, 59: 771–785.

Hare, B., Call, J., and Tomasello, M. (2001). Do chimpanzees know what conspecifics know? *Animal Behaviour*, 61: 139–151.

Heyes, C. M. (1998). Theory of mind in nonhuman primates. *Behavioral and Brain Sciences*, 21: 101–148.

Hostetter, A. B., Cantero, M., and Hopkins, W. D. (2001). Differential use of vocal and gestural communication by chimpanzees (*Pan troglodytes*) in response to the attentional status of a human (*Homo sapiens*). *Journal of Comparative Psychology*, 115: 337–343.

Kamamar, D., and Olson, D. R. (1998). Theory of mind in young human primates: Does Heyes' task measure it? *Behavioral and Brain Sciences*, 21: 122–123.

Kamamar, D., and Olson, D. R. (1998). Theory of mind in young human primates: Does Heyes' task measure it? *Behavioral and Brain Sciences*, 21: 122–123.

Kaminski, J., Call, J., and Tomasello, M. (2004). Body orientation and face orientation: Two factors controlling apes' begging behavior from humans. *Animal Cognition*, 7: 216–223.

Leslie, A. M. (1987). Pretense and representation: the origins of 'theory of mind'. *Psychological Review*, 94: 412–426.

Povinelli, D. J. (1996). Chimpanzee theory of mind? The long road to strong inference. In: P. Carruthers and P. Smith, eds. *Theories of Theories of Mind*, pp. 293–329. Cambridge: Cambridge University Press.

Povinelli, D. J. (2000). *Folk Physics for Apes: The Chimpanzee's Theory of How the World Works*. Oxford: Oxford University Press.

Povinelli, D. J. (2001). On the possibilities of detecting intentions prior to understanding them. In: B. Malle, D. Baldwin, and L. Moses, eds. *Intentionality: A key to Human Understanding*, pp. 225–248. Cambridge, MA: MIT Press.

Povinelli, D. J., and Bering, J. M. (2002). The mentality of apes revisited. *Current Directions in Psychological Science,* 11: 115–119.

Povinelli, D. J., Bierschwale, D. T., and Cech, C. G. (1999). Comprehension of seeing as a referential act in young children, but not juvenile chimpanzees. *British Journal of Developmental Psychology*, 17: 37–60.

Povinelli, D. J., Dunphy-Lelii, S., Reaux, J. E., and Mazza, M. P. (2002). Psychological diversity in chimpanzees and Humans: New longitudinal assessments of chimpanzees' understanding of attention. *Brain, Behavior and Evolution*, 59: 33–53.

Povinelli, D. J., and Eddy, T. J. (1996a). What young chimpanzees know about seeing. *Monographs of the Society for Research in Child Development*, 61, No. 2, Serial No. 247.

Povinelli, D. J., and Eddy, T .J. (1996b). Chimpanzees: Joint visual attention. *Psychological Science*, 7: 129–135.

Povinelli, D. J., and Eddy, T. J. (1996c). Factors influencing young chimpanzees (*Pan troglodytes*) recognition of attention. *Journal of Comparative Psychology*, 110: 336–345.

Povinelli, D. J., and Eddy, T. J. (1997). Specificity of gaze-following in young chimpanzees. *British Journal of Developmental Psychology*, 15: 213–222.

Povinelli, D. J., and Giambrone, S. (2000). Escaping the argument by analogy. In: D. Povinelli, ed. *Folk Physics for Apes*, pp. 9–72. Oxford: Oxford University Press.

Povinelli, D. J., Nelson, K. E., and Boysen, S. T. (1990). Inferences about guessing and knowing by chimpanzees (Pan troglodytes). *Journal of Comparative Psychology*, 104: 203–210.

Povinelli, D. J., and Prince, C. G. (1998). When self met other. In: M. Ferrari and R. J. Sternberg, eds. *Self-awareness: Its Nature and Development*, pp. 37–107. New York: Guilford.

Povinelli, D .J., Reaux, J. E., Bierschwale, D. T., *et al.* (1997). Exploitation of pointing as a referential gesture in young children, but not adolescent chimpanzees. *Cognitive Development*, 12: 423–461.

Povinelli, D. J., Theall, L. A., Reaux, J. E., *et al.* (2003). Chimpanzees spontaneously modify the direction of their gestural signals to match the attentional orientation of others. *Animal Behaviour*, 66: 71–79.

Povinelli, D. J., and Vonk, J. (2003). Chimpanzee minds: Suspiciously human? *Trends in Cognitive Science*, 7: 157–160.

Premack, D., and Woodruff, G. (1978). Does the chimpanzee have a theory of mind? *Behavioral and Brain Sciences*, 4: 515–26.

Reaux, J. E., Theall, L. A., and Povinelli, D. J. (1999). A longitudinal investigation of chimpanzees' understanding of visual perception. *Child Development*, 70: 275–290.

Sackett, G. P. (1966). Monkeys raised in isolation with pictures as visual input: Evidence for an innate releasing mechanism. *Science*, 154: 1468–1473.

Theall, L. A., and Povinelli, D. J. (1999). Do chimpanzees tailor their attention-getting behaviors to fit the attentional states of others? *Animal Cognition*, 2: 207–214.

Tomasello, M., and Call, J. (1996). The effects of humans on the cognitive development of apes. In. A. E. Russon, K. A. Bard and S. T. Parker (Eds.). *Reaching into Thought: the Minds of the Great Apes*. p. 371–403. Cambridge University Press.

Tomasello, M., and Call, J. (1997). *Primate cognition*. New York: Oxford University Press.

Tomasello, M., and Call, J. (2004). The role of humans in the cognitive development of apes revisited. *Animal Cognition*, 7: 213–215.

Tomasello, M., Call, J., and Hare, B. (1998). Five primate species follow the visual gaze of conspecifics. *Animal Behaviour*, 58: 769–777.

Tomasello, M., Call, J., and Hare, B. (2003a). Chimpanzees understand psychological states—the question is which ones and to what extent. *Trends in Cognitive Science*, 7: 153–156.

Tomasello, M., Call, J., and Hare, B. (2003b). Chimpanzees versus humans: it's not that simple. *Trends in Cognitive Science*, 7: 239–240.

Wellman, H. M. (1990). *The Child's Theory of Mind*, Cambridge, MA: Bradford.

Chapter 19

Belief attribution tasks with dolphins: What social minds can reveal about animal rationality

Alain J.-P. C. Tschudin

Abstract

This paper explores the question, 'Are animals rational?' Some may well be, demonstrating what different authors consider to be rationality of various forms. Other authors, however, continue to deny animal rationality, arguing that they lack language or merely demonstrate what appears to be 'clever behaviour'. Approaching the problem from a Darwinian reading of social evolution, my chapter focuses on evidence for rationality in dolphins and addresses associated methodological issues. Findings from neuroanatomical and behavioural research appear to support the Social Intellect Hypothesis in gregarious non-human animals, including dolphins and primates. Recent research in comparative social cognition has investigated the capacity for 'theory of mind' in apes and dolphins by means of non-verbal tasks, including tests for the capacity to attribute false beliefs to others. In this context, I shall focus on the results of non-verbal false belief tasks with captive bottlenose dolphins and compare these to other species. These findings are consistent with ascribing the evolution of a 'social mind' to dolphins, although further work is necessary to rule out alternative interpretations. If dolphins can continue to succeed on false-belief tasks, with increased controls against learning and cueing, this would raise intriguing further questions and have striking implications for the topic of rationality in animals.

Magpies and parrots can utter words as we do, and yet they cannot speak as we do [i.e.] by showing that what they are saying is an expression of thought; whereas men, born deaf and dumb...habitually invent...certain signs, by means of which they make themselves understood by those who...have the time to learn their language. And this shows not only that animals have less reason than men, but that they have none at all. Rene Descartes, *Discourse on Method*, 1637 (Descartes, 1968, pp. 74–75)

19.1 **Background: Animal rationality, evolution, and social intelligence**

Descartes' statement provoked the original title of this paper 'Dumb animals, deaf humans?' and his contention slots into a broader historical debate concerning animal rationality. Since classical times, the issue has been divided largely into two camps: those in favour of rationality in animals (e.g. Plato, who, following Socrates, credited animals with access to belief) and those against (e.g. Aristotle, who largely denied animals any form of reason or belief, although he did speculate on a gradation of intellect). The view opposing animal rationality predominated into medieval times, largely under the influence of St Thomas Aquinas, who followed Aristotle in arguing that animals could not transcend sensation, since they were incapable of understanding or reason. Kant and significant others adopted related lines of thinking, which remain to the present day. The most extreme representative of this school remains Descartes, who, with his conception of animals as machines, could be classified as a 'protobehaviourist' about animals. An alternative school of thought endured, however, which gave some credence to both sentience and rationality in animals. Bentham recognized the importance of animal sentience and Hume conceded that animals and humans shared a common reasoning underlying 'vulgar actions', although he held that animals were incapable of belief and inferential reasoning. Darwin was influenced by such thinkers and, although he also regarded animals as sentient, his views on animal rationality were more radical than his predecessors. In *The Descent of Man*, he posited that no fundamental difference existed between the 'mental faculties' of humans and higher mammals. He laid down the challenge to determine when animals became capable of abstract reasoning, and may be regarded as a 'protorationalist'.

Since Darwin, who acted as a major catalyst for further investigations into animal rationality, progress on the question of animal rationality has been difficult, because, as MacIntyre (1988) recognizes, 'there are rationalities rather than rationality.' This pluralism of meanings, while problematic, needs to be engaged and has recently been discussed elsewhere (see Kacelnik, Dretske, Millikan, Bermúdez, Hurley, this volume, for discussion). My aim in this paper, however, is not to explore the distinction between specific forms of rationality, but rather to engage the general question, 'Are animals rational?'

In order to explore this possibility, I focus on the concept of *belief* which links the ground of experience and the emergence of rational structures (Audi 2001) and which therefore occupies a central place in the study of rationality, of psychological complexity, and of consciousness.[1] In particular, I focus on whether animals have the capacity to attribute beliefs to others—a topic to which much attention has been directed in the study of social cognition. My paper relies on experimental evidence from comparative

---

[1] See Baars, 1988; Dennett, 1978; Neisser, 1988; Rosenthal, 1993; Tschudin 2001a; Wimmer and Perner, 1983.

behavioural tasks and develops a preliminary argument that some animals, such as bottlenose dolphins (*Tursiops*), have evolved a 'social mind' and may be capable of mental state attribution, and hence that rationality can be accredited to them. Competing explanations are addressed in a discussion of the comparative results.

In the field of social cognition, the capacity to have false and true beliefs about reality and to attribute these to others forms an important aspect of what Premack and Woodruff (1978) first defined as 'theory of mind'; namely, the imputation or attribution of mental states, including belief, knowledge, pretence, and intention, to one's self and others. Such 'mind reading' is dependent upon second-order intentionality, or the capacity to mentally represent another individual's mental representation (Leslie 1987).[2]

Tests for 'theory of mind' have tended to focus on establishing the capacity of an agent to attribute to another individual beliefs that the agent knows are false. Since true beliefs correctly represent reality, it is difficult to distinguish between behaviour that expresses the agent's own belief about reality and behaviour that expresses the agent's attribution of a corresponding true belief to another, or genuine 'mind reading'.

By contrast, since false beliefs do not correctly represent reality, behaviour that expresses the agent's attribution of what it regards as a false belief to another can more readily be distinguished from behaviour that merely expresses the agent's own belief. For this reason, some authors argue that it is only the attribution of a false belief state to another that indicates the capacity for 'mind reading', or reflection on the mental state of another (Dennett 1978; Suddendorf 1999). Hence, the importance of the false belief task as an indicator of 'theory of mind' (see Call and Tomasello 1999).

The significance of 'theory of mind' can be understood as an evolutionary adaptation, in the context of the Social Intellect Hypothesis (Humphrey 1976; see also Jolly 1966), which proposes that it is social complexity—rather than merely ecological complexity—that primarily drives the evolution of primate intelligence.[3] Life in a social group confers an adaptive benefit to individual primates, such as a stable, secure environment for mothers and their offspring and the possibility for the young to learn from older, experienced animals. However, there is a trade-off between such benefits and related

----

[2] Editors' note: This volume generally follows the widespread current practice of using 'mind reading' rather than 'theory of mind' to refer to the generic capacity to attribute mental states to others (see S. Nichols and S. Stich (2003), *Mindreading*, Oxford, Clarendon Press, p. 2), in order to avoid confusion between this generic capacity and a specific account, the theory theory, about how this is accomplished, namely, by inference from a theory about mental states, as opposed to by simulative use of one's own psychological equipment, as postulated by simulation theory. However, before the distinction between theory theory and simulation theory became current, 'theory of mind' was used in the generic sense. Tschudin continues to use 'theory of mind' in this generic sense that does not imply the correctness of the theory theory, which is how it should be understood in this chapter.

[3] For alternative views, see Clutton-Brock and Harvey, 1980; Gibson, 1986; Milton, 1988; Wrangham, 1980.

costs of sociality, such as competition for access to mates and food. The challenges of managing the complex trade-offs of social life and resulting selection pressure for social manipulation are thought to explain the evolution of high intelligence.[4]

The Social Intellect Hypothesis draws support from neuroanatomical studies showing that relative brain size in primates is correlated with social complexity (Sawaguchi and Kudo 1990). Moreover, Dunbar (1992) demonstrated that there is a significant relationship between neocortex size and social group size in anthropoid primates but not prosimians, which prompted him to suggest that the necessary selection pressure only arose recently in primate social evolution; and Byrne (1993) correlated neocortex size with tactical deception. Given these findings, it might be suggested that 'theory of mind' gives an evolutionary advantage to an individual, in the sense that it can better outwit its conspecifics during competition (see and compare Byrne and Whiten 1988; Tomasello and Call 1997). On the other hand, it increases the potential for an individual to take the perspective of others, and may therefore enable increased co-operation, empathy, and peacemaking.

19.2 The big issue: who has beliefs about beliefs? Background and pilot study

The capacity to attribute false beliefs, as part of full-blown 'theory of mind', has a seemingly critical role to play in understanding primate and human evolution. A central question is whether this capacity for advanced social cognition arose after the split in the lineage between apes and humans some 6–8 million years ago (Tomasello and Call 1997; Call and Tomasello 1999; but see Tomasello and Call, Chapter 17, this volume). If not, we might expect apes and other species to demonstrate a capacity to attribute false beliefs or exhibit other aspects of 'theory of mind'.

By age 5, human children not only display the cognitive precursors of 'theory of mind', but also actively demonstrate this capacity by succeeding on the demanding false-belief task, a 'benchmark' for 'theory of mind' (Baron-Cohen *et al.* 1985; Call and Tomasello 1999; Wimmer and Perner 1983; but see Sections 19.2.2 and 19.8 below). The task involves changing the location of an object in the absence of a naïve participant, who then has a resultant—false—belief that the object remains in its original location (Call and Tomasello 1999). To succeed on the false belief task, the subject must recognize that this naïve participant may well hold a belief that the subject knows, via inferential reasoning, to be false.

Premack and Woodruff (1978) extended the debate to non-human primates when they posed the question, 'Does the chimpanzee have a theory of mind?' While subsequent findings have tended to support a negative answer to this question (see Heyes 1993; Povinelli 1994), the debate concerning the 'mind-reading' capacities

---

4 See also Byrne and Whiten (1988) on what they call the *Machiavellian Intelligence Hypothesis*.

of animals has continued. Although evidence for the capacity to attribute belief states in non-human primates remains equivocal (see Call and Tomasello 1999; Premack 1988; but see also O'Connell 1995), some recent evidence suggests that chimpanzees are capable of discriminating between what others know and do not know, and see and do not see (see Hare *et al.* 2001; and see Tomasello and Call, Chapter 17, this volume).[5] Nonetheless, the contrasting performances between human and non-human primates on social cognition tasks has sustained the belief that 'theory of mind' remains elusive in species other than humans.

19.2.1 Darwin's dolphins, intense sociality, and convergent cognitive evolution

Primatologists have been awkwardly placed. Respect for Darwinian continuity seemingly compels them to expect to find precursors of 'mind reading' in social non-human primates, but their observations and empirical findings on false belief tasks have tended to indicate otherwise. But should the investigation of 'theory of mind' and rationality in animals be discouraged merely because non-human primates fail related tasks? Certainly not. From a Darwinian perspective, it would be wrong to assume that, because certain primate relatives of human beings do not demonstrate belief attribution, the phenomenon does not exist in other species (Tschudin 2001a). Particularly worth investigating would be other species—perhaps radically different from the human species—that have also been subjected to complex selection pressures and trade-offs in respect of foraging and sociality, which might be expected to result in potentially similar cognitive capacities.

Dolphins make good candidates for comparison. They have large brains (Worthy and Hickie 1986), a diverse and complex social organisation and foraging ecology (see Tschudin 1999; Connor and Mann, Chapter 16, this volume), and a demonstrated capacity for advanced forms of cognition (Herman, Chapter 20, this volume; Herman *et al.* 1994; Marten and Psarakos 1995; Pack and Herman 1995; Reiss and Marino 2001; Xitco and Roitblat 1996).

An initial examination of the relative neocortex size of several dolphin species, using the neuroradiological techniques of computed tomography (CT) and magnetic resonance imaging (MRI), led to the conclusion that dolphins possess extremely large neocortex ratios (volume of neocortex to rest of the brain) (Tschudin 1995). Tschudin *et al.* (1996) found that the neocortex ratio was correlated with group size in dolphins. In a more in-depth analysis, with a larger sample size and increased number of species,

---

5 Authors such as Povinelli and Vonk (this volume) argue that the non-verbal tasks used in this recent work cannot discriminate between the capacity for genuine 'mind reading'—such as knowledge of what others perceive—and a capacity for clever behaviour reading—such as knowledge of what others are looking at—and call for different experimental methods. In contrast, see Tomasello and Call, this volume.

the dolphin neocortex was studied in relation to behavioural ecology variables representing both foraging ecology and sociality. This study confirmed that the neocortex ratio predicted mean social group size and also maximum aggregate size ($P < 0.05$) in dolphins (Tschudin 1999). This is complemented by the finding that the encephalization quotient (EQ), a general measure of relative brain size, is correlated with pod (school) size in dolphins (Marino 1996). Based on such findings I argued that species other than humans, such as dolphins, should be viewed as having the potential for social intelligence, possibly via convergent evolution (Tschudin 1999).

Evidence from the field supports the suggestion that dolphins have a 'social mind'. Dolphins, like humans (see Harcourt 1992), are capable of forming second-order alliances, or alliances of alliances (Connor *et al.* 1999; and see Connor and Mann, Chapter 16, this volume). Notably, a recent study documents culture in whales and dolphins (Rendell and Whitehead 2001).

19.2.2 Evidence for a 'social mind' in dolphins? The false belief pilot study and its precursors

Given the support for the hypothesis of social intelligence in dolphins from neuroanatomical and behavioural ecology research, the focus moved on to behavioural tasks investigating social cognition in dolphins. It has recently been discovered that dolphins display precursors to 'theory of mind', such as the capacity for joint attention in the form of the comprehension of referential pointing (Herman *et al.* 1999; Tschudin *et al.* 2001), gaze following, and replica usage (Tschudin *et al.* 2001). This is particularly significant, given that on a comparable task, chimpanzees (unlike children) neither appeared to grasp novel communicative signs, nor improved after training (Tomasello *et al.* 1997). In this context, Tschudin *et al.* (1999) conducted a non-verbal false belief task pilot study, modified from the design of Call and Tomasello (1999), with four bottlenose dolphins at Sea World, Durban, South Africa.

In the design of the pilot study (see Appendix 19.1 for detailed information about our procedures), a single dolphin was stationed in the water between two empty, identical, opaque boxes that were located on runners at the poolside. The experimenter (serving as both the baiter and the presenter) placed a fish in one of the boxes in the presence of another human being, the communicator. On all trials, the communicator could observe the baiting of the boxes by the experimenter and observe the dolphin. In contrast, the dolphin could only observe the communicator, while its view of the boxes during baiting was always obscured by a screen. The screen was always removed after baiting. The aim was for the dolphin to choose the box containing the fish reward, based on a tap signal from the communicator in the context of the experimental environment. During all phases of the experiment (training, pretest controls, test, and post-test controls), the communicator indicated the supposed location of the reward by tapping on the box that she 'believed' to contain the food. The dolphin indicated its choice by 'pointing' its head/rostrum towards, or by moving towards, one of the boxes.

Briefly, on training trials, no switching of the boxes occurred after baiting. On the pretest trials, however, either the boxes were switched in the presence of the dolphin, after the tap signal had been provided by the communicator (displacement control), or the fish was visibly moved from one box to the other in the presence of the dolphin and before the tap signal by the communicator (ignore communicator control). In the false belief test trials, the experimenter switched the boxes after baiting, in the presence of the dolphin and while the communicator was absent; the latter then returned and tapped on the box that she (now falsely) 'believed' to contain the reward. On the post-test control, the communicator was stationed between the boxes after baiting and did not provide any tap signal to indicate the location of the fish reward.

Remarkably, all of the dolphins who participated in the pilot study experiment passed the false belief task, as well as all the pretest and post-test controls (see Appendix 19.2). This is especially noteworthy, given the performance of children and great apes on this task (Call and Tomasello 1999).

The finding that the dolphins had passed this task in the pilot study provided the first preliminary evidence that several members of a non-human species may possess the advanced social cognitive capacity to attribute false beliefs. However, several potential limitations in the pilot study were also identified, which are addressed in the current study. It may be argued that the previous experimental design was not fully adequate for at least two reasons.

First, based on the training and pretest control phases, the animals may have used a learning strategy to solve the experimental test condition. That is, they may have learned the rule: whenever there is a switch of boxes, choose the box opposite to that indicated by the communicator. Although all animals passed the pretest controls, demonstrating that they could comprehend spatial displacement and also ignore the communicator when they had seen that she had been misinformed because the fish had been moved (see Appendix 19.1), these controls may have generated a potential confound in the form of conditional discrimination learning.

Some recent findings may be informative in this regard. On a modified non-verbal false belief task run with two groups of 4-year-old children (Tschudin *et al.* 2000), the first group (n = 14) proceeded directly from training to test, while the second (n = 10) received, as an intermediate stage, the pretest controls used in the pilot study with dolphins described above. All of the children in the first group failed the false-belief trials and passed the true belief (including 'dud') trials, while 5/10 of the second group passed the false-belief trials ($\chi^2 = 7.786$, $P < 0.01$). This difference suggests that the possibility of learning a conditional discrimination should be taken seriously.

Second, since the same experimenter acted as baiter and as presenter, it is possible that experimenter bias may have skewed the performance of the animals. That is, the baiter-presenter always knew the location of the fish reward and hence may have unwittingly given cues to the dolphin about the correct location.

19.3 **Subsequent experiments: methods**

The series of experiments described below seeks to address these shortcomings. First, they introduce a series of true-belief trials, interspersed with false-belief trials, to prevent learning. Second, the roles of the different experimenters have been modified substantially, to deal more adequately with problems relating to experimenter bias and cueing.

19.3.1 **Subjects**

Experiment 1 was conducted during January 2000 and the subjects comprised two female (Affrika and Khanya, 6 and 5 years old respectively) and two male (Jula and Kani, 10 and 9 years old respectively) captive bottlenose dolphins (*Tursiops sp.*) at Sea World, Durban. These were the same subjects as were used in the pilot study (Tschudin *et al.* 1999). Experiment 2 was conducted during May–June 2000 with one male (Domino, 9 years old) captive bottlenose dolphin at Bayworld, Port Elizabeth, South Africa. All dolphins were trained to respond to all trainers and performed in public displays, where they executed tasks based on various acoustic and hand signals. All of the Sea World dolphins were born at the facility and lived in sex-segregated groups with several other dolphins, while Domino, captive-born in the facility at Bayworld, lived with his mother, Dolly. Whereas all trainers at Sea World, Durban, provided equal attention to all the animals, so as to prevent the formation of intense bonds with particular individuals, Domino appeared to be heavily reliant on his chief trainer.

19.3.2 **Materials**

Experiments 1 and 2, as on the previous pilot study, used a fabric screen to hide two boxes from the dolphin's view. The boxes were identical in size, shape, and colour. Both boxes had air spaces between the outer and inner walls and were packed with crushed ice to minimize sound transmission. For Experiment 1 (as for the pilot study), the boxes were mounted on two runners—representing left and right—clipped onto the poolside. For Experiment 2, the boxes were secured behind a specifically manufactured mount, with two polystyrene extensions to be used as touch-pads for 'left' and 'right' by the subjects, in order to reduce any possible ambiguity of response. Fish were used as rewards throughout all phases of the studies. To prevent visual cueing, all experimenters wore dark glasses throughout all phases of Experiments 1 and 2.

19.3.3 **Procedure**

In order to better understand the detailed procedures for the current experiments, readers are strongly encouraged to first consult the details of the pilot study procedures found in Appendix 19.1.

19.4 **Experiment 1: Sea World, Durban**

Since this series was conducted 10 months after the pilot study of March–April 1999, testing was reinitiated with a training phase similar to that described in training phase (i) of the pilot study (see Appendix 19.1). However, since we had decided that it had been problematic to have the same experimenter as both baiter and presenter during the pilot study, we modified the experimental design by giving these roles to different individuals. On all trials, the naïve presenter was located in an area totally isolated from the experimental area and was therefore free of any experimenter bias. After success on training, the dolphins skipped the pretest controls (because of the potential learning problems associated with these) and proceeded directly to the test phase.

19.4.1 **Training phase**

The baiter placed a fish in one of two empty, identical boxes, in full view of the communicator, who always witnessed where the fish was located. While the screen always prevented the dolphin from seeing which box was being baited, the dolphin could still observe the communicator, who demonstrated by her stance and head position that she was visually attending to the baiting. After baiting, the baiter removed the screen and the communicator tapped the box containing the fish before leaving the experimental area. The naïve presenter then entered the experimental area and presented the boxes to the dolphin, which indicated its choice by orienting towards or approaching one of the boxes.

19.4.2 **False belief/ true belief test phase**

In the light of the concerns expressed about the possibility of conditional discrimination learning in the design of the pilot study, we countered this problem by interspersing false-belief trials with true-belief trials.

In the false-belief trials, baiting occurred as per training (above). After the removal of the screen, the communicator left the area. In her absence, and in full view of the dolphin, the experimenter switched the boxes. Subsequently, the communicator returned and tapped on the box at the location that she had originally seen baited, after which the presenter entered, presented the boxes and the dolphin chose, as per training (above).

In the interspersed true-belief trials, after baiting and the removal of the screen (as above), the communicator left the area. Upon her return, the boxes were switched in her presence. As she had witnessed the switch (and consequently had a true belief), she tapped the originally baited box, now in its new location. Following this, the presentation of the boxes and choice occurred (as above).

19.5 **Experiment 2: Bayworld, Port Elizabeth**

Subsequent to Experiment 1, further potential confounds were identified with respect to cueing. First, although the issue of possible presenter–experimenter bias in the presenter had been adequately addressed, the communicator still remained indirectly knowledgeable of the fact that there were only two trial types, that is when the boxes had been switched in her presence (true-belief trials) or not (by her inference, false-belief trials). Our concern was that such knowledge could possibly unwittingly bias her tap signal to the subject, thus potentially cueing the subject and confounding the results.

Second, concern about potential baiter cueing was raised following videotape analysis of the trials filmed from head-height, across the pool, with a downward view over the experimental area. Detailed post-test video analysis of recordings filmed from the perspective of the dolphin, in a specially housed camera in the water, however, indicated that it was highly improbable for the dolphin to sight potential baiter cues from the water.

Finally, another source of potential confound may have been that the animals tested on Experiment 1 were the same animals tested 10 months previously in the pilot study, who may have remembered a conditional discrimination or who may have been in some way influenced by their acquaintance with the previous task.

The design of Experiment 2 sought to address these issues and was conducted with a naïve animal at a different facility. This test series comprised the random interspersal of false-belief trials and two types of true-belief trials, including a 'dud' belief trial. The purpose of this modification (adapted from Tschudin *et al.* 2000) was to prevent the communicator from knowing whether the boxes had been switched or not. Finally, an experimental manager was located opposite the baiter, to ensure that the fish reward was placed exactly behind the centre of the screen, erected to occlude the dolphin's vision of the two boxes. In this way, the final experiment avoided the potential experimenter bias or cueing in all three role-players, as well as avoiding the concerns about learning.

19.5.1 Training phase

This resembled Experiment 1, except that we attempted to train the animal to tap the target corresponding to the baited box, as opposed to accepting its natural body orientation and head or rostrum 'pointing' as indicating its choice of box.

19.5.2 False belief/true belief/dud belief test phase

In the light of our concerns regarding the potential for inadvertent cueing on the part of the communicator, we designed a series of randomly alternated false belief and true-belief trials, with a novel trial type, referred to as a 'dud' belief trial (Tschudin *et al.* 2000). The false-belief and true-belief trials were as described above for Experiment 1. In the dud-belief trials, after baiting and the removal of the screen, the communicator would leave the area. No switch would occur, either in her absence or upon her return. She would tap the box that she had originally seen baited, following which the naïve presenter would present the boxes to the dolphin. While a dud trial is actually a form of true-belief trial, including dud trials prevents the communicator from knowing whether or not the trial is a false belief or true-belief trial and hence avoids communicator bias or cueing.

19.6 Results of Experiments 1 and 2

On Experiment 1, Affrika, Kani, and Khanya proceeded to the test phase, the results of which are summarized in Table 19.1. Jula, however, failed to pass the training phase (he continually persevered to the left-hand side). Affrika, Kani, and Khanya succeeded on first trial for the false belief (FB) trials; whilst only Affrika and Kani succeeded on first trial for true belief (TB) trials (Table 19.1).

There is a significant relationship between the value of the belief examined (true versus false) and the dolphins' response to the signal provided by the communicator (in accord with versus against) ($\chi^2 = 6.74$, $P < 0.01$). In the light of this interesting finding, which indicates some interaction between the belief state and the signal communicated, the results were pooled across animals for individual true belief and false-belief trial performances. On the one-tailed binomial test, this yielded significance for true belief performance ($P = 0.046$) and a trend towards significance for false belief performance ($P = 0.073$). On the whole, however, no individual performances on different test conditions reached significance, although the performance of Kani on true-belief trials represented a trend in this direction (binomial test: $P = 0.062$, 4/4 trials correct, one-tailed significance). This may be an effect of the small number of trials run with each individual subject and can be addressed in the future by using an increased sample size.

The above performances stand in stark contrast to the findings for Domino during Experiment 2. We undertook a total of 144 training trials with Domino, of which he succeeded on a total of 65 trials. Although achieving 8/8 trials correct on pretraining, when we carried out additional 'refresher' trials the next morning (before proceeding to the test phase), Domino failed continuously. During continued training sessions, he attempted to use a variety of rules to solve the task, for example persevering with

Table 19.1 Experiment 1: performance of dolphins on the false-belief true-belief task

Subject	Task sequence and outcome
Affrika	TB, FB, FB, TB, TB, FB, [FB], TB, [TB], FB, FB, [TB]
Kani[a]	FB, TB, FB, TB, TB, [FB], TB
Khanya	FB, [TB], TB, [FB], FB, TB[b]

TB, true belief task; FB, false belief task; [ ] indicates incorrect choice by dolphin.

[a] Two trials scratched due to communicator error and invasion of experimental area respectively.

[b] Subsequent trials excluded due to misinformation concerning the location of the animal's reward at the end of the trial (i.e. trial number 6).

the left-hand box or the right-hand box. To avoid causing any distress to the animal and following his lack of success on numerous trials, we terminated our investigation.

19.7 **Possible interpretations**

What can be gleaned from the performance of dolphins on these belief tasks? The combination of results from Affrika, Kani, and Khanya shows that dolphins are capable of passing true-belief trials and indicates a definite trend towards successful perform-ance on false-belief trials. It may be tempting to ascribe 'theory of mind' to dolphins on the basis of their performance on Experiment 1, especially when taken in conjunction with their unanimous success in the previous non-verbal false belief task in the pilot study (Tschudin *et al.* 1999). However, this would be premature; several plausible alter-native explanations of their performance should be considered, including learning, cueing, and chance.

19.7.1 **Learning?**

Learning represents a possible alternative explanation for the success of the dolphins; this was identified as a potential problem for the pilot study design. In the subsequent series of tests, a concerted effort was made to control for the dolphins making any simple conditional discrimination by excluding the pretest control trials from the design. As indicated, we interspersed true and false-belief trials to counter this possibility. The results from the test phase of Experiment 1 represent a marked contrast with the perseverant and unsuccessful use of simple rule-based strategies by Jula and Domino during the training phases of Experiments 1 and 2. Importantly, in Experiment 1, all three dolphins were successful on the first, as well as most subsequent, false-belief trials, and two of the three were successful from the first, as well as most subsequent, true-belief trials. Affrika and Kani only recorded failures towards the end of their test trial runs, which might be explained by limitations in attention span, distraction, or fatigue. Despite the design problems with the pilot study and with Experiment 1, discussed above, their results are consistent in suggesting that the animals generally respond well, from the first trial, on false and true-belief trials.

First trial success is important because it counters the 'one-trial learning' explana-tion, namely, that an animal who fails on first trial can succeed on subsequent trials by learning from the failed first attempt. A larger sample size and naïve dolphins are needed to address this issue more thoroughly.

19.7.2 **Attributing lack of visual access?**

Another interpretation is that the dolphins in Experiment 1 may have solved the false belief task by attributing lack of visual access to the communicator (suggested in Tschudin *et al.* 1999). Attributing lack of visual access to the communicator need not require the dolphin to make a mental state attribution, but merely to track whether or not they had had visual access to the contents of the box; some might regard this to constitute a kind of 'behaviour reading' rather than 'mind reading'.[6]

Such an explanation of the results of Experiment 1 should be rejected, however, because it cannot account for the dolphins' largely consistent choice of the correct box, especially on first false-belief trials. Perhaps a behavioural rule could outperform chance in Experiment 1: when the communicator lacks visual access, choose the opposite box; when the communicator has visual access, choose the box indicated. But this rule would have to be learned. In the absence of both a mental state attribution to the communicator and learning, the best performance possible is that of chance. If learning is not present on first trials, then first trial performance should be at chance in the absence of mental state attribution. But it is not. Recall Experiment 1, where Affrika, Kani, and Khanya all succeeded on their first attempt at false-belief trials. Indeed, below chance level performances were also recorded during the post-test controls of the pilot study, where the animals appeared to choose randomly across trials. While the animals had responded almost immediately on false-belief trials, there was a significant time delay on post-test cue control responses (see Appendix 19.2), during which the animals vocalized repeatedly, attempted to interact with—and prompt—the experimenter, and were markedly ambivalent in their choices.

19.7.3 Cueing?

The possibility of inadvertent cueing by the experimenters, as in the Clever Hans phenomenon (Pfungst 1911), or other experimenter bias, is consistently problematic for the interpretation of comparative behavioural research. While we have attempted to minimize the possibility of its occurring by successive modifications to our experimental design, we can never entirely rule out cueing as the explanation of our results. Unfortunately, Domino was unsuccessful on training in Experiment 2, with the design best able to exclude cueing, and hence did not proceed to test. Nevertheless, given the number of different experimenters and conditions in the pilot study and Experiment 1, we think it unlikely that the dolphins' performances were the result of cueing. One way in which baiter cueing could be definitively ruled out in the future would be to use an automated baiting system, although this may introduce new complexities and could affect the interpersonal nature of the task.

19.7.4 A 'social mind'?

A final possibility is that dolphins succeeded on the 'theory of mind' tasks in our pilot study and on Experiment 1 because they have the capacity to attribute either a

---

6   Editors' note: consider this point in light of the challenge pressed by Povinelli and Vonk, this volume, to design tests for mind reading success on which cannot be equally well explained by behaviour reading.

true or a false belief about the location of an object (the fish reward) to another individual (here, the human communicator). Unlike Povinelli and Giambrone (2001), I argue that the continued viability of this hypothesis is supported by a range of neuroanatomical and behavioural evidence (discussed above). Indeed, the brains of dolphins are critical to the understanding of social evolution in cetaceans (Connor *et al*. 1998). The 'social mind' hypothesis is particularly supported by experimental data indicating that dolphins display certain abilities usually regarded as precursors to full-blown 'theory of mind': they comprehend (human) communicative signs such as referential pointing (Herman *et al*. 1999; Tschudin *et al*. 2001), gaze following, and replica use (Tschudin *et al*. 2001). The importance of the comprehension of such signs to the understanding of communicative partners as intentional agents has been recognized elsewhere (Tomasello *et al*. 1997). Furthermore, despite the design problems in the pilot study and Experiment 1 discussed here, the success of dolphins on false-belief tasks in these studies, and on social cognition tasks in general, contrasts with the documented literature concerning non-human primates, where the evidence for elaborate social cognition remains more equivocal.[7]

This interpretation requires some explanation for the divergent performances of Jula and Domino. There may simply be individual differences, although the rule-based strategy of Jula was surprising, especially given his good performance on the pilot study (see Appendix 19.2) and is not readily interpretable. Domino presents a different case. Whereas the animals at Sea World are part of a wider social group, with structured hierarchies and regular social interaction, at the time of this experiment Domino lived alone with his mother. As a result of this isolated existence and his heavy reliance upon his trainers, he might be considered as 'socially isolated'. Within his life context, therefore, it is unlikely that he has any regular experience of third-party social relationships, which has been regarded as important to the development and expression of social intelligence (Tomasello and Call 1997); his interaction is either with his mother or his trainers. Domino's failure is thus consistent with the hypothesis that the development and manifestation of a 'social mind' may depend heavily upon the social environment in which an individual resides.

While some authors suggest that human enculturation improves the potential for the expression of social cognition in chimpanzees (Tomasello *et al*. 1993), this may not be the case in dolphins. In our research, Affrika (one of the younger subjects) obtained more consistent results than several of her older counterparts. The performance of our subjects relative to their age, to the intensity of their human contacts, and to their social

---

7 Povinelli, 1994; Povinelli, Nelson and Boysen, 1990; Povinelli, Nelson and Boysen, 1992; Povinelli, Parks and Novak, 1992, Povinelli and Vonk, this volume; but see Hare *et al*., 2001; O'Connell, 1995; Tomasello and Call, this volume.

circumstances (consider Domino) indicates the possibility that sustained human intervention within a captive setting may not have the effect described by Tomasello *et al*. (1993, p. 1690), of 'encourag[ing] human-like skills of social cognition and learning' (see also Savage-Rumbaugh 1990).[8] Rather, sustained human interaction, when substituted for social interaction with conspecifics, may have the opposite effect and may potentially retard the cognitive functioning of otherwise spontaneously social animals. Alternatively, perhaps younger animals display more plasticity in understanding novel social situations. This would not accord, however, with our understanding of the development of social cognition in children, where a recent meta-analysis (Wellman *et al*. 2001) confirms an age-related effect with respect to capacity for conceptual change (see Wellman and Cross 2001), as opposed to the hypothesis of early competence (see Scholl and Leslie 2001).

19.8 **A window for comparison and continuing challenges**

Given the value of the comparative method to our understanding of the evolution of psychological capacities (Darwin 1859), we ran our modified non-verbal belief attribution task using the design in Experiment 2 described above with two other species: seals and human children. We did this in recognition of the learning and cueing concerns identified in our dolphin research. The two captive Cape fur seals (*Arctocephalus pussilus*) that we tested failed the false-belief trials, although one succeeded on true- and dud-belief trials (Tschudin and Peddemors 2000).

Next we tested 42 children (14 subjects per age group), 4, 5, and 6 years old. Our findings may prove informative. On test, all age groups demonstrated success on true belief (including 'dud') trials. Notably, however, none of the 4-year-olds succeeded, who proceeded directly from training to test on false-belief trials. Only three 5-year-olds passed the false-belief trials on first trial (a significant improvement compared to 4-year-olds [$P < 0.05$]) and although six 6-year-olds passed on first trial, their overall performance did not depart significantly from the performance of the 5-year-olds (Tschudin *et al*. 2000).

Compare these findings with the related child development literature. Some authors protest that false-belief task designs are too complex and suggest that 3-year-olds may be capable of passing modified tasks (for reviews, see Suddendorf and Whiten 2001; Wellman, *et al*. 2001). Documented tasks, however, credit children with first attributing false belief states to others (and to themselves) from 4 to 5 years of age (Call and Tomasello 1999; Wimmer and Perner 1983; but see Baron-Cohen *et al*. 2000 on children with autism). Our Experiment 2, however, indicates just how difficult young children find these non-verbal belief tasks: 4-year-olds fail, while 5- and 6- year-olds struggle.

---

[8] Editors' note: See and cf. Savage-Rumbaugh *et al*, this volume.

Significantly, young children who succeed on non-verbal false belief attribution tasks have been credited with 'theory of mind' (see Call and Tomasello 1999). Yet dolphins, who initially passed a similar task (see our pilot study) to the one conducted with children by Call and Tomasello, are merely considered as gifted learners or perceptive cue-readers. This view needs to be reconsidered, especially given that dolphins subsequently succeeded on our modified Experiment 1, which has greater controls than the Call and Tomasello version that was used with children.

The success of dolphins on the false-belief trials, from first trial, in both the pilot study and Experiment 1—with cautious interpretations concerning the potential for learning and cueing—differs markedly from demonstrating that they failed. Granted, Domino did not reach the test phase of the most modified design, Experiment 2, but, as discussed above, his behaviour may be explicable, given his limited tertiary social interactions, in contrast with the dolphins at Sea World, Durban.

Our comparative results and previously anticipated methodological issues indicate the need for cautiously designed future research, which could control for competing interpretations and have considerable implications. Non-invasive experimental evidence remains necessary to counter alternative explanations such as learning, which are difficult to challenge using non-controlled 'field data' (Heyes 1993; Tschudin 2001b). Such evidence also addresses the scepticism that follows certain experimental outcomes in the domain of related primate research (see Povinelli and Giambrone 2001; Povinelli and Vonk, Chapter 18, this volume).

19.9 Concluding reflections and future directions

While a 'clever behaviour reading' interpretation might be invoked to explain the success of subjects on training trials and true-belief trials (for reasons discussed above), it does not adequately explain false-belief trial performances. This remains the case because success on first trial in false-belief trials means that the subjects do not have the opportunity to acquire and apply a behavioural rule, in this case, a conditional discrimination, to pass the trial.

Alternatively, from a 'mentalist' interpretation, the attribution of a state of ignorance to the communicator is insufficient to explain false-belief trial success, because one would then expect the subjects to perform at chance. Experiment 1 dismisses this possibility, because of the success of all dolphins on false belief first trials and the significant interaction between the belief state of the communicator and the animals' response to her signal. They would then appear to be making an attribution regarding the content of the communicator's mental state, that is it is a false representation of reality.

Specifically, the performance of dolphins on a previous non-verbal false belief task and on Experiment 1—if taken in conjunction with the findings for precursors of 'theory of mind' in dolphins—supports a case for rationality in animals. Exactly what form of rationality this indicates and the extent to which it prevails in different species remains open to debate, as noted by Hurley, this volume.[9]

Mounting comparative evidence appears to support the view that animals' behaviour displays rationality, ranging from representational states in chimpanzees (Hare *et al.* 2001; Suddendorf and Whiten 2001), to culture in cetaceans and chimpanzees (Rendell and Whitehead 2001; de Waal 1999; Whiten *et al.* 1999) and socially complex cognition in birds (Emery and Clayton 2001; Bugnyar and Kotrschal 1999).

In this paper I have focussed on sociable animals and 'social minds' and I have argued elsewhere that socially advanced animal species are prime candidates for higher-order consciousness (Tschudin 2001a). This does not imply that non-social animals or animals who fail social cognition tasks should not be considered as rational. They may demonstrate rationality in other guises, some of which remain to be discovered. In dolphins, for example, the possibility of expressing certain kinds of rationality may be contingent upon their social environments, variation in which can exercise significant effects on them; but this may differ for other species. Continued investigation is thus critical.

Finally, Descartes appears to have been misguided in his denial of animal rationality: the 'absence' of language does not preclude the possibility of animal rationality. Rather, animals appear to be capable of demonstrating rational behaviour, with or without a language that humans perceive. In recognition of this, future investigations could explore potential linguistic and conceptual capacities in animals and assess their significance to the expression of rationality. Doing so may be remarkably challenging, precisely because language is regarded as a necessity for most forms of advanced cognition, including 'theory of mind' (see Astington 2001). Mindful of Darwin's call for a new, comparative psychology, however, I conclude this reflection by proposing that there are indeed 'open fields' for research into animal rationality.

---

[9] Editors' note: If a mind reading account of dolphin success on false belief tests were preferred over a behaviour reading account, the further question would remain whether such evidence for dolphin rationality should be understood as an expression of theoretical rationality (as in theory theory accounts of mind reading) or of practical rationality (as in simulationist accounts of mind reading). According to 'theory theory' views, mind reading is accomplished by means of a kind of protoscientific theory. This theory represents laws about mental states and behaviour, which are either known innately or discovered by testing hypotheses against evidence, and by means of which specific mental states can be inferred from other mental states and behaviours. According to simulation theories, by contrast, the mind reader takes someone else's perspective and uses his own decision-making processes, adjusted to the other's perspective, to generate mental states that are then attributed to the other. Such simulation is a decentred extension of practical rationality rather than an exercise in theoretical rationality (Davies and Stone 1995a,b; Carruthers and Smith 1996). Since practical rationality may be more readily attributed to animals (Hurley, this volume), it would be of particular interest to devise tests for theory-based rationality in animals.

Acknowledgements

I remain deeply indebted to my coworkers, especially to Vic Peddemors, friend and mentor, and to Tim Harkness, to Gabby Harris, Linda Clokie, to the directors, staff and animals at Sea World, Durban, Bayworld, Port Elizabeth and the Natal Sharks Board. Sincere thanks to the following: Josep Call for the suggested true-belief trials and Robin Dunbar for early comment; the Departments of Experimental Psychology and Psychiatry, University of Cambridge and especially Tony Dickinson, Nick Mackintosh, Ed Bullmore, and Simon Baron-Cohen for guidance and continued support; Kristin Hagen, Anastasia Christakou, Dom Dwyer, Clive Basson, Mike Aitken, and Frits Rijkenberg for comments on earlier part-drafts and aspects of this manuscript, to Laura Basell and Lila Koumandou for their comments on the final version, to Marcelle Tschudin and Friederike Subklew for editorial assistance; to Ian Cannell, Tim Kay, and Pete McGregor for technical support. To the Association of Commonwealth Universities, the National Research Foundation, Advanced Imaging Services, Natal and Liverpool Universities, who have formerly sponsored the work and in particular to the Swiss Academy, who sponsored me during this research. To my current sponsors, Corpus Christi College, Cambridge, and to the Burney Fund, Faculty of Divinity, Cambridge. Finally, to my family, friends, D.G., thank you.

Appendix 19.1 Pilot study procedure: Sea World, Durban, April 1999 (reproduced from Tschudin, Call, and Dunbar 1999)

Training phase

The experimenter (who acted as both baiter and presenter) placed a fish in one of two boxes, in full view of the communicator, who always witnessed where the fish was located. The screen, however, always occluded the dolphin's view of which box was being baited, although the dolphin could still observe the communicator. During this phase, dolphins learnt to associate the communicator's tap on one of the boxes with the location of the hidden reward, that is they learnt that the communicator represented the only reliable information source concerning the location of the fish. Two different sequences were used in training, to ensure that the dolphins were capable of understanding the different variations on subsequent trials: (i) the communicator tapped before leaving the experimental area; or (ii) tapped upon her return. Dolphins indicated their choice by orienting towards or approaching one of the boxes.

Pretest controls

On completion of the training phase, subjects received (in counterbalanced order across subjects) two types of pretest controls, displacement control and ignore-communicator control.

1. The displacement control served to establish that the dolphin could comprehend spatial displacement. Following the baiting of one of the boxes behind the screen, the screen was removed and the communicator tapped on the baited box before leaving the area. In the absence of the communicator and in full view of the dolphin, the experimenter switched the boxes before presenting them for the animal to make a choice.

2. The ignore-communicator control was critical in demonstrating that the dolphin could ignore the communicator when it knew—as a consequence of seeing the fish switched between boxes in her absence—that she was wrong. After baiting and the removal of the screen, the communicator left the area. At this time the experimenter opened both boxes, exaggerating the transfer of the fish from its original location to the other box, in full view of the dolphin. Upon her return, the communicator tapped on the box that she had originally observed being baited. Following this, the experimenter presented the boxes to the dolphin.

False-belief task

Following satisfactory performance on the above pretest controls, the dolphin was subjected to the false-belief test trials. To succeed on test, the dolphin had to infer that the communicator had a false belief about the location of the reward, even though the dolphin had no knowledge of where the reward was located. After baiting and the removal of the screen, the communicator left the area without indicating the location of the fish to the dolphin. In her absence, and in full view of the dolphin, the experimenter switched the boxes. Subsequently, the communicator returned and tapped on the box *at the location* that *she had originally seen baited*, after which the experimenter presented the boxes to the dolphin. Of critical importance is that on test trials, unlike ignore-communicator control trials, the dolphin did not know where the fish was located until the communicator tapped on a box. Each dolphin received four false-belief trials.

Post-test cue control

This control was devised in order to establish that the dolphins were not passing the false belief task (a) without regard for information provided by the communicator, for example by independent use of sonar, or (b) by using inadvertent cues from the communicator or experimenter during baiting. After baiting and the removal of the screen, the communicator left the area without indicating the location of the fish. She returned after a short delay and positioned herself between the two boxes, stood still and gazed directly ahead without providing any signal to indicate the location of the fish. After 15 seconds she left again, following which the experimenter presented the boxes to the dolphin.

Appendix 19.2

Table 19.2 Performance of dolphins on pretest controls, false-belief task and post-test controls (successful trials/total number of trials); reproduced from Tschudin, Call, and Dunbar (1999)

Name	Displacement control	Ignore trainer control	False-belief task[d,e]	Post-test cue control[d,e]
Jula	10/12	10/12	7/8[a]	2/4
Affrika	7/9	8/8	4/4	1/4
Khanya	8/12[c]	10/12	4/4	2/4
Kani	11/12	9/12	4/4	2/4
χ^2 [b]	27.64	32.13	23.49	1.19
P	<0.001	<0.001	<0.01	>0.99

[a] Jula failed on the first false-belief trial after an apparatus malfunction; he was given another set of 4 trials after completing the first series. If Jula's first set of trials alone are considered (3/4 correct), $\chi^2 = 18.96$ ($P < 0.02$) for the combined results.

[b] Fisher's procedure for combining P-values from k independent tests; df = 2, k = 8.

[c] This subject scored only 5/6 correct simultaneously.

[d] All false-belief trials and cue control trials were scored blindly (i.e. without knowledge of the outcome) on videotape by two independent observers, in order to assess interobserver reliability. Cohen's kappa = 0.88 and 0.87 respectively for concurrence with the experimenter, affirming that the choices could be reliably scored.

[e] Mean of subject means for false belief task: 5.3 ± 1.72 s vs. 14.5 ± 6.06 s on post-test control trials: ANOVA with subjects as covariate, $F_{1, 24} = 4.53$, $P = 0.044$.

References

Aquinas, T. St. (1956). *Summa Contra Gentiles*. New York: Doubleday.

Aquinas, T. St. (1970). *Summa Theologiae*. London: Blackfriars.

Aristotle (1976). *The Ethics of Aristotle: the Nicomachean Ethics*. London: Penguin.

Aristotle (1995a). On the soul. In: J. Barnes, ed. *The Complete Works of Aristotle*, vol. 1, Bollingen Series LXXI·2, pp. 641–692. Princeton, NJ: Princeton University Press.

Aristotle (1995b). Parts of animals. In: J. Barnes, ed. *The Complete Works of Aristotle*, vol. 1, Bollingen Series LXXI·2, pp. 994–1086. Princeton, NJ: Princeton University Press.

Astington, J. W. (2001). The future of theory-of-mind research: understanding motivational states, the role of language, and real-world consequences. *Child Development*, **72**: 685–687.

Audi, R. (2001). *The Architecture of Reason: the Structure and Substance of Rationality*. Oxford: Oxford University Press.

Baars, B. J. (1988). *A Cognitive Theory of Consciousness*. New York: Cambridge University Press.

Baron-Cohen, S., Leslie, A. M., and Frith, U. (1985). Does the autistic child have a 'theory of mind'? *Cognition*, **21**: 37–46.

Baron-Cohen, S., Tager-Flusberg, H., and Cohen, D. J., eds. (2000). *Understanding Other Minds: Perspectives from Developmental Cognitive Neuroscience*. Oxford: Oxford University Press.

Bentham, J. (1789/1982). *An Introduction to the Principles of Moral and Legislation*. In: J. H. Burns and H. L. A. Hart, eds. London: Methuen.

Bugnyar, T. and Kotrschal, K. (1999). *The raiding of conspecific food caches in common ravens: is it a clue for deceptive abilities and intentionality?* Paper presented at the Association for the Study of Animal Behaviour. Winter Meeting, London.

Byrne, R. W. (1993). Do larger brains mean greater intelligence? *Behavioral and Brain Sciences*, **16**: 696–697.

Byrne, R. W., and Whiten, A., eds. (1988). *Machiavellian Intelligence: Social Expertise and the Evolution of Intellect in Monkeys, Apes and Humans.* New York: Oxford University Press.

Call, J., and Tomasello, M. (1999). A nonverbal theory of mind test: the performance of children and apes. *Child Development*, **70**: 381–395.

Carruthers, P., and Smith, P. K., eds. (1996). *Theories of Theories of Mind*, Cambridge: Cambridge University Press.

Clutton-Brock, T. H., and Harvey, P. H. (1980). Primate brains and ecology. *Journal of Zoology*, **190**: 309–323.

Connor, R. C., Heithaus, M. R., and Barre, L. M. (1999). Superalliances of bottlenose dolphins. *Nature*, **397**: 571–572.

Connor, R. C., Mann, J., Tyack, P. L., and Whitehead, H. (1998). Social evolution in toothed whales. *Trends in Ecology and Evolution*, **13**: 228–232.

Darwin, C. R. (1859). *On the Origin of Species by Means of Natural Selection.* London: John Murray.

Darwin, C. R. (1981). *The Descent of Man, and Selection in Relation to Sex.* Princeton: Princeton University Press.

Davies, M., and Stone, T. (1995a). *Folk Psychology.* Oxford: Blackwell.

Davies, M., and Stone, T. (1995b). *Mental Simulation.* Oxford: Blackwell.

Dennett, D. C. (1978). *Brainstorms: Philosophical Essays on Mind and Psychology.* Cambridge, MA: MIT Press.

Descartes, R. (1968). *Discourse on Method and the Meditations*, translated by F. E. Sutcliffe. London: Penguin.

de Waal, F. B. M. (1999). Cultural primatology comes of age. *Nature*, **399**: 635–636.

Dunbar, R. I. M. (1992). Neocortex size as a constraint on group size in primates. *Journal of Human Evolution*, **20**: 469–493.

Emery, N. J., and Clayton, N. S. (2001). Effects of experience and social context on prospective caching strategies in scrub jays. *Nature*, **414**: 443–446.

Gibson, K. R. (1986). Cognition, brain size and the extraction of embedded food resources. In: J. Else and P. C. Lee, eds. *Primate Ontogeny, Cognition and Social Behaviour*, pp. 93–104. Cambridge: Cambridge University Press.

Gilson, E. (1993). *The Philosophy of St. Thomas Aquinas.* New York: Barnes and Noble.

Harcourt, A. H. (1992). Coalitions and alliances: are primates more complex than non-primates. In: A. H. Harcourt and F. B. M. de Waal, eds. *Coalitions and Alliances: are Primates More Complex Than Non-primates?*, pp. 445–471. New York: Oxford University Press.

Hare, B., Call, J., and Tomasello, M. (2001). Do chimpanzees know what conspecifics know? *Animal Behaviour*, **61**: 139–151.

Herman, L. M., Abichandani, S. L., Elhajj, A. N., Herman, E. Y. K., Sanchez, J. L., and Pack, A. A. (1999). Dolphins (*Tursiops truncatus*) comprehend the referential character of the human pointing gesture. *Journal of Comparative Psychology*, **111**: 347–364.

Herman, L. M., Pack, A. A., and Wood, A. M. (1994). Bottlenose dolphins can generalize rules and develop abstract concepts. *Marine Mammal Science*, **10**: 70–80.

Heyes, C. M. (1993). Anecdotes, training, trapping and triangulating: do animals attribute mental states? *Animal Behaviour,* **46**: 177–188.

Hume, D. (1985). *A Treatise of Human Nature.* London: Penguin.

Humphrey, N. K. (1976). The social function of intellect. In: P. P. G. Bateson and R. A. Hinde, eds. *Growing Points in Ethology,* pp. 303–317. Cambridge: Cambridge University Press.

Jolly, A. (1966). Lemur social behaviour and primate intelligence. *Science,* **153**: 501–506.

Kant, I. (1999). The metaphysics of morals. In: M. J. Gregor, ed. *Practical Philosophy,* pp. 355–603. Cambridge: Cambridge University Press.

Kerner, G. C. (1990). *Three Philosophical Moralists: Mill, Kant, and Sartre: an Introduction to Ethics.* Oxford: Clarendon.

Leslie, A. M. (1987). Pretence and Representation: the origins of 'theory of mind'. *Psychological Review,* **94**: 412–426.

MacIntyre, A. (1988). *Whose justice? Which rationality?* Notre Dame: University of Notre Dame Press.

Marino, L. (1996). What can dolphins tell us about primate evolution? *Evolutionary Anthropology,* **5**: 81–85.

Marten, K., and Psarakos, S. (1995). Using self-view television to distinguish between self-examination and social behavior in the bottlenose dolphin (*Tursiops truncatus*). *Consciousness and Cognition,* **4**: 205–225.

Milton, K. (1988). Foraging behaviour and the evolution of primate intelligence. In: R. W. Byrne and A. Whiten, eds. *Machiavellian Intelligence: Social Expertise and the Evolution of Intellect in Monkeys, Apes and Humans,* pp. 285–305. Oxford: Oxford University Press.

Morrel-Samuels, P., and Herman, L. M. (1993). Cognitive factors affecting comprehension of gesture language signs: a brief comparison of dolphins and humans. In: H. R. Roitblat, L. M. Herman, and P. Nachtigall, eds. *Language and Communication: Comparative Perspectives,* pp. 211–222. Hillsdale, NJ: Lawrence Erlbaum.

Neisser, U. (1988). Five kinds of self-knowledge. *Philosophical Psychology,* **1**: 35–59.

O'Connell, S. (1995). *Theory of Mind in Chimpanzees.* PhD thesis, University of Liverpool.

Pack, A. A., and Herman, L. M. (1995). Sensory integration in the bottlenosed dolphin: immediate recognition of complex shapes across the senses of echolocation and vision. *Journal of the Acoustical Society of America,* **98**: 722–733.

Pfungst, O. (1911). *Clever Hans: the Horse of Mr von Osten.* New York: Holt.

Plato (1999a). Philebus. In: E. Hamilton and H. Cairns, eds. *Plato: the Collected Dialogues,* Bollingen Series LXXI, pp. 1086–1150. Princeton, NJ: Princeton University Press.

Plato (1999b). Thaetetus. In: E. Hamilton and H. Cairns, eds. *Plato: the Collected Dialogues,* Bollingen Series LXXI, pp. 845–919. Princeton, NJ: Princeton University Press.

Povinelli, D. J. (1994). Comparative studies of animal mental state attribution: a reply to Heyes. *Animal Behaviour,* **48**: 239–241.

Povinelli, D. J., and Giambrone, S. (2001). Reasoning about Beliefs: A human specialization? *Child Development,* **72**: 691–695.

Povinelli, D. J., Nelson, K. E., and Boysen, S. T. (1990). Inferences about guessing and knowing by chimpanzees. *Journal of Comparative Psychology,* **104**: 203–210.

Povinelli, D. J., Nelson, K. E., and Boysen, S. T. (1992a). Comprehension of social-role reversal in chimpanzees: evidence of empathy? *Animal Behaviour,* **43**: 633–640.

Povinelli, D. J., Parks, K. A., and Novak, M. A. (1992b). Role reversal by rhesus monkeys, but no evidence of empathy. *Animal Behaviour,* **43**: 269–281.

Premack, D. (1988). 'Does the chimpanzee have a theory of mind?' revisited. In: R. W. Byrne and A. Whiten, eds. *Machiavellian Intelligence: Social Expertise and the Evolution of Intellect in Monkeys, Apes and Humans*, pp. 160–179. Oxford: Oxford University Press.

Premack, D., and Woodruff, G. (1978). Does the chimpanzee have a theory of mind? *Behavioral and Brain Sciences*, 1: 515–526.

Reiss, D., and Marino, L. (2001). Mirror self-recognition in the bottlenose dolphine: A case of cognitive convergence. *Proceedings of the National Academy of Sciences* 98(10): 5937–5942.

Rendell, L., and Whitehead, H. (2001). Culture in whales and dolphins. *Behavioral and Brain Sciences*, 24: 309–382.

Rosenthal, D. M. (1993). Thinking that one thinks. In: M. Davies and G. W. Humphreys, eds. *Consciousness*, pp. 197–223. Oxford: Blackwell.

Savage-Rumbaugh, S. (1990). Language as a cause-effect communication system. *Philosophical Psychology*, 3: 55–76.

Sawaguchi, T., and Kudo, H. (1990). Neocortical development and social structure in primates. *Primates*, 31: 283–290.

Scholl, B. J., and Leslie, A. M. (2001). Minds, modules and meta-analysis. *Child Development*, 72: 696–701.

Sorabji, R. (1993). *Animal Minds and Human Morals*. London: Duckworth.

Suddendorf, T. (1999). The rise of the metamind. In: M. Corballis and S. E. G. Lea, eds. *The Descent of Mind: Psychological Perspectives on Hominid Evolution*, pp. 218–260. London: Oxford University Press.

Suddendorf, T., and Whiten, A. (2001). Mental evolution and development: evidence for secondary representation in children, great apes, and other animals. *Psychological Bulletin*, 127: 629–650.

Tomasello, M., and Call, J. (1997). *Primate Cognition*. New York: Oxford University Press.

Tomasello, M., Call, J., and Gluckman, A. (1997). Comprehension of novel communicative signs by apes and human children. *Child Development*, 68: 1067–1080.

Tomasello, M., Savage-Rumbaugh, S., and Kruger, A. C. (1993). Imitative learning of actions on objects by children, chimpanzees, and enculturated chimpanzees. *Child Development*, 64: 1688–1705.

Tschudin, A. (1995). *An Assessment of the Non-invasive Techniques Available for Cetacean Brain Research*. BSocSc Hons dissertation, University of Natal, Durban, South Africa.

Tschudin, A. (1999). *Relative Neocortex Size and its Correlates in Dolphins: Comparisons with Humans and Implications for Mental Evolution*. PhD thesis, University of Natal, Pietermaritzburg, South Africa.

Tschudin, A. (2001a). 'Mind-reading' mammals: attribution of belief tasks with dolphins. *Animal Welfare*, 10: S119–127.

Tschudin, A. (2001b). Getting at animal culture. *Behavioral and Brain Sciences*, 24: 357–358.

Tschudin, A., and Peddemors, V. (2000). Unpublished data.

Tschudin, A., Call, J., and Dunbar, R. (1999). *Theory of mind in dolphins*. Paper presented at the Association for the Study of Animal Behaviour. Winter Meeting, London.

Tschudin, A., Call, J., Dunbar, R. I. M., Harris, G., and van der Elst, C. (2001). Comprehension of signs by dolphins (*Tursiops truncatus*). *Journal of Comparative Psychology*, 115: 100–105.

Tschudin, A., Daji, K., Henzi, S., Peddemors, V., and Royston, D. (1996). Relative brain size and social structure in dolphins. In: Zoological Society of Southern Africa. *Integrating Zoology*, p. 152. Pretoria: University of Pretoria.

Tschudin, A., Harkness, C., and Harkness, T. (2000). Unpublished data.

Wellman, H. M., and Cross, D. (2001). Theory of mind and conceptual change. *Child Development*, **72**: 702–707.

Wellman, H. M., Cross, D., and Watson, J. (2001). Meta-analysis of theory-of-mind development: the truth about false belief. *Child Development*, **72**: 655–684.

Whiten, A., Goodall, J., McGrew, W. C., Nishida, T., Reynolds, V., Sugiyama, Y., Tutin, C. E., Wrangham, R. W. and Boesch, C. (1999). Cultures in chimpanzees. *Nature*, **399**: 682–685.

Wimmer, H., and Perner, J. (1983). Beliefs about beliefs: representation and constraining function of wrong beliefs in young children's understanding of deception. *Cognition*, **13**: 103–128.

Worthy, G. A. P., and Hickie, J. P. (1986). Relative brain size in marine mammals. *The American Naturalist*, **128**: 445–459.

Wrangham, R. W. (1980). An ecological model of female kin-bonded groups. *Behaviour*, **75**: 262–300.

Xitco, M. J. Jr, and Roitblat, H. L. (1996). Object recognition through eavesdropping: Passive echolocation in bottlenose dolphins. *Animal Learning and Behavior*, **24**: 355–365.

Part VI

Behaviour and cognition in symbolic environments

Chapter 20

Intelligence and rational behaviour in the bottlenosed dolphin

Louis M. Herman

Abstract

A rational animal is defined as one that can perceive and represent
how its world is structured and functions, and can make logical
inferences and draw conclusions that enable it to function effectively and
productively in that world. Further, a rational animal is able to incorporate
new evidence into new perspectives of the world and can then modify its
behaviours appropriately—in effect creating a new or revised model of
the world in which it is immersed. Rational behaviour is necessarily built
on the bedrock of general and specific intellectual capacity. Intelligence,
a multidimensional trait, may appear to various degrees in various
behavioural, cognitive, or social domains. Data and observations are
presented on dolphin cognitive performance and on apparent rational
responses within four intellectual domains within the context of a variety
of empirical studies that we have conducted. These domains are:
(a) the declarative (semantic or representational) domain (does the
dolphin display knowledge or understanding about things?); (b) the
procedural domain (does the dolphin exhibit competency in means,
operations, or methods?); (c) the social domain (does the dolphin reveal
social awareness and appropriate responsiveness in social interactions or
relations?); and (d) the domain of the self (does the dolphin exhibit
knowledge or awareness of itself?). In each case, the particular
experimental paradigms are briefly outlined and instances of apparent
inferential or creative acts within each paradigm are given.

20.1 Introduction

The bottlenosed dolphin (*Tursiops truncatus*) is a highly social, cosmopolitan species
found throughout the temperate and tropical zones of the world's oceans. It is a
compelling subject for study of cognitive attributes and abilities. Early motivations for

such study were reports of the remarkable development of the bottlenosed dolphin brain (e.g. Kruger 1966; Lilly 1961), including its large size and the extensive fissurization of the cerebral cortex. The surface area of the cortex exceeds that of the human brain, although the total cortical volume is only about 80 per cent of the human volume (Ridgway 1990). In absolute size, the bottlenosed dolphin brain is somewhat larger than the human brain (Ridgway 1990). However, because larger animals tend to have larger brains, in part to control the greater bulk of somatic tissue, comparative brain studies focus on *relative* size, expressed as a ratio of brain mass to body mass (Jerison 1973; Lashley 1949). An 'encephalization quotient' (EQ) (Jerison 1973) may be computed from the mean values for a species, expressed mathematically as EQ = brain weight/0.12 (body weight)$^{0.67}$. The exponent 0.67 refers to the general relationship between mammalian brain mass (Mbr) and body mass (Mbo), or Mbr = aMbo^b. An exponent, b, of 0.75 is sometimes offered instead of 0.67. Values of EQ greater than 1.0 indicate a brain mass greater than that expected for the corresponding body mass. The largest EQ values (~7.0) occur for humans; second are four dolphin species, including the bottlenosed dolphin, with EQs ranging from 4.14 to 4.56 (Marino 2002). These values surpass those of non-human primates, including the great apes, the latter with EQs ranging from 2.09 to 3.86 (Jerison 1973; Marino 1998; Ridgway 1990).

Developmental and social factors add to the interest in bottlenosed dolphin cognitive abilities. These animals are long-lived and enjoy a protracted period of nurturance and development, not unlike the human. In the wild or in captivity, male dolphins may live into their 40s and females into their 50s (Reynolds *et al.* 2000). Physical growth continues until 15 to 20 years of age. Age at sexual maturity differs somewhat among different populations of bottlenosed dolphins, but, in general, female sexual maturity occurs between 7 and 9 years of age (range 5–12) and males between 10 and 13 years. This protracted period of development, together with the close sustained affiliation of the calf with the mother and her associates, allows the youngster the time and opportunity to learn and develop the multitude of behavioural and social skills necessary for effective functioning within its world. The social world, in particular, can be highly intricate and complex (see for various chapters in Conner *et al.* 2000; Reynolds *et al.* 2000).

Early work with bottlenosed dolphins housed in aquaria or in marine parks revealed their ready trainability and their high degree of sociability, furthering interest in the study of their behaviours, sensory capabilities, and cognitive skills. Much of the early research focussed on sensory features, especially the hearing and sound production systems, and was principally carried out at US Navy facilities or was Navy sponsored (e.g. see Wood 1973). These studies documented the dolphin's[1] remarkably well developed echolocation (sonar) system, capable of detecting differences among objects

---

[1] Unless otherwise specified, henceforth the term 'dolphin,' when used alone, refers to the bottlenosed dolphin *T. truncatus*.

in size, shape, material composition, thickness, and many other factors (see various chapters in Busnel and Fish 1980, and later reviews in Au 1993; also see Herman *et al.* 1998). Based on these competencies and on the large size of the auditory area of the brain and the eighth (acoustic) cranial nerve, some insisted that the large size of the dolphin brain is merely an evolutionary response to the advantages of echolocation and not an indicant of general processing power or intelligence. This specious argument re-emerges frequently. For example a recent article in Smithsonian magazine states:

> ... not all marine biologists agree that dolphins and other cetaceans are especially smart. Though a dolphin has an impressive ability to be trained to perform tricks, skeptics say that this behavior reflects not intelligence—the capacity to make choices based on weighing possible consequences—but conditioning, a programmed response to a stimulus like food. In that view, dolphins are no more intelligent than dogs, horses, or for that matter parrots. In addition, notions about the dolphin's exceptional intelligence have been based on the observation that they have disproportionately large brains. Again, some scientists point out that the animal's brain is wired chiefly for sonar processing and motor control, not 'thinking' (Dowling 2002, p. 49).

In contrast to these types of statements, Pabst *et al.* (1999) state strongly that 'there is insufficient evidence to assert that odontocetes (toothed whales, including dolphin species) use relatively large areas of the cerebral cortex for auditory processing' (p. 59). Further, if the large dolphin brain were primarily for auditory processing, we would expect to find that there was little remarkable about dolphin cognitive abilities. As I stated in an early review of dolphin cognitive characteristics '... descriptions (of the dolphin brain) hint at the intellectual potential of the species, which ultimately depends on brain structure and organization... However, it is behavior, not structure, that measures the intellectual dimensions and range of the species...' (Herman 1980, pp. 363–364). Behavioural studies we carried out during the years subsequent to 1980, as well as many completed earlier, have catalogued a wealth of data on dolphin cognitive traits and abilities that together provide compelling evidence for a level of intelligence commensurate with expectations based on the size, structure, and development of the brain.

The concept of 'intelligence' is thus highly germane to any discussion of dolphin cognitive capabilities and of rational behaviour, but there is far from a consensual definition of it, especially for animals. In my view, however, intelligence manifests itself in *behavioural flexibility*, the ability to modify or create behaviour adaptively in the face of new evidence or changes in world conditions (cf. Herman and Pack 1994). The intelligent animal, in principle, can go beyond the boundaries of its familiar world, and beyond its biologically programmed or learned repertoire of behaviours that enable successful responding in that familiar world, to function effectively in new worlds or in new world conditions.

Intelligence is a multidimensional trait that may appear to various degrees in various behavioural, cognitive, or social domains. Effective functioning in multiple domains would suggest rich behavioural flexibility. In this chapter, I consider four domains in which we have obtained data on dolphin cognitive performance and rational responding. These are: (a) the declarative (semantic or representational) domain (does the dolphin display knowledge or understanding about things?); (b) the procedural domain (does the dolphin exhibit competency in means, operations, or methods?); (c) the social domain (does the dolphin reveal social awareness and appropriate responsiveness in social interactions or relations?); and (d) the domain of the self (does the dolphin exhibit knowledge or awareness of itself?). I use these domains to categorize our findings and to illustrate the behavioural flexibility of the dolphin that extends even to unusual innovative behaviours. Creating new effective responses or new solutions to problems provides strong evidence for behavioural flexibility and for an understanding of the conditions existing at the moment within a particular domain. These four domains are of course not entirely independent of one another. For example knowledge of oneself may influence how one responds in particular social settings. Knowledge of how something works may define or enhance one's concept of the thing itself.

Rational behaviour is, then, necessarily built on the requisite bedrock of general and specific intellectual capacity. To understand what I mean by rational behaviour, I begin with the following premise: *a function of a mind is to create a model of the world.*[2] This model then influences our perceptions and our behaviours, and may allow us to function effectively in that world, provided our model reasonably reflects reality. However, there is not just one world, but many worlds within our life experiences. For the human, the model of the world may differ with context, situation, or culture; as a simple example, consider the different world models of a teenager for home, work, school, sports, and peer group relationships. A model that yields effective behaviour in one situation may be inappropriate or even counterproductive in another. When we bring the dolphin from the wild into the laboratory situation, it enters a radically different world. The ability of the dolphin to function effectively, even creatively, within that new world is contingent on its learning how that world is structured, how it operates, who the actors are, what features are significant, what rules and contingencies apply, and much more; effective functioning, in turn, provides inferential evidence that the dolphin has created an accurate model of that world.

A rational animal may then be defined as one that can perceive how the current world it occupies is structured and how it functions, and can then make logical inferences and draw conclusions that enable it to function effectively and productively in that world. Further, a rational animal is able to incorporate new evidence into new perspectives of the world and can then modify its behaviours appropriately—in effect creating a new or revised model of the particular world in which it is immersed.

---

[2] Jerison (1973) has argued similarly that 'the construction of the model (of the world) ... is the work of the brain.' (p. 17).

Can we find evidence, in our structured laboratory settings, for these types of rational responding by the dolphin? To answer this question, I return to the different domains described earlier and examine the evidence for flexible and appropriate responding within those domains. Much of the evidence is taken from our published literature, but some is taken from observations or procedures that may not have appeared in these reports. In each case, the essentials of the paradigm in effect are summarized briefly to provide an appreciation of that context, and then specific instances are given in which rational responding seems evident within that paradigm.

20.2 Declarative and procedural domains

Intelligence within the declarative domain can be revealed by the ability to perceive the nature, properties, or characteristics of things and to understand symbolic or direct references to those things. Intelligence within the procedural domain can be revealed by the capacity to understand how things function or how to do things (see Anderson 1978 for a discussion of declarative and procedural knowledge). Declarative and procedural knowledge may exist independently or may intertwine. We understand, for example, what the word 'car' refers to, but we may not know how to drive. These two domains are discussed together here, because for some of the paradigms we have employed, both declarative and procedural knowledge is required—knowing what things are and knowing how to manipulate those things. Five paradigms requiring declarative or procedural knowledge, or both, are presented in the following sections.

20.2.1 Understanding of the semantic and syntactic components of a symbolic system: knowing what things are and how to manipulate them

We constructed a language-like symbolic system to test the dolphin's understanding of complex instructions delivered through sequences of gestures, with the sequences governed by both semantic and syntactic rules (Herman *et al.* 1984; Herman 1986). The most complex instructions were expressed through so-called relational sequences, which required the dolphin to construct a relationship between two objects, by transporting one to the side of another (indicated by use of the action term Fetch) or by placing one object on top of or inside of another (indicated by use of the action term In/On). To insure that the dolphin was processing the entire sequence as a cohesive unit, and not simply responding to each gesture in turn ('word-by-word' processing), we constructed an 'inverse' grammar. In this grammar, the order in which the gestural symbols appear is uncorrelated with the order in which responses to those symbols must occur. The dolphin cannot, therefore, respond to each symbol in turn as it occurs, as it might in a linear (left-to-right) grammar, but must wait until the entire sequence is completed before it can know what to do. For example consider the symbol sequence and the corresponding response sequence illustrated below:

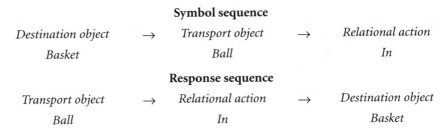

Symbol sequence

Destination object	$\rightarrow$	*Transport object*	$\rightarrow$	*Relational action*
Basket		*Ball*		*In*

Response sequence

Transport object	$\rightarrow$	*Relational action*	$\rightarrow$	*Destination object*
Ball		*In*		*Basket*

The gestural symbol sequence *Basket Ball In* requires the dolphin to place the floating ball in the floating basket. *Basket* is the first symbol given, but it signifies the destination object, the last item in the response sequence. *Ball* is the second symbol presented, but it signifies the transport object, the first object that must be responded to. *In*, the final symbol, defines the type of relationship to construct, but is the second operation in the response sequence. In the sequence illustrated, *Basket* functions grammatically as the indirect object of the verb *In*. However, in other sequences, the occurrence of *Basket* in the first position does not necessarily signal that it will have a grammatical function as indirect object. Thus, a symbol sequence might be, for example, *Basket Over*, in which case *Basket* functions grammatically as the direct object of the verb and the dolphin would be required to act directly on the basket by leaping over it. Thus, the grammatical function of an initial object symbol is not clear until additional symbols appear. In addition to the constraints on interpretation already listed, the symbol sequence *Basket Ball In* might instead have terminated with *Fetch* rather than *In*, in which case the dolphin would be required to bring the Frisbee to the side of the basket, rather than put it inside. The sequence might also have terminated with the gestural symbol *Erase*, which nullifies the preceding terms and requires the dolphin to do nothing other than attend again to the experimenter.

All of this is understood functionally by the dolphin as implied by its appropriate responding to a variety of relational sequences. To carry out the instruction conveyed by a relational sequence of symbols, the dolphin must take account of both the semantic value or referent of the symbol and the order in which symbols occur, as governed by the syntax inherent in the inverse grammar. For example the two three-symbol sequences illustrated below consist of the same three terms but the order of the destination and transport objects are reversed in the second sequence relative to the first. The first sequence instructs the dolphin to bring the swimmer to the surfboard while the second requires that the surfboard be brought to the swimmer.

Surfboard swimmer fetch
Swimmer surfboard fetch

20.2.1.1 Evidence for rational responding

There were many occasions when the dolphin demonstrated apparent inferential reasoning to achieve a solution to a new semantic proposition or new syntactic structure. Three examples can be given. First, many semantically reversed sequences of the type just illustrated were given to the dolphin Akeakamai ('Ake') and many had not been experienced by her previously. That is, they were novel instructions. Nonetheless, in the majority of cases, she was able to carry out the instructions correctly, in effect inferring that various semantic entities could be inserted into the slots within familiar syntactic frames (Herman *et al.* 1984; Herman 1986).

Second, we probed Ake's responses to anomalous sequences that violated either the semantic rules or the syntactic structure of sequences (Herman *et al.* 1993). For example, the sequence *Water Phoenix Fetch,* although structured correctly syntactically, is a semantic anomaly in that it asks Akeakamai to transport the dolphin Phoenix to the stream of water entering the pool. Phoenix cannot or will not be transported. Ake responded to this request, and to most other semantic anomalies of this type, by not taking any action at all, instead remaining at station 'staring' at the experimenter. In effect, she rejected the instruction. When given a gestural sequence that violated a syntactic rule, Ake typically decomposed the sequence to find a subset of items that were syntactically and semantically proper and acted on that. For example the sequence *Speaker Water Pipe On* is a syntactic anomaly in that there is no syntactic structure that allows for three object names in a row. Ake responded to this sequence by essentially conjoining the items *Speaker Pipe On,* a proper instruction, and placed the length of pipe floating in her pool on top of the speaker affixed to the pool wall. Note that in this case, as well as in many other similar anomalous sequences of this type, she conjoined nonadjacent items.

Third, Ake was not only familiar with three-item relational sequences of the type illustrated above, but was also familiar with a *non-relational* three-item sequence consisting of a locative term (*Left* or *Right)* followed by an object term, and then by an action term. For example, *Left Hoop Through* instructs Ake to swim through the hoop to her left (and not through the one to her right). To probe Ake's ability to infer solutions to new syntactic structures, for the first time, and without any training, we gave her a four-item sequence combining the relational and non-relational structures (Herman *et al.* 1984). Her understanding was immediate. Examples of four-item structures are *Right Hoop Frisbee Fetch* and *Hoop Right Frisbee Fetch.* The first sequence instructs Ake to take the Frisbee to the hoop on her right while the second instructs her to take the Frisbee on her right to the hoop. She also immediately understood, without training, five-item semantically contrasting sequences, such as *Right Basket Left Ball In* (put the ball on your left in the basket on your right) and *Left Basket Right Ball In* (put the ball on your right in the basket on your left) (Herman 1986). Ake therefore correctly inferred the properties of these new four- and five-word sequence structures, and made the appropriate response, strictly from her knowledge of shorter structures.

Fourth, Ake was able to rearrange objects in her pool spontaneously and appropriately in order to carry out the instruction given to her. Examples of spontaneous rearrangements innovated by Ake include: (a) lifting a hoop lying flat on the bottom of the tank into an upright position and then darting through it to complete the instruction *Hoop Through* ('swim through the hoop'); all previous instructions of this type were to hoops floating at the surface or suspended midway in the water column; (b) moving a surfboard resting against the wall of the pool to the centre of the pool and then leaping over it, in response to the instruction *Surfboard Over* ('jump over the surfboard'); and (c) removing a ball already in a basket and then quickly replacing it in the basket in response to the instruction *Basket Ball In* ('put the ball in the basket'). These examples illustrate the dolphin's ability to rearrange the physical objects of her world in order to effect a solution to a problem. They also illustrate that the various items of her language system were understood at the level of a concept—for example, what *through, over,* and *in* mean in a general sense.

20.2.2 Improvising an efficient strategy: fetching multiple objects at once

At the conclusion of a session examining competency in the gestural language paradigm, there may be as many as 11 objects floating about in the dolphin's pool, most of which may have been referred to during the course of testing. To end the session, the tankside experimenter, using the single gesture *Fetch*, asks Ake to retrieve the objects so they may be removed from the pool. Ake understands that in the absence of a specified destination object preceding the *Fetch* gesture, all objects are to be brought to the experimenter.

20.2.2.1 Evidence for rational responding

Initially, we expected that Ake would bring back one object at a time, as *Fetch* had been trained in that manner, and that after each retrieval the experimenter would simply repeat the *Fetch* gesture to request another object, until all were returned. Instead, Ake spontaneously developed a strategy of bringing back multiple objects at once. Typically, she 'rounds up' several objects and then carries the group back at once, perhaps wearing a hoop around her head, pushing a Frisbee with her rostrum, pushing a pipe with a pectoral fin, and carrying the surfboard on her back. Apparent planning can be seen in her retrieval strategy, in that she will usually first swim to the most distant object, carry it closer to another object, and then carry the two to a third object, and so forth. On her first retrieval attempt, as many as three or four objects may be gathered and brought *en masse* to the experimenter. As the number of objects is reduced, fewer objects are available for retrieval, and fewer are brought back at once.

An interesting extension of her fetching behaviour may occur between testing sessions or in the evenings. During these free intervals, Ake sometimes can be heard producing a distinctive, loud whistle, her head lifted high out of the water. We have learned that she is calling to us to take a piece of debris she is holding in the tip of the mouth (e.g. it may be a leaf that blew into her outdoor pool). She holds the item gently; it is clearly visible to us. When someone responds by coming to the edge of the pool Ake will swim to that person and offer the debris. The person takes the offering and then goes to the fish kitchen, returning after a few minutes with a fish for Ake, as well as one for each of the other dolphins who are waiting patiently at poolside next to Ake. The behaviour is obviously rewarded but was not explicitly taught. Also, Ake's distinctive calling behaviour was of her own design.

20.2.3 Tests of referential understanding: understanding references to absent objects

A strong indicant that a symbol elicits the concept or the properties of an object is an understanding of a reference to the *absent* object. We understand the referent of the word 'car,' although no car may be immediately present. To test Ake's understanding that symbols refer to things, we constructed a paradigm in which she was required to report whether an object referred to by a symbolic gesture was present or not in her pool (Herman and Forestell 1985). Two paddles were placed along the wall of the pool, one to the left and one to the right of the dolphin, who was positioned facing the experimenter. A press of the right paddle meant *Yes (present)* and a press of the left paddle meant *No (absent)*. The experimenter, who stood outside the pool wall facing Ake, showed her from one to three objects from an available set of nine different objects, throwing each in turn over the dolphin's head and into the pool. The experimenter than gave a symbolic gesture referring to an object (which may or may not be one of those shown to Ake) and then gave a second gesture glossed as 'question.' For example the gestural sequence *Ball Question* meant, 'Is there a ball in your pool?' Ake could respond *Yes* or *No*. Overall, Ake was as accurate (80–83 per cent correct for 182 queries) at reporting absence, as she was at reporting presence, providing compelling evidence for declarative knowledge—that the gestural symbols we used for the objects represented those objects.

20.2.3.1 Evidence for rational responding

The ability of Ake to create a reasonable and informative response to a problem that went beyond the boundaries of what she had been explicitly taught was illustrated when we began to give relational sequences within the reporting paradigm (Herman *et al.* 1993). For example, instead of asking *Hoop Question,* we might give her a relational instruction such as *Hoop Frisbee Fetch* (= bring the Frisbee to the hoop). However, both objects might be present, or only one of them, or neither. If both were present, Ake simply carried out the instruction—taking the Frisbee to the hoop, in conformance with the inverse grammar. If both were absent, or if the transport object was absent (e.g. the Frisbee), she directly pressed the No paddle, indicating that the action could not be completed or initiated. But, if the destination object was absent (e.g. the hoop) she created, on the very first occasion, an innovative response that greatly surprised us. Our expectation was that she would again press the No paddle, to indicate that the instruction could not be done, or that an object was absent. Instead, she swam to the transport object (the Frisbee) and brought it the *No* paddle, in effect reporting that the Frisbee was present but its destination, the hoop, was not. Ake has used that same type of response ever since if given a missing destination object within this paradigm, regardless of the particular destination or transport objects specified. For example on the first occasion that we gave her the gestural instruction *Surfboard Person Fetch* (bring the person floating in the pool to the surfboard), with the person present but the surfboard absent, Ake swam to the person and pushed her from the centre of the pool directly to the *No* paddle.

20.2.4 Understanding of representations of the real world: television scenes

Savage-Rumbaugh (1986) provided an extensive review of responses of home-reared chimpanzees or language-trained chimpanzees to television scenes. All of the chimpanzees received extensive exposure to television, but, uniformly, all showed, at best, only fleeting interest in the television scenes. In the case of the home-reared chimpanzee Lucy, even scenes of familiar humans, dogs, or children failed to elicit a response. Apparently, the images or scenes were not processed as representations of the real world. Similarly, Savage-Rumbaugh's own language-trained apes, Sherman and Austin, failed to show initial interest in television, even to scenes portraying their own familiar surroundings. To develop her chimps' appreciation of television scenes, Savage-Rumbaugh exposed them daily, 30 minutes per day for months, to television scenes or film of other chimps. During this time, the experimenters would exclaim and vocalize when interesting scenes appeared. Sherman and Austin's attention to the television scenes gradually increased, and eventually they began to exhibit appropriate overt responses to the scenes.

20.2.4.1 Evidence for rational responding

Despite this history of initial non-responsiveness of chimpanzees to television scenes (and, in my experience, similar non-responsiveness of dogs and cats), we nevertheless decided to probe the initial interest of the dolphins Ake and Phoenix to television scenes (Herman *et al.* 1990). Neither dolphin had been exposed to television of any sort previously. Given the results with apes, we expected to see an initial lack of response, which we would then follow with a long period of exposure and training to achieve appropriate responding to television scenes. To our astonishment, however, no training was needed. Both dolphins attended immediately and appropriately to the televised scenes we presented. For these initial scenes, we placed a 13-inch (diagonal measurement) black-and-white television set behind an underwater window and projected on it an image of a trainer being filmed live in a remote studio. We also placed a video camera next to the television set, looking out through the window. The image from this camera was fed back to a television screen in the studio, enabling the 'TV trainer' to see the dolphin peering into the window. The dolphin and trainer were thus linked electronically. We first directed Ake (gesturally) to swim down to the window, where she saw the image of the TV trainer, about 8 inches high, giving her an instruction in the familiar gestural language. Ake turned from the window and carried out the instruction immediately and correctly. She was given 14 different instructions altogether and completed all but two correctly. We then directed Phoenix to the window and gave her 19 different gestural instructions. She completed all but one correctly.

Subsequently, we tested Ake's responses to degraded images of the trainer, to examine how abstract gestural information could be and still be understood. In a series of steps, we first showed only the gesturing trainer's arms and hands on the television screen, then only the hands, and, finally, only two circles of white light moving about the screen. Two 3-inch-diameter sponge balls held by the gesturing trainer, whose body parts were totally obscured, produced the moving circles. Almost no degradation in performance was observed under the first two conditions. In the final condition, Ake remained significantly above chance, although below the level of the other conditions. Nonetheless, when we tested our staff with this latter display, only the most experienced staff outperformed the dolphin in interpreting the instructions embedded within the moving circles of light. This result supported the hypothesis that Ake had developed rich representations of the gestural symbols and that she used those representations to make sense of stimuli that bore little physical similarity, other than movement pattern, to the gestures displayed by the full-body image of the trainer.

Hiapo and Elele, who had arrived at our facility some 9 years after Ake and Phoenix, showed a similar responsiveness to television scenes as did Ake and Phoenix. However, Hiapo and Elele may have observed television images informally while we were testing Ake and Phoenix. In later studies, however, we showed that all four dolphins were not only able to carry out gestural instructions conveyed by a televised person, but could imitate that person's behaviours (see later section on imitation). Further, Elele was able to carry out match-to-sample tests as accurately as she did in the real world when the sample object was displayed on the television screen and only the alternative objects appeared in the real world (Pack and Herman 1995). No special training was required for Elele to use the television displays in this manner.

20.2.5 Knowing how to integrate behaviours: combining multiple discrete behaviours holistically

All of the dolphins understand many individual gestural signs (ca. 80) that are not part of the language vocabulary and that elicit particular behaviours. Some examples of behaviours elicited by a single gestural sign are *back dive, blow bubbles, tail wave, open mouth, spiral swim, spit water,* and *pirouette.* In addition, sequences of these discrete signs may be given, in which case the experimenter is asking the dolphin to perform each of the behaviours in the sequence.

20.2.5.1 Evidence for rational responding

Initially, only a few sign combinations were used, consisting of the single-sign instruction *spit water*, or *open mouth*, or *wave pectoral fins* followed by the generic sign for *swim*. In response, the instructed dolphin (all dolphins were exposed to this sequential signing) swam at the surface while spitting water, or while keeping its mouth open, or while wiggling its pectoral fins. These responses may have been shaped, but our records on this early training are incomplete. However, all dolphins spontaneously extended the behaviour from swims to jumps and from two discrete behaviours to three or more. For example in response to a sequence of three gestures glossed as *pectoral fin wave, tail wave,* and *jump,* Phoenix will leap into the air in a porpoising dive while simultaneously wiggling her pectoral fins and waving her tail. Another example is the sequence of three gestures *spit, pectoral fin slap,* and *swim*. In response, Phoenix spits water from her mouth while swimming on her back and slapping a pectoral fin on the water surface. Specific combinations such as these are not trained; new combinations are carried out reliably as an integrated response in almost all cases. As the number of individual elements increases above three, one or more may drop out, possibly reflecting forgetting of some of the instructed behaviours. Clearly, however, the dolphins' ability to execute or attempt to execute an integrated behavioural response to new sequences of action signs implies that each has developed a concept of 'combinations.'

20.2.6 Knowing how to improvise behaviours: understanding the concept of 'create'

All four dolphins are familiar with a gestural sign that we gloss as *create*. No specific behaviour is required in response. Instead, the gesture asks the dolphin to create a behaviour on its own—any behaviour or even multiple behaviours. In formal tests, the dolphin Elele's responses to the *create* sign were tested over the course of 18 sessions of 24 trials each (Braslau-Schneck 1994). For eight of these 24 trials, the trainer gave the *create* sign, so that there were 144 trials of this type altogether over the 18 sessions. For the remaining 16 'filler' trials in a session, the trainer interacted socially with the dolphin at tankside, giving specified gestures only for behaviours that maintained the dolphin in a relatively simple activity such as allowing herself to be stroked, or engaging the dolphin in some game such as putting rings on a stick. The eight *create* trials were organized into two blocks of two consecutive trials and one block of four consecutive trials, with blocks separated from each other and from the beginning of the session by at least two filler trials. Responses to filler trials were reinforced with fish reward and social praise. To avoid training of responses to *create* trials, these responses were not reinforced.

20.2.6.1 Evidence for rational responding

Elele most often offered multiple behaviours per *create* trial. Over the 144 *create* trials, Elele offered 323 behaviours, 72 of which were unique. Of these 72, 38 were novel, in that they had not been seen before or were not under stimulus control. An important component of this demonstration was the understanding of the *create* gesture itself. Elele only offered self-initiated behaviours if given this gesture. Clearly then, Elele understood 'create' at the level of a concept.

20.3 **Social intelligence**

Dolphins live in complex, fluid societies, with individual affiliations ranging from close bonding to more fleeting transient associations (e.g. Reynolds *et al.* 2000). The dolphin social network is often described as a 'fission–fusion' society because of the interplay of relatively permanent and temporary associations (Connor *et al.* 2000). The closeness of an affiliation is often seen, for example, in pairs or small groups of animals performing highly synchronous swimming and leaping behaviours. Social knowledge for a dolphin seems to be about attending to others and knowing others, both categorically in terms of gender and age class as well by one's history of interactions with particular individuals. Social knowledge also entails knowing how to collaborate or compete with others, as in foraging, feeding, or reproductive activities. Additionally, social knowledge involves interpreting other's behavioural and vocal signals as well as producing such signals for others. Here, I consider three aspects of social knowledge or social interactions that we have studied in our laboratory: (a) joint attention; (b) behavioural synchrony; and (c) imitation.

20.3.1 **Joint attention: understanding human pointing and gaze**

Manual pointing or gazing by humans serves to call another's attention to an object, event, or place of interest. Human pointing or gazing is thus a social, triadic transaction, involving the co-ordination of the attention of the informant and the observer to the same target or event, and may be termed referential pointing or gazing. Implicit in adult human pointing or gazing is not only the intent to manage the attention of another individual but also the expectation that the observer understands the referring function of the action.

Wild chimpanzees have not been observed using pointing gestures to direct another's attention (Tomasello and Call 1997). In laboratory conditions characterized by extensive contact with humans, chimps can learn to use pointing to direct a human's attention to something desired, such as a piece of food or a place to go. Paradoxically, however, they fail to understand the human's intention when the human points at something. For example in the Povinelli *et al.* (1997) study, the experimenter sat between two boxes, one of which contained food, and pointed at the baited box. The chimp chose that box reliably only when the experimenter's outstretched arm and hand almost touched the box. However, when the experimenter was positioned away from the boxes by 3 meters or so, the chimp no longer reliably approached the indicated box. The chimp thus failed to understand referential pointing. Understanding of human gaze by chimpanzees is also only marginally effective. Povinelli *et al.* (1997) found that none of the eight chimps he tested performed above chance in initial trials using human gaze (or gaze plus pointing) to direct attention to the baited box. Similar results were reported by Call *et al.* (1998).

In contrast to these negative findings with chimpanzees, domestic dogs can select the object pointed to or gazed at by a human handler (Hare *et al.* 1998; Soproni *et al.* 2001). The dog's understanding of these human communicative signals is most likely the product of selective breeding and domestication, as human-reared wolves do not show this capability (Hare *et al.* 2002).

Dolphins, like chimpanzees, are of course not domesticated animals. We tested Ake's understanding of referential pointing by placing three objects in the pool, one to her left, another to her right, and a third behind her, each at approximately 3 metres distance (Herman *et al.* 1999). We could then refer to these objects by pointing at them or by using the familiar gestural language symbols. In either case, the dolphin was required to choose the referenced object and take the particular action to it (for example, leaping over it, swimming under it, touching it with the tail, etc.) as directed by a gestural action symbol that immediately followed the object indicator. For example, we can instruct Ake to swim under a hoop either through the symbolic gestural sequence *Hoop Under* or by pointing at the hoop and then signing *Under*. We also tested gaze comprehension by Ake and Phoenix by turning the head left or right to look at an object, and asking the dolphin to take a signalled action to that object (Pack and Herman 2004).

20.3.1.1 Evidence for rational responding

Surprisingly, Ake not only understood a direct point to a distal object, made with arm and finger extended, but also immediately understood a cross-body point (Herman *et al.* 1999). For the cross-body point, the left arm is extended across the body to indicate an object to the pointer's right, or the right arm is extended across the body to indicate an object to the left. Although Ake had been exposed to human direct pointing previously during informal sessions with trainers, she had never before been exposed to cross-body pointing. Her immediate understanding of this novel form of pointing reinforces the idea that the dolphin understood pointing, in either form, as a reference to an object. In addition, both Ake and Phoenix showed spontaneous understanding (without any prior training) on their first trials of human-directed gaze (also see Tschudin *et al.* 2001, for an additional demonstration of spontaneous understanding of human gaze by dolphins). The immediacy of understanding of gaze by Ake and Phoenix extended to static as well as to dynamic forms of gaze. In static gaze, the dolphin sees only the terminal position of the head, left or right, while in dynamic gaze the dolphin views the active head movement from straight ahead to either the left or right side. Static pointing was also immediately understood by both dolphins.

Furthermore, Ake demonstrated spontaneous understanding of complex pointing sequences. She immediately responded correctly (without training) to sequences of two points, each to a different object, and then followed by the symbolic gesture *fetch*, by taking the object pointed to second to the object pointed to first. Thus, she applied the same inverse grammatical rule to a sequence of two points that she normally did to a sequence of two symbolic object gestures followed by the symbolic gesture *fetch*. For example, in response to the sequence *point at surfboard, point at pipe, fetch*, Ake swam to the floating pipe and transported it to the floating surfboard, the same type of response she takes if given the wholly symbolic gestural sequence *surfboard pipe fetch*.

It may seem surprising, given the dolphin's armless anatomy, that it would so easily understand the human pointing gesture, while chimps do not. Elsewhere (Herman *et al.* 1999), we speculated that the dolphin may understand pointing because it possesses an acoustic analogue of pointing in its natural world, through its highly focused echolocation beam. It has been shown, for example, that one dolphin can detect what another is inspecting through echolocation, simply by listening to the echoes returning from the emitter's beam (Xitco and Roitblat 1996). The 'eavesdropping' dolphin, positioned next to the echolocating dolphin, is able to identify the object of interest to the echolocating dolphin, in effect sharing attention with the echolocator. That the dolphin also understands the referring function of gaze as directed by human head movements may signal that it has developed a profound appreciation for human attentional signals in general, and the value of sharing in that attention.

20.3.2 Behavioural synchrony: carrying out behaviours together

We devised a gestural sign we termed 'tandem.' This sign, when followed by a second sign denoting an action, directs a pair of dolphins to carry out that action together, synchronously. For example, the gestural sequence *tandem backdive* directs the pair to join together and perform a back dive in close synchrony. Two different pairs of dolphins, Ake and Phoenix and Hiapo and Elele, were separately exposed to the requirements of the tandem sign and both pairs have responded to it reliably and roughly equivalently.

20.3.2.1 Evidence for rational responding

Once the tandem sign was taught, using only a few actions, both pairs carried out a variety of additional behaviours synchronously, as requested gesturally by the experimenter. An example of a complex sequence carried out synchronously is the three-item gestural sequence *tandem kiss jump*. In response, the pair swim away together and then leap out of the water touching each other's rostrum (beak) in mid-air. Of particular interest was how a pair might respond to the sequence *tandem create*. In effect, this sequence asks the pair to select or create together a behaviour of their own choosing, and carry it out synchronously. Typically, the pair will first swim about side-by-side, generally for a longer time than when given a specific behaviour to perform synchronously, then apparently select some behaviour in common and execute it in close synchrony. The selected behaviours may range from simple types, such as synchronous tail waves, to a complex spinning leap while spitting water from their mouth. We have not been able to determine how the dolphins manage this task. Their apparent joint performance may be a case of near-simultaneous mimicry, one following the other's action closely, but we have not been able to confirm this through detailed video analyses. Alternatively, underwater intention movements by one dolphin may guide the second dolphin to select that same behaviour. Clearly, there are other possibilities as well, and further study of mechanisms is needed.

Importantly, behavioural responses to the *tandem create* sequence were not taught. Instead, after a pair had been exposed to many multiple instances of the tandem sign followed by a sign for a specific behaviour, they were exposed for the first time to the *tandem create* sequence. Each executed the same self-selected behaviour, in close synchrony with the other. Their spontaneous response was apparently accomplished through generalization of their knowledge of expected responses to the *create* gesture and their knowledge of the synchronous requirement of the *tandem* gesture.

20.3.3 Mimicry: copying sounds and copying the motor behaviours of others

Both the definition of imitative behaviour and which animals may exhibit such behaviour have been areas of considerable discussion and debate (e.g. Whiten and Ham 1992, Zentall 1996). Three premises have guided our approach to the study of imitation by dolphins: (i) evidence for imitative ability is strongest if imitation can be demonstrated to a variety of behaviours and in a variety of contexts; (ii) although the genesis of imitative ability may derive from social factors, its expression may also appear in non-social contexts; and (iii) if imitation is understood conceptually by the subject, it should generalize to many situations, and be controllable by abstract symbols (Herman 2002). Here I consider three different areas in which we have demonstrated imitative abilities of dolphins: (a) vocal mimicry; (b) motor imitation of dolphins and humans; and (c) imitation of behaviours viewed on a television screen.

20.3.3.1 Evidence for rational responding

Bottlenosed dolphins are highly vocal animals and use sounds for echolocation, communication, and emotional expression (Herman and Tavolga 1980). We trained Ake to imitate vocally a variety of electronically generated 'model' sounds broadcast into her pool through an underwater speaker (Richards *et al.* 1984). In response to the model sound, Ake vocalized into an adjacent hydrophone. She mimicked a variety of different waveforms, including pure tones (sine waves), triangle waves, and slow frequency modulation. Some models were imitated accurately on Ake's first attempt, illustrating her development of a generalized concept of mimicry. Reliability of mimicry over successive exposures to a given model was excellent. There were two instances when model sounds were played that were out of Ake's preferred vocal range, one above and one below that range. In each case, Ake accurately reproduced the sound contour, but at an octave below the high frequency model and an octave above the low frequency model.

Octave generalization of the type displayed by Ake is a rare phenomenon in the animal world. Tests for octave generalization of pitch contours in starlings, cowbirds, and a mockingbird yielded negative results (Hulse and Cynx 1986; Hulse 1989). The birds instead focussed on the absolute frequencies of the pitch series and could not recognize a tonal series as the same when it was shifted by an octave. Rats and monkeys also fail to evidence octave generalization (D'Amato and Salmon 1984). In contrast to these frequency constraints of birds, rats, and monkeys, Ralston and Herman (1995) demonstrated excellent octave generalization of pitch contours in the dolphin Phoenix. Thus, dolphins appear to possess a particularly robust form of frequency contour perception, as do humans. This may serve dolphins well in recognizing variants of the different communicative whistle sounds produced by familiar dolphin associates (Herman and Tavolga 1980).

In addition to these demonstrations of dolphin vocal imitation, our studies have shown that dolphins are exceptional behavioural imitators of other dolphins and, surprisingly, of humans (Xitco 1988; Herman 2002). In formal tests of imitation, we use a gesture we call 'mimic' that directs the dolphin to imitate the behaviour of a live model, either another dolphin or a human. In the case of a human model, the dolphin must relate its body image to the body plan of the human, and draw analogies where necessary. For example if the human raises a leg in the air, the dolphin raises its tail. If the human waves an arm, the dolphin waves its pectoral fin. These correspondences were not explicitly taught. Opaque correspondences occur as well, in that movements of the human's head, nodding or shaking, are copied by corresponding movements of the dolphin's head. The dolphin, of course, cannot see its own head. Within the formal mimicry context, the dolphins attempt imitation only if the *mimic* sign follows the behavioural demonstration. If some other sign is given, for example, *spiral swim*, the dolphin will respond to that sign and ignore the behaviour of the model. That imitation of novel behaviours has been carried out successfully, and that imitation, or an attempt at it, will occur only in the presence of the mimic gesture, implies an understanding of mimicry as a general concept applicable to any behaviour observed.

As a natural follow-up to the dolphins' earlier success in responding to televised gestural instructions (Herman *et al.* 1990), we explored their ability to imitate television scenes of humans or dolphins performing behaviours. For our first attempt, we filmed a human or dolphin performing a behaviour in one of two interconnected tanks, while another dolphin watched the ongoing behaviours on a television screen located behind an underwater window in the second tank (Herman *et al.* 1993). The television camera was located almost 30 metres distant from the performing models, yielding a rather small image on the television screen, reflecting the limits of the zoom lens we were using. In addition, we did not use the mimic gesture after a behaviour, waiting to see whether imitation would occur spontaneously. Not surprisingly, perhaps, mimicry was not immediate. Imitation did occur, however, once we increased the size of the image (by using a different lens) and, additionally, presented the mimic sign after the behavioural demonstration. The mimic sign was given either by the person demonstrating a behaviour, or by switching to a person on screen giving the mimic sign after a dolphin model completed a behaviour. With these changes, the dolphins (Phoenix and Ake) responded to the televised images in the same way as they did to live models, for example by nodding their head up and down or side to side in imitation of a human doing the same, or by pirouetting when the human or dolphin model pirouetted. The dolphins thus correctly inferred that they could respond to the television models in the same way as they did to live models.

20.4 Self-knowledge: awareness of oneself

Here we asked what a dolphin may know or understand about itself. Traditionally, self-awareness in animals has been tested through the mirror self-recognition (MSR) test. Chimpanzees and at least one gorilla and one orang-utan have passed this test, showing evidence that they recognized the image in the mirror as their own (see various chapters in Parker *et al.* 1994). Recently, Reiss and Marino (2001) demonstrated mirror self-recognition in bottlenosed dolphins. Self-awareness may, however, exist in many forms, not just self-recognition. We tested two additional forms: (a) awareness of one's own behaviours; and (b) awareness of one's own body parts.

20.4.1 Awareness of one's own behaviours

We developed a procedure that required the dolphin Phoenix either to repeat the behaviour she had performed last or choose a different behaviour, depending on whether she observed a gesture we glossed as '*repeat*' or a gesture we glossed as '*any*' (Cutting 1997; also summarized in Herman 2002). For this paradigm, Phoenix was restricted to five different behaviours: jump over an object (*over*), swim under it (*under*), touch it with the tail (*tail-touch*), touch it with the pectoral fin (*pec-touch*), or mouth it (*mouth*). A typical trial began with Phoenix given a gesture requiring her to perform a particular one of the five behaviours to an object in the water, such as *tail-touch* or *under*. After completing the behaviour to the object, Phoenix was given either the sign *repeat* or the sign *any*. The *any* sign instructed her to choose any of the five behaviours other than the one she had just completed. The *repeat* sign directed her to perform the same behaviour again (see also Mercado *et al* 1998, 1999). This procedure was repeated three times and Phoenix was rewarded only if she completed the entire four-item sequence correctly. An example of a sequence given was *Mouth, Repeat, Any,*

Repeat. Here, in succession, Phoenix correctly mouthed the object, mouthed again, leaped over the object, and finally leaped over again.

20.4.1.1 Evidence for rational responding

To successfully complete a sequence, Phoenix had to maintain a mental representation of the behaviour she last performed, update that with each succeeding behaviour, semantically process the *repeat* or *any* gesture, and then either self-select the same behaviour for *repeat,* or choose a behaviour from the remaining four if given *any.* Phoenix was highly successful at this task completing correctly between 79 and 95 per cent of 160 different four-item sequences, depending on the particular sequence type given. An interesting example of apparent inferential reasoning occurred when, for the first time, we began a sequence not with a specific directed behaviour, but with the *any* sign. Here, Phoenix was being asked, in effect, to begin a sequence herself by choosing any of the five behaviours. Phoenix waited in front of the experimenter for approximately 12 seconds, watching and seemingly waiting for a specific instruction. The experimenter remained motionless. Phoenix then turned toward the object and leaped over it *(over),* apparently concluding that she could initiate the sequence on her own. She then successfully completed the remainder of the four-item sequence as successively instructed by the experimenter.

20.4.2 Awareness of one's own body parts: displaying named body parts and using them in unique ways as instructed by symbolic gestures

Children, even at the relatively young age of two, can point to as many as 20 of their different body parts in 'show me' interchanges with a caregiver (MacWhinney *et al.* 1987; Witt *et al.* 1990). These young children can also carry out a variety of simple verbal instructions referring to their body parts (e.g. 'wash your hands,' 'brush your hair,' 'push the ball with your foot,' 'push the ball with your hand'). Clearly, we credit the young child with an understanding that the words we use for body parts refer to or represent those body parts.

Can an animal understand symbols as references to or as representing its own body parts? To conclude that it can, at least three conditions must be met: (a) different body parts must be reliably associated with different symbols (e.g. 'pectoral fin' vs. 'tail' vs. 'rostrum'); (b) different symbolically referenced body parts must be used in the same or in analogous ways (e.g. 'touch the ball with your pectoral fin' vs. 'touch the ball with your tail' vs. 'touch the ball with your rostrum'); and (c) the same symbolically referenced body part must be used in different ways (e.g. 'shake your rostrum' vs. 'touch the Frisbee with your rostrum' vs. 'toss the Frisbee with your rostrum'). In addition, an inference that the animal understands symbols as representations of its body parts is strengthened under Conditions *b* and *c* if the animal is able to use its body parts in novel ways in response to a symbolic reference, especially if it occurs the first time such novel use is requested.

Thus, a dog that offers its paw in response to a verbalization such as 'give me your paw' is not of itself evidence that the words are understood as representing its paw. Instead, it can only be concluded that the dog is making a learned instrumental response to a discriminative stimulus. To decide otherwise would require fulfilling conditions *a* to *c*.

We examined whether the dolphin Elele could develop referential understanding of gestural symbols for her different body parts, by fulfilling conditions *a* to *c* (Herman *et al.* 2001). The ability of Elele to carry out these different body-part tasks reliably would provide strong evidence for her understanding of symbols as representing her body parts.

20.4.2.1 Evidence for rational responding

Elele's responses revealed both semantic and topographical knowledge of her body parts. She not only understood symbolic gestural references to any of nine different body parts (*rostrum, mouth, melon, dorsal fin, pectoral fin, side, belly, genitals,* and *tail*), but also successfully used the same body part to carry out different symbolically referenced actions (e.g. she was able, on instruction, to *display* her rostrum, or *shake* it, or use it to *touch* a basket or *toss* a Frisbee). She was also able to use different body parts to carry out the same action (e.g. touch a basket with her pectoral fin, side, tail, or mouth). Two examples of novel body-part requests completed successfully on the first occasion given were *Frisbee dorsal fin touch* (touch the floating Frisbee with your dorsal fin—a response not in the natural or learned repertoire of Elele) and *surfboard genital toss* (toss the surfboard with your genital region—completed by swimming under the surfboard in an inverted position and then thrusting her pelvic region upwards against the board). These and other novel responses strengthened the conclusion that the gestural symbols we used for those body parts represented those body parts to Elele.

In humans, the identification and conscious control of one's own body parts is dependent on the development and maintenance of a body image (e.g. Gallagher 1995; Sirigu *et al.* 1991). In carrying out novel body-part instructions, such as those illustrated above, Elele, in effect, had to conceive of (ideate) those body-part responses that would satisfy the instruction. The results of these dolphin body-part studies (Herman *et al.* 2001), as well as the findings reviewed earlier, of versatile motor-mimicry capabilities (Herman 2002), testify to the development in the dolphin of a well-articulated body image that can be consciously accessed for effective body-part responses to even highly abstract instructions or observed motor events.

20.5 General discussion

This chapter began with the argument that rational behaviour is a derivative of intelligence manifested in one or more domains, and that a fundamental function of intellect is to allow the animal to construct a model of its world, or of multiple worlds as it encounters new or changed conditions or contexts. To construct a representative model, the animal must perceive and interpret the structural and functional components of the world in which it is currently immersed. Rational behaviour, then, involves responding adaptively to the perceived elements and dynamics of that world, inferring, extrapolating, deducting, concluding, or creating behaviour as necessary. I presented evidence of rational responding by bottlenosed dolphins that met one or more of these characteristics of rationality, as exhibited in our various behavioural studies at the Kewalo Basin Marine Mammal Laboratory. The major points of evidence may be recapped as follows.

20.5.1 Declarative and procedural domains

The question posed about the declarative (semantic or representational) domain of intellect was does the dolphin display knowledge or understanding about things. The question posed about the procedural domain was does the dolphin exhibit competency in means, operations, or methods. The evidence gave strong affirmative answers to both questions, through the results of a variety of experiments and observations. Both questions addressed the issue of the intellectual flexibility of the dolphin in responding to new or changing world conditions.

Ake's inferential and extrapolative abilities were demonstrated within the gestural language system we developed for communicating instructions to her. After exposure to a limited number of exemplars of sentence frames, Ake was able to reliably interpret and act on new instructions given to her when new semantic elements were inserted into the various semantic slots of sentence frames (i.e. object, action, relationship, and locative slots). Further evidence for semantic processing was found in her rejection of semantic anomalies and by her apparent understanding of the intent of an instruction by rearranging objects to enable her to complete an instruction, such as lifting a hoop off the bottom of the pool in order to swim though it. Syntactic processing was convincingly illustrated by her appropriate responding to semantically reversed sequences, by her immediate ability to interpret new and more complex syntactic structures that combined elements of simpler structures, and by her extraction of semantically and syntactically correct subsets of instructions embedded within longer syntactically anomalous sequences. This suite of abilities required both declarative and procedural knowledge—understanding of the referents of the gestures and understanding how to respond to them appropriately.

Ake created an ostensibly efficient strategy for returning all floating objects to the experimenter after being given the generic 'fetch' command. Her strategy was to gather multiple objects together and return the group to the experimenter, rather than retrieving one at a time. The strategy was not always time efficient, however, as multiple objects might be gathered together, some then escaping and requiring regathering perhaps several times, but it did minimize the number of discrete swimming trips required and seemed to us, admittedly anthropomorphically, to be more interesting and enjoyable to the dolphin than fetching single objects.

Ake's understanding of the referents of the gestures we used for the various semantic entities is illustrated by some of the examples cited above. More stringent evidence was obtained through her accurate responses in experiments that specifically asked her to report, through pressing one of two paddles, whether a 'named' object was present or not in her pool. Within these experiments, she created a new type of response that was unanticipated by us when she was asked, for the first time, to take an object that was present to an object that was absent. Here, she transported the object that was present to the paddle signifying 'absent.' Further, and rationally, if both objects were absent, or if the object to be transported was absent, she simply pressed the 'absent' paddle immediately.

Ake and Phoenix, on the first occasion that each was exposed to a TV image of a person signing to them, immediately carried out the person's gestural instructions. It was as if each dolphin inferred that the television scenes were representations of the real world, despite radical differences in image size, clarity, dimensionality, and context, and that she could therefore respond in the same manner as she did to real-world scenes. Further, Elele inferred immediately that objects shown on a television screen could be matched with objects displayed in the real world.

All four dolphins understand a concept of 'combinations' when given a sequence of gestures, each calling for a different action. Rather than selecting and producing one of the actions only, or carrying out several actions successively, the dolphins construct a response that integrates the multiple behaviours into a single act.

Creativity was formally demonstrated in studies that asked the dolphins, through a specific gesture we called 'create,' to vary their behaviour in any way, as long as each successive behaviour was different from the immediately preceding behaviour. Elele demonstrated impressive behavioural variability, exhibiting 323 behaviours, 72 of which were unique, during 144 trials, each asking her to create her own behaviour. Elele only offered self-initiated behaviours if given the 'create' gesture. Clearly then, Elele understood 'create' at the level of a concept in that she only offered variable behaviours if given the create gesture.

These various results illustrate declarative and procedural knowledge in several different contexts, and provide evidence of inferential reasoning and innovative responding.

20.5.2 The social domain

Here, the question asked was does the dolphin reveal social awareness and appropriate responsiveness in social interactions or relations. Again, for answers, I referred to the results of studies at our laboratory and, again, the answers were in the affirmative.

Dolphins and dogs, but not chimpanzees or other great apes, understand the referring function of the human pointing gesture. Ake and Phoenix both understood and acted on direct points to objects as a reference to that object. Furthermore, both dolphins, on their first exposure to cross-body pointing, responded appropriately, in essence inferring that the direction of the arm's extension, rather than which arm it was, called attention to an object. In addition, Ake, when presented with a sequence of points immediately understood that she could respond in the same manner as she did to sequences of symbolic gestures, by applying the same inverse grammatical rule. Both Ake and Phoenix also immediately understood the referring function of human-directed gaze. The complexity of the dolphins' responses and interpretations of human pointing or gazing at objects, and that none of these responses were specifically taught, signal that they have developed a profound appreciation for human attentional signals in general, and the value of sharing in that attention.

Pairs of dolphins (e.g. Ake–Phoenix or Hiapo–Elele) understood the concept of acting together synchronously in response to a particular gestural instruction to do so. In addition to carrying out specifically requested behaviours in close synchrony, they were proficient at choosing their own behaviours to carry out together in response to the gestural sequence *tandem create*. The mechanism by which the dolphins achieve this co-ordinated and inventive response is not clear. Even if it involves real-time mimicry or exceptional sensitivity to the intention movements of the other, rather than some more abstract form of communication, it still demands an understanding of the requirements of a highly abstract task, the close monitoring and anticipation of each other's movements, and the maintaining of an inventory of most recently performed behaviours in order to satisfy the requirement for variability in response.

Dolphins are clearly the most facile vocal and motor imitators of any non-human mammal. In addition to vocal imitation of arbitrary computer generated sounds, Ake also demonstrated octave generalization, recognizing sound sequences ('tunes') that were transposed by an octave and producing sounds an octave removed from a model sound that was outside of her tonal range. Fundamentally, octave generalization of tunes further illustrates sensitivity to sequential structures, which was illustrated previously by Ake's sensitivity to the syntactic structure of her gestural language. All four dolphins demonstrated not only an ability to copy observed motor behaviours of other dolphins, but also of humans. Imitation of human motor behaviour required forming structural relationships between the dolphin's perception of the human body plan and its own body image. Analogies were formed, such as the correspondence of the human leg and the dolphin's tail. The dolphins' understanding of the television world was also forcefully illustrated by the effective copying of motor behaviours of dolphins or humans observed on a television screen.

These various findings illustrate the dolphins' acute social awareness, including sensitivity to the attentional mechanisms of others and to the nuances of their behaviours, sufficient to create replicas of those behaviours.

20.5.3 The domain of the self

The question asked was what may a dolphin know or understand about itself. Recognizing that self-awareness is a multidimensioned concept, we chose two facets of self-awareness for study that appear not to have been formally studied previously in any animal: awareness of one's own behaviours and awareness of one's own body parts.

Phoenix demonstrated conscious awareness of her own behaviours, as illustrated by her ability to repeat her previous behaviour or to perform a different behaviour, contingent on which one of two abstract gestural instructions was given, 'repeat' or 'any' (= do not repeat). She chose behaviours from a set of five. She was able to reliably self-select each successive behaviour in sequences of four gestural instructions. The first instruction required a specific behaviour, and the remaining three each asked her to repeat or not repeat her prior act. She faithfully followed the rule embedded within each instruction.

Elele demonstrated conscious awareness and conscious control of her own body parts, as illustrated by her understanding of symbolic gestural references to those body parts and by her ability to use those body parts in four different ways as directed by gestural instructions. Some instructions called for novel (first-time) uses of a body part, and were successfully carried out by Elele.

These results illustrate aspects of self-awareness that address the multidimensionality of this concept whose study has previously been limited primarily to mirror self-recognition. Recent findings of mirror self-recognition by dolphins (Reiss and Marino 2001), together with the findings presented here, reveal what appears to be a self-concept in this species.

20.5.4 Evolutionary perspectives

Finally, what pressures might select for the evolution of the levels of intellectual flexibility and the type of innovative and rational responding demonstrated? The often-heard argument that the exceptional size and development of the dolphin brain is accounted for by the requirement for extensive auditory processing, especially echolocation, can be dismissed on at least two grounds. The first is based on comparisons with insectivorous bats, the group of bats feeding on flying insects and having the most sophisticated echolocation system among all bats. Examination of the brain and body weights of 20 species of insectivorous bats listed by Hutcheon *et al.* (2002) reveals that all 20 have EQs falling below 1.0. These bats do have relatively larger auditory processing areas (cochlear nuclei complex and superior olivary complex) than do fruit eating and nectar sipping bats, which do echolocate but rely more on vision and olfaction. Thus, using insectivorous bats as a model of a sophisticated echolocating group, it is clear that echolocation and attendant auditory capabilities can be supported through specialized areas without overall enlargement of the brain or extensive enlargement of the cerebral cortex. In contrast, the dolphin brain has obviously expanded greatly in many areas, especially the cerebral cortex, the site implicated in much of human cognitive functioning.

Second, the question of the intelligence of the dolphin, as I stated earlier, can best be resolved through the results of empirical behavioural study rather than through examination of brain structures. In our studies, those reviewed here as well as others (see e.g. Herman *et al.* 1993), the diversity, depth, and breadth of cognitive skills demonstrated by the dolphins revealed exceptional behavioural flexibility in several different domains of intellectual functioning: declarative, procedural, social, and self. These domains encompass aspects of the physical world (knowing what things are and how they function or may be utilized), the social world ('reading' and interpreting the behaviours of others, and engaging in beneficial activities with others), and the world of self (being consciously aware of one's own physical being and one's own actions).

In many of the cases described in this chapter, the responses of the dolphins to the challenges posed appeared to require logical inferences and innovative responding, as well as adaptive responding to new or changed 'world' conditions. These capabilities seem reflective of the exceptional adaptability of the bottlenosed dolphin in its natural world. This species exhibits remarkable diversity in its natural habitats, including all of the oceans within the temperate and tropical zones, as well as inshore and offshore ranges within those habitats, and migration between the two ranges. The different habitats may each demand unique foraging, feeding, antipredator, and social strategies, suggestive of an ability of bottlenosed dolphins to adapt behaviourally to diverse worlds or to changes in world conditions, including movement from the natural world to the world of the laboratory or marine park.

Although it is difficult to cleanly isolate a single root cause for the evolution of intelligence in a species, in the case of the dolphin the major determining factor may be social pressure—the requirement for integration into a social order having an extensive communication matrix for promoting the well-being and survival of individuals. I first advocated this idea in my early discussion of the cognitive characteristics of dolphins (Herman 1980), inspired in part by Humphrey's (1976) treatise on the social basis of intellect. The idea finds much support today, as expressed, for example, in many of the chapters within the earlier and recent volumes on 'Machiavellian intelligence' (Byrne and Whiten 1988; Whiten and Byrne 1997; also see e.g. Tomasello and Call 1997). Dunbar (1992) has shown that within primate societies the ratio of neocortex to the remainder of the brain increases with increases in typical group size. Within large, complex societies, individuals benefit from assessing the physical and behavioural characteristics of others in the community, and from recruiting others into alliances or collaborative activities. Within bottlenosed dolphin societies, recognition of others through unique individual signature whistles has been reported as well as mimicry of whistles of others, as an apparent affinitive act (Janik 1997; Tyack 1999). Collaborative foraging, feeding, and predator defence occurs regularly (Reynolds *et al.* 2000) and the formation of complex alliances among male dolphins has been noted among the dolphin population of Shark Bay, Australia (Conner *et al.* 1992).[3]

---

[3] Editors' note: see also Connor and Mann, this volume.

Ecological pressures may additionally select for intelligence. In particular, the advantage of learning, storing, and utilizing information about such things as prey identification and habits, feeding sites, feeding strategies, predator characteristics and defences, topographical features of the habitat (cognitive mapping), and environmental cycles can select for the type of intelligence that supports declarative and procedural knowledge and skills. Both ecological and social pressures may act to select for intelligence, not necessarily exclusively or independently but more likely in parallel or in some sequential steps. Tomasello and Call (1997) concluded that, for primates, both social and ecological pressures select for intelligence, but each influences the development of different types of cognitive skills, possibly at different periods in the evolution of the species. This analysis seems compelling for dolphins as well, particularly given the convergence of dolphins and apes in so many of their cognitive attributes and skills (see e.g. Herman and Morrel-Samuels 1990).

Acknowledgments

None of this work would have been possible without the help, insights, and creativity of the many students and colleagues who worked together with me on these many projects. Foremost, I am grateful for the opportunity to have interacted for so long with four such amazing animals. I dedicate this chapter to them—all four are now gone: Elele passed away on December 16, 2000, Akeakamai on November 2, 2003, Phoenix on January 10, 2004, and Hiapo on February 24, 2004. Their accomplishments are legend and are preserved in many of the citations in this chapter.

References

Anderson, J. R. (1978). *Language, Memory, and Thought*. Hillsdale, NJ: Erlbaum.

Au, W. W. L. (1993). *The Sonar of Dolphins*. New York: Springer-Verlag.

Busnel, R-G., and Fish, J.F., eds. (1980). *Animal sonar systems*. New York: Plenum Press.

Braslau-Schneck, S. (1994). *Innovative Behaviors and Synchronization in Bottlenosed Dolphins*. Unpublished master's thesis, University of Hawaii, Honolulu.

Byrne, R. W., and Whiten, A., eds (1988). *Machiavellian Intelligence: Social Expertise and the Evolution of Intellect in Monkeys, Apes and Humans*. New York: Oxford University Press.

Call, J., Hare, B., and Tomasello, M. (1998). Chimpanzee gaze following in an object choice task. *Animal Cognition*, 1: 89–100.

Connor, R. C., Smolker, R. A. and Richards, A. F. (1992). Two levels of alliance fromation among bottlenose dolphins (*Tursiops* sp.) *Proceedings of the National Academy of Sciences USA*, 89: 987–990.

Connor, R. C., Wells, R. S., Mann, J., and Read, A. J. (2000). The bottlenose dolphin: social relationships in a fission-fusion society. In: J. Mann, R. C. Connor, P. L. Tyack, and H. Whitehead, eds. *Cetacean Societies: Field Studies of Dolphins and Whales*, pp. 91–126. Chicago: University of Chicago Press.

Cutting, A. E. (1997). *Memory for Self-Selected Behaviors in a Bottlenosed Dolphin (Tursiops truncatus)*. Unpublished master's Thesis, University of Hawaii, Honolulu.

D'Amato, M. R., and Salmon, D. P. (1984). Processing of complex auditory stimuli (tunes) by rats and monkeys (*Cebus apella*). *Animal Learning and Behavior*, 12: 184–194.

Dowling, C. G. (2002). Incident at Big Pine Key (the dolphin wars). *Smithsonian*, **33**: 45–51.

Dunbar, R. I. M. (1992). Neocortex size as a constraint on group size in primates. *Journal of Human Evolution*, **20**: 469–493.

Gallagher, S. (1995). Body schema and intentionality. In: J. L. Bermúdez, A. Marcel, and N. Eilan, eds. *The body and the Self*, pp. 225–244. Cambridge, MA: MIT Press.

Hare, B., Brown, M., Williamson, C., and Tomasello, M. (2002). The domestication of social cognition in dogs. *Science,* **298**: 1634–1636.

Hare, B., Call, J., and Tomasello, M. (1998). Communication of food location between human and dog (*Canis familiaris*). *Evolution of Communication,* **2**: 137–159.

Herman, L. M. (1980). Cognitive characteristics of dolphins. In: L. M. Herman, ed. *Cetacean Behavior: Mechanisms and Functions*, pp. 363–429. New York: Wiley Interscience.

Herman, L. M. (1986). Cognition and language competencies of bottlenosed dolphins. In: R. J. Schusterman, J. Thomas, and F. G. Wood, eds. *Dolphin Cognition and Behavior: A Comparative Approach*, pp. 221–251. Hillsdale, NJ: Lawrence Erlbaum Associates.

Herman, L. M. (2002). Vocal, social, and self-imitation by bottlenosed dolphins. In: C. Nehaniv and K. Dautenhahn, eds. *Imitation in Animals and Artifacts*, pp. 63–108. Cambridge, MA: MIT Press.

Herman, L. M., Abichandani, S. L., Elhajj, A. N., Herman, E. Y. K., Sanchez, J. L., and Pack, A. A. (1999). Dolphins (*Tursiops truncatus*) comprehend the referential character of the human pointing gesture. *Journal of Comparative Psychology*, **113**: 1–18.

Herman, L. M., and Forestell, P. H. (1985). Reporting presence or absence of named objects by a language-trained dolphin. *Neuroscience and Biobehavioral Reviews*, **9**: 667–691.

Herman, L.M., Kuczaj, S. III, and Holder, M. D. (1993). Responses to anomalous gestural sequences by a language-trained dolphin: Evidence for processing of semantic relations and syntactic information. *Journal of Experimental Psychology: General,* **122**: 184–194.

Herman, L., Matus, D., Herman, E. Y. K., Ivancic, M., and Pack, A. A. (2001). The bottlenosed dolphin's (*Tursiops truncatus*) understanding of gestures as symbolic representations of its body parts. *Animal Learning and Behavior*, **29**: 250–264.

Herman, L. M., and Morrel-Samuels, P. (1990). Knowledge acquisition and asymmetries between language comprehension and production: Dolphins and apes as a general model for animals. In: M. Bekoff and D. Jamieson, eds. *Interpretation and Explanation in the Study of Behavior, Vol. 1, Interpretation, Intentionality, and Communication*, pp. 283–312. Boulder, CO: Westview Press.

Herman, L. M., Morrel-Samuels, P., and Pack, A. A. (1990). Bottlenosed dolphin and human recognition of veridical and degraded video displays of an artificial gestural language. *Journal of Experimental Psychology: General*, **119**: 215–230.

Herman, L. M., and Pack, A. A. (1994). Animal intelligence: Historical perspectives and contemporary approaches. In: R. Sternberg, ed. *Encyclopedia of Human Intelligence*, pp. 86–96. New York: Macmillan.

Herman, L. M., Pack, A. A., and Hoffmann-Kuhnt, M. (1998). Seeing through sound: Dolphins perceive the spatial structure of objects through echolocation. *Journal of Comparative Psychology*, **112**: 292–305.

Herman, L. M., Pack A. A., and Morrel-Samuels, P. (1993). Representational and conceptual skills of dolphins. In: H. R. Roitblat, L. M. Herman, and P. Nachtigall, eds. *Language and Communication: Comparative Perspectives*, pp. 273–298. Hillside, NJ: Lawrence Erlbaum.

Herman, L. M., Richards, D. G., and Wolz, J. P. (1984). Comprehension of sentences by bottlenosed dolphins. *Cognition,* **16**: 129–219.

Herman, L. M., and Tavolga, W. N. (1980). The communication systems of cetaceans. In: L. M. Herman, ed. *Cetacean Behavior: Mechanisms and Functions*, pp. 149–209. New York: Wiley Interscience.

Hulse, S. H. (1989). Comparative psychology and pitch pattern perception in songbirds. In: R. J. Dooling, and S. H. Hulse, eds. *The Comparative Psychology of Audition: Perceiving Complex Sounds*, pp. 331–352. Hillsdale: NJ: Lawrence Erlbaum Associates.

Hulse, S. H., and Cynx, J. (1985). Relative pitch perception is constrained by absolute pitch in songbirds (*Mimus, Molothrus, Sturnus). Journal of Comparative Psychology*, **99**: 176–196.

Humphrey, N. K. (1976). The social function of intellect. In: P. P. G. Bateson and R. A. Hinde, eds. *Growing points in ethology*. Cambridge: Cambridge University Press.

Hutcheon, J. M., Kirsch, J. A. W., and Garland Jr., T. (2002). A comparative analysis of brain size in relation to foraging ecology and phylogeny in the chiroptera. *Brain Behavior and Evolution*, **60**: 165–180.

Janik, V. M. (1997). Whistle matching in wild bottlenose dolphins. *Journal of the Acoustical Society of America*, **101**: 31–36.

Jerison, H. J. (1973). *Evolution of the Brain and Intelligence*. New York: Academic Press.

Kruger, L. (1966). Specialized features of the dolphin brain. In: K. S. Norris, ed. W*hales, Dolphins, and Porpoises*, pp. 232–254. Los Angeles: University of California Press.

Lashley, K. S. (1949). Persistent problems in the evolution of mind. *Quarterly Review of Biology*, **24**: 28–42.

Lilly, J. C. (1961). *Man and Dolphin.* New York: Doubleday and Co.

MacWhinney, K., Cermak, S. A., and Fisher, A. (1987). Body part identification in 1- to 4-year-old children. *American Journal of Occupational Therapy*, **41**: 454–459.

Marino, L. (1998). A comparison of encephalization between odontocete cetaceans and anthropoid primates. *Brain, Behavior and Evolution,* **51**: 230–238.

Marino, L. (2002). Brain size evolution. In: W. F. Perrin, B. Wursig and J. G. M. Thewissen, eds. E*ncyclopedia of Marine Mammals*, pp. 158–162. New York: Academic Press.

Mercado, E. III, Murray, S. O., Uyeyama, R. K., Pack, A. A., and Herman, L. M. (1998). Memory for recent actions in the bottlenosed dolphin (*Tursiops truncatus*): Repetition of arbitrary behaviors using an abstract rule. *Animal Learning and Behavior,* **26**: 210–218.

Mercado, E. III, Uyeyama R. K., Pack, A. A., and Herman, L. M. (1999). Memory for action events in the bottlenosed dolphin. *Animal Cognition,* **2**:17–25.

Pabst, D. A., Rommel, S. A., and McMellan, W. A. (1999). The functional morphology of marine mammals. In: J. E. Reynolds III, and S. A. Rommel, eds. *Biology of Marine Mammals*, pp. 15–72. Washington DC: Smithsonian Institution Press.

Pack, A. A., and Herman, L. M. (1995). Sensory integration in the bottlenosed dolphin: immediate recognition of complex shapes across the senses of echolocation and vision. *Journal of the Acoustical Society of America*, **98**: 722–733.

Pack, A. A., and Herman L. M. (2004). Dolphins (*Tursiops truncatus)* understand the referent of both static and dynamic human gazing and pointing in the object choice task. *Journal of Comparative Psychology*, **118**:160–171.

Parker, S. T., Mitchell, R. W., and Boccia, M. L., eds. (1994). *Self-awareness in Animals and Humans: Developmental Perspectives.* Cambridge: Cambridge University Press.

Povinelli, D .J., Reaux, J. E., Bierschwale, D. T., Allain, A. D., and Simon, B. B. (1997). Exploitation of pointing as a referential gesture in young children, but not adolescent chimpanzees. *Cognitive Development*, **12**: 423–461.

Ralston, J. V., and Herman, L. M. (1995). Perception and generalization of frequency contours by a bottlenose dolphin (*Tursiops truncatus). Journal of Comparative Psychology,* **109**: 268–277.

Reiss, D., and Marino, L. (2001). Mirror self-recognition in the bottlenose dolphin: A case of cognitive convergence. *Proceedings of the National Academy of Science,* **98**: 5937–5942.

Reynolds, J. E., Wells, R. S., and Eide, S. D. (2000). *The Bottlenose Dolphin: Biology and Conservation.* Gainesville: University Press of Florida.

Richards, D. G., Wolz, J. P., and Herman, L. M. (1984). Vocal mimicry of computer generated sounds and vocal labeling of objects by a bottlenosed dolphin, *Tursiops truncatus. Journal of Comparative Psychology,* **98**: 10–28.

Ridgway, S. H. (1990). The central nervous system of the bottlenose dolphin. In: S. Leatherwood and R. R. Reeves, eds. *The bottlenose Dolphin*, pp. 69–97. New York: Academic Press.

Savage-Rumbaugh, E. S. (1986). *Ape Language: From Conditioned Response to Symbol.* New York: Columbia University Press.

Sirigu, A., Grafman, J., Bressler, K., and Sunderland, T. (1991). Multiple representations contribute to body knowledge processing: Evidence from a case of autotopagnosia. *Brain,* **114**: 629–642.

Soproni, K., Miklosi, A., Topal, J., and Csanyi, V. (2001). Comprehension of human communicative signs in pet dogs (*Canis familiaris*). *Journal of Comparative Psychology,* **115**: 122–126.

Tomasello, M., and Call, J. (1997). *Primate Cognition.* New York: Oxford University Press.

Tschudin, A., Call, J., Dunbar, R. I. M., Harris, G., and van der Elst, C. (2001). Comprehension of signs by dolphins (*Tursiops truncatus*). *Journal of Comparative Psychology,* **115**: 100–105.

Tyack, P. L. (1999). Communication and cognition. In: J. E. Reynolds III and S. A. Rommel, eds. *Biology of Marine Mammals*, pp. 287–323. Washington, DC: Smithsonian Institution Press.

Whiten, A., and Byrne, R. W., eds. (1997). *Machiavellian Intelligence II: Extensions and Evaluations.* Cambridge: Cambridge University Press.

Whiten, A., and Ham, R., (1992). On the nature and evolution of imitation in the animal kingdom: Reappraisal of a century of research. *Advances in the Study of Behavior,* **21**: 239–283.

Witt, A., Cermak, S., and Coster, W. (1990). Body part identification in 1- to 2-year-old children. *American Journal of Occupational Therapy,* **44**: 147–153.

Wood, F. G. (1973). *Marine Mammals and Man: The Navy's Porpoises and Seas Lions.* Washington, DC: Luce.

Xitco, M. J. Jr. (1988). *Mimicry of Modeled Behaviors by Bottlenose Dolphins.* Unpublished master's thesis, University of Hawaii, Honolulu.

Xitco, M. J. Jr, and Roitblat, H. L. (1996). Object recognition through eavesdropping: Passive echolocation in bottlenose dolphins. *Animal Learning and Behavior,* **24**: 355–365.

Zentall, T. R. (1996). An analysis of imitative learning in animals. In: C. M. Heyes and B. G. Galef Jr, eds. *Social Learning in Animals: The Roots of Culture*, pp. 221–243. New York: Academic Press.

Chapter 21

Intelligence and rationality in parrots

Irene M. Pepperberg

Abstract

Studies both in the field and the laboratory demonstrate that the capacities of non-human animals to solve complex problems form a continuum with those of humans. Such measures of intelligence often imply the ability to choose the solution that human beings facing the same task would rationally choose. However, animals that are deemed intelligent by human standards may not always be deemed rational by these same human standards. Conversely, sometimes they display elements of rationality that go beyond what might be expected simply on the basis of documented cognitive ability. This chapter examines several such divergences that arise in the study of Grey parrots.

21.1 Intelligence vs. rationality: convergences and divergences

For over 35 years, beginning with the so-called 'cognitive revolution' (Hulse *et al.* 1968), researchers have been demonstrating, through tests both in the field and in the laboratory, that the capacities of non-human animals to solve complex problems form a continuum with those of humans (see Balda *et al.*1998; Shettleworth 1998). When studying animals' abilities, however, researchers sometimes find, in subjects that otherwise appear quite intelligent, behavior that doesn't seem rational—at least by standards usually applied to human behavior. Clearly, 'intelligence' and 'rationality' are not entirely interchangeable. We should thus ask what the relationship is between intelligence and rationality in non-human animals, and whether rationality can be examined through studies of intelligence.

Although 'intelligence' as applied to animals may suggest sophisticated forms of cognitive processing (something like Kacelnik's 'B-rationality', Chapter 2, this volume), which may approach or even surpass comparable human cognitive capacities,[1]

---

[1] For example, animals' sometimes superior perceptual capacities or memory may enable them to perform more accurately or quickly than humans on a given task, such as memory for cached food (see Balda *et al.* 1998).

nevertheless animal intelligence does not always lead to patterns of behavior that would be regarded as fully rational in a human being. Generally, this divergence occurs because human rationality is based not only on cognitive processes, but also on human social, psychological, economic, moral, and any other human-centered values, some or all of which may differ for any given non-human species (and, of course, may sometimes differ for different human cultures). Sometimes an animal may respond differently from human beings on a task because of a different evolutionary or social history; behavior patterns different from those that human beings would exhibit in such a situation may have increased fitness or been reinforced in the past. Thus an animal may react in a way that, by standards normally applied to human behavior, seems irrational. The inability of Boysen's chimpanzees (*Pan troglodytes*) to inhibit their choice of a larger array of candy, even after learning that the candy array they choose will always go to their partner and not to themselves, is such an example (see Boysen, Chapter 22, this volume; Boysen and Berntson 1995). How then can we judge animal rationality? How can we devise experiments and collect data without relying on human standards that may be inappropriate?

This process is simplified when the intelligent or cognitively sophisticated response for the animal coincides with behavior that would be regarded as rational for human beings. When this is the case, researchers can base studies of animal intelligence—and expectations of rationality—on criteria of human rationality. Although many objections exist to evaluating animal cognition using human tasks (see Pepperberg 2001a), when animals' tests are fully comparable to those given to humans, with respect to both ecological validity and experimental control, then animal and human abilities can indeed be compared on this basis (Pepperberg, 2005a). Intelligence, studied in this manner, usually implies the ability to choose the solution that would be regarded as the rational choice for human beings facing a comparable task. Nevertheless, we should keep the question in mind: is this solution also rational for the animal?

In many instances, this is indeed the case. For example, countersinging marsh wrens (*Cistothorus palustris*) learn not only their own repertoire of songs and song order but also their neighbors' approximately 200-song repertoires and their song orders, so as to choose the *particular* song in each neighbor's repertoire that can be used to jam the neighbor's order and dominate that neighbor (Kroodsma 1979; Kroodsma and Byers 1998). Another example: female great tits (*Parus major*) appear to decide whether to enter a male neighbor's territory based on eavesdropping upon experimentally manipulated interactions between a stranger and her mate and between the same stranger and said male neighbor. Thus, she apparently makes her decision by inferring the ranking of the two resident males based on their respective abilities in dealing with the same intruder; she is much more likely to enter the territory of the neighboring male if he can be inferred to be dominant to her mate (Otter *et al.* 1999). Her behavior is thus rational in the sense that it conforms to the requirements of transitive inference, a complex and rational cognitive process (Pepperberg, 2005a).[2] Interestingly, the information thus derived might even counter what she knows about previous interactions between her mate and her neighbor. Standards of biological intelligence and of human rationality converge in application to such behavior patterns by wrens and great tits.

Intelligence also often converges on human standards of rationality in the Grey parrots (*Psittacus erithacus*) that I study (Pepperberg 1999). The cognitive capacities that my research documents have been surveyed elsewhere (Pepperberg 1999; Pepperberg *et al.* 1995). Very briefly, my oldest subject, Alex, labels more than 50 different objects, seven colors, five shapes, quantities to six, and three categories (material/color/shape); he uses '*no*', '*come here*', '*wanna go X*', '*want Y*' ('X', 'Y' are location or item labels) appropriately. He combines labels to identify, classify, request, or refuse approximately 100 items and to alter his environment, and he comprehends these labels. He processes queries to judge category, relative size, quantity, presence or absence of similarity and difference in attributes, and he responds correctly to recursive, conjunctive questions. He, like some other 'language-trained' subjects, can use symbols dispassionately, that is, can separate identification of an object from a request for that item (Pepperberg 1988; i.e. he separates illocutionary force from propositional content).[3] Moreover, he requests absent objects or an action not currently being performed (i.e. he demonstrates elements of 'displacement', according to Hockett, 1959), and will accept that object or action and no other. He uses labels to refer to known aspects of similar but non-identical items, for example, to identify the material ('wood') of a piece of wood of a novel color or shape without additional training. Similarly, he understands that the label 'green' expresses the concept of greenness, which applies to beans as well as to training objects, and that the arbitrary label 'green' is subsumed into a category whose arbitrary label is 'color' (i.e. he understands that labels can be grouped into hierarchies based on human coding; Pepperberg 1996). He also adds new categories and concepts to his repertoire fairly easily. He and other Greys understand object permanence (i.e. that a hidden item does not cease to exist), and are able to track complex movements of hidden items. Grey parrots can also use mirrors to find hidden objects. Such abilities were once presumed to be limited to humans and apes (Premack 1978), but Alex is not unique: other Greys are replicating some of his results (Pepperberg 1999). His responses to the tasks described here would be regarded as rational responses by a human child to similar tasks.

However, the behavior patterns of Alex and at least two of our juveniles, Griffin and Arthur (aka Wart), sometimes exhibit characteristics that, although clearly intelligent,

---

[2] Editors' note: see also and compare Allen's discussion, this volume, of behavior that suggests transitive inference, and alternative interpretations.

[3] The content is 'this is a green grape' and the force comes in the form of a demand 'I *want* green grape.'

cannot be understood merely as advanced or complex forms of cognitive processing, and may or may not be rational by human standards. Two aspects of such behavior are notable: first, the birds' learning of labels outside of the standard training procedure, where their behavior may suggest more human-like rationality than we might expect if we based intelligence only on cognitive processing ability, and, second, the mistakes they make on certain types of tasks, where their behavior may at first appear to fall short of human standards of rationality despite their intelligence. In the remainder of this chapter I review these two categories of behavior for the light they may shed on the relationships between intelligence and rationality.

21.2 Label acquisition

To appreciate how my birds learn outside of standard training, one must first understand the standard procedure. My model/rival training system (Pepperberg 1981) is built upon studies by Todt (1975) and Bandura (1971) of how social modeling affects learning by, respectively, parrots and human beings. Model/rival training in my laboratory uses three-way *social* interactions with two human beings and a parrot to demonstrate a targeted vocal behavior. Typically, a parrot observes two human beings talking about one or more objects in which it has already shown an interest: the trainer presents the object or array of objects and queries the other human being about it (e.g. 'What's here?', 'What color?'). Responses are given in English, and the trainer rewards correct answers *referentially*, by giving the other human the object(s) as well as praise. Incorrect responses by the human being, similar to those a bird might make, are punished by scolding and by temporarily removing the object(s) from sight. Thus the second human being is a model for the parrot's responses and its rival for the trainer's attention. This human model/rival also illustrates effects of an error, and s/he tries again or articulates the required words more clearly after a (deliberately) incorrect or garbled response, thereby demonstrating corrective feedback and the reason for learning the specific sounds of the label. The bird is included in interactions, being asked about the object(s) in turn, and is rewarded for its successive approximations to the correct response in English speech; training is thereby adjusted to its level. An initial response by the bird might, for example, be 'ay-er' for a large piece of paper; /p/ is particularly difficult for a parrot, lacking lips, to produce. Sometimes the bird spontaneously answers before the model in an attempt to obtain the object.

Unlike Todt's (and other researchers') modeling procedures (see Pepperberg and Sherman 2000), our procedure also *reverses* roles of human trainer and model and includes the parrot, in order to emphasize that one being is not always the questioner and the other the respondent, and that the procedure causes environmental change. Role reversal also counteracts a feature of Todt's work: his birds, whose trainers maintained their respective roles, responded only to the particular human being posing questions; our birds, however, respond to, interact with, and learn from all trainers.

Model/rival training uses only *intrinsic reinforcers*: the reward for uttering 'X' is the object X, in order to ensure closest possible correlations between labels or concepts to be learned and their referents. Earlier unsuccessful programs for teaching birds to communicate with humans used *extrinsic* rewards (e.g. Mowrer 1950): on the few occasions when those subjects correctly labeled any items, or responded appropriately to specific commands, they received the same favored food, which neither related to nor varied with the labels or concepts being taught, thereby delaying acquisition by confounding the targeted label or concept with that of the food (Greenfield 1978; Miles 1983; Pepperberg 1981; for comparable data on intervention programs for children with various dysfunctions, see Pepperberg and Sherman 2000, forthcoming).[4] Furthermore, *de facto* use of labels as requests demonstrates *functionality*.

The positive results of model/rival training showed that our parrots, like young children (e.g. Hollich *et al.* 2000), learn most easily when input is referential, clearly demonstrates the functionality of the material to be learned, and emphasizes modeling and social interaction. But the results described so far do not show whether these input elements, either individually or together, were *necessary and sufficient* for allospecific acquisition (i.e. learning from different species). What if training lacked some of these elements? Answering that question required parrots uninfluenced by prior experience; Alex might cease learning because training changed, not because of how it changed. Thus I added the juveniles Kyaaro, Alo, and Griffin to test the importance of reference, context/function, modeling, and social interaction. In brief, by designing and conducting a large number of experiments that carefully eliminated, separately and in combination, various elements of reference, functionality, social interaction, and modeling, my students and I learned that *all* these elements were crucial to establishing referential use of English labels by Grey parrots (reviewed in Pepperberg 2002).

However, we also learned that model/rival training was not the entire story. First, we found that our parrots' initial learning of labels, although not stimulus-bound, was still slow and difficult (like that of children; see Hollich *et al.*, 2000), while their later label acquisition was much faster and involved interesting types of transfer and concept formation (also the case for children). Second, we found that considerable learning was occurring outside of sessions, and that some of this learning was initiated by the birds, much like children playing the 'naming game' (Brown 1973). The point is that we observed a transition from learning that is slow and difficult to learning that is relatively fast, involves complex processing, and is self-initiated. Thus this transition, although not as impressive in parrots as in human beings (e.g. McCarthy 1954; but see P. Bloom 2000), is not limited to human beings. How this transition occurs is not well understood even for human beings (e.g. Golinkoff *et al.* 2000), nor is the role of rationality therein, but this issue can now also be explored with animal subjects.

....

[4] Editors' note: for comparisons on this point, see Savage-Rumbaugh *et al*, this volume.

21.2.1 **Parallels with initial learning by children**

For both our parrots and for children, first labels are qualitatively different from later labels. First labels may be acoustically biased by and based on prenatal (DeCasper and Spence 1986; Querleu *et al*. 1981) or prehatching exposure to sounds (e.g. Gottlieb 1982). First labels also have a clear, probably evolutionarily based, predisposition to refer to whole objects in a general way rather than to particular aspects of objects (Macnamara 1982; Markman and Wachtel 1988; Pepperberg and Wilcox 2000). For example, for birds and monkeys a 'hawk' alarm call may not initially refer to that specific predator or even to some aspect of a bird, but simply to any big object overhead with a certain general shape (Cheney and Seyfarth 1990). First labels are often mimetic, indexical in that they refer to a specific item rather than a class, and may lack true meaning and communicative intent (de Villiers and de Villiers 1979). As an initial approach to learning, however, use of such labels is biologically 'rational': 'better safe (to react to something that *could* eat me, like a hawk) than sorry'.

Nevertheless, at least for humans—and I suspect for my parrots—even first label use may express some relatively high level of rationality; these labels' meanings are not based exclusively on low-level, simple one-to-one associations. If that were the case, children could be as easily trained to use tones rather than labels (Colunga and Smith 2000), or to ignore the whole-object concept of early labeling and learn to focus exclusively on attributes—and they can't (Macnamara 1982; Markman and Wachtel 1988). Moreover, my birds easily transfer labels to related objects; for example, without training, they used 'paper' to refer to an old-fashioned huge sheet of computer output as well as to the piece of index card used as the original exemplar. Even so, some of my birds' earliest utterances lacked full reference; thus, for example, one bird began to say 'Hello' each time any phone rang. Their use of early labels obviously would not count as rational for an *adult* human being, but some aspects of rational label usage may be beginning to emerge.

What seems to be lacking in the functioning of first labels is the use of them as representations. That is, first labels may: (a) refer only to something immediately present; or (b) incompletely encode all the elements of the items to which they refer, such that label use may sometimes be overextended to irrelevant objects or underextended to only one or two exemplars. This behavior may occur because subjects are unable to hold images in memory long enough to form a full representation, or because they are unable to sort early labels into categories used to form representations. Conceivably, both these factors might be involved.

Concerning memory: studies have shown that both children under a year, as well as somewhat younger parrots, lack full object permanence; that is, if an object disappears for more than a few seconds before the subject is allowed to search for it, retrieval presents considerable difficulty (Diamond 1985; Pepperberg *et al*.1997). The standard explanation of such failure is that the subject is not able to store a representation of the object. Whether this ability is dependent upon neural development (Chugani 1999; Diamond 1990) or some other effect (Wellman *et al*. 1986; Pepperberg *et al*. 1997) is unclear.

Concerning categorization: we know that adults use images of basic categories in representation. When we adults hear 'car', we have a generic image of a car and do not generally think of some other type of vehicle, such as a truck, nor do we think of the particular jalopy that we first owned (Rosch *et al.* 1976). Children—and probably birds—initially seem to lack these basic categorical images; for them, 'turtle' is specifically (and solely) the squeak toy in the bath. But later these underextensions, and sometimes the specific labels, are completely lost. For children, 'ur-ul' then refers to the class of critter (de Villiers and de Villiers 1979); for birds, 'key' can now refer to something of different colors, shapes, or materials (Pepperberg 1999).

21.2.2 Social, emotional, and neural mediation of the transition from early to later, more representational uses of labels

What is occurring as the label-learning process continues in children (Hollich *et al.* 2000)—and I believe in my birds—is that they begin to process information within an entirely different context that arises as their understanding of social systems and rationality develop in concert. Various levels of neural development probably underlie these changes in understanding, which may or may not be reflected in overt behavior patterns that are apparent to an observer. Although learning still appears to be self-directed in the sense of being driven by a need to influence others and to have basic needs met, learning now advances because the subject is able to attend to the intentions of others and to recognize others as information sources separate from self. That is, responses begin to show elements of the social rationality we find in human adults. My suggestion is that these rudiments of social rationality support the transition to representational label use. Let's explore how such advances might occur and lead to concept formation, representation, and complex cognitive processing.

Interestingly, most studies that involve both labeling and concept formation deal with older children (18 to 24 months; for example, L. Bloom 2000; Tomasello 2001). Only a few studies attempt to examine the transitional stages either in human beings (Hollich *et al.* 2000) or animals (Cheney and Seyfarth 1990; Pepperberg 1999). By 18 months, a child can be playing with one toy and notice that an experimenter is playing with and labeling another toy; the child will then change its focus to view the experimenter's toy when it hears the experimenter's label, rather than continue to be absorbed with the one that has captured its own interest. But at only 12–14 months, the child is so self-centered that, in the same situation, it will prefer to look at its own toy when it is given a choice between toys and hears the experimenter's label (Hollich *et al.* 2000), and at about 10 months is actually likely to associate the experimenter's label with its own toy (Hirsh-Pasek *et al.* 2000). Thus, during a quite brief period, a transition occurs in which the child begins to lose its completely self-centered bias and begin to act in a way that approaches adult behavior.[5]

---

[5] Editors note: on the capacity for decentering and its relationship to rationality, see also the chapters by Millikan and by Currie, this volume.

Moreover, data show that it is at the end of this transitional stage that autistic behavior and its associated lack of communication often becomes evident in humans (Tager-Flusberg 2000)—that is, at the point at which self-awareness and the need to understand that self is separate from others and that others are information sources become critical for learning. Some autistic children never move beyond the self-centered 10-month-old stage in the way they associate labels and objects (Baldwin and Tomasello 1998); that is, they fail to acquire the rudiments of adult rationality.

Given that many researchers argue against full consciousness or self-awareness in animals but admit that animals exhibit extensive cognitive processing,[6] the question arises: what level of awareness might be necessary for the kind of abilities that we see in Grey parrots, and how might it develop? In particular, what elements of rationality might be necessary? It may be that most creatures learn how to generalize and how to make the initial separation of self from other by first categorizing and generalizing emotions with respect to environmental events (Humphrey 2000) and then with respect to intersubject interactions (Damasio 1999). Damasio argues that 'core' consciousness (the basic form, which involves total awareness of the present, but not of the future or the past) emerges when we interact with an object (including other beings); this feeling accompanies the making of a mental image—even one that is retained for less than a minute. His theory might explain why social interaction so handily assists learning, in both children and parrots. The mental image allows for categorization of emotions and events with respect to their emotional content and in normal individuals eventually leads to relevant categorization of the objects and actions. Note that a child, for example, initially does not label an emotion, but talks about objects about which it cares and expresses the relevant emotion by displaying positive or negative affect (L. Bloom 2000).[7] It is still unclear how these categorizations lead to representations of objects and actions that can be manipulated to allow for advanced learning. But at some basic level, self-hood (not necessarily full self-consciousness), beginning in the *emotional* domain prior to the emergence of language, seems to lead to the ability to categorize, which then leads to the understanding and use of representation.[8] Development seems to require the rudiments of rationality on the part of the subjects in order for them to advance.

Of course, one might argue that sensitivity to input, separation of self from other, and emergent aspects of rationality are simply the outcome of brain maturation, and that the transition to advanced learning is mediated by neural development. Many neural connections that we have at birth die off early (Changeux and Danchin 1976).

---

[6] Blumberg and Wasserman 1996; Heyes 1993; see however Griffin 2001; Lovibond and Shanks 2002.

[7] Might a child's so-called 'terrible twos' occur because the child does not yet have the ability to express verbally not only what s/he wants, but also how s/he *feels*?

[8] Editors' note: see and compare Savage-Rumbaugh *et al*, this volume, on the emotional roots of language.

Do systems used in early, simple label learning die off? Many new connections are formed in the first few years of life. Given that neural categorization occurs when a neural ensemble provides the same output from different inputs, is this type of connectivity largely absent in year-old babies but progressively present in older ones? Is the failure to form the new connections as the old ones die off responsible for the emergence of autistic behavior and/or the failure of rationality to develop (see Chugani 1999)? Might analogous neural explanations hold for animals?

Relevant neural development may include the maturation of mirror neurons and neural mirror systems, which have been studied in non-human primates and human beings since their discovery in the 1990s. A mirror neuron fires in the same way when an agent carries out a manual or oral action as when it observes another individual carry out that same action. Mirror neurons are found in human Broca's area and in the homologous brain area in monkeys; researchers have thus argued that mirror systems are involved in the evolution of language (see the reviews in Arbib 2002, 2005). The monkey mirror neuron system, however, differs significantly in certain respects from the human mirror neuron system. Observed actions associated with the firing of mirror neurons in monkeys (but not in human beings) must already be in the observer's behavioral repertoire and must be goal-oriented (Chaminade *et al.* 2001, 2002; Rizzolatti *et al.* 2001). Moreover, mirror neurons need not be involved in imitation. Although monkeys can emulate observed goal-oriented behavior by using actions already in their own repertoire, they cannot directly imitate a novel action (Visalberghi and Fragaszy 2002). I have suggested (Pepperberg, 2005 b; Pepperberg and Sherman, forthcoming) that human mirror systems probably evolved in order to analyze and *developmentally* to recreate actions to which their bearers are exposed—even actions such as speech actions, to which neurons react which activate muscles that are not directly observed (Craighero *et al.* 2002; Sundara *et al.* 2001). Thus, I propose that *exposure* to complex speech patterns and observation of everyday combinatorial acts in human beings (and possibly apes, but not monkeys) recruits neurological paths including mirror neurons for use in combining simple patterns in specific, hierarchical ways to form various types of complex behavior, including communicative actions using simplistic grammar as well as physical actions. The human mirror neuron system, more flexible than a monkey's, allows different syntactic/phonemic patterns to arise from input. The emergence of linguistic behaviors that will eventually become genuinely representative then involves both achieving physical competence for execution and rational intentions (choice). Such behavior may lack *ostensible* training, but 'training' begins at birth. Interestingly, autistic human beings, whose mirror neurons are, like the monkey's, functional at the motor cortex level (Avikainen *et al.* 1999), seem similarly unable to integrate information from mirror neurons into higher level cognitive processes, and thus show deficits in imitation, communication, mind reading (Heavey *et al.* 2000), and in solving hierarchical problems (Yirmiya and Shulman 1996).

What is important is that hierarchical, combinatory behavior of the type that seems correlated with the development of mirror systems is a cornerstone of communication, intelligence, and rationality. It is likely that different types of mirror systems exist in different species and may support behavior patterns of varying levels of complexity and possible rationality. Capuchin monkeys (*Cebus apella*), for example, behave in many ways that appear both intelligent and rational, but still fail certain tests that they might be expected to solve by means of experience and choices that would seem rational from a human standpoint (Visalberghi and Limongelli 1994; Addessi and Visalberghi, Chapter 15, this volume). However, if changes in neural connections are indeed what produces the transition to a different form of learning and corresponding shifts in behavior, such neural reorganization is unlikely to be *specific* to human beings or even to primates, because this transition also occurs in parrots, and parrot brain architecture differs significantly from that of humans (Medina and Reiner 2000). Mirror neuron systems have not yet been discovered in birds, though analogies may exist (see the review in Pepperberg, 2005b).

21.2.3 Parrot transitions: referential mapping, sound play

Evidence for our birds' transitions away from self-centered learning comes from three forms of vocal actions that they use in very similar ways. First, although our birds' new labels usually emerge in sessions initially in a modified, rudimentary pattern—first as a vocal contour, then with vowels, and finally with consonants—the birds occasionally utter completely-formed new labels after minimal or no training and without overt preliminary 'practice'. Second, our birds—like young children (Kuczaj 1998)—often engage in a form of sound play outside of sessions, in which they spontaneously recombine labels or label parts. In both cases, these labels quickly become part of their repertoire if we provide a corresponding object—that is, if we *referentially map* the label (Pepperberg 1990). Thus, after learning 'grey', Alex produced and was rewarded for 'grape', 'grate', 'grain', 'chain', and 'cane'. Another bird, Wart, after learning 'spoon' produced 'spool' and 'school'. 'Spool' remained in the repertoire because it could easily be rewarded; 'school' dropped in frequency because it could not (Pepperberg, unpublished data). What appears to happen in both cases, and also in a third (see below) is that the birds begin to test out the possibility that humans are indeed good information sources for the reference for these novel labels (Pepperberg 2001b). They see humans in this context during training; then, in what could be considered a rational process, they take the situation a step further. In the third form of vocal action, they not only play with label phonetics, but they take a label that they have seen used in a very specific context, such as 'wool' for a woollen pompon, and, for example, pull at a trainer's sweater while uttering that label. The probability of such action happening by chance or simply as an attention-getting mechanism is slim; at that stage the bird usually has three to four other labels. The birds—like children (Brown 1973)—seem to be probing the situation to test an hypothesis in what appears to be a rational manner.

We respond in all three cases with high affect and excitement, which stimulates the birds further, shows them the power of their utterances, and reinforces their early attempts at categorization. Even if the birds err in initial categorizations, they still get positive reinforcement, in that we provide a correct, new label for something; for example, we tell them that the almond isn't a 'cork', but suggest the term 'cork nut' (Pepperberg 1999).

These vocal actions do not display trial-and-error learning, but rather a form of guided invention (Lock 1980): from the initial label mapping to the generalization to what might be considered imaginal syntax. Parrots, like children, have a repertoire of desires and purposes that drive them to form and test ideas in dealing with the world—in a manner that would be regarded as rational in children. The ideas they test in these ways can amount to the first stages of categorization and representation in cognitive processing. Their manipulation of representations may amount to a syntax of imagery, which Damasio argues requires some level of self-awareness (Pepperberg 2001b). Such processes may demonstrate the emergence of rudimentary forms of rational behavior in these birds

If so, what training procedures might foster such emergent behavior? In particular, does model/rival training, with its emphasis on interaction and adjustment of the learning situation, do so? Arguably, subjects must observe and identify with a *model* who responds to the command 'do X' in order to determine the behavior patterns to be learned and to connect those patterns with 'X'. By contrast, typical single-trainer instruction of the 'do as I do' form might actually prevent the subject from separating the targeted behavior pattern or the target of the command from the instantiation of the command, and thus might inhibit building a representation of the required response (Pepperberg and Sherman, forthcoming): think about a command such as 'touch your nose' that a single trainer demonstrates on *her* nose. Specifically, what we are initially doing via our two-trainer modeling system is to teach the subject to *imitate* the response of the model—to identify with the model, to take the model's point of view so as to recreate the model's actions in the subject him/herself—and thus to build a representation by seeing the model's actions. Does model/rival training work because of the *rational* decision of the bird to reproduce the actions of the model?

Preliminary data on children with various dysfunctions, including autism, suggest that they are also sensitive to the elements of input of the model/rival protocol (Pepperberg and Sherman 2000, forthcoming). Children in these studies have not yet achieved fully age-appropriate behavior patterns, but their communication and social abilities improve significantly after model/rival training. This is *not* to suggest that the mental processes of these children resemble that of our parrots, but rather that both may profit from procedures that engender exceptional learning—learning that does not occur under what are considered normal input conditions (Pepperberg 1985). Although we recognize that the actual role of mirror systems in imitation is still not clear and that several different types of mirror systems probably exist (Pepperberg, 2005b), we speculate that our model/rival procedure may help in forming or strengthening connections in whatever mirror system may exist in our subjects (note Wolf *et al.* 2001). This suggestion is consistent with Gordon and Barker's (1994) argument that what is lacking in autistic children (and perhaps also in untrained animals) is not a *theory* of behavior, but a *skill*. If action planning (that is, hierarchical combination of simple actions to form complex behavior) depends on selection (whether conscious or unconscious) and combination of appropriate neurons to produce patterns of appropriate temporal activation (Arbib and Rizzolatti 1996), then skilful action planning indeed might be engendered or trained by our model/rival procedure (Pepperberg and Sherman, forthcoming). Work in progress (Sherman, personal communication) suggests that children with particularly severe disabilities may need prior training to prepare them to accept model/rival input. Data from such studies may provide additional understanding of the transitional processes of learning, help us determine what if any parallels exist in non-humans, and examine the extent to which rational thought processes or decision making may be involved.

21.3 'Irrational' but intelligent behavior?

The above data suggests that parrots develop allospecific communicative competence in a way that may be related to children's very early language acquisition, and that the birds' pattern of development has rudiments of rationality in a way that again parallels the development of children. Our parrots, however, occasionally respond to our tasks in ways that, from a strictly human viewpoint, appear to fall short of rationality. Examples occurred during tasks involving object permanence and recursive, conjunctive comprehension.

21.3.1 Piagetian object permanence: surprise and anger versus prolonged search in response to being 'tricked'

Students and I have shown that parrots can not only solve a complex Piagetian object permanence task, but also demonstrate knowledge of the specific item that was hidden (Funk 1996; Pepperberg *et al.* 1997). In such tasks, an object is placed in a small container; the container is then passed under successive screens until the item is hidden in the designated site, whereupon the researcher shows the empty initial cover, then passes it under another screen, and finally leaves the initial cover in an accessible site that varies in each trial. An additional, untouched screen is present to see if the subject examines only screens handled by the researcher. The order of movement varies among trials. A subject shows it is not using a 'go to last place item was seen' or 'go to last place researcher touched' rule by ignoring the initial cover. On occasion, a researcher tricks the subject by showing that a particularly desirable item is being hidden, but hides something else. Griffin and Alex reacted with surprise and anger (for example, with odd yips, beak-banging, cup tossing) when the hidden item was other than the expected one (Pepperberg *et al.* 1997). Subjects must have some level of awareness of and memory for the identity of the hidden item in order to respond in these ways when a less desirable one is found. They must also actively track the item's movement in order to infer when and where it was hidden and to remember the site and how to extract the item from the site; simple rules will not provide the answer. A standard argument would be that the subject has a representation of what is hidden, and it reacts to the difference (the 'cognitive dissonance') between the observed item and its representation. The birds' reactions also suggest they do not expect that item A can routinely 'turn into' B. In responding in these ways, a subject may not be aware of its *use* of these multiple representations, but it must be aware *of* these representations and the cognitive dissonance arising when its represented expectations are not matched by reality, demonstrating something akin to Natsoulas' (1978) 'perceptual consciousness' (Pepperberg and Lynn 2000).[9] Birds that remember not just where they cached an item, but when and the specific nature of the item (Clayton and Dickinson 1998; Clayton *et al.*, Chapter 9, this volume) are also good candidates for such awareness.

However, we should next ask: why do our birds respond in these cases with surprise and anger, rather than continued search or even immobility (Pepperberg and Lynn 2000)? And how should we assess the rationality of this response? Would it be rational or irrational to continue searching, when logic says that the item must be 'there'? The birds' behavior differs from that of a very young child, who predictably looks longer at a situation that violates its expectations (e.g. Baillargeon *et al.* 1985), and from the standard, expected dishabituation response (Mishkin and Petri 1984). As far as we know, a computer—(the ultimate rational information processor), assuming it can be programmed to perform the inferential searching task, would (unless programmed for this specific eventuality) react with an 'error message', and freeze. Perhaps anger and surprise are 'error messages' emitted by the birds. We do not suggest that dissonance in cognitive processing requires or must be indicated by emotions; indeed, emotional responses often short-circuit cognitive processes in human beings. Perhaps parrots' anger short-circuits the more rational response to continue searching—if that indeed is the more rational response. What we do suggest is that the animal must be aware of the cognitive dissonance and the immutability of items, at the level of perceptual consciousness, or it would not exhibit such emotion-driven surprise at the outcome (Pepperberg and Lynn 2000).

Of course, lack of a specific surprise reaction might not be very telling. Allen (1997) would argue that our birds' reactions indeed indicate awareness (and possibly rationality?) because they differ, for example, from the slow extinguishing of lever pressing in an animal

---

[9] Natsoulas defines perceptual consciousness as encompassing awareness of one's sensory perceptions, e.g. how information provided by the senses is acknowledged, processed, and integrated such that it can be used for several purposes.

whose expectations are violated in an operant task. However, given operant experimental design, researchers would not be able to observe an initial surprise reaction in the animal. Moreover, the slow extinguishing of an animal's operant behavior is not so different from repeated button pressing by human beings when food does not come, as expected, from a vending machine. Yet we may be reluctant to describe such continued 'searching' behavior by human beings as either rational or irrational.

What therefore can we infer about the rationality of our birds, when they do not continue to search but instead display anger and surprise (Pepperberg and Lynn 2000)? Would their behavior be more rational if they prolonged their search? (One bird did continue to search, but only after displaying anger.) Can we argue that our birds are either rational or irrational, given that our expectations of human behavior are unclear, and may even differ with the age of the human involved?

Would it help to answer that question if we know more about what our birds are aware of and their conscious mental processes? For some researchers, a specific feature of consciousness is the existence of '... noncomputable, seemingly random, conscious choices with an element of unpredictability...' (Hameroff 1997; also Barinaga 1996; Allen and Bekoff 1997); our birds' behavior in the above example appears to fit this description (Pepperberg and Lynn 2000). This linkage suggests, surprisingly, that a subject may be conscious when normal cognitive processes fail so that it must access something else to decide how to proceed (Allen and Bekoff 1997). This 'something else' is not necessarily logical or rational, and can be an emotional state. Have we circled back to Mishkin and Petri's 'habit state'? Connected back to the ideas of Damasio? The answers to these questions are not clear. I suggest that positing rationality in this situation should lead us only to expect a reaction indicating cognitive dissonance, not the nature of the reaction or any more specific explanation of it. We are not yet in a position to say more.

21.3.2 **Playing games?**

Let's examine one more instance of problematic behavior. Consider a parrot who is given seven items of various colors, shapes, and materials and is asked to label the color of the one that, for example, is wood and square, rather than the shape of one that is green and paper (a conjunctive, recursive question). On occasion, he successively provides each of six possible wrong color answers, and then repeats the wrong answers, thus avoiding the correct answer on 12 out of 12 trials (Pepperberg 1992). A chance explanation is not supported by statistics. Moreover, recursive tasks cannot be solved by responding with respect to a fairly simple set of criteria, such as performing color-based oddity-from-sample by making a match and then avoiding the match, as do pigeons (Zentall *et al.* 1981); in such a task errors are explained by the bird simply forgetting to avoid the match. Rather, the parrot sees seven items, and must recursively decode a symbolic query in order to recall the designated (color) attribute of the one item defined by conjunction of two other attributes. In turn, he must encode this attribute symbolically: to produce the color label of the one item symbolically labeled 'wood' *and* 'square' within an array of other items that are wood *or* square, but not both. To perform this task successfully to obtain a reward—as indeed he can (Pepperberg, 1992)—the bird must represent the labels of all the attributes, integrate that information with a search for the appropriate item (which requires combining representations of two attributes), and encode the correct attribute label. To systematically avoid success, as he occasionally does, he must specifically *avoid* uttering the correct label and instead produce other relevant labels (in this case, color) for 12 trials.

The bird's behavior is intelligent, no doubt; but it does not appear to be rational (Pepperberg and Lynn 2000)—at least if we assume his goal is to obtain his reward. Because he can respond correctly on such tasks, one would expect him to do so in order to obtain his reward. Yet the bird chooses to respond in what at first appears (to his human trainers) to be an intentionally illogical manner.

However, this example should prompt us to reconsider the goal we are attributing to the animal. Perhaps the bird knows that the task requires a certain response that would yield a specific reward, yet opts to inhibit that response in order to elicit some other caretaker reaction? Perhaps the bird is aware not only of how to answer *in*correctly, but also of how to use its knowledge to affect its trainers' behavior, for its own ends? The trainer's frustration and escalation of affect as the answers continue to be erroneous, an increase in the trainer's vocal volume and in the intensity of interaction, may on occasion provide the parrot (an admittedly social creature; Pepperberg 1999) with a greater reward than the chance to chew on a familiar green square of wood. If so, the parrot's apparent irrationality in relation to an assumed goal may actually be a rational choice in relation to a different goal. The parrot may here be approaching a level of social understanding and rationality at which it actively manipulates the trainer's behavior for its own reasons.

21.4 **Conclusions**

Studies of non-human animals both in the field and the laboratory demonstrate that their capacities to solve complex problems—that is, to behave intelligently—form a continuum with those of humans. These measures of intelligence usually require the animal to choose the answer that, to humans, is logical—that is, rational—and that a human in the same situation would choose. I have shown, however, that Grey parrots who have consistently succeeded on tasks designed to measure complex cognitive processing may sometimes engage in behavior that doesn't seem rational, at least by human standards.

Previously, Lynn and I have posited that the evolutionary homologies/convergences across taxa in brain function that lead to continuity (but not necessarily isomorphism) in cognitive processing also allow for (but do not necessarily lead to) convergence (but not necessarily isomorphism) with respect to consciousness (Pepperberg and Lynn 2000).

I now make a parallel suggestion with respect to rationality. Especially for behavior patterns that at first seem irrational, our focus should not necessarily be on the goal that is assumed relevant to human researchers who are concerned exclusively with cognitive processing, but rather should encompass goals that might be meaningful/useful/of interest to the animal subject within a broader perspective. Does a chimpanzee care that much about a slice of banana that it can obtain by using a rake in a designated manner (Povinelli 2000), or is it more interested in seeing other ways in which the rake can be handled? Do our parrots provide themselves with the reward of increased interaction by deliberately erring? Is a vervet monkey (*Cercopithecus aethiops*) irrational because it ignores a carcass that implies that a leopard is near (Cheney and Seyfarth 1990), or is it making a rational inference that the leopard already has food and won't be hunting again soon? In order to determine in what senses and to what extents animals are rational, we have much to learn not only about animal intelligence, but also about experimental design and interpretation.

References

Allen, C. (1997). Animal cognition and animal minds. In: P. Machamer and M. Carrier, eds. *Philosophy and the Sciences of the Mind*, pp. 227–243. Pittsburgh, PA: Pittsburgh University Press and Konstanz, Germany: Universitätsverlag Konstanz.

Allen, C., and Bekoff, M. (1997). *Species of Mind*. Cambridge: MIT Press.

Arbib, M. A. (2002). The mirror system, imitation, and the evolution of language. In: K. Dautenhahn and C. Nehaniv, eds. *Imitation in Animals and Artifacts*, pp. 229–279. Cambridge, MA: MIT Press.

Arbib, M.A. (2005). From monkey-like action recognition to human language: An evolutionary framework for neurolinguistics. *Behavioral and Brain Sciences*, **28**: 105–124.

Arbib, M. A., and Rizzolatti, G. (1996). Neural expectations: A possible evolutionary path from manual skills to language. *Communication and Cognition*, **29**: 393–424.

Avikainen, S., Kulomaeki, T., and Hari, R. (1999). Normal movement reading in Asperger subjects. *Neuroreport*, **10**: 3467–3470.

Baillargeon, R., Spelke, E. S., and Wasserman, S. (1985). Object permanence in five-month-old infants. *Cognition*, **20**: 191–208.

Balda, R. P., Pepperberg, I. M., and Kamil, A. C., eds. (1998). *Animal Cognition in Nature*. London: Academic Press.

Baldwin, D. A., and Tomasello, M. (1998). Word learning: a window on early pragmatic understanding. In: E. V. Clark, ed. *Proceedings of the of the Twenty-Ninth Annual Child Language Research Forum*, pp. 3–23. Chicago, IL: Center for the Study of Language and Information.

Bandura, A. (1971). Analysis of modeling processes. In: A. Bandura, ed. *Psychological Modeling*, pp. 1–62. Chicago: Aldine-Atherton.

Barinaga, M. (1996). Neurons put the uncertainty into reaction times. *Science*, **274**: 344.

Bloom, L. (2000). The intentionality model: how to learn a word, any word. In: R. M. Golinkoff, K. Hirsh-Pasek, L. Bloom, L. B. Smith, A. L. Woodward, N. Akhtar, M. Tomasello, and G. Hollich, eds. *Becoming a Word Learner: A Debate on Lexical Acquisition*, pp. 124–135. New York: Oxford University Press.

Bloom, P. (2000). *How Children Learn the Meaning of Words*. Cambridge, MA: MIT Press.

Blumberg, M. S., and Wasserman, E. A. (1996). Animals have minds? *American Psychologist*, **51**: 59–60.

Boysen, S. T., and Bernston, G. (1995). Responses to quantity: perceptual vs. cognitive mechanisms in chimpanzees (*Pan troglodytes*). *Journal of Experimental Psychology and Animal Behavior Processes*, **21**: 82–86.

Brown, R. (1973). *A First Language: the Early Stages*. Cambridge, MA: Harvard University Press.

Chaminade, T., Meary, D., Orliaguet, J-P., and Decety, J. (2001). Is perceptual anticipation a motor simulation? A PET study. *Brain Imaging*, **12**: 3669–3674.

Chaminade, T., Meltzoff, A. N., and Decety, J. (2002). Does the end justify the means? A PET exploration of the mechanisms involved in human imitation. *NeuroImage*, **15**: 318–328.

Changeux, J.-P., and Danchin, A. (1976). Selective stabilization of developing synapses as a mechanisms for the specification of neuronal networks. *Nature*, **264**: 705–721.

Cheney, D. L., and Seyfarth, R. M. (1990). *How Monkeys See the World*. Chicago: University of Chicago Press.

Chugani, H. T. (1999). Metabolic imaging: a window on brain development and plasticity. *Neuroscientist*, **5**: 29–40.

Clayton, N. S., and Dickinson, A. (1998). Episodic-like memory during cache recovery by scrub jays. *Nature*, **395**: 272–274.

Colunga, E., and Smith, L. B. (2000). Learning what is a word. *25th Annual Boston University Conference on Language Development*.

Craighero, L., Buccino, G., and Rizzolatti, G. (2002). Speech listening specifically modulates the excitability of tongue muscles: A TMS study. *European Journal of Neuroscience*, **15**: 399–402.

Damasio, A. (1999). *The Feeling of What Happens*, San Diego, CA: Harcourt.

DeCasper, A. J., and Spence, M. J. (1986). Prenatal maternal speech influence newborn's perception of speech sounds. *Infant Behaviour and Development*, **9**: 133–150.

deVilliers, P. A., and deVilliers, J. G. (1979). *Early Language*. Cambridge, MA: Harvard University Press.

Diamond, A. (1985). Development of the ability to use recall to guide action, as indicated by infants' performance on AB tasks. *Child Development*, **56**: 868–883.

Diamond, A. (1990). The development of neural bases of memory functions as indexed by the AB and delayed response tasks in human infants and infant monkeys. In: A. Diamond, ed. *The Development and Neural Bases of Higher Cognitive Functions*, vol. 608, pp. 267–309. New York: Annals of the New York Academy of Sciences.

Funk, M. S. (1996). Development of object permanence in the New Zealand parakeet (*Cyanoramphus auriceps*). *Animal Learning and Behavior*, **24**: 375–383.

Golinkoff, R. M., Hirsh-Pasek, K., Bloom, L., Smith, L. B., Woodward, A. L., Akhtar, N., Tomasello, M., and Hollich, G., eds (2000). *Becoming a Word Learner: A Debate on Lexical Acquisition*. New York: Oxford University Press.

Gordon, R. M., and Barker, J. A. (1994). Autism and the 'theory of mind' debate. In: G. Graham and G.L. Stephens, eds. *Philosophical Psychopathology*, pp. 163–181. Cambridge, MA: MIT Press.

Gottlieb, G. (1982). Development of species identification in ducklings: IX. The necessity of experiencing normal variations in embryonic auditory stimulation. *Developmental Psychobiology*, **15**: 507–517.

Greenfield, P. M. (1978). Developmental processes in the language learning of child and chimp. *Behavioral and Brain Sciences*, **4**: 573–574.

Griffin, D. R. (2001). *Animal Minds: Beyond Cognition to Consciousness*. Chicago: University of Chicago Press.

Hameroff, S. R. (1998). Did consciousness cause the Cambrian evolutionary explosion?. In: S. R. Hameroff, A. W. Kazniak, and A. C. Scott, eds. *Toward a Science of Consciousness II: The Second Tucson Discussions and Debates*, pp. 421–437. Cambridge, MA: MIT Press.

Heavey, L., Phillips, W., Baron-Cohen, S., and Rutter, M. (2000). The Awkward Moments Test: A naturalistic measure of social understanding in autism. *Journal of Autism and Developmental Disorders*, **30**: 225–236.

Heyes, C. M. (1993). Imitation, culture and cognition. *Animal Behaviour*, **46**: 999–1010.

Hirsh-Pasek, K., Golinkoff, R. M., and Hollich, G. (2000). An emergentist coalition model for word learning. In: R. M. Golinkoff, K. Hirsh-Pasek, L. Bloom, L. B. Smith, A. L. Woodward, N. Akhtar, M. Tomasello, and G. Hollich, eds. *Becoming a Word Learner: A Debate on Lexical Acquisition*, pp. 136–164. New York: Oxford University Press.

Hockett, C. (1959). Animal 'languages' and human language. *Human Biology*, **31**: 32–39.

Hollich, G. J., Hirsh-Pasek, K., Golinkoff, R. M. (2000). Breaking the language barrier: an emergentist coalition model for the origins of word learning. *Monographs of the Society for Research in Child Development*, **262**: 1–138.

Hulse, S. H., Fowler, H., and Honig, W. K., eds (1968). *Cognitive Processes in Animal Behavior*. Hillsdale, NJ: Erlbaum.

Humphrey, N. (2000). The privatization of sensation. In: C. Heyes and L. Huber, eds. *The Evolution of Cognition*, pp. 241–252. Cambridge, MA: MIT Press.

Kroodsma, D. E. (1979). Vocal dueling among male marsh wrens: evidence for ritualized expressions of dominance/subordinance. *Auk*, **96**: 506–515.

Kroodsma, D. E., and Byers, B. E. (1998). Songbird song repertoires: an ethological approach to studying cognition. In: R. P. Balda, I. M. Pepperberg, and A. C. Kamil, eds. *Animal Cognition in Nature*, pp. 305–336. London: Academic.

Kuczaj, S. A. (1998). Is an evolutionary theory of language play possible? *Cahiers Psychologie Cognitive*, **17**: 135–154.

Lock, A. (1980). *The Guided Reinvention of Language*. London: Academic Press.

Lovibond, P., and Shanks, D. R. (2002). The role of awareness in Pavlovian conditioning: Empirical evidence and theoretical implications. *Journal of Experimental Psychology: Animal Behavior Processes*, **28**: 3–26.

McCarthy, D. (1954). Language development in children. In: L. Carmichael, ed. *Manual of Child Psychology*, pp. 476–581. Oxford: Wiley.

Macnamara, J. (1982). *Names for Things: A Study of Human Learning*. Cambridge, MA: MIT Press.

Markman, E. M., and Wachtel, G. F. (1988). Children's use of mutual exclusivity to constrain the meanings of words. *Cognitive Psychology*, **20**: 121–157.

Medina, L., and Reiner, A. (2000). Do birds possess homologues of mammalian primary visual, somatosensory and motor cortices? *Trends in Neurosciences*, **23**: 1–12.

Miles, H. L. (1983). Apes and language. In: J. de Luce and H. T. Wilder, eds. *Language in Primates*, pp. 43–61. New York: Springer-Verlag.

Mishkin, M., and Petri, H. L. (1984). Memories and habits: Some implications for the analysis of learning and retention. In: L. Squire and N. Butters, eds. *Neuropsychology of Memory*, pp. 287–296. New York: Guildford Press.

Mowrer, O. H. (1950). *Learning Theory and Personality Dynamics*. New York: Ronald Press.

Natsoulas, T. (1978). Residual subjectivity. *American Psychologist*, **33**: 269–283.

Otter, K. A., McGregor, P. K., Terry, A. M. R., Burford, F. R. L., Peake, T. M., and Dabelsteen, T. (1999). Do female great tits *Parus major* assess males by eavesdropping? A field study using interactive song playback. *Proceedings of the Royal Society, London B*, **265**: 1045–1049.

Pepperberg, I. M. (1981). Functional vocalizations by an African Grey parrot (*Psittacus erithacus*). *Zeitschrift für Tierpsychologie*, **55**: 139–160.

Pepperberg, I. M. (1985). Social modeling theory: A possible framework for understanding avian vocal learning. *Auk*, **102**: 854–864.

Pepperberg, I. M. (1988). An interactive modeling technique for acquisition of communication skills: Separation of 'labeling' and 'requesting' in a psittacine subject. *Applied Psycholinguistics*, **9**: 59–76.

Pepperberg, I. M. (1990). Referential mapping: attaching functional significance to the innovative utterances of an African Grey parrot. *Applied Psycholinguistics*, **11**: 23–44.

Pepperberg, I. M. (1992). Proficient performance of a conjunctive, recursive task by an African Grey parrot (*Psittacus erithacus*). *Journal of Comparative Psychology*, **106**: 295–305.

Pepperberg, I. M. (1996). Categorical class formation by an African Grey parrot (*Psittacus erithacus*). In: T. R. Zentall and P. R. Smeets, eds. *Stimulus Class Formation in Humans and Animals*, pp. 71–90. Amsterdam: Elsevier.

Pepperberg, I. M. (1999). *The Alex Studies*. Cambridge, MA: Harvard University Press.

Pepperberg, I. M. (2001a). Evolution of avian intelligence. In: R. Sternberg and J. Kaufman, eds. *The Evolution of Intelligence*, pp. 315–337. Mahwah, NJ: Erlbaum Associates.

Pepperberg, I. M. (2001b). Lessons from cognitive ethology: Animal models for ethological computing. *Proceedings of the First Conference on Epigenetic Robotics, Lund, Sweden*.

Pepperberg, I.M. (2002). In search of King Solomon's ring: Cognition and communication in Grey parrots. *Brain, Behavior and Evolution*, **59**: 54–67.

Pepperberg, I. M. (2005a). Cognitive aspects of networks and avian capacities. In: P. K. McGregor, ed. *Animal Communication Networks*, pp. 568–582. Cambridge: Cambridge University Press.

Pepperberg, I. M. (2005b). Insights into vocal imitation in Grey parrots (*Psittacus erithacus*). In: S. Hurley and N. Chater, eds. *Imitation* vol. 1, pp. 243–262. Cambridge, MA: MIT Press.

Pepperberg, I. M., and Lynn, S. K. (2000). Perceptual consciousness in Grey parrots. *American Zoologist*, **40**: 893–901.

Pepperberg, I. M., and Sherman, D. V. (2000). Proposed use of two-part interactive modeling as a means to increase functional skills in children with a variety of disabilities. *Teaching and Learning in Medicine*, **12**: 213–220.

Pepperberg, I. M., and Sherman, D. V. (2002). A two-trainer modeling system to engender social skills in children with disabilities. *International Journal of Comparative Psychology*, **15**: 138–153.

Pepperberg, I. M., and Sherman, D. V. (forthcoming). Training behavior by imitation: from parrots to people ....to robots? In: K. Dautenhahn and C. Nehaniv, eds. *Models and Mechanisms of Imitation and Social Learning in Robots, Humans and Animals: Behavioural, Social and Communicative Dimensions*. Cambridge: Cambridge University Press.

Pepperberg, I. M., and Wilcox, S. E. (2000). Evidence for a form of mutual exclusivity during label acquisition by Grey parrots (*Psittacus erithacus*)? *Journal of Comparative Psychology*, **114**: 219–231.

Pepperberg, I. M., Willner, M. R., and Gravitz, L. B. (1997). Development of Piagetian object permanence in a Grey parrot (*Psittacus erithacus*). *Journal of Comparative Psychology*, **111**: 63–75.

Pepperberg, I. M., Garcia, S. E., Jackson, E. C., and Marconi, S. (1995). Mirror use by African Grey parrots (*Psittacus erithacus*). *Journal of Comparative Psychology*, **109**: 182–195.

Povinelli, D. J. (2000). *Folk Physics for Apes: The Chimpanzee's Theory of How the World Works*. Oxford: Oxford University Press.

Premack, D. (1978). On the abstractness of human concepts: why it would be difficult to talk to a pigeon. In: S. Hulse, H. Fowler, and W. Honig, eds. *Cognitive Processes in Animal Behavior*, pp. 423–451. Hillsdale, NJ: Erlbaum.

Querleu, D., Renard, X., and Versyp, F. (1981). Les perceptions auditives du foetus humain. *Médecine et Hygiène*, **39**: 2101–2110.

Rizzolatti, G., Fogassi, L., and Gallese, V. (2001). Neurophysiological mechanisms underlying the understanding and imitation of actions. *Nature Review Neurology*, **2**: 661–670.

Rosch, E. H., Mervis, C. B., Gray, W. D., Johnson, D. M., and Boyes-Braem, P. (1976). Basic objects in natural categories. *Cognitive Psychology*, **8**: 382–439.

Shettleworth, S. J. (1998). *Cognition, Evolution, and Behavior*. Oxford: Oxford University Press.

Sundara, M., Kumar Namasivayam, A., and Chen, R. (2001). Observation-execution matching system for speech: A magnetic stimulation study. *Neuroreport: For Rapid Communication of Neuroscience Research*, **12**: 1341–1344.

Tager-Flusberg, H. (2000). Language and understanding minds: connections in autism. In: S. Baron-Cohen, H. Tager-Flusberg, and D. J. Cohen, eds. *Understanding Other Minds: Perspectives from Developmental Cognitive Neuroscience*, pp. 124–149. Oxford, UK: Oxford University Press.

Todt, D. (1975). Social learning of vocal patterns and modes of their applications in Grey parrots. *Zeitschrift für Tierpsychologie*, **39**: 178–188.

Tomasello, M. (2001). Perceiving intentions and learning words in the second year of life. In: M. Bowerman and S. C. Levinson, eds. *Language Acquisition and Conceptual Development*, pp. 132–158. Cambridge: Cambridge University Press.

Visalberghi, E. and Fragaszy, D. (2002) Do monkeys ape?—Ten years after. In: K. Dautenhahn and C. Nehaniv, eds. *Imitation in Animals and Artifacts*, pp. 471–499. Cambridge, MA: MIT Press.

Visalberghi, E. and Limongelli., L. (1994). Lack of comprehension of cause-effect relations in tool-using capuchin monkeys (*Cebus apella*). *Journal of Comparative Psychology*, **108**: 15–22.

Wellman, H. M., Cross, D., and Bartsch, K. (1986). Infant search and object permanence: Meta-analysis of the A-not-B error. *Monographs of the Society for Research in Child Development*, **214**: 1–51.

Wolf, N. S., Gales, M. E., Shane, E., and Shane, M. (2001). The developmental trajectory from amodal perception to empathy and communication: the role of mirror neurons in this process. *Psychoanalytic Inquiry*, **21**: 94–112.

Yirmiya, N., and Shulman, C. (1996). Seriation, conservation, and theory of mind abilities in individuals with autism, individuals with mental retardation, and normally developing children. *Child Development*, **67**: 2045–2059.

Zentall, T. R., Edwards, C. A., Moore, B. S., and Hogan, D. E. (1981). Identity: the basis for both matching and oddity learning in pigeons. *Journal of Experimental Psychology: Animal Behavior Processes*, **7**: 70–86.

Chapter 22

The impact of symbolic representations on chimpanzee cognition

Sarah T. Boysen

Abstract

Two decades of studies with chimpanzees from the Comparative Cognition Project at the Ohio State University Chimpanzee Center suggest that the enculturation process, including the immersion of chimps in an artifact- and symbol-laden human culture and long-term, stable social relationships with human beings, affects the animals' access to attentional resources in dramatic ways. Such changes, in turn, can facilitate acquisition of complex cognitive skills by the chimpanzees and/or override behavioral predispositions that would reduce their capacity to comprehend task demands and respond 'rationally' to them. We here survey recent findings on: (1) the effects of numerals on chimps' evaluative dispositions; (2) the effects of numerals and symbols for 'same' and for 'different' in reaction time tasks; and (3) the abilities of some (female) chimps to employ scale models and other representations of location to solve problems.

22.1 Introduction

The ability of symbolic representations to encompass selective features of their referents may constitute a critical advantage of such representations. Symbols may permit an organism to process selected information efficiently and respond rationally and adaptively based on that knowledge structure. At the same time, symbols can minimize potential interference from lower-level or more primitive evaluative dispositions (Boysen and Berntson 1995).

Over the past two decades, we have been investigating numerical skills and the comprehension of scale models by chimpanzees raised in an enriched, enculturated captive environment (e.g. Boysen 1993; Boysen and Berntson 1989; Kuhlmeier *et al.* 1999; Kuhlmeier and Boysen 2001; Kuhlmeier *et al.* 2001). An impressive array of cognitive accomplishments has been reported for chimpanzees, in work from our research program as well as other primate cognition laboratories (e.g. Biro and Matsuzawa 2001; Hare *et al.* 2001; Tomonaga 2001; Tomasello and Call 1997). At the same time, some rather surprising limitations for acquiring or processing some types of information by our chimps have come to light (Boysen and Berntson 1995; Boysen *et al.* 1996; Boysen *et al.* 1999). Given the significant demands imposed by their dynamic social structure and co-operative living, it is likely that highly sophisticated cognitive capacities in chimpanzees were selected by evolution (Jolly 1966). Nonetheless, clear differences between the cognitive capacities of human beings and apes challenge our understanding of the evolution of cognition in both.

22.2 Evaluative processes and symbolic representations

The relationship between numerical processing and behavioral action that can be viewed in a broader framework of evaluative processes is of particular interest (Berntson *et al.* 1993). The evaluative dimension of behavior has been widely recognized as a pervasive organizing principle that captures behavior's bivalent, directional component (Konorski 1948; McLaren *et al.* 1989). Evaluative dispositions, characterized by approach–avoidance tendencies or reactions, are widely represented phylogenetically, including in human beings. Evaluations of the adaptive significance of environmental objects and events are so central to survival that all species have biological mechanisms for approaching, acquiring, or ingesting certain classes of stimuli, avoiding or rejecting others, and establishing enduring response predispositions toward classes of stimuli. Evaluative processes are apparent across levels of neurobehavioral organization, and range from simple pain-withdrawal reflexes, conditioned approach or avoidance responses, and to higher-level attitudinal predispositions toward broad classes of stimuli.

The significance of the multiple levels of evaluative mechanisms lies in the potential for several evaluative dispositions to be expressed within a given context. Although these multiple dispositions may in some cases lead to concordant behavioral actions, in other cases they may result in conflict. Our work with candy arrays and numeral arrays illustrates the latter possibility.

22.2.1 Summary of earlier work with candy arrays and with numeral arrays

In our studies of quantity judgment during the 1990s, chimpanzees were unable to select a smaller collection of candy in order to achieve a larger reward (Boysen and Berntson 1995; Boysen *et al.* 1996; Boysen *et al.* 1999) (Fig. 22.1).

These data suggest that response biases arising from the incentive features of candy were directly opposed to the associative disposition based on the reinforcement contingency. The chimpanzees clearly had acquired implicit knowledge of the food-distribution rule ('select the smaller array of candy in order to obtain the larger array'), as evidenced by their immediate correct performance when numerals symbols were substituted for candy arrays as choice stimuli. However, when candy arrays served as choice stimuli, the inherent incentive features of the candy arrays may have introduced a powerful and conflicting dispositional bias (Fig. 22.1).

It is likely that the relative potency of this interference effect was related, in part, to the perceptual immediacy of the candy arrays, relative to the delay inherent in postresponse reinforcement (Forzano and Logue 1994; Logue 1988). Despite their poor performance, the animals' selections were sensitive to size disparities between the candy arrays, and hence to the reward differential. Increases in disparity, however, were associated with progressively poorer performance (Fig. 22.2). Thus, performance was poorest under the very conditions where the animal stood to benefit the most by the optimal selection of the smaller array.

These findings are reminiscent of the self-control literature in animals and children (Eisenberger *et al*. 1989; Logue 1988; Mischel *et al*. 1989; Tobin and Logue 1994). Young children, for example, have difficulty inhibiting a direct response to food in order to achieve a larger but delayed reward (Mischel *et al*. 1989). Similarly, adult humans are more impulsive and show poorer self-control when food reinforcers are delivered immediately after a response than when delayed (Forzano and Logue 1994). The improved performance of the chimpanzees when Arabic numerals served as choice stimuli is further reminiscent of the self-regulation literature, just as children who were unable to inhibit suboptimal approach tendencies to food items in the Mischel *et al*. (1989) study were able to respond more adaptively when the food stimuli were presented more abstractly as two-dimensional photographs.

The improvement in performance of our chimpanzees with Arabic numbers suggested that symbolic stimuli could accurately represent the requisite numerical information, without encompassing, among others, the incentive properties that trigger the interfering response bias. Indeed, the ability of symbols to represent some aspects or features of their real-world referents, but not others, may constitute an important advantage of

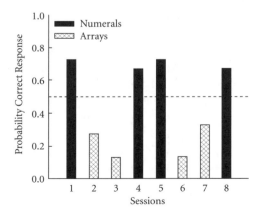

Fig. 22.1 Performance for numerals vs. candy arrays.

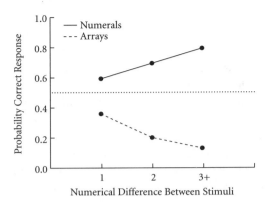

Fig. 22.2 Performance and disparity between stimuli.

abstract representations. The chimpanzees in our studies had extensive training in the labeling of arrays with Arabic numerals across a wide variety of contexts and their numerical symbols were not consistently associated with any specific reinforcer. Rather, the common dimension across training was the numerical significance of the symbols. The training approach we used approximated the acquisition and application of number symbols by young children, and this may account for the apparent lack of interference effects with Arabic numerals.

We were not convinced, however, that only the presentation of candy arrays would trigger the interference effect. To address this issue more directly, we created arrays composed of small rocks taken from the parking lot of the Chimp Center, and presented competing collections of rocks to the chimps. The reward contingencies for rocks were similar to those we had used with Arabic numbers; the chimps' behavior was reinforced based on the number of rocks in the array they did *not* choose. We quickly discovered that rocks were treated precisely the same way as the candy arrays, demonstrating that the perceptual feature of array mass, regardless of whether it was rocks or candy, was more likely to elicit the interference effect in the chimps (Fig. 22.3). Thus, their performance with rocks or candies never reached success with statistical significance, while their performance with numerals did.

Our studies of numerical skills and constraints on rational judgments between quantities have included experiments that explore the way the number of food items or the numerals were integrated during the task, how these factors were assessed across array elements, and the impact of numerical symbols on choice behavior. We have also studied the mechanisms underlying quantity judgments, particularly the subitizing and counting processes that produce numerosity judgments, and their relationship to evaluative dispositions. *Subitizing* is defined as the rapid apprehension of a small number of items, typically between one and four or five items, whose cardinal value can be immediately reported. Once the size of an array exceeds five to six items, counting is required to arrive at the correct total number of items. In addition, the time

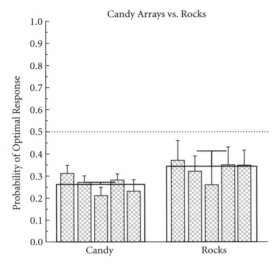

Fig. 22.3 Candy arrays vs. rocks.

required to count the items is much longer than the response time for subitizing smaller arrays (Gallistel 1990; Kaufman *et al.* 1949).

Among the various experimental approaches to these questions, we have investigated the determinants of performance within the food-choice task by systematically manipulating the reward value, reward size, and the subjects' motivational state, to identify possible determinants of performance and interference in the quantity judgment task. We have also examined processing of quantity across object arrays and number symbols. That is, what differences in performance might emerge if the chimps were required to choose between two numerical representations that differed, two objects arrays that were different, or two choice stimuli that included one numeral and one object array? In addressing this empirical question we explored the parameters of the incentive value of food arrays, the contributions of numerosity and incentive value, and the manner by that these factors were assessed across array elements. Mechanisms of quantity judgment and evaluative processing provoked numerous hypotheses for study. Consequently, we have explored mechanisms underlying quantity judgments and links with evaluative processes using reaction time studies that evaluated quantity judgments using small array sizes, as well as additional studies that examined possible differences in modes of number processing (Himes 1998; unpublished master's thesis).

Our results from these studies are consistent with a model of evaluative processes according to which several response dispositions may be invoked when chimpanzees are faced with such tasks. One of these dispositions may be primarily associative, with the smaller array serving as a conditioned stimulus (S+) in a stimulus–response– reinforcement relationship established by the task contingencies. The second response disposition may be a direct incentive-based disposition toward the choice stimuli, arising either from the intrinsic incentive value of candy, or by association with the process of reinforcement, as with the rocks and number stimuli. A third contribution may come from predispositions to the perceptual schema representing mass, as the size of the quantity increases (Boysen *et al*. 2001; Menzel 1962). The demonstration that interference was minimal with Arabic numerals indicates that symbolic representations can subsume numerosity, without the associated incentive or perceptual properties. Additional findings support this interpretation. Performance with candy arrays progressively deteriorated with larger disparities between the two arrays. With Arabic numerals as choice stimuli, however, performance was significantly above chance, and was largely insensitive to the disparity between the two numerals (Fig. 22.2). An alternate explanation for the performance with Arabic numerals might be that the subjects were not responding based on number representations, but had acquired a broad set of specific stimulus–response associations based on the differential probabilities of reinforcement across the Arabic symbols. This is not a plausible explanation for several reasons. For example all responses were reinforced in the present paradigm.

22.3 Processing of quantity across object arrays and numerical symbols: Do mixed stimuli evoke the interference and facilitation effects?

Up to this point in our research, the quantity judgment task had entailed the presentation of candy arrays with other candy arrays, and numerals with other numerals, with only one stimulus type used in a given test session. In view of the reduced interference effect with Arabic numerals, the question arose as to performance in the conflict task if one of the choice stimuli was a candy array and the other a numeral. Of interest was whether the presence of an Arabic numeral as one choice might foster an abstract mode of processing in that the potential incentive interference from the candy array could be overcome when these differing modes of representation were employed.

We have consistently found that using the quantity judgment approach, performance with candies is poor, presumably due to an interfering incentive bias toward the larger candy array. When Arabic numerals were presented as comparisons, however, quantity judgments must have been based upon abstract number representations, since the direct incentive effects were minimized. Thus, an abstract mode of numerical processing appeared to free the animals from previously counter-productive and a partial incentive-based response biases. The question we sought to address with the present study was whether the presentation of a numeral as one choice might foster an abstract mode of processing, even if the second comparison stimulus was a candy array. Although preliminary, pilot data suggested this as a possibility, and supported the general feasibility of the study.[1]

---

[1] Editors' note: See also Hurley, this volume, on whether the results of the mixed array feasibility study are better interpreted in terms of the 'scaffolding' of rationality by symbols or in terms of

22.3.1 **Mixed candy–numeral array experiment: design and methods**

The experiment followed the general design of our standard interference task (see Boysen and Berntson 1995), except that one Arabic numeral and one food array (presented in separate trays) served as the comparison stimuli. Although detailed predictions were not possible *a priori*, the feasibility data indicated that animals would not simply select the candy arrays when given a choice between numerals and food arrays. In fact, the overall probability of selections of the candy arrays and numerals were approximately equivalent, and were related to numerosity.

The same five chimpanzees participated in each experiment, including two females (ages 15 and 37 years at the beginning of the study), and three males (ages 10, 17, and 17.5 years). All animals had extensive experience and training on a variety of cognitive–behavioral tasks, including counting and Arabic number skills. All subjects had also demonstrated the requisite competence with numerals '0' through '6' required for the present studies. Testing was completed with individual animals in a home cage area that was equipped with a large polycarbonate window and a shelf outside the window where stimulus items (candy arrays or numeral placards) could be displayed.

In the experiment, arrays of candy, arrays of numerals, and mixed arrays of both candy and numerals were presented to the animals; the animals pointed to stimuli in full view of a video camera. A blind experimenter presented stimuli while a second experimenter, positioned out of sight, monitored the subjects' choices on a color monitor, recorded the data, and verbally informed the first experimenter of the animals' selections. The chimpanzees were then reinforced, depending upon their choices, by the first experimenter.

For each trial, two arrays of food items (M and M peanut candies), two Arabic numeral symbols (black numerals on a white 7.5 × 12 cm background, affixed to clear Plexiglas placards), or a combination of a candy array and numeral placard were placed in two separate small trays. Trays were placed on the window shelf in front of the subject, approximately 30 cm apart, and the chimpanzee selected one of the arrays by pointing. The experiment entailed a reversed reinforcement contingency: after the chimpanzee made it choice, the experimenter removed the contents of the selected tray, and the animal was given the remaining (non-chosen) candy array, if candy arrays were compared. If numerals were the comparison stimuli (or a numeral and a candy array), the chimpanzees received the number of candies corresponding to the unselected Arabic numeral as their reward for that trial. It was thus to the subject's advantage to select the dish containing the

---

their inhibition of non-instrumental responses. Recall also the work by Call, this volume, on the greater difficulty that apes find in making arbitrary as opposed to causal associations. Given that symbols are arbitrarily related to what they represent, does Call's work bear on the hypothesis of symbolic facilitation?

smaller quantity of candy or representative numeral symbol, in order to receive the larger number of candies as a reward. Test sessions consisted of 16 trials, including four trials of candy–candy comparisons, four number–number comparisons, and eight mixed array pairs (candy–numeral); subjects completed six sessions. Stimulus pairs were comprised of a subset of number pairs selected to sample a range of array sizes and disparities. Order and positions of stimulus pairs were counterbalanced across trials and sessions. The chimps were maintained on their normal food rations throughout the study.

22.3.2 Results from previous feasibility study with mixed arrays

One potential complication with this experiment was the possibility that subjects could fail to respond based on numerosity under these conditions, but would instead simply select the candies on each trial. During an earlier feasibility study with this task, three animals were tested over six sessions, with a mix of candy–candy, numeral–numeral, and candy–numeral stimulus pairs. Six trials with each stimulus-combination type were tested within each session, in counter-balanced order and position. For candy–numeral trials, we also counterbalanced, across animals and sessions, the assignment of the number combinations to the types of stimuli (i.e. for the number combination 2–4, 2 candies and the numeral 4 were used on half the trials, and 4 candies and the numeral 2 on the other half). The results from the feasibility study indicated that the animals did not adopt a simple strategy of selecting the candy array on each candy–numeral trial. Rather, their selection of the candy array or the numeric stimulus was sensitive to the quantities represented.

Results of the feasibility study for the numeral–numeral and candy–candy pairs were similar to those obtained in our earlier studies with these non-mixed arrays (mean proportion of optimal selections with numerals = 0.69, with candies = 0.41). Moreover, results of the feasibility study for mixed array candy–numeral trials were only slightly lower than with numerals (0.61), consistent with the hypothesis that presentation of an Arabic numeral as one of the choices might help to minimize incentive or perceptual-based interference from the comparison candy array. Additionally, the animals' performance on candy–numeral trials with higher disparity ratios was more improved over choices between stimulus pairs that had lower disparity ratios (split half analysis; higher ratios = 0.67, lower ratios = 0.53), suggesting a flat or slightly positive slope of the disparity ratio function.

22.3.3 Results of mixed candy–numeral array experiment

The overall results of the mixed arrays study revealed several interesting relationships that had not been predicted by the feasibility study. While the trends across all animals for the candy–candy and numeral–numeral comparisons were in the direction towards significance, as in our previous studies that showed an interference effect when candy arrays were presented, in the current experiment, the tendency for the candy–candy arrays to produce interference (compared to the numeral–numeral arrays) did not reach significance. Moreover, the novel mixed candy–numeral array comparisons did not show any facilitation effect via the symbol, regardless of disparity or specific comparisons (for example, larger candy array vs. symbol for smaller number). Instead, the combined trial types (candy–numeral) within a single session reduced the interference effect such that, across all animals, the disparity ratio regression functions for all three trial types showed essentially flat slopes. Thus, the mixed array tasks did not provide any evidence for interference that might have been based on the types of stimuli used, nor did the regression slopes provide a sensitive quantitative measure that would account for variability attributable to the stimuli used for quantity judgments.

Subsequent analyses did shed additional light on the results and supported an interpretation that the subjects were unable to decipher the comparison rules within the novelty of the mixed-array task. Instead, they appear to have adopted several types of response strategies that did not permit them to optimize their choices when symbols were available among the comparison stimuli. For example, when responses across all five subjects were analyzed for position bias within a two-choice framework, analysis of left–right responses revealed a significant right-hand bias; the animals chose the response tray closer to the food cup more often by a factor of 2 to 1. That is, they were twice as likely to select the tray situated to their immediate right, that was located nearer the reward cup. This was a particularly striking bias since no response biases had been observed in the previous five experiments run with the same quantity judgments tasks and the same subjects. This suggests that in their confusion over the task parameters, the chimps invoked a lower-level response strategy that provided some consistent reinforcement but did not necessarily match the best response strategy dictated by the reverse-contingency reward option.

Specific responding to the three trial types (candy, numeral, and candy–numeral pairs) shed additional light on the strategies the animals attempted, in an effort to respond despite their apparent confusion when faced with the mixed-array task. For both candy–candy and numeral–numeral comparisons, the animals responded randomly. That is, they were just as likely to choose the smaller numeral as the larger, and to choose the smaller array of candy as the larger array. Thus, no interference effect was evident for candy–candy comparisons, nor was there any positive effect toward correct responses when symbolic (numeral–numeral) comparisons were available. For the combined stimulus pairs, however, when a candy array and a number were available as choices, a significant difference did emerge in the animals' decisions. In this case, they were more likely to choose the numerical representation, regardless of the symbolic quantity represented, when compared with the candy array. Such choices in the mixed array trials were statistically significant, and thus the subjects chose the number symbol more often during the mixed array trials. Again, this response strategy was not optimal, and reflected less understanding of the general rules governing the mixed-array experiment, compared with previous studies that did not mix array types within a given session.

22.4 **Replication of original interference and symbolic facilitation results with unmixed arrays**

Additional explanations for the minimal demonstration of both the symbolic facilitation and interference effect observed in the mixed array task might include the possibility that, over the course of a series of completed studies with the quantity judgment approach, the animals were no longer motivated to attend as readily to the task demands. Another possibility was that the conflict paradigm no longer evoked choices based upon competing predispositions from several differing sources (e.g. perceptually based evaluative schema or incentive-related relationships inherent in the task demands). Given the results of the mixed array study, we completed an additional study whereby the parameters of the original quantity judgment task were reinstated, to insure that both phenomena (interference; facilitation) were available for subsequent study, and that results obtained prior to the mixed array study could be replicated with these subjects.

In this study, we presented one session of candy–candy arrays, a session of numeral–numeral arrays, and one session of mixed array types (candy–numeral) arrays, with no trial types mixed within a given test session. As in previous tasks using this approach, all stimuli were counter-balanced for left–right position, and trial sequences were randomly selected. The animals were reinforced with the non-chosen candy array, if candies were presented, or with the number of candies represented by the non-selected numeral, for number–number comparisons. For number–candy comparisons, they were rewarded with the corresponding number of candies for either the non-chosen candy array or numeral.

As seen in all prior quantity judgment tasks, our findings revealed a similarly powerful interference effect when the chimpanzees were required to choose between candy arrays of competing sizes. This interference effect was seen across the group and evident in the individual performance of the five subjects. Additionally, the presentation of number symbols as response options revealed precisely the same facilitation effects previously seen with the quantity judgment paradigm, with the chimps now able to optimize their responses by choosing the numeral that represented the smaller quantity and thus were rewarded with the larger number of candies corresponding to the remaining numeral. Like the interference effect, the symbolic facilitation effect was observed both at the group level and with individual subjects.

The results of the mixed-array session, during which only mixed candy–numeral arrays were presented and no candy–candy or numeral–numeral arrays, revealed no consistent group pattern. This suggested that future experiments were needed to clarify the conditions under which interference and symbolic facilitation effects can be found. Overall, these results suggested several new experiments that might clarify the contributions of the testing context and response set during the task, and potentially offer other approaches for assessing the effects of stimulus mode on the evaluative choice.

22.5 **Mechanisms of quantity judgment and evaluative processing**

To examine possible mechanisms underlying the animals' quantity judgments under the evaluative dispositions model, we next assessed the chimps' ability to respond to same–different discriminations of Arabic number pairs, arrays of randomly-distributed dots, or combinations of dot arrays and numbers. A new dependent measure for this version of the task was reaction time, in hopes of teasing apart potential processing differences among the three categories of comparison stimuli (dot–dot; number–number; number–dot). One of our interests was to evaluate potential mechanisms that might subserve a counting-like process, compared with other hypothesized perceptually based schema for rapid assessment of quantity such as subitizing. Again, subitizing is understood as an apprehension-type process for making rapid, accurate numerosity judgments of small-sized arrays, while counting requires a slower, serial enumeration process.

22.5.1 **Background and motivation**

Though viewed by some investigators as a more primitive cognitive process, subitizing has been suggested by Davis and Pérusse (1988) to underlie most quantity judgments by non-human animals. While this latter view is not widely held (Miller 1993; Gallistel 1990), since the range of counting-like or numerical discriminations by animals does not often exceed the array size suggested for subitizing (Capaldi and Miller 1988; Meck and Church 1983; Miller 1993), little empirical support for either position has been forthcoming. Some authors (e.g. Gelman and Gallistel 1978; Gallistel 1990) have suggested that subitizing is a high-level emergent process representing a similarly rapid counting mechanism. Thus, the subitizing literature suggests that the underlying cognitive support for quantity judgments by the chimpanzees may differ depending upon array size. Similarly, the symbol facilitation effect observed in the quantity judgment tasks with our animals suggested the possibility that comparisons between stimulus types (arrays vs. numerals) might also be supported by differing processing mechanisms. Since definitive data were not available, we proposed to explore this question.

One of the primary lines of evidence in support of subitizing in humans comes from reaction times studies that generally reveal a relatively flat slope of the reaction-time–set-size function for small sets, with a positive inflection in the slope at the putative subitizing boundary, where number judgments of larger arrays require a serial counting process (see Mandler and Shebo 1982). We were interested in examining the potential for similar reaction time differences in chimpanzees with previous counting experience and an understanding of representational symbols for quantities. Therefore, we sought to explore potential relations between the mechanisms that subserve quantity judgments and the evaluative processes that underlie choice, with our first efforts at examining the possibility of distinct modes of quantity judgment in chimps and possible relationships between these modes of number processing and task interference. Based on both the human and animal literature, reaction-time studies were chosen to evaluate potential differences in the mechanisms for quantity judgments across arrays sizes.

22.5.2 Same–different experiment with numerals: design and methods

Our animals were previously trained and highly experienced with the use of Arabic symbols as representations for quantities of edibles or objects, and this capability allowed us to adapt a conceptual same–different task within a reaction-time paradigm to evaluate potential processing differences. All animals had demonstrated competence with Arabic symbols from '0' to '6' that extended across the purported subitizing boundary of 3 to 5 items, while two of them were experienced with a broader numeral repertoire including '0' to '8'. All the chimps were also accomplished in the use of a computer-based touch-screen interface by which choice stimuli could be presented and responses registered. The animals also had previous experience on same–different tasks with color and shape stimuli, and thus had a working knowledge of the association of two distinct black-and-white non-iconic symbols that represented 'same' and 'different'. For the current task, the 'same' and 'different' symbols appeared in the lower left and right corners of the computer touch-frame monitor. These symbols served as response options to indicate a numerical match ('same') or mismatch ('different') between two different arrays of filled circles (0.5 in dots), a pair of Arabic numerals (from '0' to '6'), or a dot array and a single numeral, presented in the middle display portion of the screen.[2]

Each trial was self-initiated by the subject who contacted a three in central white 'start' square on an otherwise black CRT screen. Touching the start square triggered the stimulus display and response choices for each trial, and helped insure that the chimps were attending to the screen at the time the stimuli were presented. The computer system recorded all responses and reaction times (to the closest millisecond), and a single food reward (from an assortment of small candies) was provided by the experimenter for a correct response. The response data would permit determination of reaction-times and error rates as a function of array size, for both correct matches ('same' responses) and correct mismatches ('different' responses). All dot stimulus arrays were variable in configuration, and randomly sequenced, since fixed configurations would likely facilitate a subitizing strategy with array sizes that might otherwise require counting (e.g. 5 or 6 dots).

The procedures of the reaction time study differed from previous quantity-judgment tasks in that all stimuli were presented on a computer touch-frame system, as two-dimensional arrays of dots or numerals, or a combination of dots–numerals, for

[2] Editors' note: Note the combination in this task of symbols for sameness and difference with symbols for number. See also Herman, this volume, on tasks with dolphins that combine symbols in novel ways.

comparison. Response options became available simultaneously with the presentation of the comparison stimuli following a subject-paced initiation response (touching a central white square that then disappeared). Once the subjects had made a same–different judgment of the paired stimuli, the screen turned black, and the initiation square was redisplayed in preparation for the next trial. Only correct trials were reinforced by the experimenter, who provided a single candy reward per trial; no correction procedures were employed.

In the first phase of training, the same–different task was introduced with a limited number of comparison stimuli (all combinations of dot–numeral arrays composed of quantities 2 and 5). Thus, the animals were presented with dot–dot arrays (e.g. 2 dots vs. 2 dots for 'same' trials; 2 dots vs. 5 dots for 'different'), numeral–numeral comparisons, and numeral–dot comparisons (e.g. an array of 5 dots vs. the numeral 2 for 'different'), using all possible combinations of two and five dots. Comparison stimuli were counterbalanced for stimulus type (dot–numeral), position (left–right on CRT monitor), and presented quasirandomly in 16-trial blocks distributed across two sequences per session. Animals were tested daily and typically completed two to three trial blocks per session for a total of 32 to 48 trials per day.

Following criterion performance with the initial stimulus pairs (two successive sessions of >80% correct responses), in the next phase of training novel stimuli were introduced that included analogous combinations of dot, numeral, and mixed stimuli using 1 and 6 (dots and numerals). Training with the new stimuli continued until the animals' performance stabilized at criterion levels, followed by the introduction of trial blocks that included all possible combinations of comparison stimuli from the first and second phases (quantities 2, 5, 1, 6), presented in four sequences. The test sequences were presented in quasirandom order within a session with two to three trial blocks completed within a given test session for a total of 32–48 trials per day per subject.

After reaching criterion performance with combinations of the four quantities, novel probe trials were presented embedded among the training trials used for the previous phase (all combinations of 1, 2, 5, and 6). Novel probes included all possible quantity combinations that had not been previously used in training (0, 3, and 4, in novel combinations with all other quantities between 0 and 6).

22.5.3 Results

Definitive trends were apparent from the data and offer an intriguing look into processing relationships among trial types (same–different) and stimulus modes (dots, numerals, and mixed arrays). Of particular interest were the relative patterns of errors for array pairs, numeral pairs, and array–numeral combinations. Analyses of the first phase of training for the same–different reaction time study covered the final 12 training sessions (from a total of 20), including the final two sessions during which the animals achieved criterion performance. Data were analyzed for reaction time differences and overall correct responses to the same–different discriminations across all animals. A repeated measures

2×3 ANOVA (reaction time for same–different comparisons $\times$ stimulus type (dot, numeral, and dot–numeral comparisons) demonstrated a significant difference in both reaction time and performance levels for the same–different discriminations. Thus, latencies to respond to 'different' comparisons across all possible comparisons of dots and numerals were significantly more rapid than those associated with 'same' comparisons.

The longer response time to 'same' comparisons also reflected processing differences that were sensitive to array size in the dot–dot comparisons. For example latency to respond to comparisons of six dots vs. six dots was approximately 200 millisecond longer than the same comparison of arrays composed of one dot vs. one dot. Both latency measures to these 'same' comparisons, however, were significantly longer than the 'different' trial types that included comparisons of arrays composed of one dot vs. six dots. In the latter trials, latency to respond correctly ('different') was significantly faster, with a median reaction time of 1550 millisecond, compared to 2400 millisecond to respond correct ('same') in the six dot vs. six dot comparisons.

Numerical comparisons for 'same' trials also revealed some intriguing differences in reaction time for same–different discriminations, with latencies to respond to 'same' trials significantly longer than those for 'different' trials, regardless of stimulus type (dot vs. numeral). Although the effects of stimulus type did not reach significance, specific trends were evident in all comparisons. Latencies for the dot–dot judgments were more rapid than for numeral–numeral comparisons, and reaction times to dot–numeral mixed pairs were more rapid than numeral-numeral, but slower than dot–dot judgments.

22.6 Problem solving by chimpanzee using a scale model and photographic representations of location

In addition to our series of quantity judgments studies, we have also been investigating the potential for comprehension of scale models by chimpanzees (Boysen and Kuhlmeier 2001, 2002; Kuhlmeier *et al.* 1999, 2002; Kuhlmeier and Boysen 2001). These studies have focused on the nature of cognitive representations used by chimps for coding the topographic relationship between a 1:7 scale model and its real-world referent. The unique results of the scale model studies had not been reported previously for non-human primates and have expanded our understanding of the type and complexity of abstract representations that might fall within the cognitive domain of the chimpanzee.

22.6.1 Background and motivation

In these experiments, novel two- and three-dimensional stimuli (scale models and photographs) were presented in a task initially examined in children (DeLoache 1987, 1989). In addition to comparisons of the chimps' performance on the quantity judgment task with the self-control results for children, the developmental literature offered other potential paradigms for examining the nature of representational systems in the chimpanzee. Work by DeLoache and co-workers (e.g. 1987, 1989, 1991, 1992, 1995, 2000) represents one creative approach.

In a series of highly innovative studies, DeLoache investigated the difficulty that very young children (aged 2½) had in understanding the representational features of a scale model of a playroom in her laboratory. As the child watched, the experimenter hid a miniature toy in the model and then asked the child to enter the real playroom and find the full-size toy that had been hidden in the analogous site. Although many of the younger children were not successful with the task, children who were only 6 months older, at 3 years of age, could locate the toy in the full-size room quite successfully. However, when photographs were substituted for the scale model, children of both age groups performed well, demonstrating what DeLoache termed the 'picture superiority' effect (DeLoache 1987). She and her colleagues argued that the conceptual ability to manipulate dual representations is necessary to appreciate the symbolic qualities of the scale model. Her results with younger children suggested that they were unable to see the model as both: (1) a separate object, with toy-like features; and (2) as a representation or symbol for its real-world referent. Photographs require only the second capacity and thus they were more readily interpreted as representations of the actual room by both groups of children.

These studies suggested an intriguing question to us. Would chimpanzees also recognize the spatial and perceptual relationships among the features of a scale model depicting a familiar room, using DeLoache's hiding game?

22.6.2 Scale model experiment: design and methods

To address these issues, a 1:7 scale model of the chimps' indoor playroom was constructed (Fig. 22.4). Two adult subjects, Sheba, a 16-year-old female (100 lbs), and Bobby, an 11-year-old male (181 lbs), despite their age and size, were both quite tractable and were able to move freely among several secured rooms in the laboratory with the author, who had raised both of them from 2 years of age. The animals were tested individually, with the scale model placed in a hall just adjacent to the real playroom.

Fig. 22.4 Scale model of chimps' indoor playroom.

Each chimp watched as the experimenter hid a miniature soda can in one of four possible locations within the model (Fig. 22.5). The chimps were verbally instructed to wait by the model (the animals are extremely well behaved and under verbal control) while the experimenter entered the playroom briefly to hide a real can of soda. When she returned, the chimp entered the playroom to search for the hidden soda (Fig. 22.6). Once it was retrieved, the subject was asked to return to the scale model and required to point to the site where the miniature can had originally been hidden just minutes before. This second retrieval insured that failures to retrieve the real item were not due to memory failure.

Both chimps were tested for their ability to locate the soda can in the real room under five conditions. These were: (1) hiding the miniature can among the four possible locations in the scale model; (2) presentation of the individual items from the model (e.g. the miniature chair) to indicate each hiding place; (3) presentation of a color panorama photograph of the real room, with the experimenter pointing to individual hiding places within the photo; (4) 8 inch × 10 inch color photographs of

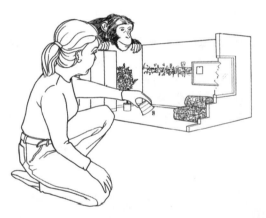

Fig. 22.5 The hiding event.

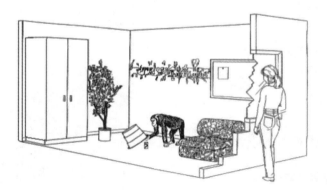

Fig. 22.6 Retrieval 1 for indoor scale model study.

the hiding locations in the room such as a photograph of the chair, behind which the soda could be hidden; and (5) prerecorded, videotape scenarios of the experimenter hiding the soda can in the four locations in the playroom. Two color video cameras recorded all events. One camera was located in the hall and recorded all interactions related to the model, and a second camera was positioned outside the playroom positioned in front of a large Lexan window and filmed all responses by the chimps in the playroom.

22.6.3 Results and discussion

Although both chimpanzees initially grasped the critical features of the 'hiding game' and were correct on two orientation trials of the task, only Sheba's performance reached statistical significance. Sheba was successful at retrieving the hidden item, regardless of test conditions (Fig. 22.7).

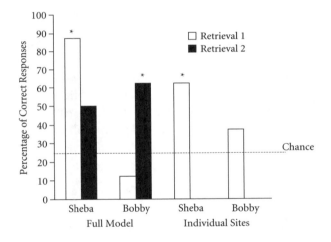

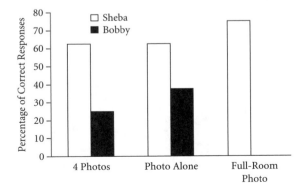

Fig. 22.7 (a) Per cent correct response for standard scale model task and presenting individual miniature sites; *$p < 0.05$. (b) Percentage of correct responses for Sheba and Bobby during the three conditions of the photograph task; *$p < 0.05$

Neither the scale model, miniature items, photographs, nor videotape tasks proved difficult for her. Bobby, on the other hand, failed to locate the hidden soda can in the playroom, regardless of how the information was presented. That is, he was equally unsuccessful with the model, photographs, and videotape, and instead, adopted a rigid spatial strategy whereby he either ran immediately to the last location where he had eventually found the hidden item or he moved in clock-wise fashion around the room, from site to site, until he located the hidden soda (Fig. 22.8).

Bobby's performance was similar to that we found in the quantity judgment task, in which none of the chimpanzees tested could inhibit attending to and choosing a larger candy array despite consistent reinforcement contingencies that dictated choosing the smaller one. Bobby perseverated with response strategies in the scale model-task that resulted in consistent failure, regardless of the test condition. In fact, Bobby failed nearly every trial of every condition, and in the case of the videotape testing, failed to get even one trial correct (N = 8); his performance was below chance. Both the level of failure and the persistence of the maladaptive strategies he used immediately brought to mind the rigid (and incorrect) response predispositions with candy arrays during the quantity judgment task.

Bobby was unable to meet the attentional demands of the task, which were similar to the immediate attentional requirements of the quantity judgment choices; his behavioral impulsivity could not be overridden by a higher-level cognitive strategy that Sheba seemed readily able to apply to the scale-model task. Instead, he used a spatial response strategy to solve the task with remarkably similar persistence that, to us, was reminiscent of the interference phenomena observed in the quantity-judgment task where the chimps seemed unable to inhibit incorrect responses in the face of reward parameters that mandated otherwise.

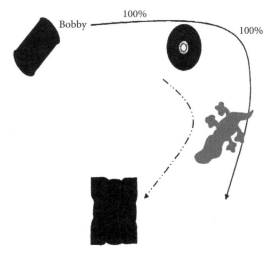

Fig. 22.8 Moving sites: search pattern for unsuccessful subject.

22.6.4 **Further experiments: individual differences in use of scale models**

To address the question of individual differences in performance, an additional study was designed and completed so that all adult chimpanzees in the Ohio State colony could be tested with the scale model paradigm. Since the other five animals could not safely be tested outside their cages, a 1:7 scale model of one of their outdoor play yards was constructed, using four hiding sites indicated by heavy-duty play items, including a large blue polystyrene barrel, a tire, a child's red plastic sandbox, and a plastic teeter-totter shaped like an alligator (Fig. 22.9). Miniature versions of each site were made for the model and testing procedures were similar to the original indoor scale model task.

Following testing of all seven chimpanzees (including Bobby and Sheba, who were retested with the outdoor model), we found that performance by three adult females (ages 16, 25, and 38 years) and one adult male (18.5 years), achieved statistical significance if the positions of the sites in the model and the outdoor enclosure remained constant throughout testing (Fig. 22.10). However, two other adult males (ages 11 and 18 years) and one adolescent male (8 years) failed the task.

However, if the sites in the model and the test enclosure were changed on every trial, only the adult females could achieve optimum performance (Fig. 22.11). All the males, including three adults and one adolescent, performed poorly, showing the same stereotyped, clock-wise search pattern exhibited by Bobby during the original indoor scale model task.

These results replicated and extended the findings from the first scale model study with the animals and clearly demonstrated that chimpanzees have the requisite cognitive sophistication to manipulate novel two- and three-dimensional representations, in the form of photographs or scale models representing actual real-world object–space relationships. Failure to inhibit powerful, purportedly lower-level response dispositions

Fig. 22.9 Scale model of chimps' outdoor play yard.

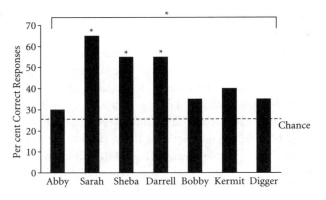

Fig. 22.10 Percentage of correct responses for all chimpanzees when hiding sites remained in constant positions; *$p < 0.05$.

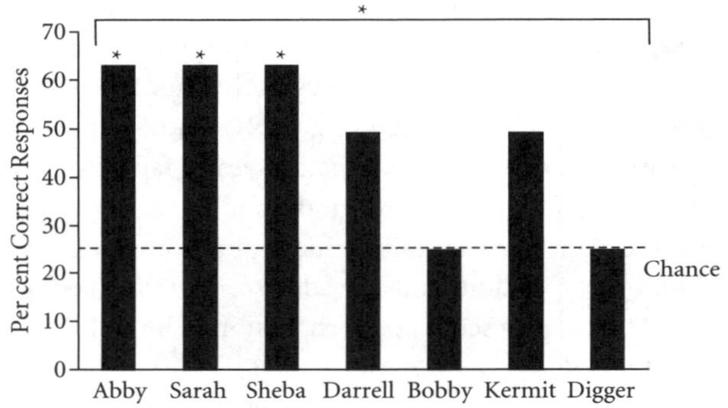

Fig. 22.11 Significant performance by female chimpanzees only when hiding sites were moved in model and enclosure on every trial; *$p < 0.05$.

expressed as stereotyped spatial or perseverative object-oriented responding was evident in three of the seven animals tested and suggests that the attentional requirements of such representational processing may be difficult for some individuals. This was particularly the case for the male chimps.

The failure of these three males suggested that future studies examining the attentional demands imposed by the scale model task could shed light on possible developmental differences observed with younger and older children in interpreting scale models. Given the degree of impulsivity and rigidity of incorrect responding by some of the chimps, and the symbolic facilitation provided by numerals we had found in the earlier quantity judgment studies, these results also suggested that animals who failed the scale-model task might benefit from a metarepresentation that could form a bridge between the scale model and the actual room. Such a metarepresentation could potentially provide an inhibitory, symbolic link between the two spaces. Might a metarepresentation linking and relating the separate three-dimensional spaces serve the same intervening function as the numerals did for quantity judgments, thus allowing the animals to inhibit competing, lower-level response biases?[3] It is possible that such use of a representation

---

[3] Editors' note: See also and compare the discussion of metarepresentation by Proust, this volume.

would result in the attenuation of the non-productive, spatially biased searching observed with the males, and allow for more optimal search strategies based on topographic analogy. This is of particular interest to us since similar error patterns have been reported for children who had difficulties with DeLoache's task, and thus additional studies addressing this question are planned for the near future with the chimps.

22.7 **Conclusion**

The present studies have evaluated the effects of representations on the cognitive capacities of chimpanzees in three experimental paradigms. The quantitative comparison experiments reveal the way in which instrumentally 'rational' responses are facilitated, and/or interfering incentive properties of stimuli are avoided, by the use of numerals. By contrast, the same–different reaction time experiments, using symbols for numbers as well as for sameness and difference, do not show a similar symbolic facilitation effect by numerals. Finally, some chimps show a consistent ability to derive real-word locations from representations of location in a scale model, photographs, or videos, while others consistently fail to do so. Our earlier results prompt the question of whether a metarepresentation linking representation of location to actual locations might facilitate performance of these last tasks for those animals who fail on it.

In summary, clear effects of representation type, including photographs, models, or abstract symbols, on information processing and cognition by chimpanzees have been demonstrated by these experiments. Indeed, in the case of the quantity judgments, a fairly simple task has revealed a dramatic difference in how the animals were able to best respond. These results illustrate the significant and exponential impact that symbolic communication, including writing, must have had on our early hominid ancestors. The power of a symbol to override and inhibit strong biological and behavioral predispositions likely represents one of its most adaptive functions. There is little doubt that the continued study of the extant species with whom we have 23 million years of shared evolutionary history promises to further elucidate our own cognitive evolution.

References

Balakrishnan, J. D., and Ashby, F. G. (1992). Subitizing: Magical numbers or mere superstition? *Psychological Review*, **54**: 80–90.

Berntson, G. G., Boysen, S. T., and Cacioppo, J. T. (1993). Behavioral organization and the cardinal principle of evaluative bivalence. *Annals of the New York Academy of Sciences*, **702**: 75–102.

Biro, D., and Matsuzawa, T. (2001). Chimpanzee numerical competence: Cardinal and ordinal skills. In: T. Matsuzawa, ed. *Primate Origins of Human Cognition and Behavior*, pp. 199–225. Tokyo: Springer-Verlag.

Boysen, S. T. (1993). Counting in chimpanzees: Nonhuman principles and emergent properties of number. In: S. T. Boysen and E. J. Capaldi, eds. *The Development of Numerical Competence: Animal and Human Models*, pp. 39–59. Hillsdale, NJ: Lawrence Erlbaum Associates.

Boysen, S. T., and Berntson, G. G. (1989). Numerical competence in a chimpanzee (*Pan troglodytes*). *Journal of Comparative Psychology*, **103**: 23–31.

Boysen, S. T., and Berntson, G. (1995). Responses to quantity: perceptual vs. cognitive mechanisms in chimpanzees (*Pan troglodytes*). *Journal of Experimental Psychology and Animal Behavior Processes*, **21**: 82–86.

Boysen, S. T., Berntson, G., Hannan, M., and Cacioppo, J. (1996). Quantity-based inference and symbolic representation in chimpanzees (*Pan troglodytes*). *Journal of Experimental Psychology and Animal Behavior Processes*, **22**: 76–86.

Boysen, S. T., Berntson, G. G., and Mukobi, K. L. (2001). Size matters: The impact of size and quantity on judgments by chimpanzees (*Pan troglodytes*). *Journal of Comparative Psychology*, **115**: 106–110.

Boysen, S. T., and Kuhlmeier, V. A. (2001). Representational capacities in chimpanzees: Numerical and spatial reasoning. In: L. Sheeran, N. Briggs, G. Shapiro, J. Goodall, and B. Galdikas, eds. *Great Apes at the Crossroads: Chimpanzees, Bonobos, Gorillas and Orang-utans, Proceedings of the Third International Conference on Great Apes of the World*, pp. 131–147. New York: Plenum Press.

Boysen, S., and Kuhlmeier, V. (2002). Representational capacities for pretense with scale models and photographs in chimpanzees (*Pan troglodytes*). In: R. W. Mitchell, ed. *Pretending and Imagination in Animals and Children*, pp. 210–228. Cambridge: Cambridge University Press.

Boysen, S. T., Mukobi, K. L., and Berntson, G. G. (1999). Overcoming response bias using symbolic representations of number by chimpanzees (*Pan troglodytes*). *Animal Learning and Behavior*, **27**: 229–235.

Capaldi, E. J., and Miller, D. (1988). Counting in rats: Its functional significance and the independent cognitive processes that constitute it. *Journal of Experimental Psychology and Animal Behavior Processes*, **14**: 3–17.

Davis, H., and Pérusse, R. (1988). Numerical competence in Animals: Definitional issues, current evidence, and a new research agenda. *Behavioral and Brain Sciences*, **11**: 561–615.

DeLoache, J. S. (1987). Rapid change in the symbolic functioning of very young children. *Science*, **238**: 1556–1557.

DeLoache, J. S. (1989). Young children's understanding of the correspondence between a scale model and a larger space. *Cognitive Development*, **4**: 121–139.

DeLoache, J. S. (1991). Symbolic functioning in very young children: Understanding of pictures and models. *Child Development*, **62**: 736–752.

DeLoache, J. S. (1995). Early understanding and use of symbols: The model model. *Current Directions in Psychological Science*. **4**: 109–113.

DeLoache, J. S. (2000). Dual representation and young children's use of scale models. *Child Development*, **71**: 329–338.

DeLoache, J. S., Kolstad, D. V., and Anderson, K. N. (1991). Physical similarity and young children's understanding of scale models. *Child Development*, **62**: 111–126.

DeLoache, J. S., and Marzolf, D. P. (1992). When a picture is not worth a thousand words: Young children's understanding of pictures and models. *Cognitive Development*, **7**: 317–329.

Eisenberger, R., Weier, F., Masterson, F. A., and Theis, L. Y. (1989). Fixed-ratio schedules increase generalized self-control: Preference for large rewards despite high effort or punishment. *Journal of Experimental Psychology and Animal Behavior Processes*, **15**: 383–392.

Forzano, L. B., and Logue, A. W. (1994). Self-control in adult humans: Comparison of qualitatively different reinforcers. *Learning and Motivation*, **25**: 65–82.

Gallistel, C. R. (1990). *The Organization of Learning*. Boston, MA: MIT Press.

Gelman, R., and Gallistel, C. R. (1978). *The Child's Understanding of Number*. Cambridge, MA: Harvard University Press.

Hare, B., Call, J., and Tomasello, M. (2001). Do chimpanzees know what conspecifics know? *Animal Behaviour*, **61**: 139–151.

Jolly, A. (1966). Lemur social behaviour and primate intelligence. *Science*, 153: 501–506.

Kaufman, J., Lord, E. L., Reese, M. W., and Volkmann, T. W. (1949). The discrimination of visual number. *American Journal of Psychology*, 6: 498–525.

Konorski, J. (1948). *Conditioned Reflexes and Neuron Organization*. Cambridge, MA: Cambridge University Press.

Kuhlmeier, V. A., and Boysen, S. T. (2001). The effect of response contingencies on scale model task performance by chimpanzees (*Pan troglodytes*). *Journal of Comparative Psychology*, 115: 300–306.

Kuhlmeier, V. A., Boysen, S. T., and Mukobi, K. (1999). Comprehension of scale models by chimpanzees (*Pan troglodytes*). *Journal of Comparative Psychology*, 113: 396–402.

Kuhlmeier, V. A., Mukobi, K., and Boysen, S. T. (2002). Chimpanzees (*Pan troglodytes*) recognize spatial and object correspondence between a scale model and its referent. *Psychological Science*, 13: 60–63.

Logue, A. W. (1988). Research on self-control: An integrating framework. *Behavioral and Brain Sciences*, 11: 665–709.

Mandler, G., and Shebo, B. J. (1982). Subitizing: An analysis of its component processes. *Journal of Experimental Psychology*, 111: 1–22.

Marzolf, D. P., and DeLoache, J. S. (1994). Transfer in young children's understanding of spatial representations. *Child Development*, 65: 1–15.

McLaren, I. P. L., Kaye, H., and Mackintosh, N. J. (1989). An associative theory of the representation of stimuli: Applications to perceptual learning and latest inhibition. In R. G. M. Morris, ed. *Parallel Distrubuted Processing: Implications for Psychology and Neurobiology*, pp. 102–130. Oxford: Oxford University Press.

Meck, W. H., and Church, R. M. (1983). A mode control model of counting and timing processes. *Journal of Experimental Psychology and Animal Behavior Processes*, 9: 320–334.

Menzel, E. W., Jr (1962). Individual differences in the responsiveness of young chimpanzees to stimulus size and novelty. *Perceptual and Motor Skills*, 15: 127–134.

Menzel, E. W., Jr, and Davenport, R. K., Jr (1962). The effects of stimulus presentation variables upon chimpanzees' selection of food by size. *Journal of Comparative Physiology and Psychology*, 55: 235–239.

Miller, D. (1993). Do animals subitize? In: S. T. Boysen and E. J. Capaldi, eds. *The Development of Numerical Competence: Animal and Human Models*, pp. 149–169. Hillsdale, NJ: Lawrence Erlbaum Associates.

Mischel, W., Shoda, Y., and Rodriguez, M. L. (1989). Delay of gratification in children. *Science*, 244: 933–937.

Murofushi, K. (1997). Numerical matching behavior by a chimpanzee (*Pan troglodytes*): Subitizing and analogue magnitude estimation. *Japanese Psychological Research*, 39: 140–153.

Peterson, S., and Simon, T. J. (2000). Computational evidence for the subitizing phenomenon as an emergent property of the human cognitive architecture. *Cognitive Science*, 24: 93–122.

Tobin, H., and Logue, A. W. (1994). Self-control across species (*Columba livia, Homo sapiens, and Rattus norvegicus*). *Journal of Comparative Psychology*, 108: 126–133.

Tomasello, M., and Call, J. (1997). *Primate Cognition*. New York: Oxford University Press.

Tomonaga, M. (2001). Investigating visual perception and cognition in chimpanzees (*Pan troglodytes*) through visual search and related tasks: From basic to complex processes. In: T. Matsuzawa, ed. *Primate Origins of Human Cognition and Behavior*, pp. 55–86. Tokyo: Springer-Verlag.

Chapter 23

Language as a window on rationality

E. Sue Savage-Rumbaugh, Duane M. Rumbaugh,
and William M. Fields

Abstract

The question of whether 'animals' share aspects of rational thought with
human beings has been discussed from many different perspectives in
this volume. We suggest that the distinction between human and animal
is an overly simplistic one. This may have been missed because the
important underlying distinction is between the presence and the absence
of language. As Bermúdez (Chapter 5, this volume) states: '...logic
requires language. Language offers the possibility of *intentional ascent*—
of thinking about thoughts. A thought can 'only be held in mind' in such
a way that it can be the object of further thought if it has a linguistic
vehicle.' It has been assumed that all humans possess language and that
all animals do not, regardless of the complexity of their nervous systems,
their vocal systems, their relationships to our own species, or the fact that
some apes acquire human language without training when exposed to it
from early infancy (Savage-Rumbaugh *et al.* 1986). Some species may
utilize natural languages that we do not understand (Savage-Rumbaugh
et al. 1996).

When captive animals acquire language, they also begin to display
evidence of rationality that we are not able to discern or measure in their
non-linguistic companions (Savage-Rumbaugh and Fields, in press). These
differences are sufficiently great as to suggest that truly linguistic animals
are capable of holding a thought in mind and reflecting upon it.
Nonetheless, caution is warranted before making a claim that an animal
has learned language. The ability to 'name' an exemplar, or to respond to
a display of items, may be acquired as a learned association rather than
as a context-free symbol. To operate independently as 'mind stuff,'
symbols must be freed from their context of acquisition.

23.1 **Introduction**

Because of the widespread assumption that only humans possess language, most of the tests for 'rational thought' included in this volume (except those posed by Herman and Pepperberg) are based on non-linguistic behaviors. Often these tests require complex paradigms and multiple levels of inference, making it difficult to differentiate between behaviors produced by prior experience/learning and those produced by computing the mental world of the other. Such difficulties, for example, lead Povinelli (Chapter 18, this volume) to conclude that the non-linguistic tests for mind reading in apes employed to date are inadequate. Povinelli argues that any successful test must adequately address past experience of the subject and that virtually all extant test results can be accounted for in terms of probabilistic learning about how other individuals behave in certain circumstances ('S_b'), without the need to postulate understanding of what other individuals think ('S_{b+ms}').

Tests of non-human mind reading and inferential capacities have become popular ever since Terrace argued that animals cannot learn language (Terrace *et al.* 1979). In response to Terrace's critique of ape language, many researchers turned to non-verbal tests of inferential reasoning and/or mind reading to elucidate the presence and form of rational thought in non-human animals. We agree with Povinelli (Chapter 18, this volume) that many such tests are questionable because they utilize learned rather than novel behaviors. Unlike Povinelli, however, we suggest that the clearest way to determine whether and what animals think is simply to ask them. To do so requires that animals use language to understand our questions and to answer them, and, importantly, *that the questions and answers are not trained behaviors*. They must be novel questions and novel responses, based on a symbolic ability to encode and decode linguistic information on a first trial basis.

We provide an overview of language research conducted with four different groups of apes during the last four decades to the end of delineating the conditions that promote, or discourage, various linguistic capacities. We argue that true language acquisition occurs only through social rearing in environments that are language mediated. Ideally, these environments should include other apes as well as human beings. It is essential that the social environments provide a narrative for the language learner about what things are, what is about to happen, what has previously occurred, and the implications of various events and/or utterances. Real life symbolic communication involves constantly informing the other's mind about ones' own thoughts and intentions. This information supports the mutual co-ordination of behavior and the establishment of rules of social–symbolic engagement.

Additionally, we observe that structured training situations that require the subject to use a symbol to label an object (for a reward), in the absence of a communicative context that meaningfully embeds that symbol in life, result in behaviors that only superficially resemble representational language. These behaviors can be accounted for by what Povinelli calls 'S_b'. Such behaviors are essentially learned associations between 'quasisymbols' and some movement, drawing, or sound such, that re-presentation of a part or portion of the original context (or some subset of behaviors associated with that original context) re-evoke the quasisymbol. These behaviors lack the defining characteristic of displacement, and, thus, they cannot be employed to communicate information about things not present or displaced in time. We suggest that a fully representational linguistic system arises only through emersion in a linguistic community in which comprehending the utterances of others becomes central to everyday life. Through constant exposure to novel context-appropriate utterances, individuals become able to understand things they are told. They employ this understanding to build up a common mental representation of the social/physical world in which they exist. It is through that social/physical representation of the world, constructed by a common language, that symbols and syntax operate to communicate information. In the absence of such a world, they are meaningless. Symbol–object association skills limited to training sessions do not result in a commonly shared mental world. Consequently, no matter how complex the tasks or how elaborate the behavior, what is learned is unable to function as language. Such tasks may show that an organism has the potential for various aspects of language, but the skills themselves remain aloof and apart from the essence of language. The blossoming of language carries with it the realization that the contents of minds of others differ from those of one's own. This realization is inherent within the comprehension and construction of a linguistic system, for one is constantly exposed to symbolic information regarding the intentions of others—which generally differ from one's own intentions. One cannot acquire a representational linguistic system, and effectively employ it to communicate novel information, without a basic understanding that the function of such communications is to bring into behavioral alignment differing views and/or states of knowledge and intention.

23.2 Basic scientific findings: apes, dolphins, parrots

The last half of the 20th century witnessed a major shift in how the foundations and character of animal behavior were viewed. This shift was, in large part, the product of new research paradigms designed to test whether or not animals have the capacities for what we human beings term 'sentience', 'consciousness', 'awareness of self', 'language', and 'understanding of other minds'. Particularly the last two of these topics might be regarded as studies of rationality. Does any animal form, apart from the human, have the potential to acquire language, in part or whole, or the capacity to attribute knowledge to another being? A number of researchers employing very different methodologies and different species began to address the question of linguistic capacities in non-human beings (see Appendix; and see: Aano *et al.* 1982; Gardner and Gardner 1969; Premack 1976; Rumbaugh 1977; Fouts and Mills 1997; Savage-Rumbaugh 1986; Hillix and Rumbaugh 2004; Savage-Rumbaugh and Lewin 1994; Patterson and Linden 1981; Patterson and Cohn 1990; Miles 1983; Matsuzawa 1985; Pepperberg 1999; Herman *et al.* 1993; Idani, in press; Terrace 1979; Shumaker 2003).

As reports from this work were published, the behavior of the various animal subjects and many aspects of the method were translated into 'words.' These translations inevitably obscured the actual procedures employed to produce the behaviors and made it appear that training procedures were much more similar than was in fact the case. This translation of behavioral procedures into vocabulary content has been responsible for considerable confusion in the field of animal language research. It has also made it possible for the skeptic to dismiss, by caveat, many different abilities without seriously evaluating or understanding important competency distinctions. Appendix 1 attempts to alleviate some of this confusion by procedurally describing the different methods and competencies displayed across the different projects. When modern ape language studies began in the late 1960's, it was thought that only human beings could use symbols referentially—that is, to refer to objects and events not necessarily present for the purpose of communication. Nine separate projects arose, each of which employed different methodologies.

1. Gardner, Gardner, and Drum (1989) employed American Sign Language for the deaf and emphasized symbol training and symbol production. Their first chimpanzee, Washoe, was reared alone, but later chimpanzees had exposure to peers. Their chimpanzees acquired between 70 and 160 signs, which were employed to communicate basic wants and desires. Their training program emphasized cultural immersion plus teaching by molding the chimpanzees' hands into the proper sign configuration. The Gardners note that they taught language production, but not language comprehension. They reported that comprehension occurred spontaneously, but no tests of comprehension (such as: 'give X', 'give Y', 'do X and Y') are reported. Their chimpanzees learned signs rapidly and formed many combinations, some of which were clearly novel. Gardner and Gardner (1998) also described the conditions that tended to elicit signing from their subjects: signing was nearly always preceded by questions such as 'Where go?', 'What now?', 'What do we do,?' 'How many?', 'What size?', 'What that make from?', 'Who is that?', 'Want what?', 'Where potty?', and so on. The rate of signing is a function of the rate of questions directed toward the ape. If signing rarely occurred except when solicited by a question, it seems that such questions must have been posed very frequently, since the Gardners note that Dar signed over 479 times per hour! It is difficult to see how anyone could have kept appropriate contextual notes at that rate while also working with the chimpanzee.

Similar sign language methodologies were utilized by: Terrace (1979); Fouts and Mills (1997); Miles (1983, 1990, 1994); Patterson and Cohn (1990); and Idani (in press). (Terrace 1979 with Nim, a chimpanzee) and Miles (with Chantek, an orang-utan) reared their subjects alone, while Patterson reared Koko alone during her formative years and later added a male. Idani is raising his apes as age-related peers. Except for Terrace, the additional investigators who employed sign language also utilized spoken language. These additional projects report similar results regarding sign acquisition. It is quite clear that apes so reared can and do use symbols to express basic needs and desires. It is not completely clear that these apes use symbols in a manner that is detached from immediate personal desires, such as to report events of which others lack knowledge. Miles states that Chantek understands thousands of words, and Patterson reports that both Koko and Michael sign about past events, the most dramatic example being Michael's report of his capture at an early age (see Hillix and Rumbaugh 2004). However, data-based verification of these comments is needed.

Terrace interpreted his results with Nim, and consequently all of the results of the other signing projects, as the result of imitation of caretakers' signs (Terrace 1979; Terrace *et al.* 1979). Many suggestions have been offered regarding the reasons that Nim's signing seemingly differed from that of other apes (Hillix and Rumbaugh 2004). The most salient procedural difference was that Nim, unlike the others, was taught to produce signs in the absence of any referent. Nim had only the signed input of his teachers to depend upon. These circumstances would force a heavy reliance upon imitation. It is also clear that at least some of the other linguistically competent apes did not imitate their trainers nearly as frequently as did Nim (Miles 1983; Brakke and Savage-Rumbaugh 1996).[1]

2. Premack and Premack (1972) employed plastic chips and emphasized the training of specific language-like skills, such as the ability to determine whether a given plastic chip was the 'name of' or 'not the name of' a presented item. The Premacks' methodology was combined conditional match-to-sample tasks, associative learning tasks, and conditional discrimination tasks. They generally presented only a few plastic chips on any given trial, thereby limiting the possible number of choices that could be made. His chimpanzee subject Sarah was not permitted to employ her plastic tokens as a means of either productive or receptive linguistic communication. Therefore she did not use her symbols to express her basic needs and desires, nor did she evidence comprehension, or possess the ability to comment on past events, or to form novel utterances. She learned to associate plastic symbols with different fruits, to solve same/different match-to-sample problems, and to answer 'name of,' 'color of', and relational questions. Premack reported that Sarah 'comprehended' her symbols (Hillix and Rumbaugh 2004) but no tests of comprehension are reported (such as 'Give X', 'Give Y', 'Put X or Y').

3. Rumbaugh (1977) employed a computer-based touch panel of printed symbols. These symbols or 'lexigrams' were geometric forms employed for communicative purposes; all known symbols were available at all times. Lana, like the apes who learned signs, could employ symbols to request her basic needs and desires at any time. Unlike all other animal language subjects, Lana was required to produce sentences with a specific phrase structure grammar. She learned new sentences very quickly and recomposed their

---

[1] Project Nim should not have been the deathblow to ape language research that many assumed it was. Terrace certainly did not imagine his report in Science to be the last word. Indeed, Terrace's anecdotal report on Project Nim, published in the book Nim, reflects accomplishments in non-human language acquisition that Terrace's epistemology and methods would not allow as an empirical report. We are reminded of Merlin Donald's comment that 'Behaviorists should not be surprised to find that there is much more data on animals capacities to be found in ethology than could ever be found in Skinner's boxes' (2001, p. 90).

various parts appropriately (Pate and Rumbaugh 1983) As with Washoe, and other apes discussed above, it was not clear that her receptive capacities equaled her productive capacities, or that she could use symbols in a manner that was detached from immediate personal desires—such as the ability to report past events.

4. Savage-Rumbaugh (1986) employed a methodology similar to Rumbaugh's, although single words rather than stock sentences were taught. In addition, human companions were with the chimpanzees throughout the day, employing many different symbols and engaging them in tasks such a food preparation, cleaning, games, drawing, etc. The teaching of symbols emphasized comprehension, as well as production, and the subjects were immersed in a language environment throughout the day. This study provided the first clear test data to demonstrate that chimpanzees could comprehend symbols and give items as requested by the experimenter. It also provided the first data to show that chimpanzees could employ symbols to communicate with each other. These chimpanzees did not form sentences using a phrase structure grammar but they did combine symbols into meaningful requests. They could use their symbols to express their basic needs and wants as well as their future intentions, and recent past events. They spontaneously reported these events to others who had not seen them, and these events were verified.

5. Asano, Kojima, Kubota, and Murofusi (1982), Matsuzawa (1985), and Biro and Matsuzawa (2001) employed a combination of the methods utilized by the Premacks and by Rumbaugh: a touch-sensitive panel composed of lexigrams was utilized, but only those symbols related to a specific task were available at any one time, limiting the answers that could be given. Ai was peer reared with other chimpanzees. She learned 14 object names and 11 color names. Like Lana, she could give the color and the name of any object upon request. Matsuzawa (1985) later concentrated specifically on teaching counting to Ai, who can count up to nine. Ai also learned to trace her symbols. She did not use her symbols to communicate her wishes and desires, and no data on symbol comprehension has been presented (although there are reports of a receptive task with numbers; see Biro and Matsuzawa 2001).

6. Herman (1986, and Chapter 20, this volume) focused on symbol comprehension rather than production. Using hand signs with the dolphin Akeakamai (Ake) and computer-generated sounds with Phoenix, Herman and coworkers directed dolphins to engage in a variety of behaviors. The dolphins had no means of communicating their desires and needs to the experimenters. Herman employed a phrase structure grammar in sentences given to the dolphins, and they displayed great sensitivity to that grammar (Herman and Uyeyama 1999; Herman, Chapter 20, this volume). They also responded appropriately to novel combinations, understood 'yes' and 'no,' and were able to report the presence or absence of objects and to respond to televised images of hand signals. They were able to imitate movements by human beings, to perform previously acquired behaviors in synchrony with another dolphin, and to produce novel behaviors in response to the request 'create.' When the sign for 'create' was combined with that for 'synchrony,' they were able to create novel behaviors and immediately to perform them in synchrony

with another dolphin. No other animal subjects have achieved this particular skill and there is no understanding of how the dolphins are were able to co-ordinate their behaviors in this manner. There is no evidence that one dolphin was following another. This suggests that they may be employing a communication system of their own as they plan these synchronous novel actions.

7. Savage-Rumbaugh (1993) employed a lexical system similar Rumbaugh's in order to expose bonobos to language. This was the first and only investigation to date to eschew training (Savage-Rumbaugh and Lewin 1994). Like the Gardners, Savage-Rumbaugh emphasized total cultural immersion, but unlike them provided no explicit teaching of, or practice with, the symbols. The caretakers pointed to symbols as they spoke about events of interest, such as travel in the forest, play, food, startling occurrences, and so on. All symbols were available at all times and these apes could use them to express their wants and needs. They could state their future intentions, carry on a dialogue about those intentions with others, and report past events. Comprehension and production both emerged spontaneously, with comprehension of spoken English preceding use of the equivalent lexical symbol as determined by standardized blind tests procedures.

By age 7, these apes were able to demonstrate comprehension of a wide variety of syntactical structures and to understand clearly the grammatical roles of subject, object, recipient, action, location, and the indicative. They understood the semantics and syntax of novel utterances; these abilities were confirmed using blind test procedures (Savage-Rumbaugh *et al.* 1993; Williams *et al.* 1997). Savage-Rumbaugh suggested that training actually inhibited language—the realization of its true nature and its application to all aspects of daily life.

8. Pepperberg (1999) employed a verbal 'model–rival' procedure with her subject, the parrot Alex. Two people talked about and exchanged items in Alex's presence. Pepperberg's work with Alex (1999) emphasizes production. Alex can name a large number of items and produce labels for their color and their shape. Her testing methods entail presenting an item or group of items to the parrot and then asking a question ('What matter?', 'What color?', 'What shape?', and so on). The parrot acquired productive symbol use, but does not respond to commands to give items or to engage in behaviors upon objects, so in this sense has not acquired comprehension. While the parrot employs words to express his desires and needs, it is not completely clear whether these spontaneous utterances hold meaning in the same way that human utterances do. In a recent book, Hillix observed the following:

> When everyone, including the three small aliens, had settled down, Alex launched into repeated requests: 'Want corn.' 'Grain.' 'Grapes.' 'Want cork.' Over and over and over he repeated the requests. When Nicole and Irene were not otherwise occupied, they treated each request as real, and offered the requested item. I soon got into the act as well, and after all initial requests had been honored, when Alex asked for items that he already had, we pointed out to Alex that each item was already on his table. Repeated offering of the items nearly always led to rejection. (Hillis and Rumbaugh 2004, p. 239)

Hillix and Rumbaugh (2004) suggest that Alex was emitting these utterances in order to get people to move about the room, open cabinets, and so on. However, since parrots are known to produce spontaneous utterances for no apparent reason, more data on this issue is needed. Alex is unable to name items if they are presented in a box or as photographs. He must be permitted to taste them or have them rotated in front of him. The reasons for this difficulty are not clear.

9. Shumaker and Beck (2003) have begun a study with two orang-utans (Indah and Azy) using a touch-sensitive screen and symbol system, similar to that first employed with Lana. These orang-utans are being trained to label foods and objects with single lexigrams. They have produced a few symbol combinations. They are permitted to make requests of their human caretaker. Training sessions are limited to an hour per day. Data have not yet been presented regarding symbol acquisition, comprehension skills, or their capacity to report present or past events.

The Appendix further delineates similarities and differences in the procedures employed in these various studies and in the competencies achieved by different subjects. Their skills include associatively learned pairings between objects, complex conditional discriminations, communication of basic desires, use of symbol independently of context, apparent grammatical processing, reporting of past events, and many other complex skills that make up language as we typically employ it in everyday living (Segerdahl, Fields, and Savage-Rumbaugh, in preparation). As the Appendix reveals, the field is rife with different approaches to the topic of language. These differences currently confound any comparative conclusions about different species, except when co-rearing studies are done with different species (discussed below).

23.3 Lexigram studies at the Language Research Center

In the remainder of this chapter we focus upon four distinct phases of research conducted at the Language Research Center. While the lexigram keyboard has remained a constant across all four phases, methodologies have differed significantly, with each phase informed by previous experience. No other animal language studies have experienced this metamorphism across time, with previous work informing how each new set of subjects goes about the process of acquiring language. These differing methodologies have, at times, puzzled readers who have not realized that the changes across time were driven by the research results within our laboratory.

23.3.1 Phase 1: Lana

The first use of the lexigram system began with a single ape, Lana (LANA stood for the 'Language Analogue Project'). Lexigrams were embossed on keys, which initially had to be touched in sequence to form a stock sentence. When Lana was shown a particular food or object, a computer record was made of every utterance by the experimenter and ape. This record made it possible to go back and see what had been said without having to transcribe video tapes and make difficult decisions about whether the ape produced a specific lexigram or not. The lexical system was interfaced with a computer that could respond to the lexical utterances by giving requested items, such as coffee, or by making things happen in the environment, such as opening a window or playing music. The computer was also used to place certain grammatical requirements on the structure and order of lexigrams selected by the ape.

Because the lexical system consisted of distinctive geometric symbols, it was possible to portray them visually in real time, enabling the experimenter literally to read what the ape uttered and *vice versa*. The symbols were displayed through the use of small projectors above the keyboard. One row portrayed what the experimenter had 'said' while the other row portrayed what the ape had 'said.'

At the beginning of work with Lana, there was little confidence that it would be possible to produce true language in an ape. The initial intent was to use the computer to teach Lana language by making her environment change in response to her phrases. However, Lana learned essentially nothing until her caregiver, Timothy V. Gill, went into her chamber and began to talk to Lana using her keyboard. Lana's progress through interacting with Tim provided strong evidence that apes can learn highly complex things, but through the processes of social observation and learning rather than through the rewarding of motor responses (Rumbaugh 1977).

Lana was about one and half years old when her formal training began. She was taught a large number of sentence strings, called 'stock sentences' because their structure followed the specific rules of a phrase-structured grammar. Some examples of such sentences include: 'Please machine give M and M', 'Please machine pour milk', 'Please machine give piece of bread', and so on. Over the course of 3 years, Lana made many novel sentences through the selections of parts of stock sentences. For example, one day Tim was standing outside Lana's room drinking a coke. Lana had learned the word for coke, but had previously used this only in the context of obtaining coke from the machine, by using the stock sentence 'Please machine give coke.' However, at this point coke was not in her vending machine, a fact she could plainly discern. Lana looked at Tim and formed the novel sentence 'Lana drink this out of room?'. Lana had previously learned the term 'this' in the stock sentence 'X name of this' and she had learned her name in the stock sentence 'Tim give Lana X'. She had also learned the stock sentences 'Lana drink milk' and 'Lana move out of room', which she used in order to get Tim to take her out of the room. Elements of all of these different stock phrases were recombined to form the appropriate new sentence 'Lana drink this out of room?', which resulted in Tim's sharing of his coke with Lana. The following day Tim intentionally repeated the behavior of drinking a coke outside of her room. This time Lana asked, 'Please Lana drink coke this room.'

Thus Lana was able to form two completely novel sentences to obtain coke from Tim. Because all of her previous experiences with coke had been limited to using the stock sentence 'Please machine give coke', this example meets Povinelli's requirement as applied here: no previous experience could lead to an S_b account of Lana's behavior, so that an S_{b+ms} account is needed. That is, Lana's novel sentences about coke suggest that Lana had some understanding of Tim's state of mind. Tim was drinking coke outside the room and was not aware that Lana wanted his coke until Lana 'spoke' through her lexigrams. Lana must have realized that she needed to move beyond her stock sentences if she wished to communicate a novel idea to Tim.

Like Herman's dolphins, Lana could work with novel combinations of sentence elements and utilize the appropriate grammatical order as she did so. However, in contrast to the dolphins, it was Lana, not the experimenter who was making up the novel sentences. Like the dolphins, Lana was also taught 'yes' and 'no.' However, instead of using 'no' to indicate that an object was absent and 'yes' to indicate that it was present, Lana learned to answer questions about the state of window or the door (whether it was open or shut). Lana also spontaneously extended the meaning of 'no.' She began to use it when Tim stood outside her room eating or drinking something that was not available to her in the machine. She even stamped her feet to indicate her feelings as she employed 'no.' She also spontaneously expanded 'no' to comments such as 'No chow in machine' when her vending device was empty. Moreover, if Tim represented the state of affairs incorrectly, for example by asserting that coke was in the machine when it was not, Lana would repeatedly respond 'No coke in machine' and then ask 'Tim move behind room?' so that Tim could see for himself that there was no coke in the machine. Thus the term 'no' was spontaneously employed as a comment about the state of affairs in her world and as a statement about Lana's feelings.

When she was first shown an object that she did not know how to name, and no one bothered to tell her its name, Lana soon spontaneously asked 'What name of this?' This behavior could not be explained in terms of any past situations in which Lana might have learned to use this sentence by trial and error. Rather, she must have understood Tim's use of the sentence when it was directed toward her. When the appropriate circumstances occurred, she directed the same sentence toward Tim.

Are these kinds of linguistic usages indicative of rational thought? They certainly indicate that Lana understood the communicative limitations of her stock sentences. Neither Lana's feelings nor her knowledge of things in her world were facts that—apart from Lana's expressions—Tim was privy to. Lana's direct and novel expressions (as opposed to a blind reliance on stock sentences) suggest that she possessed an understanding of the contents of Tim's mind as differing from her own.

However, even though Lana formed many appropriate novel combinations, she also formed sentences that were not semantically interpretable. When this occurred, she was not able to recognize the nature of her error or correct it appropriately and severe confusion could ensue. Moreover, Lana's receptive skills remained limited. She could look at an array of six objects of differing colors and answer questions about the name of the object or its color. However she could not look over an array of objects or photographs and hand over the one requested by the experimenter.

It seemed that Lana's training had given her a piece of the language pie—but not all of it. Therefore, when we began again with new ape subjects, we decided to employ single words rather than stock sentences and to try to achieve two-way communication between our ape subjects. Up to this point all investigations of animal language had focused upon human/animal interactions and upon teaching the subjects to produce symbols or responses. Human researchers tended to scaffold all of the communicative interactions and to assist their subject's every communication. Clearly, language could not have evolved in such circumstances. We hoped that investigations of communications between apes would help us to understand more about language evolution, as well as normal language use.

23.3.2 Phase II: Sherman and Austin

The second phase of the research began in 1975 with the introduction of two additional young chimpanzees, Sherman and Austin, to the keyboard system. Instead of focusing on words and syntax, this research emphasized the training of both sides of the language coin, production and comprehension. Sherman and Austin were taught to name items, to request items and to give items in response to requests from the experimenter. Initially they did not differentiate between 'naming' and 'requesting.' As with Alex the parrot (Pepperberg, Chapter 21, this volume), when they named an item such 'corn' they were given corn as a reward, thereby confounding communication of the name of an item with communication of a desire for that item. However, unlike Alex, when they did request an item, they did not repeat their requests over and over if they already the item they were requesting. Additional training required them to 'name' the item without receiving it as a reward (Savage-Rumbaugh 1986).

Once Sherman and Austin had acquired the skills of naming and requesting, we attempted to determine whether they possessed the receptive competence to give specific foods to the experimenter upon request. Between five and ten foods were placed on a table in front of them and they were asked to give any food item symbolically requested by the experimenter. They were not initially predisposed to do so. Boysen (Chapter 22, this volume) has demonstrated how difficult it is for chimpanzees to ignore quantities of real items and to select the smaller pile of candy for another chimpanzee, while they have no difficulty using numerals to do this task. A similar difficulty appeared when Sherman and Austin were presented with the real foods in a receptive task. There was strong bias for them to attend to the foods that they wanted to eat rather than the symbolically encoded food which the experimenter requested. Thus the experimenter might ask for 'chow' but be ignored as the chimpanzee pointed to a banana and made food barks to indicate that he wanted to eat the banana. The chimpanzees had to learn to set aside their own wishes and to attend to those of the experimenter. Unlike Herman's dolphins whose comprehension tasks dealt with inert objects such as hoops, balls, and pipes, this task dealt with M and Ms, bananas, juice, and many other items that chimpanzees wanted to consume.

Once they achieved the ability to inhibit their immediate desires in order to listen and respond to the experimenter's requests, Sherman and Austin were invited to communicate with one another. This was accomplished by placing foods on the table in front of one chimpanzee, while the other set beside the keyboard ready to ask for the foods he wanted. They initially failed miserably. They lacked the non-verbal co-ordination that co-ordinates symbolic communication. They had to be shown how to listen, how to share, how to hand food to one another without eating it themselves, and how to take turns—cultural behaviors we as human beings take for granted. They also had to learn that the speaker initiates conversations through the use of glances, body posture, timing, and symbolic expression and that it was the role of the listener to use all of these signals to decode the symbolic expressions of the speaker. Decoding also required translating symbolic information into non-symbolic actions, and acting on behalf of the speaker at the speaker's request. When Sherman and Austin did begin to communicate with each other, a variety of spontaneous communicative gestures arose to augment symbolic communication. These gestures indicated that they paid close attention to the visual regard of the other. For example if Austin were looking away when Sherman selected a symbol, Sherman would wait until Austin looked back. He would then point to the symbol that he had used and if Austin still hesitated, Sherman would point to the food itself. If Austin's attention wandered even more, Sherman would take Austin's head and turn it toward the keyboard. If Sherman were not attending to Austin's request, Austin would gaze steadfastly at the symbol until Sherman took note. They recognized that the speaker had to monitor the listener, watch what he was doing and make judgments about his state of comprehension. Depending upon these judgments, the speaker had to decide how to proceed with conversational repair.

The fact that Sherman and Austin had to learn how to co-ordinate their communications at the non-verbal level complements Povinelli's findings that chimpanzees do not spontaneously employ or understand didactic gestures or gaze (Bering and Povinelli 2003). However, when Sherman and Austin were placed within a setting that required communicative co-ordination, these skills rapidly appeared. Although they gave every appearance of clearly understanding and appropriately utilizing symbol, gaze, and gesture, it could nonetheless be argued that S_b could account for these behaviors, and that they did not understand the need to tell each other anything.

Hence a new test for two chimpanzees was devised and data were collected from the first trial. This test did not require the use of lexical symbols, thereby eliminating any effects of training associated with lexigrams and the keyboard. In place of the lexigrams, 14 manufacturer's brand labels were glued onto plastic plaques (for example the 'M and M' label on the M and M bag, or the 'Coca cola' label on a package of Coke). These labels were completely new and different symbols. They did not light up or produce a sound, as did the symbols on the keyboard. The chimpanzees were familiar with the packaging of foods and had seen the labels on the containers many times, but no one had ever used these labels for any communicative purpose. The keyboard was rendered non-functional for chimpanzee A during this test. The test required chimpanzee A to communicate the contents of a container to chimpanzee B without using a keyboard. Chimpanzee B could then employ his keyboard to ask for the food. Only chimpanzee A knew the contents of the container. It was his job to tell the other chimpanzee, in whatever way he could, the name of the hidden food. It was B's job to translate this information back into a lexical request. They were not shown how to do this, they were merely provided with the means to do so as the food plaques were left on the floor in chimpanzee A's room. The task required that they recognize:

- That labels such as 'Coca cola' and 'M and M' were symbols.
- That chimpanzee A, who was shown the contents of the container, needed to communicate this information to chimpanzee B.
- That chimpanzee A could not use the keyboard.
- That the novel plaques on the floor could be employed as symbols.
- That chimpanzee B could understand the information chimpanzee A was trying to convey if chimpanzee A showed chimpanzee B the appropriate novel plaque.

Thus, if they understood the importance of communicating specific information, the tools were present for them to do so. If, however, their previous 'communications' via the keyboard were simply trained responses and they did not understand the differential knowledge states of speaker and listener, they would not know what to do. This task was essentially a version of mind-reading tasks in which one chimpanzee has knowledge that another does not have. In order to share that knowledge, the fact of differential knowledge states must be understood. This test meets Povinelli's criteria for S_{b+ms} because no previous experience or training could lead to successful performance on this task without an understanding of mental states.

Both chimpanzees used these food labels, from the first trial, to tell the other individual the type of food that was hidden in the container. They did this despite having had:

- No training to use these labels as symbols.
- No training to associate a specific label with a specific food.
- No training to show these labels to the other chimpanzee.
- No training to associate a label with a lexigram.

The behavior observed on trial 1 with each of the 14 food labels indicated that individual A realized that his state of knowledge differed from that of B and that this could be changed through of symbolic communication. This test verified that Austin and Sherman possessed an understanding of the mental state of the other (Savage-Rumbaugh 1986).

As critical as this test was, it was not as important as the spontaneous emergence of statement skills. With the emergence of the ability to state their intentions, Sherman and Austin arrived on the doorstep of true symbolic function. Not only did they recognize differential knowledge states between themselves and others when required to do so, they began to realize that differential knowledge states existed generally and that the way to co-ordinate with others was to announce one's intentions. Thus they began to say 'Go outdoors' and then to head for the door, or 'Apple refrigerator' and then to take an apple from the refrigerator (rather than any of the other foods that were also located in the refrigerator).

These statements of intended action differed from the requests of Lana in that they were not uttered for the purpose of getting an experimenter to do something for them (such as open the cabinet, take them outdoors, play with them, etc.). Sherman and Austin were instead attempting to co-ordinate their intended behavior with that of others by explaining what they were going to do before they did so. In order to be able to produce statements about intended action, for the purpose of co-coordinating future actions with others, one must be able to form a thought and hold it in mind. Could they also reflect upon that thought?

We tested for this capacity by teaching them to sort six objects into the lexically marked categories of food and tool. In the initial task, three of the objects were food symbols and three were tool symbols. We then presented them with the remainder of the food and tool lexigrams present in their vocabulary and asked them whether the symbol itself belonged to the category of food or tool. Each symbol was presented for a single trial only in a blind test paradigm. In order to answer this question correctly, they had to look at each symbol, reflect upon its referent, and make a categorical decision about it. Their ability to do this correctly was nearly perfect, indicating that they could hold a symbol in mind, reflect upon it and assign it to category, without ever having seen others so classify these symbols (Savage-Rumbaugh *et al.* 1980).

Along with ability came the spontaneous reporting of events. For example when Sherman was outdoors he saw an anaesthetized chimpanzee being carried on a table past our building. He immediately rushed inside to the keyboard to state, 'outdoors scare', and gestured in the direction of the anaesthetized ape. Another day while he was outdoors, someone lit a sparkler and Sherman rushed in to announce, 'Straw give scare outdoors.' Given that no training paradigm had required Sherman and Austin to tell us what they saw outdoors, they must have realized they were privy to information that we were not. Indeed, only those events that were unusual or frightening or posed possible danger were brought to our attention. Such selectivity lends weight to the conclusion that they were reporting things to us that they felt were important to them, and that they assumed we did not know, but needed to know.

Sherman and Austin also began at this time to attend to television. Herman (Chapter 20, this volume) reports that, unlike the dolphins, Sherman and Austin had to be trained to watch television. It is true that when Sherman and Austin were first exposed to television broadcasts of old entertainment programs they did not watch with interest. However, just as with the dolphins, when the television was employed to depict relevant information to them, they responded immediately. For example when they saw a televised picture of a table of food in an adjacent room, they specified, at the keyboard, that they wanted to go to that room and they reported which food they wanted to eat. After this they began to watch various shows and appeared to follow aspects of the narration. Austin for example, enjoyed King Kong. He would watch the scenes of Fay Ray in Kong's hand with rapt engagement. If someone moved in front of the TV, Austin would become upset and begin to rock. When there were scenes of Kong in a cage, Sherman would go and sit in a similar cage that was located nearby.

They also began to differentiate between live and taped images of themselves. When they saw a TV with their image on the screen they would wave their hand, stick out their tongue, wiggle their feet, and so on, and watch the TV to determine whether there was a correspondence between what they did and the image on the TV. If not, they generally lost interest in their image. However, if it was live, they engaged the image. Sherman would find a cloth and make a cape of it and puff up the hair on his shoulders so that he looked very large and parade up and down in front of the television. Austin would find a spoon and pretend to eat imaginary food while watching himself on television. He would also look into the camera's lens and shine a flashlight down his throat, so that he could get a television perspective of his throat. This was much better than what he could see by looking in a mirror. No one demonstrated this behavior for him, yet he recognized that the perspective offered to him by a camera was distinct from the perspective offered to him by a mirror. How did he come to realize that he could see further down his throat with a TV camera than with a mirror? Their interest in their own live images on television seemed to arise from the fact that the camera, unlike a mirror, could show them many different perspectives of themselves.

Spontaneous (or indeed trained) differentiation of live and taped self-images has not been reported for any other non-human species. Further detailed study of this behavior (Menzel *et al.* 1985) also revealed that Sherman and Austin could compensate for 180 degree rotation of an image. No training was given to establish this skill; it emerged in both of them at about the same time that spontaneous statements and reporting of out of sight events began to occur. At the very least, this skill illustrates a cognitive capacity to reflect upon their own self-image from many different perspectives. It also illustrates that Sherman and Austin know what they know about whether the image they see is live or taped, for they are clearly assessing and testing this knowledge (see Shettleworth and Sutton, Chapter 11, this volume).

Sherman and Austin also engaged in other forms of pretense during this period. For example, a favorite game was to pretend that small gorilla dolls were biting their fingers. After the pretend bite, they would nurse and care for the finger as though it were hurt, showing—as Currie notes (Chapter 13, this volume)—that they were able to extend the initial pretense of biting to a logical consequence, that of having a hurt finger that needed special attention. Both apes would also pretend to talk on the telephone. Austin became entranced with his shadow. When movies were projected on the wall, he would, during chase scenes, spontaneously intersperse his shadow into the picture in such a manner that the shadow would appear to be chasing the chimpanzee on the screen. It took some considerable planning for him to place himself at the right point between the projector and screen to accomplish this effect. Austin and Sherman also played an interesting game of light chase in the evening. If Austin were given a flashlight he would shine it around different locations of the room and Sherman would chase the light beam. This game was not introduced by human beings, but was devised by the chimps.

These playful antics suggest that something was occurring in Sherman and Austin that goes beyond what was observed in the chimpanzee subjects described by Povinelli, Call, or Tomasello and Call (all in this volume). Of special interest was the Sherman and Austin's desire to determine whether a televised image of themselves was live or taped. The development of such a desire appears to require a capacity for self-reflection.[2] Austin's use of the camera to gain a new perspective on his throat suggests that he understood that there are differential perspectives to begin with. And why would they employ (with no training) a new symbol system to communicate, if they did not understand that differential knowledge was provided to the one who saw the container being baited?

We believe that Sherman and Austin were making statements, reporting events, using didactic gestures, identifying live versus taped images, playing games with images, interacting with television narratives, pretending to feed themselves and others imaginary food, and so on because, through language, they were developing mental worlds sufficiently rich to permit an increased degree of self-reflection, understanding of the mind of the other and rationality.

23.3.3 **Phase III: Kanzi**

Phase III began with a simple question—could an ape from another species learn to use a keyboard, as the chimpanzees Sherman, Austin, and Lana had? The other species of interest was the bonobo, *Pan paniscus*. This work began in 1980, at a time when there was limited data available about this species either in the wild or in captivity; many scientists still questioned whether it was a distinct species. However, the data that did exist suggested that bonobos were more vocal and more gestural than other apes during their natural communications (Savage-Rumbaugh *et al.* 1977).

One of these apes, Matata, an adult female, along with her 6-month-old son Kanzi, were assigned to the Language Research Center. Rather than separate Kanzi from his mother and attempt to teach him language, it was determined that he would be left with her even during sessions intended to teach language to her. This ape language study was the first attempt to teach communicative symbols to an adult ape and the first to work with a mother whose child was left with her during all teaching sessions. The training method with Matata was essentially the same as that utilized with Sherman and Austin.

---

2 The term 'self-reflection' argues for an assessment of self beyond Call's notion of 'reflection' in apes.

Matata was taught to name objects; however her progress was much slower than that of Sherman and Austin, and was fraught with many more difficulties. She never advanced to the level where she could give specific lexically encoded items in response to the requests of others. She could hand over a requested item that she already had in her possession; but if she was repeatedly asked to look through an array of foods and give a specific one, such as an orange or grapes, she could not do so. Her training proceeded for 2 years, by which time she could name eight foods as they were shown to her, request the foods she desired, and request simple social actions and games. Thus, she lagged far behind the Lana, Sherman, and Austin.

When her son Kanzi was 2 years of age he was separated for several months from Matata so that she could breed with the *Pan paniscus* male at the Yerkes Field Station. From this point on, Kanzi received care around the clock by a group of four caretakers who rotated through day and night shifts. At the time of the separation, we believed that Kanzi did not know any lexigrams, except 'chase', used for a game that he tended to request four or five times a day. However, during the first day of Matata's absence he used the keyboard over 300 times to request foods, actions, and objects. Moreover, he formed spontaneous lexical combinations such as 'chase apple', meaning that he wanted to play 'keep-away' with the apple. It quickly became apparent that for the first time a non-human being had acquired a symbolic system of communication without specific training.

Kanzi's abilities in that first month were determined to be equal to the combined capacities described above for Sherman and Austin. He made statements of intent, he shared food, he combined his symbols spontaneously, and he replied to requests both symbolically and behaviorally. Every skill that Sherman and Austin had been trained to execute we found extant in Kanzi by 3 years of age, even though the only 'training' he received was through observing what we attempted to teach his mother (Savage-Rumbaugh *et al.*1986).

Phase III therefore raised a most interesting question: Had our attempts to train Sherman, Austin, and Lana really been needed? Kanzi was with his natural mother during the first two and a half years of his life while people were attempting to teach her language. He was played with and entertained in those sessions, yet he silently learned far more than she did. He did not display this capacity while she was around. Possibly he felt no need to use the keyboard to communicate with his mother, for she served as the intermediary between him and the human beings in his world. However, once she was no longer there and it became important to employ the lexical system that the human caretakers had employed, these latent skills appeared at once.

Once we understood that Kanzi did not require training sessions in order to acquire symbols we abandoned all plans to train him. Instead we concentrated upon using language around him to discuss what we, and he, were doing, had done or were going to do. Kanzi responded well to this approach and began rapidly to acquire new symbols that Matata had never seen. In addition, Matata produced another baby,

Mulika, who was initiated into this new mode of lexigram usage from birth. Mulika acquired symbols even more rapidly than Kanzi. Like Kanzi, in order to acquire a large lexicon, she needed only to observe others using the keyboard in the context of daily events. Like Kanzi, she was able to recognize individual symbols off and on the keyboard, to name things, to request things, to comprehend and respond appropriately to the statements of others, and to employ lexigrams for items that were removed in space and time. Like Kanzi, she always first demonstrated her knowledge of any lexigram by recognition of the spoken word for that lexigram (Savage-Rumbaugh *et al.* 1986).

By the age of 8 years, Kanzi's understanding of spoken language had advanced to the point that he could interpret novel English utterances that were syntactically quite complex and semantically unusual, such as 'Get the toy gorilla and slap him with the can opener' (p. 115) or 'Feed the doggie some pine needles' (p. 117). Formal tests of this ability indicated that Kanzi had broken the syntax barrier and that he was simultaneously processing both syntax and semantics (Savage-Rumbaugh *et al.* 1993). Unlike Herman's dolphins, and unlike Lana, Sherman, and Austin, Kanzi was never trained to comprehend words or phrases. Normal spoken English was employed simply to carry on the business of living (Segerdahl, Fields, and Savage-Rumbaugh, in preparation). Kanzi's tests of comprehension did not intersperse novel trials with trained ones, as was the case for the dolphins. Each one of more than 600 test trials was novel for Kanzi.

Tomasello (1994) argues that although Kanzi appeared to be listening to complete sentences, it was possible for him to infer what to do in response to only one or two words and that no syntactical device was needed for such inferences. This critique ignores the demonstrations of sensitivity to word order and to embedded constructions that require an understanding of recursion. For example Kanzi was sensitive to word order, responding appropriately, and differentially, to sentences such as 'Pour X in Y' versus 'Pour Y in X.' Comprehension limited to processing only 'X' and 'Y' would result in picking up or pointing to X and Y, in any order. It would not account for the correct response to the verb 'pour' or for the correct subject–object relationship dictated by the verb. Since sentences given were the precise inverse of one another (for example 'Pour the juice in the egg' versus 'Pour the egg in the juice'), any explanation that implies Kanzi could guess what to do simply from the items themselves is insufficient.

It has also been suggested that Kanzi was simply responding to the first item spoken first and to the second spoken item second—without processing syntax. However, this would not account for his differentiation of 'Pour X in Y' from 'Pour Y in X.' Moreover, when presented with sentences such as 'Give X and Y' Kanzi often gave only one item and it could be either the first or second item. Often the experimenter had to pose another request for the remaining item. Thus sentences that required syntactical processing were easier for Kanzi than those that required him to recall a short list of items. The holistic aspect of a syntactically integrated sentence placed more of a processing load, though less of memory load, upon Kanzi.

Equally important is the fact that context often could not serve to help Kanzi understand what to do. When he was presented with a wide array of objects, which happen to include juice in a bowl and an egg in bowl (along with a knife, a mask, an apple, a toy dog, a balloon, a cup of water, and a TV) the correct action was not at all obvious. In order to know how to appropriately respond to 'Pour the egg in the juice' or to its inverse, Kanzi had to comprehend all three words and their ordered relationship.

In addition to word order, Kanzi (and Panbanisha—see Phase IV, below) demonstrated sensitivity to sentences requiring recursion, such as 'Get the X that is in Y', where numerous exemplars of X are located in alternative locations. Only if Kanzi recognized that a particular X is distinguished by its syntactically defined relationship to Y, could sentences containing such recursions successfully be comprehended. Thus if Kanzi is asked to get the tomato that's in the microwave, he has to ignore the tomato in front him and all the other tomatoes located in various parts of the laboratory, in order to bring only the one that is in the microwave (other items would be placed in the microwave as well, so that Kanzi could not simply go to the microwave and return with whatever he found there). When Kanzi heard such a sentence he did not even look at the tomato directly in front of him. He headed straight for the microwave.

The importance of a syntactical structure such as recursion lies in the fact that the relationship between the content words ('X' and 'Y') is specified by a word (in this case the word 'that') whose *sole* function is syntactical. That is, the function of the word 'that' is specified by purely its location within the sentence. There exists no external semantic referent for 'that' which is defined apart from the structure of the sentence. In a sentence such as 'Get the X that is in Y,' the word 'that' serves to specify which of a number of Xs is the X to be acted upon.

An understanding of the two words 'tomato' and 'microwave' might result in the ape looking at the microwave and eating or pointing to the tomato. It would not result in the ape bringing the tomato from the microwave. This action requires that the syntactical structure of the sentence be comprehended. Supporting this view is the fact that when recursion was not utilized (as in 'Go to X and get Y'), responses indicating ambiguity were significantly more probable; for example, Kanzi was likely to go the microwave but then hand the tomato in front of him to the researcher, rather than to assume that the researcher wanted a particular tomato from the microwave. Thus it was clear, from blind test data and novel trials, not only that the syntactical device of recursion was understood, but that without it sentences became ambiguous.

The powerful influence of syntax for the bonobos was initially revealed in a surprising way when the syntax of a sentence that was used to explain something to Kanzi was incongruent with what the user of the sentence had had intended to convey. At that time that the first author (SSR) believed that Kanzi's comprehension of syntax was, if extant, little more than the protosyntax attributed to him by Bickerton (1995) and Tomasello (1994) and that when bonobos responded to complete sentences, they were relying on inference and semantic reference alone. However, on a particular occasion, SSR was seated by the keyboard chopping up some food for Kanzi when she happened to cut her thumb with the knife. At moment the accident occurred, Kanzi was looking away, engaged in play with a toy. When he turned and realized that her thumb was bleeding, he assumed an expression of surprise and concern as though he were wondering how this could have suddenly happened to her. SSR might well have said in English 'I cut myself with the knife'; if she had, Kanzi would readily have understood, as she had cut herself before and had often cautioned him to be careful when he used a knife. He thus understood that a knife could slip and cut himself or someone else. However, since SSR happened to be sitting next to the keyboard, she in fact chose to use it rather than to speak. Since the word 'cut' was not on the keyboard, she expressed the situation as best she could given the available vocabulary and keyed 'Knife hurt Sue.' Although SSR said 'Knife hurt Sue,' she meant 'I cut myself with the knife' and assumed (wrongly, as it turned out) that if Kanzi only understood the words knife and Sue, he could infer that she had cut herself. He had seen this happen before and was always interested in inspecting such injuries.

However, immediately on hearing SSR's words, Kanzi looked amazed and then produced a loud sound of alarm and knocked the knife across the room. This was *not* his normal response to seeing someone injure herself with a knife. Rather, by saying 'Knife hurt Sue', SSR had *syntactically* conveyed something she had *not* intended to convey, namely, that the knife had acted with an agency of its own. And Kanzi had responded not to her intended meaning, but to her actual utterance and its syntactical structure. Had she said instead 'I cut myself with the knife', Kanzi would not, based on past experience, have shown alarm or knocked the knife away.

Kanzi's sensitivity to the implications of the specific syntax used, as opposed to what had been intended, was surprising. If Kanzi had been sensitive only to the words 'knife' and 'hurt', he would have no reason to attribute agency to the knife. The fact that Kanzi apparently did attribute agency to the knife in accordance with the implications of syntax, when the speaker had not intended to do so, made us aware that we had thus far underestimated the power for syntax in Kanzi's understanding.

A critic might elect to dismiss this observation as anecdotal. However, to do so would have been unwise for those working with Kanzi around knives, as confusion about knives and their magical powers could have ensued for sometime; Kanzi might well not have wanted that particular knife to be used again. When SSR realized what had occurred, she explained to Kanzi in English, without the keyboard, that the knife itself had not hurt her but that she had simply cut herself with it accidentally. She retrieved the knife and continued to use it. Kanzi looked suspicious for a few moments, but then acted as though the knife was no longer a threat, and was solicitous about the cut thumb.

The difference between position of someone who is actively engaged with apes moment by moment and that of someone who speculates from a detached perspective about alternative interpretations is clear in this case. While such speculations serve the purpose of argument well, they ignore the reality of the communicative situation and the critical nature of culturally appropriate actions that participants must take on the spot in order to manage evolving social situations.

During the ensuing tests that were presented to Kanzi, many sentences were employed that could not be properly processed without a sophisticated grasp of syntax. For example Kanzi was asked 'Can you feed your ball some tomato?' His 'ball' was a round sponge ball with a pumpkin face, and was on the floor along with other objects, including other balls. Kanzi had not treated this ball as a face before, or even paid attention to the fact that the ball had a 'face' on it. However, upon hearing the request, he turned the ball around until he found the face and then pretended to place the tomato on the ball's mouth, as though the ball could eat the tomato. In this instance, not only was he able to treat the ball as an animate agent capable of eating, he was simultaneously able to recognize that he and the researchers were only pretending that the ball was an animate agent. Had Kanzi understood only the words 'ball' and 'tomato' and their order, as Tomasello suggests, how could such a complex behavior have ensued? Rather, he needed to understand the verb 'feed,' and the syntactical implication that the ball was to be the object of a pretend action with a tomato, and not the other way around. He did not try to feed the ball to the tomato. Yet 'ball' is used first and 'tomato' second, and Kanzi could use words in sentences such as 'Rose chase Liz', where Rose is designated as the chasor and Liz as the chasee. A response strategy tied only to word order would thus have led him to make the ball (mentioned first) act on the tomato (mentioned second).

23.3.4 Phase IV: Panbanisha and Panzee

Phase IV was initiated in order to determine whether the ease with which Kanzi acquired symbolic and protogrammatical skills was the result of species or procedural variables. A critical question was: would Lana, Sherman, and Austin have fared differently if we had begun to work with them during infancy and had avoided the training of specific lexigrams? This was a critical question. Hundreds of trials had been required to teach Sherman and Austin to select specific lexigrams for specific for specific items. After this initial training, new food items could be acquired from a single presentation. Yet even though food names came to be easily acquired, each completely new conceptual class (object names, color names, location names, persons names) required similar initial extensive training for Sherman and Austin; moreover, the full complement of symbol competency with food items did not always transfer when new conceptual classes were introduced. These limitations did not hold for Kanzi, who had not had such training. Any conceptual class (requesting, comprehending, naming, making statements, and so on) could be utilized in combination with any vocabulary item. If training was not required for apes to become linguistically competent, our training tasks and careful data collection had been unnecessary. On the other hand, if bonobos were simply able to learn language on their own, while other ape species were not, then bonobos, like human beings, were an exceptional species.

We approached these questions by corearing a female infant of each species Panbanisha (a bonobo, or *Pan paniscus*) and Panzee (a chimpanzee, or *Pan troglodytes*) in an environment similar to that experienced by Kanzi and Mulika (Brakke and Savage-Rumbaugh 1995, 1996). We studiously avoided training the apes to point to particular symbols and rewarding them for any specific symbol usage. This corearing study began shortly after birth and lasted for 4 years. During this period of time Panbanisha and Panzee were together 24 hours a day and shared all of their living experiences in an environment of language immersion. All uses of the keyboard by Panbanisha and Panzee were recorded with the utterance, date, name of the ape, and contextual information surrounding the utterance. Gestures also were noted.

The movements of the infant chimpanzee Panzee were jerkier and less co-ordinated than those of the infant bonobo Panbanisha. In addition, the chimpanzee was significantly delayed in her ability to track visually more than one object at a time. As she grew older she remained more distractible than the bonobo. Any sound or event in her environment immediately drew her attention away from the task at hand.

Both infants began to babble at the keyboard at 6 months of age. These babbling episodes generally occurred around naptime as the infants would look over the keyboard and touch specific symbols. Sometimes they would take the caretakers hand and have her point to symbols as well. However they showed no expectancy that anything happen as result of having selected certain symbols. They seemed content simply to point to different lexigrams and note that their caretaker was watching.

Panbanisha the bonobo began using lexigrams to communicate at 1 year of age and by 30 months she was employing more than 120 different lexigrams. By contrast, Panzee the chimpanzee relied on gestural communication until she was 20 months of age, when she used her first lexigram. By 30 months, she was using 55 different lexigrams. She continued to lag behind Panbanisha in the total number of known lexigrams throughout the study period of 4 years. Formal tests to control for contextual cues were presented to both subjects at 4 years of age. At this time Panbanisha comprehended 179 words and their lexical referents. Panzee comprehended 79 words and their lexical referents. In addition, Panzee often over-generalized her symbols. For example she employed the word 'dogs' to indicate she wanted to visit the dogs, the orangs, and other chimpanzees. Panbanisha, by contrast employed different lexigrams for each of these categories. Panbanisha also distinguished between general grooming and care of wounds, while Panzee employed a single term for both items.

Panzee's comprehension was more closely linked to the here and now and to objects in full view than Panbanisha's. Panbanisha comprehended utterances about the past and future quite readily. For example if they were told that 'The gorilla was going to appear at the house later,' Panzee would display pilo erection and orient at once toward the house. Panbanisha would comment 'gorilla later' and then inquire about the gorilla at a later time. Panbanisha acquired state lexigrams, such as 'rain', 'quiet', and 'sleep', while Panzee did not. Panbanisha also acquired names for herself, Panzee, Kanzi,

Matata, Sherman, Mary Ann, Sue, Kelly, Lana, Liz, and others, while Panzee acquired no proper names of chimpanzees or persons.

Panzee also experienced difficulty in responding to complex utterances in which multiple aspects, such as action, location, and object, all had to be kept in mind. Data were kept on all direct requests made of both individuals. These data included the request itself and notes about their response. These requests were made during the course of daily living in a normal context. However, only those requests that could not be decoded based on contextual data were recorded. Panbanisha responded correctly to 90 per cent of the English-only requests, while Panzee responded appropriately to only 78 per cent of the requests. Further analysis of the data revealed that Panzee, but not Panbanisha, had particular difficulty with three-term requests, responding correctly to only 57 per cent of the three-term requests. Thus Panzee did quite well with simple requests like 'Wash Karen' or 'Go (to) A-frame', but experienced difficulty with longer three-term requests such as 'Carry the string to the group room.' Her behavior suggested that she had difficulty remembering all the components of such utterances. Panbanisha handled both two-term and three-term requests with equal facility.

As Panzee's vocabulary increased, she continued frequently to use gestures and to combine them with lexigrams. As a result, her total number of combinatorial utterances was higher than that for Panbanisha, whose combinations were mainly lexical. At 3.5 to 4 years of age we investigated their ability to form lexical and lexical/gestural combinations and found that 16.7 per cent of Panbanisha's corpus (1088 utterances) and 16 per cent of Panzee's corpus (1000 utterances), respectively, were combinations of two or more elements—a striking cross-species similarity in frequency (Greenfield, Lynn, and Savage-Rumbaugh, in preparation). Both animals produced a higher proportion of combinations as opposed to single symbols than Kanzi, Panbanisha's half-brother (10.4 per cent combinations).

Combinations of gestures and/or lexigrams were coded according to the semantic roles of each element in a relationship. Table 23.1 presents examples and frequencies of two-element meaning relations (two lexigrams, two gestures, one lexigram and one gesture) constructed by each subject. The table also presents the distribution of orders in which a relationship was expressed (e.g. action–object order vs. object–action order).

The table shows a strong similarity between the animals of the two species. Each utilizes lexigrams and gestures to construct the same range of semantic relations; both have significant ordering tendencies in their combinations. These similarities were also shared by Kanzi and by human children. Panzee and Panbanisha also utilize the same particular orders, indicating the construction of a shared communicative culture. The notion of *Pan* culture thus extends from behavioral patterns (Savage-Rumbaugh *et al.* 1992) to symbolic patterns.

For both apes, their most frequent semantic relationship is between an entity and a demonstrative (Table 23.1), where the demonstrative is gesturally indicated and the

Table 23.1 Frequency and Order of Different Proto-Grammatical Categories

Type	Frequency		Type	Frequency		Type	Frequency	
	Pb	Pz		Pb	Pz		Pb	Pz
Action agent	41***	29**	Entity location	67***	46***	Attribute entity	8	12
Agent action	4	10	Location entity	32	12	Entity attribute	5	4
Entity demonstrative	58***	89***	Affirmative action	42***	24**	Entity action	10	4
Demonstrative entity	18	14	Action affirmative	13	8	Action entity	4	2
Action object	41***	18*	Instrument object	1	7*	Instrument goal	1	10
Object action	7	7	Object instrument	0	0	Goal instrument	1	8
Goal action	34	49	Attribute action	5	11**	Repeated category	103	123
Action Goal	39	69	Action attribute	1	1	Miscellaneous	43	43
Affirmative goal	31*	19***	Affirmative entity	15*	10			
Goal affirmative	13	3	Entity affirmative	5	5	Total	642	637

Pb = Panbanishsa; Pz = Panzee

* .05 significance level

** .01 significance level

*** .001 significance level

entity is specified with a lexigram. Similar combinations of indicative gesture and word are also the earliest and most frequent combinations of hearing and deaf children (Goldin-Meadow *et al.* 1994). For all three species, this cross-modal combination of object symbol (in lexigram, speech, or sign) and indicative gesture (point or touch) creates the basic semantic relation of reference.

Both Panzee and Panbanisha (like Panbanisha's half-brother Kanzi) utilize a spontaneous ordering strategy in their cross-modal communications. All three animals create combinations in which the lexigram precedes the gesture (88 per cent; see examples in Table 23.1). Similarly, a deaf human child never exposed to sign language models also constructed his own ordering strategy for arranging a noun-like sign plus an indicative gesture, parallel to the entity–demonstrative combinations by the apes noted above (Goldin-Meadow *et al.* 1994). That is, the ordering was the same as Panzee's and Panbanisha's and, like theirs, was not learned from a caregiver.

However, there is also an important difference between *Pan* (both species) and *Homo* that is pertinent to the independent evolution of human syntax. From about age 3, the deaf child produced indicative gestures after the noun signs but before the verb signs. In contrast, Panzee and Panbanisha placed gesture last across syntactically diverse situations.

Chimpanzees are more involved with tools; bonobos are more involved with social relations (Kano 1990; Wrangham *et al.* 1994). This difference is reflected in the semantic content of our subjects' combinations. While Panbanisha's two-element combinations overwhelmingly refer to a social relationship rather than to a tool relationship, Panzee's two-element combinations are balanced between social relations and tool relations. This differential focus of interest may have emerged during the independent evolution of bonobos and chimpanzees over the last two million years.[3]

23.4 **Linguistically based assessments of ape mind reading**

Apes that are linguistically competent can be given mind-reading tests that do not require the kind of training and test trials objected to by Povinelli (Chapter 18, this volume). Instead of having to determine, on the basis of inference, whether or not they are thinking

[3] While evolutionary conclusions may be drawn from this study, we alert the reader that while Panbanisha and Panzee's postnatal experiences were as similar as possible, their prenatal experience differed. Panbanisha's mother was housed at the language research center where she experienced close contact with human companions who exposed her to spoken English. Panzee's mother was housed at the Yerkes Primate Center with 200 other chimpanzees were the sonic environment was composed of displays, banging, and screaming. Panbanisha's mother also experienced a diet rich in fats and protein, while Panzee's mother received mainly monkey chow. However both babies appeared vigorous and healthy immediately after birth. Panzee's clinging reflex was much more readily elicited and much more forceful than Panbanisha's. Panzee could support her weight from the start, while Panbanisha needed assistance.

about the contents of others' minds, one can ask them, 'What are you thinking about? For example while the first author and Panbanisha were in the forest away from the other bonobos, Panbanisha had a pensive look on her face and the SSR asked, 'Panbanisha, are you thinking about something?' Panbanisha responded vocally with an affirmative 'peep.' SSR asked, 'What are you thinking about?' Panbanisha pointed to the lexigram keyboard and uttered 'KANZI.' The concern was relevant as Kanzi had been sick that week and was not feeling well at all. On another occasion, Panbanisha was grooming with her human companion Liz. Panbanisha seemed to be preoccupied and Liz asked, 'Panbanisha are you thinking about something?' Panbanisha responded at the keyboard board 'THINK T-ROOM,' an utterance indicating that she was thinking she and Liz might visit the T-Room where toys, musical instrument, and supplies are kept. A few minutes later Panbanisha modified her statement uttering 'T-ROOM LATER GRAPES TODAY.'

Our apes can answer this question in various contexts. Panbanisha has responded to this question in many different ways at different times. Sometimes she answers that she is thinking about food, at other times about Kanzi or Matata, at other times about traveling to a specific location, and so on. She validates these expressions of thought with her ensuing actions if given an opportunity. For example one day while riding in the car she became very upset and wanted to get out. SSR inquired as to what she was thinking and she replied 'Colony room scare.' SSR then asked if she wanted to go back to the colony room and she replied 'yes.' Upon returning there, we found that a snake had entered the colony room and scared both the bonobos and the people there. Panbanisha had been able to hear their reactions to the snake from the car, though her human companions could not, and had reported her perceptions. In another case, one day in the forest Kanzi looked upset and SSR asked him what was wrong. He replied 'Austin fight,' so we traveled back to the building where Austin lived and found that Austin and Sherman had been in a serious fight. Again, Kanzi had been able hear things that humans could not and reported his perceptions.

Language competency also makes it possible to present mind-reading tests to apes that are generally reserved for children. We have conducted tests with Kanzi and Panbanisha in which they observe a person ask for M and Ms to be placed in their lunch box. This request is complied with and then the person is asked to go out of the room and retrieve something. While the person is gone a 'trick' is played on him, and pine needles are substituted for M and Ms. When the person returns, as he tries to open the lunchbox Kanzi or Panbanisha are asked to state what the person is looking for. They both replied 'M and Ms'. When this procedure was repeated and they were instead asked what was in the lunch box, they replied 'pine needles' (Savage-Rumbaugh 2000). Thus, on a first trial, they were able to produce statements of contents of the box and statements of the content of someone's mind. When this test was repeated with Panbanisha and a tick was placed in the box rather than pine needles, she had a different response. When asked what the person was looking for, she replied instead that the experimenter was bad, indicating her moral judgment of the behavior of another.

23.5 **The emotional roots of language**

In order to better appreciate the significance of the different training methodologies, test methodologies and skills reported for different animal language subjects (summarized in the Appendix), we turn to the theory of emotional and intellectual growth presented by Greenspan and Shanker (in press). This *functional emotional model* articulates the way the earliest communications between infant and caregiver emerge from emotional signals presented by the infant. Through coregulated affective exchanges, a child learns to signal her intent. As intentions are responded to the child becomes increasingly better able to signal many varied and subtle intentions without escalating to action and to engage in more and more circles of communication. Coregulated affective interactions support the development of a number of capacities, which, when combined with symbol formation, lead to language.

At the heart of this theory is the idea that non-verbal affective interactions are circular in nature, with each affective expression drawing an appropriate response from the other party as meanings are exchanged. Meanings have their deepest roots in the affective domain and are first expressed in that domain. The more that emotional expressions come to operate as intentional signals rather than merely as raw emotive actions, the more these intentions become freed from the perceptual states that generated them. With this freeing from initial affect, that expressions are transformed to function as symbols. When this process is interfered with, for either biological or environmental reasons, the child fails to progress functionally toward more complex stages of development.

Although this theory was developed with children, since it deals with non-verbal exchanges of meaning it also applies to non-human communication systems. The theory would predict that the more complex the affective non-verbal signaling, the greater the transfer into the domain of symbolic and linguistic functioning. Consequently, those paradigms that allow an animal to make its wants and desires known (initially by non-verbal means and later through a symbolic–linguistic interface) should be the ones that lead to the highest levels of competency. Most importantly, such paradigms would need to foster a close positive relationship between one or two caregivers and the organism attempting to acquire language.

The four subjects in our studies that had the deepest environmental immersion and the closest ties with the fewest caregivers (Kanzi, Panbanisha, Koko, and Chantek[4]) are the subjects who display the highest linguistic skill levels found in apes, such as understanding of spoken English conversation. They are also the only subjects reported to translate from one symbol system to another, that is between English and lexigrams or English and sign language. Three of these subjects are only ones to

----

[4] Mentioned briefly in the introduction, Penny Patterson works with gorilla Koko (b. 1971) and Lyn Miles with orang-utan Chantek (b.1977).

produce Oldowan tools (Chantek, Kanzi, and Panbanisha) and two of these subjects (Kanzi and Panbanisha) are the only ones to pass a linguistic test for mind reading (Savage-Rumbaugh and Fields 2004). All of these subjects (along with Washoe, Moja, Dar, and Tatu) also engaged in pretend play, recognized themselves in mirrors and comprehended as well as produced indicative pointing. By contrast, subjects whose training began late in life or who were not reared in an immersion environment tended to be restricted to 'islands of competency' that center around their explicit training sessions.

Greenspan and Shanker (in press) report similar observations for children who are challenged with autism. Those whose experience is limited to training sessions do not develop the non-verbal competencies that underpin and launch true symbolic reference. Such children also may fail to follow glances, and respond only when another individual turns their head or head and body in a certain direction. These limitations are similar to those reported by Bering and Povinelli (2003) with nursery chimpanzees. They are also reminiscent of how Sherman and Austin (also nursery reared) behaved before they acquired symbolic competency. By contrast, autistic children who are placed in a program that emphasizes non-verbal circles of communication and expression of affective intent with parents and other family members begin to make leaps into linguistic and other competencies similar to, and often greater than, those reported here for apes.

Also parallel to our observations for apes, Greenspan and Shanker note that explicit attempts to train non-verbal autistic subjects can actually interfere with the ability to acquire true symbolic functioning. The reason for this is that such training does not link into the child's affective system.[5] It is only through neurological activation of the subcortical affective system that the higher centers of cortical functioning become meaningfully engaged. Therefore, such training tends to induce a kind of rote learning that is divorced from true communication. It may result in echolalia, in difficulty separating errors from correct responses (because the responses are not communicatively based), and in a lack of appropriately communicative eye contact.

Thus we observe that the early rearing conditions of various subjects in animal language studies can and do have a profound effect upon the reported findings. However, it is rare that young apes have exposure to adults who employed communication systems in the wild. To our knowledge, Matata (who was wild born and reared) is the only ape able to provide a communicative model for youngsters who are also in a language program. Most young apes have no opportunity to learn non-verbal communication from anyone other than their peers or their human caregivers.

As we come to better understand the communication of intentionality at the non-verbal level, our insights into the roots of symbolic processes will expand. By looking at these processes in species as diverse as dolphins, parrots, apes, and children, researchers have

---

[5] Editors' note: for related discussion see Pepperberg, this volume.

forced more adequate definitions of language, have explored the boundaries of other minds in new ways, and have begun—just barely begun—authentically to engage other species in meaningful dialogue. If this work continues, as it surely should with second and third generation students, the future will be filled with surprises. In our view, many limitations expressed by the animal subjects to date are the limitations of our methods, not those of the subjects themselves. We predict that, as new methods are discovered and applied, these early studies will be looked back upon as simple and naïve, and our respect for the minds of other creatures will grow accordingly.

Acknowledgements

The preparation of this chapter, as well as much of our research herein referenced, was supported by grants from the National Institute of Health (NICHD-06016 and NICHD-38051). Support from the College of Arts and Sciences, Georgia State University and from the Great Ape Trust of Iowa are also gratefully acknowledged.

Appendix 23.1 Language acquisition in non-humans: comparison of procedures, tests, and skills

1 Training paradigms

1.A Subject required to '**label**' a specific object or attribute of an object by presenting the object and then displaying for the subject a **limited set** of symbols.

Premack—Sara (chimpanzee)
Matzuawa—Ai (chimpanzee)
Shumaker—Inda, Azy (orang-utans)

1.B Subject required to '**label**' a specific object or attribute of an object by presenting the object and requiring the subject to select/produce the appropriate name from the **full set** of learned symbols.

Gardner and Gardner—Washoe, Dar, Moja, Tatu (chimpanzees)
Fouts—Washoe, Dar, Moja, Tatu (chimpanzees)
Rumbaugh—Lana (chimpanzee)
Miles—Chantek (orang-utan)
Patterson—Koko (gorilla)
Savage-Rumbaugh—Sherman, Austin (chimpanzees)
Pepperberg—Alex (parrot)

1.C Subject required to **observe the sign produced by a caretaker** and then to imitate that sign.

Terrace—Nim (chimpanzee)

1.D Subject required to **'request'** specific objects by selecting or producing the appropriate symbol from the **full set** of learned symbols.

Gardner and Gardner—Washoe, Dar, Moja, Tatu (chimpanzees)
Fouts—Washoe, Dar, Moja, Tatu (chimpanzees)
Rumbaugh—Lana (chimpanzee)
Miles—Chantek (orang-utan)
Patterson—Koko (gorilla)
Savage-Rumbaugh—Sherman, Austin (chimpanzees)
Pepperberg—Alex (parrot)
Terrace—Nim (chimpanzee)
Idani—(chimpanzee)
Shumaker—Indah, Azy (orang-utans)

1.E Subject required to **produce a specific behavior** when shown a specific symbol.

Herman—Akeakamai (dolphin)
Schusterman—(sea lion)

1.F Subject required to look over an array of objects and **give the specific item** that is requested by a symbol. (Note this is a symbol-to-real-object task.)

Savage-Rumbaugh—Sherman, Austin (chimpanzees)

1.G This training paradigm required the subject to look over an array of pictures and *give the photo requested* that is requested by a symbol.

Savage-Rumbaugh—Sherman, Austin (chimpanzees)

1.H This training paradigm required the subject to **read or listen to symbolic questions and provide answers about colors, objects, numbers, or some combination thereof.** (Note this is a symbol-to-symbol task, with visual regard to objects interspersed.)

Rumbaugh—Lana (chimpanzee)
Pepperberg—Alex (parrot)
Matsuzawa—Ai (chimpanzee)

1.I Subject required to learn to produce **same/different** responses, where one or more of the presentation sets was an object.

Rumbaugh—Lana (chimpanzee)
Premack—Sarah (chimpanzee)
Pepperberg—Alex (parrot)
Herman—Ake, Phoenix (dolphins)

1.J Subject required **to produce a sequence of behaviors, in the order required by a grammatically ordered** sequence of symbols.

Herman—Ake, Phoenix (dolphins)

1.K Subjects were **immersed in a cultural context** with language being employed around them constantly in a way that was related to daily life, **and one or more** of the above training paradigms was also utilized.

Gardner and Gardner—Washoe, Moja, Dar, Tatu (chimpanzees)
Savage-Rumbaugh—Sherman, Austin (chimpanzees)
Patterson—Koko (gorilla)
Terrace—Nim (chimpanzee)
Miles—Chantek (orang-utan)

1.L No training requirements of any kind made with the subject. Subjects were **immersed in linguistic environment that allowed language to emerge spontaneously.**

Savage-Rumbaugh—Panzee (chimpanzee)
Savage-Rumbaugh—Kanzi, Panbanisha, Mulika, Nyota, Nathan (bonobos)
Fouts—Loulis (chimpanzee)

1.M Subjects are **taught to produce synchronous novel behaviors.**

Herman—Ake and/or Phoenix (dolphins)

1.N Subjects are **taught to produce novel behaviors** on command.

Herman—Ake and/or Phoenix (dolphins)

1.O Subjects are **taught to respond to presence or absence of objects using 'Yes'/'No'.**

Herman—Ake (dolphin)

1.P Subject **are taught to imitate** novel actions.

Herman—Ake, Phoenix (dolphins)
Miles—Chantex (orang-utan)

2 Testing paradigms

2.A This testing paradigm documented that the subject could **spontaneously produce novel sequences of behaviors** in response to novel combinations of symbols.

Herman—Ake, Phoenix (dolphins)
Savage-Rumbaugh—Kanzi, Panbanisha (bonobos)

2.B This paradigm tested that subjects **can make statements about intended actions.**

Savage-Rumbaugh—Sherman, Austin (chimpanzees)

2.C Subjects **name photographic representations** of objects.

Gardner and Gardner—Washoe, Dar, Moja, Tatu (chimpanzees)
Fouts—Washoe, Dar, Moja, Tatu (chimpanzees)

Rumbaugh—Lana (chimpanzee)
Miles—Chantek (orang-utan)
Patterson—Koko (gorilla)
Savage-Rumbaugh—Sherman, Austin, Panzee (chimpanzees)
Savage-Rumbaugh—Kanzi, Panbanisha, Mulika, Nyota, Nathan (bonobos)
Matzuzawa—Ai (chimpanzee)

2.D Subjects are tested for their ability to **label video representations** of learned symbols or signs.

Savage-Rumbaugh—Sherman, Austin, Panzee (chimpanzees)
Savage-Rumbaugh—Kanzi, Panbanisha, Mulika, Nyota, Nathan (bonobos)
Herman—Ake, Phoenix (dolphins)

2.E Subjects are tested for their **ability to imitate novel actions on objects.**

Savage-Rumbaugh—Kanzi, Panbanisha, (bonobos); Panzee (chimpanzee)

2.F Subjects are tested for their ability to **imitate novel actions**.

Herman—Ake, Phoenix (dolphins)

2.G Subjects are tested for their ability to select **correct object when presented with a symbol.**

Savage-Rumbaugh—Sherman, Austin, Panzee (chimpanzees)
Savage-Rumbaugh—Kanzi, Panbanisha (bonobos)
Herman—Ake, Phoenix (dolphins)

2.H **Subjects are tested for their ability to label when presented with real object.**

Gardner and Gardner—Washoe, Dar, Moja, Tatu (chimpanzees)
Fouts—Washoe, Dar, Moja, Tatu (chimpanzees)
Rumbaugh—Lana (chimpanzee)
Miles—Chantek (orang-utan)
Patterson—Koko (gorilla)
Savage-Rumbaugh—Sherman, Austin, Panzee (chimpanzees)
Savage-Rumbaugh—Kanzi, Panbanisha, Mulika, Nyota, Nathan (bonobos)
Matzuzawa—Ai (chimpanzee)

2.I Subjects are tested for their ability **to imitate novel actions on objects.**

Miles—Chantek: succeeds (orang-utan)
Matzuzawa—Ai, Chole, Popo, Pan: fail (chimpanzees)
Savage-Rumbaugh—Kanzi, Panbanisha: succeed (bonobos); Panzee: succeeds (chimpanzee)

2.J Subjects are tested for their ability **to imitate novel actions.**

Miles—Chantek: succeeds (orang-utan)

Herman—Ake: succeeds (dolphin)

2.K Subjects are tested for their ability **to translate spoken words into lexical symbols** or signs.

Savage-Rumbaugh—Sherman, Austin: fail (chimpanzees)
Savage-Rumbaugh—Kanzi and Panbanisha: succeed (bonobos); Panzee: succeeds (chimpanzee)
Miles—Chantek: data not yet reported (orang-utan)
Patterson—Koko: data not yet reported (gorilla)

2.L Tests of subjects ability to use acquired symbol system to **communicate with each other.**

Savage-Rumbaugh—Sherman, Austin (chimpanzees)
Savage-Rumbaugh—Kanzi, Panbanisha (bonobos)

3 Blind controls and testing anomalies

3.A The experimenter who presents objects, signs, or symbols cannot be seen by subject while response is being made.

Gardner and Gardner
Savage-Rumbaugh

3.B The experimenter can be seen but the response objects (photos, lexigrams) cannot be seen by the experimenter.

Rumbaugh
Savage-Rumbaugh
Matsuzawa

3.C Experimenter can be seen but wears goggles

Herman

3.D Subject must taste or touch objects before correctly naming them—or have objects rotated in front of eye.

Pepperberg—Alex (parrot)

3.E Principle trainer turns her back on subject but interprets subject's vocal behavior. Principle trainer prepares questions 4 days or more in advance of testing.

Pepperberg—Alex (parrot)

3.F Responses are score from video tape by two independent observers.

Gardner and Gardner
Savage-Rumbaugh

4 Data drawn from daily notes

4.A Data recording indicates that subject **spontaneously produces symbols to ask for things they want that are not present**. (Once subject obtains item, they do not keep asking.)

Gardner and Gardner—Washoe, Dar, Moja, Tatu (chimpanzees)
Fouts—Washoe, Dar, Moja, Tatu, (chimpanzees)
Rumbaugh—Lana (chimpanzee)
Miles—Chantek (orang-utan)
Patterson—Koko (gorilla)
Savage-Rumbaugh—Sherman, Austin, Panzee (chimpanzees)
Savage-Rumbaugh—Kanzi, Panbanisha, Mulika, Nathan, Nyota (bonobos)

4.B Data recording indicates that subject **spontaneously produces symbols to announce intended self-action.**

Gardner and Gardner—Washoe, Dar, Moja, Tatu (chimpanzees)
Fouts—Washoe, Dar, Moja, Tatu (chimpanzees)
Rumbaugh—Lana (chimpanzee)
Miles—Chantek (orang-utan)
Patterson—Koko (gorilla)
Savage-Rumbaugh—Sherman, Austin, Panzee (chimpanzees)
Savage-Rumbaugh—Kanzi, Panbanisha, Mulika, Nathan, Nyota (bonobos)

4.C Data recording indicates that subject **spontaneously produces symbols to request actions of others.**

Gardner and Gardner—Washoe, Dar, Moja, Tatu (chimpanzees)
Fouts—Washoe, Dar, Moja, Tatu (chimpanzees)
Rumbaugh—Lana (chimpanzee)
Miles—Chantek (orang-utan)
Patterson—Koko (gorilla)
Savage-Rumbaugh—Sherman, Austin, Panzee (chimpanzees)
Savage-Rumbaugh—Kanzi, Panbanisha, Mulika, Nathan, Nyota (bonobos)

4.D Subject spontaneously demonstrates **protogrammar during production.**

Rumbaugh—Lana (chimpanzee)
Savage-Rumbaugh—Kanzi, Panbanisha (bonobos); Panzee (chimpanzee)

4.E Subject **spontaneously comments on states of affairs related to specific personal desires.**

Gardner and Gardner—Washoe, Moja, Dar, Tatu (chimpanzees)
Rumbaugh—Lana (chimpanzee)
Savage-Rumbaugh—Sherman, Austin (chimpanzees)
Savage-Rumbaugh—Kanzi, Panbanisha (bonobos)
Miles—Chantek (orang-utan)

Patterson—Koko (gorillas)

4.F Subject **spontaneously (without being asked, 'what that') comments on states of affairs independent of personal desire** for object or event.

Savage-Rumbaugh—Sherman, Austin (chimpanzees)
Savage-Rumbaugh—Kanzi, Panbanisha, Nyota (bonobos)
Patterson—Koko (gorilla)
Miles—Chantek (orang-utan)

5 Additional skills

5.A Subjects able to interpret video representations of narrative events.

Savage-Rumbaugh—Sherman, Austin, Panzee (chimpanzees)
Savage-Rumbaugh—Kanzi, Panbanisha, Mulika, Nyota, (bonobos)

5.B Subjects engage in games of pretense with objects.

Savage-Rumbaugh—Sherman, Austin, Panzee (chimpanzees)
Gardner and Gardner—Washoe, Dar, Moja, Tatu (chimpanzees)
Savage-Rumbaugh—Kanzi, Panbanisha, Mulika, Nyota (bonobos)
Patterson—Koko (gorilla)
Fouts—Washoe, Dar, Moja, Tatu (chimpanzees)
Miles—Chantek (orang-utan)

5.C Subject copies lines, circles, etc. with training.

Matzuzawa—Ai (chimpanzee)
Savage-Rumbaugh—Sherman, Austin (chimpanzees)
Miles—Chantek (orang-utan)

5.D Subject spontaneously begins writing symbols.

Savage-Rumbaugh—Panbanisha (bonobo)

5.E Subject demonstrates understanding of novel spoken English conversations.

Savage-Rumbaugh—Kanzi, Panbanisha, Nyota (bonobos)
Patterson—Koko (gorilla)
Miles—Chantek (orang-utan)

5.F Subject passes linguistically based mind-reading test.

Savage-Rumbaugh—Kanzi, Panbanisha (bonobos)

5.G Subject produces Olduwan tools.

Savage-Rumbaugh—Kanzi, Panbanisha, Nyota (bonobos)
Miles—Chantek (orang-utan)

5.H Subject produces vocal symbols that are not trained.

Savage-Rumbaugh—Kanzi, Panbanisha (bonobos)

5.I Subject produces vocal symbols that are trained.

Hayes—Vickie (chimpanzee)
Pepperberg—Alex (parrot)
Herman—Ake (dolphin)

5.J Subject sorts novel items or photographs by categories.

Savage-Rumbaugh—Sherman, Austin (chimpanzees)
Savage-Rumbaugh—Kanzi, Panbanisha (bonobos)
Hayes—Vickie (chimpanzee)

5.K Comprehension and use of indicative pointing.

Hayes—Vickie (chimpanzee)
Gardner and Gardner—Washoe, Moja, Dar, Tatu (chimpanzees)
Savage-Rumbaugh—Sherman, Austin (chimpanzees)
Rumbaugh—Lana (chimpanzee)
Miles—Chantex (orang-utan)

5.L Self-awareness or mirror recognition.

Savage-Rumbaugh—Sherman, Austin (chimpanzees)
Savage-Rumbaugh—Matata, Kanzi, Panbanisha (bonobos)
Patterson—Koko (gorilla)
Miles—Chantek (orang-utan)
Reiss and Marino—(dolphins)

5.M Learned to tie knots.

Savage-Rumbaugh—Austin, Sherman, Panzee (chimpanzees)
Miles—Chantek (orang-utan)

5.N Ability to translate from one symbol system to another (sign to English, English to lexigram).

Miles—Chantek (orang-utan)
Patterson—Koko (gorilla)
Savage-Rumbaugh—Kanzi, Panbanisha (bonobos)

6   Rearing variables

6.A Subjects raised in the **presence of adults who were wild caught** and possessed whatever communicative abilities they have acquired in the wild.

Savage-Rumbaugh—Kanzi, Panbanisha, Nyota, Nathan (bonobos)

6.B Subjects were raised in the **presence of young peers** during formative years.

Gardners—Moja, Dar, Tatu (chimpanzees)
Matsuzawa—Ai (chimpanzee)

Savage-Rumbaugh—Sherman, Austin (chimpanzees)
Herman—Ake, Phoenix (dolphins)

6.C Raised **only in the presence of human companions** during formative years.

Hayes—Vickie (chimpanzee)
Gardner and Gardner—Washoe (chimpanzee)
Miles—Chantek (orang-utan)
Patterson—Koko (gorilla)
Terrace—Nim (chimpanzee)
Pepperberg—Alex (parrot)
Patterson—Koko (gorilla)
Terrace—Nim (chimpanzee)

6.D Age training begins and vocabulary size

Subject name	Age training begins	Vocabulary size
Alex	12–13 months	90 symbols
Chantek	9 months	150 symbols
Washoe	9 months	176 symbols
Nim	2 weeks	125 symbols
Lana	1.5 years	100 symbols
Moja	1 month	168 symbols
Tatu	1 month	140 symbols
Dar	1 month	122 symbols
Sherman	3 years	81 symbols
Austin	2.5 years	81 symbols
Kanzi	6 months	150 symbols
Panbanisha	1 month	179 symbols
Panzee	1 week	79 symbols
Matata	9 years	6 symbols
Azy	Adult	8 symbols
Indah	Adult	8 symbols
Aka	2–3 years	35 symbols
Phoenix	2–3 years	35 symbols
Ai	1 year	30 symbols
Sara	8 years	130 symbols
Koko	1 year	1000 symbols*

* Different criteria are employed by different investigators to count symbols. Patterson's criterion, used with Koko (use of a sign at least once spontaneously and without prompting), is probably the least strict of all, giving rise to a much higher number of symbols for Koko than for any other subject.

References

Asano, T., Kojima, T., Matsuzawa, T. Kubota, K., and Murofushi, K. (1982). Object and color naming in chimpanzees (Pan troglodytes). *Proceedings of the Japan Academy*, **58** (B): 118–122.

Bering, J. M., and Povinelli, D. (2003). Comparing cognitive development. In: D. Maestripieri, ed. *Primate Psychology*, pp. 105–233. Cambridge, MA: Harvard University Press.

Bickerton, D. (1995). *Language and Human Behavior*. Seattle: University of Washington Press.

Biro, D., and Matsuzawa, T. (2001). Chimpanzee numerical competence: Cardinal and ordinal skills. In: T. Matsuzawa, ed. *Primate Origins of Human Cognition and Behavior*, pp. 199–225. Tokyo: Springer-Verlag.

Brakke, K. E., and Savage-Rumbaugh, E. S. (1995). The development of language skills in bonobo and chimpanzee—I. Comprehension. *Language and Communication*, **15**: 121–148.

Brakke, K. E., and Savage-Rumbaugh, E. S. (1996). The development of language skills in Pan—II. Production. *Language and Communication*, **16**: 361–380.

Donald, M. (2001). *A Mind so Rare: Evolution of Human Consciousness*. New York: Norton.

Fouts, R., and Mills, S. T. (1997). *Next of Kin*. New York: William Morrow.

Gardner, R. A., and Gardner, B. T. (1969). Teaching sign language to a chimpanzee. *Science*, **165**: 664–672.

Gardner, R. A., and Gardner, B. T. (1998). *The Structure of Learning: From Sign Stimuli to Sign Language*. Lawrence Erlbaum Associates.

Gardner, R. A., Gardner, B. T., and Drumm, P. (1989). *Teaching Sign Language to Chimpanzees*. Albany, NY: State University of New York Press.

Goldin-Meadow, S., Butcher, C., Mylander, C., and Dodge, M. (1994). Nouns and verbs in a self-styled gesture system: what's in a name? *Cognitive Psychology*, **27**: 259–319.

Greenfield, P. M., Lyn, H., and Savage-Rumbaugh, E. S. (in preparation). Semiotic combinations in *Pan*: A cross-species comparison of communication in a chimpanzee and a bonobo.

Greenspand, S. and Shaler, S. (2004). The First Idea: How Symbols, Language, and Intelligence Evolved from our Primitive Ancestors to Modern Humans. De Capo Press.

Herman, L. M. (1986). Cognition and language competencies of bottlenosed dolphins. In: R. J. Schusterman, J. Thomas, and F. G. Wood, eds. *Dolphin Cognition and Behavior: A Comparative Approach*, pp. 221–251. Hillsdale, NJ: Lawrence Erlbaum Associates.

Herman, L.M., Kuczaj, S. III, and Holder, M. D. (1993). Responses to anomalous gestural sequences by a language-trained dolphin: Evidence for processing of semantic relations and syntactic information. *Journal of Experimental Psychology: General*, **122**: 184–194.

Herman, L. M., Richards, D. G., and Wolz, J. P. (1984). Comprehension of sentences by bottlenosed dolphins. *Cognition*, **16**: 129–219.

Herman, L. M., and Uyeyama, R. K. (1999). The dolphin's grammatical competency: Comments on Kako. *Animal Learning and Behavior*, **27**: 18–23.

Hillix, A., and Rumbaugh, D. M. (2004). *Animal Bodies, Human Minds: Ape, Dolphin and Parrot Language Skills*. New York: Kluwer/Academic Press.

Idani, G. (in preparation). Studies at the Great Ape Research Institute, Hayashibara. In: D. A. Washburn, ed. *Emergents in Rational Behaviorism, Essays in Honor of Duane M. Rumbaugh*. American Psychological Association.

Kano, T. (1990). The bonobos peaceable kingdom. *Natural History*, **99**: 62–70.

Matsuzawa, T. (1985). Color naming and classification in a chimpanzee (*Pan troglodytes*). *Journal of Human Evolution*, **14**: 283–291.

Menzel, E., Savage-Rumbaugh, E. S., and Lawson, J. (1985). Chimpanzee (*Pan troglodytes*), spatial problem solving with the use of mirrors and televised equivalents of mirror. *Journal of Comparative Psychology*, **99**: 211–217.

Miles, H. L. (1983). Apes and language: The search for communicative competence. In: J de Luce and H. T. Wilder, eds. *Language in Primates: Perspectives and Implications*, pp. 43–61. New York: Springer-Verlag.

Miles, H. L. (1994). Chantek: The language ability of an enculturated orangutan (Pongo pygmaeus). In: J. Ogden, L. Perkins, and L. Sheeran, eds. *Proceedings of the International Conference on Orangutans: The Neglected Ape*, pp. 209–219. San Diego: Zoological Society of San Diego.

Miles, H. L. (1990). The cognitive foundations for reference in a signing orangutan. In: S. T. Parker and K. R. Gibson, eds. *Language and Intelligence in Monkeys and Apes*, pp. 511–539. Cambridge: Cambridge University Press.

Pate, J. L., and Rumbaugh, D. M. (1983). The language-like behavior of Lana chimpanzee: Is it merely discrimination and paired-associate learning *Animal Learning and Behavior*, **11**: 134–138.

Patterson, F. G., and Cohn, R. N. (1990). Language acquisition by a lowland gorilla: Koko's first ten years of vocabulary development. *Word*, **41**: 97–142.

Patterson, F. G., and Linden, E. (1981). *The Education of Koko*. New York: Holt, Rinehart, and Winston.

Pepperberg, I. M. (1999). *The Alex Studies*. Cambridge, MA: Harvard University Press.

Premack, A. J., and Premack, D. (1972). Teaching language to an ape. *Scientific American*, **227**: 92–99.

Premack, D. (1976). *Intelligence in Ape and Man*. Hillsdale, NJ: Erlbaum Associates.

Rumbaugh, D. M., ed. (1977). *Language Learning by a Chimpanzee: The Lana Project*. New York: Academic Press.

Rumbaugh, D. M., and Washburn, D. A. (2003). *Intelligence of Apes and Other Rational Beings*. New Haven: Yale University Press.

Savage-Rumbaugh, E. S. (1986). *Ape Language: From Conditioned Response to Symbol*. New York: Columbia University Press.

Savage-Rumbaugh, E. S. (1993). *Kanzi: A Most Improbable Ape*. Tokyo: NHK Publishing Co.

Savage-Rumbaugh, E. S. (2000). *Kanzi II*. [Videotape 52 minutes.] Niio, G. (Director). Tokyo: NHK of Japan.

Savage-Rumbaugh, E. S., Brakke, K. E., and Hutchins, S. (1992). Linguistic development: Contrasts between co-reared *Pan troglodytes* and *Pan paniscus*. In: T. Nishida, W. C. McGrew, P Marler, M. Pickford, and F. B. M. de Waal, eds. *Topics in Primatology, 1, Human Origins*, pp. 51–66. Tokyo: University of Tokyo Press.

Savage-Rumbaugh, E. S., Fields, W. M., and Spircu, Tiberu. (2004). The emergence of knapping and vocal expression embedded in a *Pan/Homo* culture. *Biology and Philosophy*, **19**: 541–575.

Savage-Rumbaugh, E. S., Fields, W. M., and Taglialatela, J. (2004). Ape consciousness-human consciousness: A perspective informed by Language and Culture. *American Zoologist*, **40**: 910–921.

Savage-Rumbaugh, E. S., and Lewin, R. (1994). *Kanzi: At the Brink of the Human Mind*. New York: Wiley.

Savage-Rumbaugh, E. S., McDonald, K., Sevcik, R. A., Hopkins, W. D., and Rubert, E. (1986). Spontaneous symbol acquisition and communicative use by pygmy chimpanzees (*Pan paniscus*). *Journal of Experimental Psychology*: General, **115**: 211–235.

Savage-Rumbaugh, E. S., Murphy, J., Sevcik, R. A., Rumbaugh, D. M., Brakke, K. E., and Williams, S. (1993). Language comprehension in ape and child. *Monographs of the Society for Research in Child Development*, Serial No. 233, Vol. 58, Nos. 3–4: 1–242.

Savage-Rumbaugh, E. S., Rumbaugh, D. M., Smith, S. T., and Lawson, J. (1980). Reference: The linguistic essential. *Science*, **210**: 922–925.

Savage-Rumbaugh, E. S., Wilkerson, B. J., and Bakeman, R. (1977). Spontaneous gestural communication among conspecifics in the pygmy chimpanzee (*Pan paniscus*). In: G. H. Bourne, ed. *Progress in Ape Research*, pp. 97–116. New York: Academic Press.

Savage-Rumbaugh, E. S., Williams, S. L., Furuichi, T., and Kano, T. (1996). Language perceived: *Paniscus* branches out. In: W. C. McGrew, L. F. Marchant, and T. Nishida, eds. *Great Ape Societies*, pp. 173–184. Cambridge: Cambridge University Press.

Segerdahl, P., Fields, W., and Savage-Rumbaugh, S. (in preparation). *The Cultural Dimensions of Language*.

Shumaker, R., and Beck, B. (2003). *Primates in Question: The Smithsonian AnswerBook*. Washington: The Smithsonian Books.

Terrace, H. S. (1979). *Nim*. New York: Knopf.

Terrace, H. S., Pettito, L. A., Sanders, R. J., and Bever, T. G. (1979). Can an ape create a sentence? *Science*, **206**: 891–900.

Tomasello, M. (1994). Can an ape understand a sentence? A review of *Language Comprehension in Ape and Child* by E. S. Savage-Rumbaugh *et al. Language and Communication*, **14**: 277–390.

Williams, S. L., Brakke, K. E., and Savage-Rumbaugh E. S. (1997). Comprehension skills of language-competent and non-language-competent apes. *Language and Communication*, **17**: 301–317.

Wrangham, R. W., McGrew, W. C., de Waal, F. B. M., and Heltne, P., eds (1994). *Chimpanzee Cultures*. Cambridge: Harvard University Press.

Index